D0146055

Robert Grover Brown, Ph.D., Iowa State University, is Professor of Electrical Engineering at that institution. Professor Brown has also been Program Manager for the Themis Program, a research project sponsored by the Office of Naval Research at the University, and was formerly with North American Aviation, Inc.

Robert A. Sharpe is an engineer with Xerox Data Systems. Previously, he taught electrical engineering at Iowa State University and served with R.C.A. Laboratories.

William Lewis Hughes, Ph.D., Iowa State University, is Professor and Head of the Department of Electrical Engineering at Oklahoma State University. Professor Hughes has wide experience in a consultative capacity with professional and governmental organizations. He is the author of *Nonlinear Electrical Networks*, published by The Ronald Press Company.

Robert E. Post, Ph.D., Iowa State University, is Professor of Electrical Engineering at that institution. He also serves as a consultant to industrial organizations, and regularly reports the results of his research in microwave engineering and wave propagation in the professional journals.

LINES, WAVES, AND ANTENNAS

The Transmission of Electric Energy

ROBERT GROVER BROWN
IOWA STATE UNIVERSITY

ROBERT A. SHARPE
XEROX DATA SYSTEMS

WILLIAM LEWIS HUGHES
OKLAHOMA STATE UNIVERSITY

ROBERT E. POST
IOWA STATE UNIVERSITY

SECOND EDITION

THE RONALD PRESS COMPANY • NEW YORK

2 c

Library of Congress Catalog Card Number: 73–77859
PRINTED IN THE UNITED STATES OF AMERICA

Preface

This book is intended to serve as an introductory text in the general area of electric energy propagation. It is assumed that the student has a reasonably thorough background in circuit analysis and static field theory and also some familiarity with differential equations and vector analysis, both of which are indispensable in wave theory.

Our approach has been to present simple situations and to progress logically to the more difficult and general cases. The fundamental concepts basic to various modes of transmission, such as reflection, velocity of propagation, and characteristic impedance are first established, using the simplest possible models for illustration. Once firmly implanted, these ideas carry forward more or less automatically to new situations.

The book begins with transmission-line theory presented from a circuit theory viewpoint; that is, voltage and current play the dominant roles rather than the associated electric and magnetic fields, which are suppressed in the form of line parameters. Pulse transmission is considered early, in order to establish firmly in the student's mind the concepts of spatial propagation and reflection. Our experience has been that these concepts are more convincing to the beginning student when presented in terms of pulses rather than in terms of sinusoids. Chapters 1 through 6 deal with transmission-line theory. Wave theory is presented in Chapters 7 through 11. The basic concepts from transmission-line theory are carried over and used freely in these chapters without detailed explanation. Chapters 12 through 14 deal with elementary antenna theory. Chapter 15 is devoted to fundamental concepts in radio-wave propagation. Within these broad areas, the material is split between basic theory and application.

Readers familiar with the First Edition will notice a number of changes and additions. The chapters on lossy line analysis and lossy wave propagation have been removed and the material contained in these chapters incorporated in other chapters. New chapters on matrix representation of transmission-line circuits, resonant cavities, and elements of radio-wave propagation have been added. Material on two-section quarter-wave transformers, transmission-line filters, propagation in anisotropic media, antenna arrays, and waveguide components using ferrites has been added, where appropriate, to the other chapters.

As in the previous edition, extensive discussions of hardware have been avoided except where a device provides an illustrative example of the application of the theory presented in the text. Our solution to the hardware problem is to deal with it primarily in the laboratory periods.

The following rectangular Smith charts are reprinted by permission from the General Radio Co.: Fig. 3–1(a), Fig. 3–5, Fig. 3–6(a, b, c, d), Fig. 3–8(a, c), Fig. 4–13(a, b, c, d), Fig. 4–18(a, b, c, d), and Fig. 6–2(a, b, c, d). The following polar charts are reprinted by permission from the Hewlett-Packard Co.: Fig. 3–1(b) and Fig. 3–8(b).

We wish to thank the many users of the First Edition for their helpful suggestions and constructive criticism. The preparation of this edition has also profited immeasurably from the interest and advice of our colleagues to whom we owe a special debt of gratitude.

ROBERT GROVER BROWN
ROBERT A. SHARPE
WILLIAM LEWIS HUGHES
ROBERT E. POST

April, 1973

Contents

LINES, WAVES, AND ANTENNAS

1

Introduction to Energy Transmission

1–1. INTRODUCTION

The central theme of this text is the transmission of electric energy. Energy transmission is fundamental and ranks with energy conversion in its importance to electrical engineers. For example, of what value is it to be able to convert megawatts of energy from mechanical to electrical form at Hoover Dam, if we fail to understand how to transmit this energy to load centers in Southern California? At the other end of the frequency spectrum, the microwave engineer dealing with radar is confronted with a similar problem—that of transmitting energy to a distant target and detecting the minute value reflected from the target. These two problems in the fields of power and radar differ vastly in detail but are nevertheless similar in the fundamental sense that they both involve transmission of electric energy. It is essential that the electrical engineer understand the mechanism involved.

Like so many engineering subjects, the principles of electric energy transmission are few and the details many. The principles are simply those of classical circuit and field theories. The myriads of details come as a practical consequence of the almost infinite variety of ways in which energy transmission may be accomplished.

The various forms of energy transmission will be divided into three classes for the purposes of this text. The division is made according to the number of conductors involved:

1. Transmission lines: two or more conductors.
2. Waveguides: single, hollow conductor with energy propagation taking place inside the hollow region.
3. Antennas: zero conductors with energy propagating in the space between transmitting and receiving antennas.

1–2. EXAMPLES OF ENERGY TRANSMISSION

The mode of energy transmission chosen for a particular application depends on such factors as: (1) initial cost and maintenance; (2) frequency band and information-carrying capacity; (3) selectivity or privacy offered; (4) reliability and noise characteristics; and (5) power level and efficiency. Each mode of energy transmission will have only some of the desirable features. Consequently it becomes a matter of engineering judgment to choose the mode of energy transmission best suited for a particular application. It is the purpose of this text to present the underlying principles as clearly and concisely as possible so that you will be in a better position to make such decisions.

The material presented in this text should be considered with as unbiased an outlook as possible. In order to emphasize why this is so essential, consider the problem of transmitting an ultra-high frequency signal (say at 1000 MHz) between two points 30 miles (48.3 km) apart. Many engineers, because of their experience with the high efficiency of power lines, may erroneously jump to the conclusion that the transmission line will be most efficient. Others may incorrectly assume that waveguides will be the most efficient. However, for this particular case, the transmission of energy without any electrical conductors (antennas) will be found to excel the efficiency of waveguides and transmission lines by many orders of magnitude. If the received energy in each case were chosen to be 1 nanowatt, then for comparison it would be found that for reasonably typical installations, the required transmitted power would be of the order of:

1. Transmission lines: 10^{1575} watts (15,840-dB loss).
2. Waveguides: 10^{150} watts (1584-dB loss).
3. Antennas: 50 milliwatts (77-dB loss).

These figures were determined from an assumed transmission-line loss of 10 dB/100 ft (328 dB/km), an assumed waveguide loss of 1 dB/100 ft

(32.8 dB/km), and transmitting and receiving antennas with an area of the order of 2 sq m. As a practical consequence of these results, microwave relay links at about 30-mile (48.3 km) intervals are used for cross-country transmission of telephone and television services.

A word of caution is needed here with regard to the preceding discussion. Both transmission lines and waveguides exhibit an attenuation characteristic as a function of distance which is exponential in form, whereas antenna field patterns in general vary inversely with the square of distance. Consequently, except for fixed distances of transmission, comparison of antennas with either waveguides or transmission lines is somewhat meaningless. For example, if the transmission path length of the preceding example were shortened by a factor of 100:1 to a distance of only 1500 ft (457 m), the comparison would become:

1. Transmission line: 1 megawatt (150-dB loss).
2. Waveguide: 30 nanowatts (15-dB loss).
3. Antenna: 5 microwatts (37-dB loss).

Quite clearly the waveguide now excels either the transmission line or antenna for efficiency of energy transfer.

A comparison of the energy input required to obtain a received power of 1 nanowatt for the three modes of energy transmission is shown in Fig. 1–1. The assumed frequency is 1000 MHz, and the results are expressed on a decibel scale with a zero-decibel reference at the required receiver power level of 1 nanowatt. The dotted section of the antenna

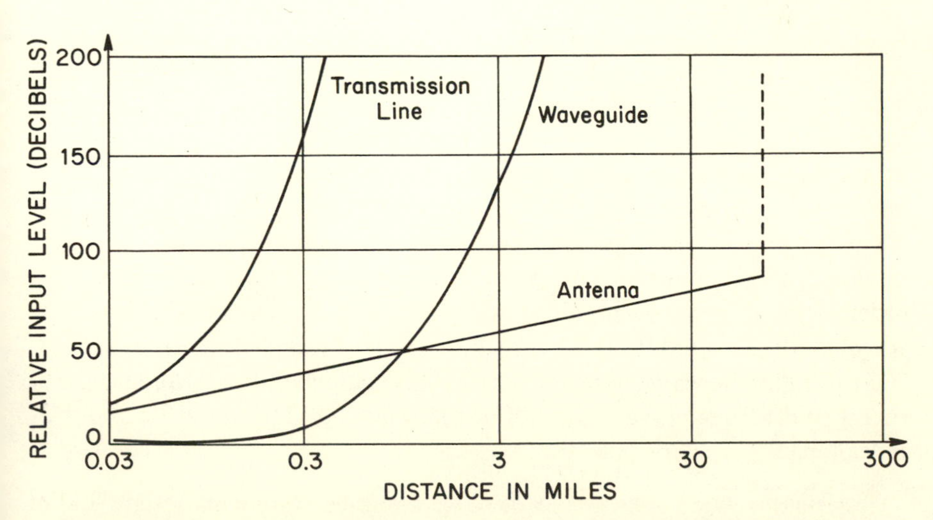

Fig. 1–1. Input power required versus distance, for a fixed receiver power.

curve, somewhat beyond 30 miles (48.3 km), indicates that the attenuation becomes very severe beyond the typical line-of-sight distance of about 50 miles (80.5 km).

The devices used in actual practice for various frequency ranges are shown in Fig. 1-2 in pictorial form. There is overlap between transmission

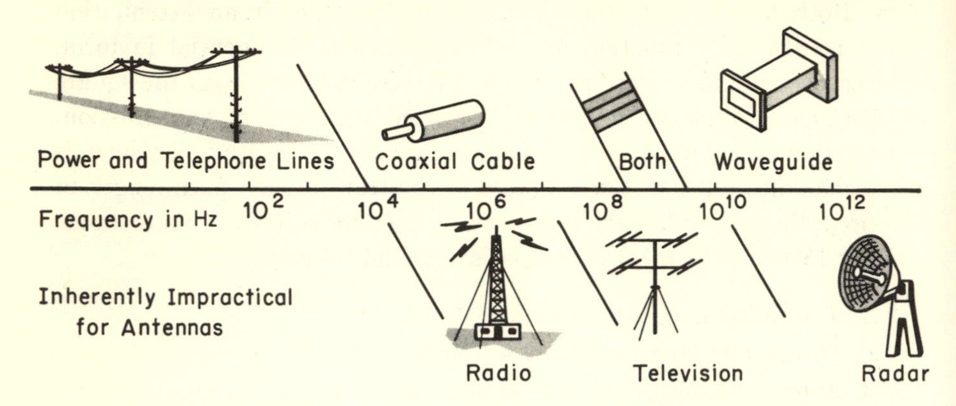

Fig. 1-2. Approximate frequency ranges for transmission lines, waveguides, and antennas.

lines and waveguides in the range of roughly 300 to 3000 MHz. It should be noted that transmission of energy down to zero frequency is practical with transmission lines, but waveguides and antennas inherently have a practical low-frequency limit. In the case of antennas this low-frequency limit is of the order of 100 kHz, but in a few cases this has been extended down to 10 kHz.[1] For waveguides the practical low-frequency limit is about 300 MHz. Theoretically the devices could be made to work at arbitrarily low frequencies, but the physical sizes required become so large as to be quite impractical. However, with lower gravity and lack of atmosphere such as found on the moon, it may be entirely feasible to have an antenna 10 miles (16.1 km) high and 100 miles (161 km) long for frequencies as low as a few hundred hertz. As an indication of the sizes involved, it may be noted that for either waveguides or antennas, the order of the dimensions involved is about one-half wavelength. One-half wavelength at 300 MHz is 0.5 m, while one-half wavelength at 30 kHz is 5000 m. Thus a waveguide for 300 MHz would be about the size of a roadway drainage culvert, and an antenna for 30 kHz would be about 3 miles long.

[1] For example, the Omega long-range navigation system operates in the 10–13 kHz range.

1–3. THE MATHEMATICS OF LINES, WAVES, AND ANTENNAS

The mathematical analysis of transmission lines, waveguides, and antennas may be approached from several different points of view. In this text the mathematical methods chosen to solve a particular problem will be just sufficiently advanced to handle the problem presented. After sophisticated mathematical methods have been developed, they will be applied to some of the problems which were solved by less elegant methods in order to tie together the material and to provide greater confidence in the processes involved. The mathematics required will range from simple addition through ordinary differential equations to vector operational equations. Appendix material will, where necessary, supplement the background of the student and permit study without recourse to other sources.

Although it may appear that the methods of analysis used in this text differ from conventional circuit theory, careful attention to details will reveal that in reality the results expressed are connected closely to conventional circuit theory and that the difference is basically only that of language. For example, the equation

$$\nabla \times \mathcal{E} = -\frac{\partial \mathcal{B}}{\partial t} \tag{1–1}$$

will appear in the analysis of waveguides. If the equation is inspected very closely and all its mathematical frills removed, it is found to be just the simple circuit equation

$$\textstyle\sum_{\text{loop}} \text{voltages} = 0 \tag{1–2}$$

A considerable amount of personal satisfaction may be found by inspecting the results of this text carefully to see how they fit into the larger picture and how the results complement those of conventional circuit theory, and vice versa.

In order to appreciate the results of this text fully, it may be necessary to discard some previous misconceptions and employ a fresh and unbiased viewpoint. Three of the most difficult barriers that may have to be overcome are: (1) the misconception that electric currents are carried only in conductors and that paths for summation of voltage drops must be through conductors; (2) the misconception that in a time-varying situation, a difference of potential may be ascribed between two physical points in space without regard to the path of measurement; and (3) a lack of appreciation of the fact that time-changing electric and magnetic fields cannot exist independently—but rather that the presence of one is responsible for the presence of the other.

The misconception that currents are carried only in conductors can be destroyed by the simple appreciation of the mechanism of current flow through a capacitor. By way of comparison, the current density in a conductor caused by a *constant* electric field may be compared with that in an insulator caused by a *time-changing* electric field by the equations

$$\mathcal{J}^c = \sigma \mathcal{E} \tag{1-3}$$

$$\mathcal{J}^d = \epsilon \left(\frac{\partial \mathcal{E}}{\partial t} \right) \tag{1-4}$$

The latter equation is the basis for explaining how energy transmission may be accomplished with either a single conductor or no conductors at all; the currents are in fact carried in space as the result of the time-changing electric fields.

The misconception that in a time-varying situation a difference of potential may be ascribed between two physical points in space comes from erroneous generalization based on observations of *slowly* varying field conditions. In a static situation a difference of potential may be ascribed between two points in space. When the static situation is modified to a very slowly changing situation, the results are, as expected, not significantly different. However, for more rapidly varying conditions (possibly as low as fractions of cycles per second, depending on the particular circumstances), it will be observed that the results depart from those of the static case.

The transformer offers the simplest example to refute the misconception that a difference of potential may be ascribed between two physical points in space without regard to path of measurement. Measurement at the terminals of the transformer will yield a certain voltage. However, rearranging the leads to the voltmeter so that they encircle the core—i.e., changing the path of measurement—can change the resultant measured voltage to virtually any desired value. One can detect the change in path in this very simple case and possibly expect a different reading. However, at much higher frequencies, where the importance of a conductor is reduced to the point of omitting it, the significance of the path is much more obscure and subtle in effect.

Dramatic evidence of the misconception about difference of potential can be had by observing a very high power microwave transmitter driving a poorly terminated waveguide. Arcing will be observed between the two sides of the hollow, rectangular metal tube serving as a waveguide. The arcing is evidence of electric fields of hundreds of thousands of volts per meter, even though the arcing is taking place between points connected by a conductive path with a resistance measured in the millionths of ohms. Study this point carefully, when it is discussed later in more detail, and you will gain a great insight into and appreciation for transformer induction.

If it were possible to summarize the results of this text in a single sentence, then "Time-changing electric and magnetic fields cannot exist independently" would probably come closest. Again, the erroneous impression that time-changing electric and magnetic fields can exist independently arises from observation of physical situations where this assumption is not invalidated. The situation is much like that of classical mechanics and Einstein's modification. Unless the particle can be forced to a velocity of about nine-tenths that of light or greater, the deviation for relativistic corrections is very minor. But that does not mean that they do not exist. At low frequencies where conduction current is predominant, the very strong connection between electric and magnetic fields is not very evident. However, at very high frequencies where conduction current is of much less consequence, the tie between electric and magnetic fields is much more evident. For example, for a plane wave traveling in space, the electric fields and magnetic fields are both present, being mutually perpendicular, and are in fact in time phase and related so that the ratio of the electric field to the magnetic field is a constant of 377 ohms. The time-changing electric field produces a current, as explained earlier, and this produces a magnetic field which in turn, in the fashion of transformer induction, produces an electric field. The connection is so strong that one *cannot* exist without the other.

The mathematics of lines, waveguides, and antennas is very brief and concise, basically consisting of a total of five equations of the general form of Eq. (1-1). The equations are simple, and they express simple ideas, but considerable care must be exercised that the reader learns to master the equations rather than find that the equations master him. Do not merely look at them as a group of symbols; look to see what they represent and mean. The mechanics of mastering them then become as simple as learning the multiplication table or an alphabet. Knowing the equations does not give you a mastery of the subject, but knowing what they mean does.

1-4. NOTATION

Notation plays an important role in this text, where a considerable number of mathematical manipulations must be performed. You should know not only precisely what the symbols mean but also the rules for choosing them, so that you need not continually consult a table.

In this text we shall have occasion to deal with actual functions of time for arbitrary functions as well as sinusoidal signals. In the case of sinusoidal signals it will be convenient to use the phasor representation from ac circuit theory. It is important to recognize from the symbols whether the functions under discussion are in time or phasor form. The general rule here is that capital script letters, such as $\mathcal{V}$, $\mathcal{I}$, $\mathcal{E}_x$, and $\boldsymbol{\mathcal{E}}$, will represent the time forms, while other capital letters—whether ordinary italic like V, I, and E_x or boldface roman like **E**—will represent phasor quantities.

A spatial vector quantity will always be indicated in boldface type. Examples are $\boldsymbol{\mathcal{E}}$ and **E**, the time and phasor symbols, respectively, for the electric-field intensity vector. If the vector function is limited in some way, such as in waveguides where the form of variation in the z-direction is assumed, then the basic symbol will be modified by a superscript or subscript symbol. Unit vectors will be denoted with a small letter **a**, with a subscript affixed to denote the direction of the unit vector.

A much abbreviated list of symbols is given here to illustrate the method of notation.

Quantity	*Time Notation*	*Phasor Notation*	*Dimensions*
Current	$\mathcal{I}$	I	amperes
Voltage	$\mathcal{V}$	V	volts
Vector electric field intensity .	$\boldsymbol{\mathcal{E}}$	**E**	volts/meter
Vector magnetic field intensity .	$\boldsymbol{\mathcal{H}}$	**H**	amperes/meter
Vector magnetic potential . .	$\boldsymbol{\mathcal{A}}$	**A**	webers/meter
Unit vector along the x-axis .	$\mathbf{a}_x$	$\mathbf{a}_x$	dimensionless

Whenever possible the symbols used will be standard. A complete list is included in Appendix H. The symbols will also be defined as they occur in the various chapters.

1–5. RELATIONSHIP BETWEEN TIME AND PHASOR NOTATION OF SINUSOIDS

Throughout the text it will be convenient to alternate between time and phasor notation for sinusoidal signals. The connection between the two is therefore rather fundamental. A basic point of difference between conventional ac circuit analysis and the analysis presented here is that the quantities involved here will for the most part be dependent on as many as three position variables. In transmission lines, waveguides, and antennas, either the quantity of interest or its effect is felt at a distant point some time after the actual change—specifically, at a distance of x meters delayed in time by x/v seconds. The quantity v is the velocity of propagation in meters/second. Thus, for a cosinusoidal driving signal, the response at a distance x from the "disturbance" will be of the form

$$\cos \omega \left(t - \frac{x}{v}\right) = \cos (\omega t - \beta x) \tag{1-5}$$

For simplicity in writing this equation, ω/v has been replaced by the standard symbol β, which has the dimensions of radians/meter.

The phasor way of representing the function given in Eq. (1–5) follows directly from conventional ac circuit theory by treating βx as the general phase angle. Consequently, if the phasor representation of $\cos \omega t$ is "chosen" to be e^{j0}, then the phasor representation of Eq. (1–5) is, by inspection,

$$e^{-j\beta x} \tag{1-6}$$

The conversion between the two representations is most conveniently stated in the direction between phasor and time. It is straightforward and involves two steps: (1) multiplying the phasor function by $e^{j\omega t}$ and (2) taking the real[2] part of the product. It is thus easily verified that a time delay of x/v seconds may be expressed as a phase lag of βx radians, as shown in Eq. (1–5) for the time expression or as shown in Eq. (1–6) for the phasor expression. The phasor expression is usually preferred in the analysis of a problem because of the resultant simplicity in performing the mathematical manipulations.

An operation such as differentiation in the time representation of sinusoidal signals is represented in phasor notation as a simple multiplication by $j\omega$. Integration is the inverse of differentiation in time notation, and so in phasor notation produces the inverse operation, i.e., division by $j\omega$.

In abbreviated form the connection between time notation and phasor notation for sinusoidal signals may be expressed as follows:

Description	*Time*	*Phasor*
Cosine wave . . .	$\cos \omega t$	e^{j0}
Sine wave	$\sin \omega t = \cos \left(\omega t - \frac{\pi}{2}\right)$	$e^{-j(\pi/2)} = -j$
Phase delay	$\cos (\omega t - \theta)$	$e^{-j\theta}$
Differentiation . . .	$\frac{\partial}{\partial t}$	$j\omega$

[2] It is also possible to use the imaginary part of the product. However, whichever is chosen must be consistently used throughout a given series of calculations.

The phasor representation of a time function is most easily found by merely thinking: "What phasor expression can we start with so that the real part of the product of exp $(j\omega t)$ and the phasor expression will in fact be the time expression?"

A problem that will arise many times throughout the text involves computation of time-averaged power for sinusoidal signals for both the time representation and the phasor representation. The definition of time-averaged power is made in terms of time, and so that must be the starting point. Consider voltage and current signals of the form

$$\mathcal{V} = |V| \cos (\omega t + \theta_1) \tag{1-7}$$

and

$$\mathcal{I} = |I| \cos (\omega t + \theta_2) \tag{1-8}$$

where $|V|$ and $|I|$ are the peak values of the two signals, respectively (and also the absolute magnitudes of the phasors V and I). The time-averaged power may be found by definition as the time average of the product of the voltage and current:

$$P_{\text{avg}} = \text{time avg of } (\mathcal{V}\mathcal{I}) = \frac{|V|\,|I|}{2} \cos (\theta_1 - \theta_2) \tag{1-9}$$

Now, quite evidently, the phasor representations of the voltage and current are

$$V = |V|e^{+j\theta_1} \tag{1-10}$$

and

$$I = |I|e^{+j\theta_2} \tag{1-11}$$

In order to obtain from these phasor expressions the correct expression for time-averaged power as given in Eq. (1–9), it is only necessary to find

$$P_{\text{avg}} = \frac{1}{2} \text{ real part of } (VI^*) = \frac{|V|\,|I|}{2} \cos (\theta_1 - \theta_2) \tag{1-12}$$

The asterisk (*) on I indicates the conjugate of a complex number. To complete the preceding table we add:

Description	*Time*	*Phasor*
Avg power	Avg of $(\mathcal{V}\mathcal{I})$	$\frac{1}{2}$ real part of (VI^*)

2

General Equations and Solution for Transmission Lines

2–1. INTRODUCTION

This chapter will introduce the general equations for the transmission line. The solution of these equations for the special case of the lossless line will be considered first, in Art. 2–3. Lossless lines are only mathematical abstractions, but for many purposes the results derived with this simplification will be of sufficient accuracy to justify their practical use. In any case the important features of wave propagation and reflection are easily illustrated for the lossless case without the cumbersome detail involved when the most general case with losses is considered. Consideration of losses is important, however, and will in fact be discussed in Art. 2–8 in considerable detail.

The transmission-line analysis presented in this portion of the text may well represent your first association with distributed parameter systems, and therefore some new attitudes toward analysis will have to be developed. If conventional lumped-constant circuit theory is to be applied, it will be necessary to attack the transmission-line problem from a differential viewpoint, since the properties of inductance, capacitance, resistance, and conductance are so hopelessly distributed and intermingled that they are not individually recognizable.

A transmission line, as shown in Fig. 2–1, may be thought of as having

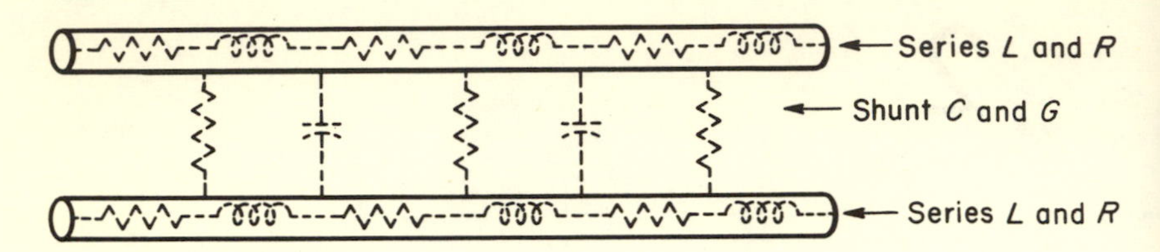

Fig. 2–1. Transmission line and parameters.

series inductance and resistance as well as shunt capacitance and conductance. A given transmission line can be characterized by giving the numerical values of these quantities per unit length of line. Evaluation of these quantities, *parameters*, as they are often called, is illustrated in Appendix A for several common transmission-line configurations. Typical values of the parameters for commercial transmission lines and power lines are given in Appendix B.

Viewing the transmission line in terms of the conventional circuit elements, even though we derive the equations on a differential basis, is very attractive but at the same time rather naive. The individual transmission-line parameters are computed on the assumption that the other quantities do not exist. Their computation proceeds under the idealized assumption that the field configuration represents actual static current flow and charge distribution, and that the parameters are going to be used for rapidly varying time situations. Fortunately, the results will be quite accurate over a rather wide range of frequencies. This fortunate situation will be discussed further in the material on waveguides, where the coaxial line is analyzed from the field viewpoint. It is interesting to note that ordinary RG–8A/U coaxial cable behaves in agreement with the results derived in the transmission-line chapters of the book up to frequencies in the range of 10 GHz.

2–2. DERIVATION OF THE GENERAL TRANSMISSION-LINE EQUATIONS

The various transmission-line parameters and their corresponding symbols are

Series $\begin{cases} \text{Inductance in henrys/meter, } L \\ \text{Resistance in ohms/meter, } R \end{cases}$

Shunt $\begin{cases} \text{Capacitance in farads/meter, } C \\ \text{Conductance in mhos/meter, } G \end{cases}$

It must be emphasized that, in general, R is not the reciprocal of G and that L, C, R, and G are per unit length of line.

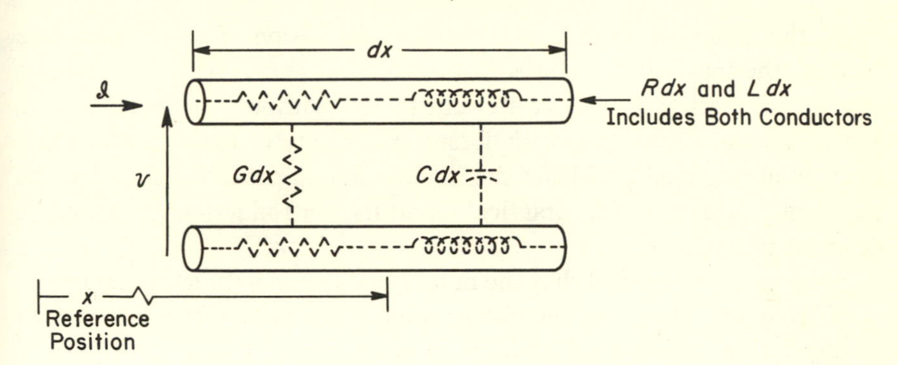

Fig. 2–2. Differential length of transmission line.

Consider now a differential length of transmission line as shown in Fig. 2–2. The two basic transmission-line equations can be written by inspection if the following pertinent facts are first noted:

1. In rationalized MKS units the actual values of the line constants for a differential length of line are $L\,dx$ henrys, $R\,dx$ ohms, $C\,dx$ farads, and $G\,dx$ mhos.
2. To a first approximation the voltage between the lines and the current along the lines are assumed constant (with respect to distance but not with respect to time) along the differential length of line.
3. One equation may be written by application of Kirchhoff's voltage law by noting that the differential voltage drop may be written either as $-(\partial \mathcal{V}/\partial x)\,dx$ or as $(L\,dx)(\partial \mathcal{I}/\partial t) + (R\,dx)(\mathcal{I})$.
4. The other equation may be written by application of Kirchhoff's current law by noting that the differential current drop may be written either as $-(\partial \mathcal{I}/\partial x)\,dx$ or as $(C\,dx)(\partial \mathcal{V}/\partial t) + (G\,dx)(\mathcal{V})$.

The two transmission-line equations obtained by cancellation of the common factor dx are:

$$-\frac{\partial \mathcal{V}}{\partial x} = L\frac{\partial \mathcal{I}}{\partial t} + R\mathcal{I} \tag{2–1}$$

$$-\frac{\partial \mathcal{I}}{\partial x} = C\frac{\partial \mathcal{V}}{\partial t} + G\mathcal{V} \tag{2-2}$$

These equations are the basis for all subsequent work dealing with transmission lines.

2–3. GENERAL SOLUTION OF THE LOSSLESS LINE EQUATIONS

At this point we shall give a detailed discussion of the lossless line because the important features of propagation and reflection of signals can be more clearly and easily illustrated without the cumbersome generalization of the line with losses. Fortunately, for a considerable number of practical problems the lossless line approach is found to be sufficiently accurate for practical analysis. Introduction to losses is deferred to Art. 2–8.

Some readers may feel that the more general case with losses should be considered first. However, the transmission-line problem, in common with all physical problems, does not have an exact answer. Consequently, since we can only hope to approximate the answer by considering equations that are approximations to the conditions of the physical problem, we are faced with the decision as to how accurate the answers must be in order to be useful. For our immediate purposes, consideration of the lossless line produces answers of satisfactory accuracy. Limitations and extensions of the results will be discussed as the occasion demands.

The equations for lossless transmission lines are obtained from Eqs. (2–1) and (2–2) by the simple expedient of setting $R = G = 0$ with the results:

$$-\frac{\partial \mathcal{V}}{\partial x} = L\frac{\partial \mathcal{I}}{\partial t} \tag{2-3}$$

and

$$-\frac{\partial \mathcal{I}}{\partial x} = C\frac{\partial \mathcal{V}}{\partial t} \tag{2-4}$$

Now, by differentiating one equation with respect to distance and the other with respect to time and substituting one expression into the other, we find the pair of equations

$$\frac{\partial^2 \mathcal{V}}{\partial x^2} = LC\frac{\partial^2 \mathcal{V}}{\partial t^2} \tag{2-5}$$

and

$$\frac{\partial^2 \mathcal{I}}{\partial x^2} = LC\frac{\partial^2 \mathcal{I}}{\partial t^2} \tag{2-6}$$

These last two equations are each in terms of a single variable and may be recognized as being in the form of the classical wave equation. The wave equation is most often seen in the form

$$\frac{\partial^2 y}{\partial x^2} = \frac{1}{v^2}\frac{\partial^2 y}{\partial t^2} \tag{2–7}$$

where v has the significance and dimensions of velocity. A solution of the wave equation, which may be easily checked by substitution for validity, is

$$y = y^+\left(t - \frac{x}{v}\right) + y^-\left(t + \frac{x}{v}\right) \tag{2–8}$$

In Eq. (2–8) y^+ and y^- are understood to be functions, completely arbitrary as yet, of the arguments $[t - (x/v)]$ and $[t + (x/v)]$, respectively. Generally speaking, the functions y^+ and y^- will be different. The superscript plus and minus have been affixed in anticipation of showing that the terms represent traveling waves, each of which travels in the direction corresponding to the superscript. It should be carefully noted that the traveling-wave behavior is entirely dependent upon the choice of arguments $[t - (x/v)]$ and $[t + (x/v)]$ but is in no way dependent on the functions themselves.

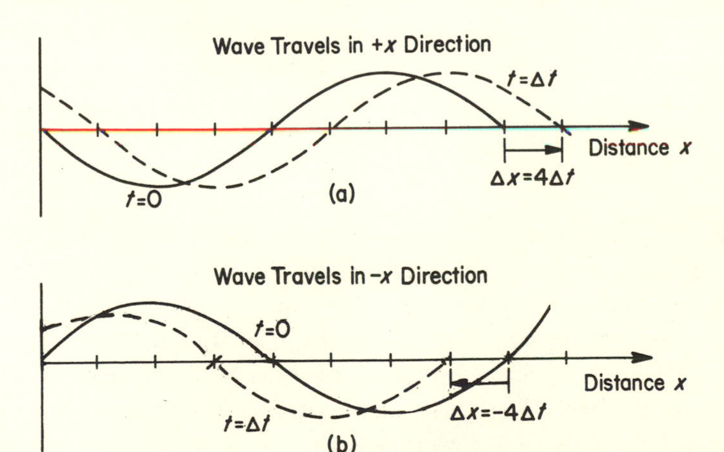

Fig. 2–3. Traveling waves.

The traveling-wave phenomenon is illustrated in Fig. 2–3 for the particular function $\sin 2\pi\theta$. In Fig. 2–3(a) the argument θ has been chosen to be $(t - x/4)$ while in Fig. 2–3(b) the argument has been chosen to be $[t + (x/4)]$. Two curves have been sketched in each case, one for $t = 0$ and a second one for $t = \Delta t$.

For the argument $[t - (x/4)]$ it is possible to "follow" the waveform (by maintaining the argument constant) by moving a distance $\Delta x = +4\,\Delta t$ in time Δt. That is, the waveform travels in the $+x$ direction with a velocity $v = 4$.

For the argument $[t + (x/4)]$ it is possible to "follow" the waveform (by maintaining the argument constant as before) by moving a distance $\Delta x = -4\,\Delta t$ in time Δt. That is, the waveform "travels" in the $-x$ direction with a velocity $v = 4$.

Returning now to Eqs. (2–3) and (2–4), armed with knowledge that both the voltage and current may be written in the general form of Eq. (2–8), we find for an assumed voltage function of

$$\mathcal{V} = \mathcal{V}^+\left(t - \frac{x}{v}\right) + \mathcal{V}^-\left(t + \frac{x}{v}\right) \tag{2–9}$$

that the corresponding current function, found by substitution into either Eq. (2–3) or (2–4), must be

$$\mathcal{I} = \frac{\mathcal{V}^+[t - (x/v)]}{R_0} - \frac{\mathcal{V}^-[t + (x/v)]}{R_0} \tag{2–10}$$

The constants v and R_0 are related to known parameters L and C by the expressions

$$v = \frac{1}{\sqrt{LC}} \tag{2–11}$$

and

$$R_0 = \sqrt{\frac{L}{C}} \tag{2–12}$$

The velocity v is found by comparison of Eq. (2–7) with Eq. (2–5). The characteristic resistance R_0 is found in the substitution process to find $\mathcal{I}$ from the assumed voltage function $\mathcal{V}$.

A very important point to note from Eq. (2–10) is that the voltage-to-current ratio in each of the traveling waves is exactly the characteristic resistance of the line. The negative sign associated with the current in the wave traveling in the $-x$-direction is the result of our positive current convention in the $+x$-direction. It may be noted that a similar sign consideration does not exist for the voltage because it is measured between the conductors rather than along them, as is done for the current.

The negative sign associated with the wave traveling the $-x$-direction may be clearly understood by reference to Fig. 2–4. In Fig. 2–4(a) a wave traveling in the $+x$-direction is illustrated. The current flow in the $+x$ direction causes the wave to propagate in the $+x$-direction by charging

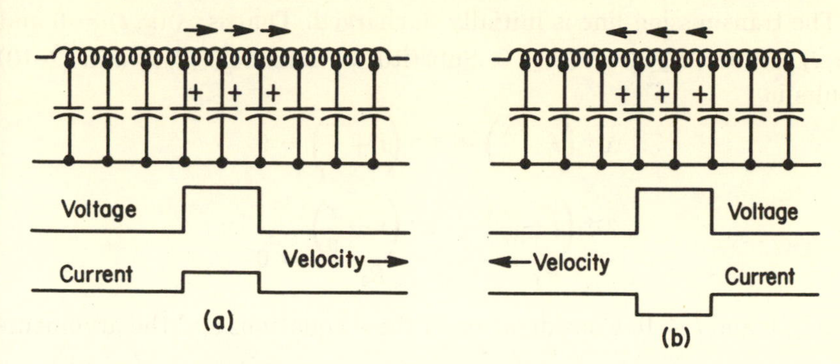

Fig. 2-4. Voltage and current distribution in traveling waves.

the capacity to the right and discharging the capacity to the left. In Fig. 2-4(b) the reverse situation is shown; current flowing to the left quite clearly causes the wave to propagate to the left. Both cases, however, are merely evidence of the same phenomenon. Propagation takes place in the direction of current flow.

It is very important that the waves traveling in both directions on the line be viewed as evidence of precisely the same phenomena and with each obeying precisely the same mathematical rules. The minus sign associated with the voltage-to-current relationship for the wave traveling in the $-x$-direction should be placed in the equations with the same attitude and confidence as signs are accounted for in simple dc circuit problems.

2-4. SATISFYING BOUNDARY CONDITIONS FOR THE GENERAL LOSSLESS LINE SOLUTION

The general solution given in Eqs. (2-9) and (2-10) is incomplete in the sense that the functions $\mathcal{V}^+$ and $\mathcal{V}^-$ are still unknown. These functions are determined by the conditions imposed by a particular problem.

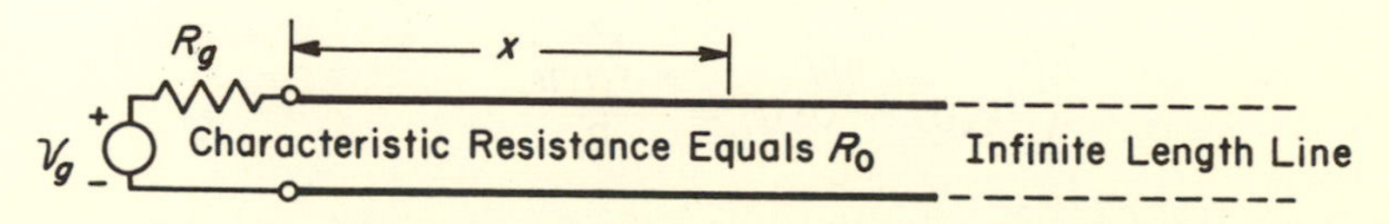

Fig. 2-5. Propagation of a signal down an infinite line.

Determination of the functions will be considered in this article.

Consider first the problem outlined in Fig. 2-5.

The transmission line is initially uncharged. That is: $\mathcal{V}(x, t) = 0$ and $\mathcal{I}(x, t) = 0$ for $x \geq 0$ and $t \leq 0$. Substituting into Eqs. (2–9) and (2–10) results in

$$\mathcal{V}^+\left(t - \frac{x}{v}\right) + \mathcal{V}^-\left(t + \frac{x}{v}\right) = 0$$

$$\frac{\mathcal{V}^+\left(t - \frac{x}{v}\right)}{R_0} - \frac{\mathcal{V}^-\left(t + \frac{x}{v}\right)}{R_0} = 0$$

for $x \geq 0$ and $t \leq 0$. Consideration of these equations and the arguments of $\mathcal{V}^+$ and $\mathcal{V}^-$ tells us that $\mathcal{V}^-\left(t + \frac{x}{v}\right)$ is zero for all values of $\left(t + \frac{x}{v}\right)$ and $\mathcal{V}^+\left(t - \frac{x}{v}\right)$ is zero for all values of $\left(t - \frac{x}{v}\right)$ less than or equal to zero. At $t = 0^+$, a voltage $V_g(t)$ is applied to the transmission line, at $x = 0$, through a source resistance of R_g ohms. We have shown that the only possible voltage on the line for the specified initial conditions is $\mathcal{V}^+\left(t - \frac{x}{v}\right)$, so we can write

$$\mathcal{V}(x, t) = \mathcal{V}^+\left(t - \frac{x}{v}\right)$$

and

$$\mathcal{I}(x, t) = \frac{\mathcal{V}^+\left(t - \frac{x}{v}\right)}{R_0}$$

for all $\left(t - \frac{x}{v}\right) > 0$.

At the source end of the line we can apply Ohm's law, with the result

$$V_g(t) - \mathcal{V}^+(t) = \mathcal{I}(t)R_g$$

But, we know that

$$\mathcal{I}(t) = \frac{\mathcal{V}^+(t)}{R_0},$$

so

$$V_g(t) - \mathcal{V}^+(t) = \frac{\mathcal{V}^+(t)R_g}{R_0}$$

or

$$\mathcal{V}^+(t) = \frac{R_0}{R_0 + R_g} V_g(t)$$

The general expressions for the voltage and current on the infinite length of transmission line can now be written as

$$\mathcal{V}(x, t) = \left(\frac{R_0}{R_0 + R_g}\right) V_g\left(t - \frac{x}{v}\right) \tag{2–13}$$

$$\mathcal{I}(x, t) = \left(\frac{1}{R_0 + R_g}\right) V_g\left(t - \frac{x}{v}\right) \tag{2–14}$$

It should be noted that the infinite length of line appears to the source as a simple resistive load of R_0 ohms; thus, one might expect that the generator voltage appears across R_g and the input to the line as though they constituted a simple resistive voltage divider. The voltage and current at other points along the line are replicas of the values at the input end of the transmission line, except for a delay of x/v in time.

Consider next the transmission line terminated in a resistance R_L, as shown in Fig. 2–6. Note that the reference point for x has been shifted

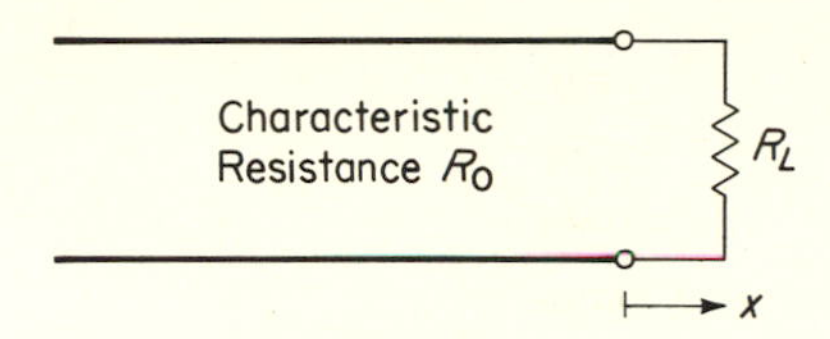

Fig. 2–6. Terminated transmission line.

to the position of the termination for convenience. Suppose that a wave traveling in the $+x$-direction is incident upon the load resistance. The voltage and current expressions for the incident wave are

$$\mathcal{V}(x, t) = \mathcal{V}^+\left(t - \frac{x}{v}\right)$$

$$\mathcal{I}(x, t) = \frac{\mathcal{V}^+\left(t - \frac{x}{v}\right)}{R_0}$$

Ohm's law tells us that the voltage and current at the load must be related by

$$\mathcal{V}(0, t) = R_L\mathcal{I}(0, t)$$

This condition cannot be met by the incident wave alone since the incident voltage and current are related by the characteristic resistance of the transmission line, R_0. To achieve the proper voltage and current relationships at the load we will postulate that some of the incident wave is reflected from the load resistance. The fraction of the voltage reflected is called the *voltage reflection coefficient* and is denoted by K_L. This means that there will now be a voltage wave traveling in the $-x$-direction and

$\mathcal{V}^-(0, t) = K_L \mathcal{V}^+(0, t)$. Now we can write the voltage and current expressions at the load as

$$\mathcal{V}(0, t) = \mathcal{V}^+(0, t) + \mathcal{V}^-(0, t) = \mathcal{V}^+(0, t)(1 + K_L)$$

$$\mathcal{I}(0, t) = \frac{\mathcal{V}^+(0, t)}{R_0} - \frac{\mathcal{V}^-(0, t)}{R_0} = \frac{\mathcal{V}^+(0, t)(1 - K_L)}{R_0}$$

You may like to think of the negative sign associated with the reflected current from the standpoint that the current reflection coefficient is always the negative of the voltage reflection coefficient. This sign change on the reflected current is the feature that permits the sum of two waves, each having a voltage-to-current ratio of exactly R_0, to produce a resultant voltage-to-current ratio different from R_0. In particular, Kirchhoff's laws are satisfied at the load if the voltage reflection coefficient K_L is chosen to satisfy Eq. (2–15) (see Fig. 2–7):

$$\frac{\text{Net load voltage}}{\text{Net load current}} = R_L = R_0 \frac{1 + K_L}{1 - K_L} \tag{2–15}$$

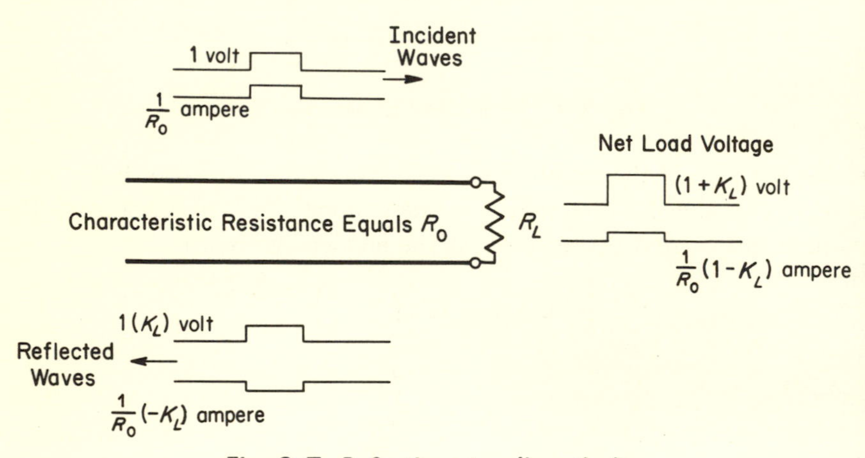

Fig. 2–7. Reflection at a discontinuity.

From this equation the voltage reflection coefficient may be easily solved for, with the result

$$K_L = \frac{R_L - R_0}{R_L + R_0} \tag{2–16}$$

An important question that may now be answered is: What load resistance, if any, may be used to terminate a finite-length transmission line without reflection? Quite clearly the value is exactly the characteristic resistance of the line R_0, since the voltage-to-current ratio everywhere along the line is R_0 for the case of the infinite line, and this situation will not be disturbed if the line is severed at *any* point and terminated in a resistance of R_0. This also explains why R_0, as defined earlier, was presumptively called the characteristic resistance of the line, since by definition the characteristic resistance of a system is of such a value that the load and input impedances are identical.

Example 2–1

As a numerical example of the use of the voltage reflection coefficient, suppose that a 50-V pulse is incident on a 30-ohm load in a line with a characteristic resistance of 50 ohms. The incident voltage and current pulses will be 50 V and 1 A. The voltage reflection coefficient at the load is −0.25. Consequently the voltage and current in the reflected pulse will be −12.5 V and +0.25 A. The net voltage and current at the load are thus 37.5 V and 1.25 A. The voltage-to-current ratio at the load is thus exactly 30 ohms—as of course it must be.

Consider finally the more general case when both the source resistance and the load resistance are different from the characteristic resistance of the line, as illustrated in Fig. 2–8. This problem is analyzed by successive

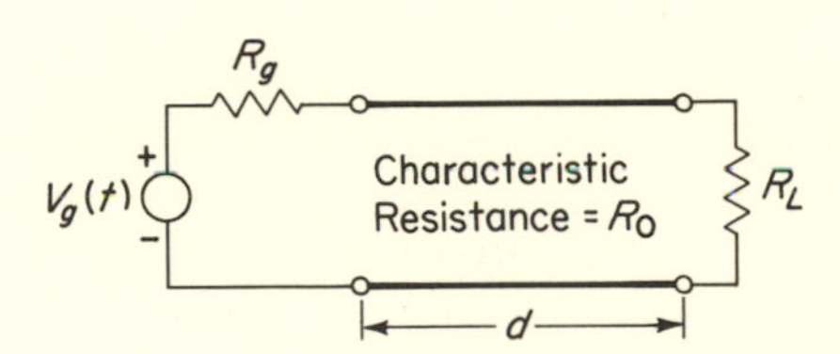

Fig. 2–8. General transmission line problem.

application of the incident properties indicated in Eqs. (2–13) and (2–14) and the reflected properties indicated in Eqs. (2–15) and (2–16). The incident wave from the source is independent of the line length and load resistance and is found by a simple divider analysis between the generator internal resistance and the characteristic resistance of the line. This signal travels to the load with a velocity v and is reflected. The reflected signal is computed from application of Eq. (2–16). This reflected signal travels back toward the generator and is reflected there by the mismatched resistance levels in the same fashion as at the load. Although the reflection at the source takes place independently of the source voltage, it must be noted that the actual signals at the input end of the line depend on the original incident values and all subsequent reflections.

This can be seen by considering the voltage at the input after the leading edge of the incident wave has traveled the length of the line, been reflected at the load, and has returned to the source end. At this time, we can write (assume $x = 0$ at the source for simplicity)

$$V_g\left(2\frac{d}{v}\right) - \mathcal{I}R_g = \mathcal{V}^+\left(2\frac{d}{v}\right) + \mathcal{V}^-\left(2\frac{d}{v}\right),$$

where $\mathcal{I} = \mathcal{I}\left(2\frac{d}{v}\right) = \frac{1}{R_0}\left[\mathcal{V}^+\left(2\frac{d}{v}\right) - \mathcal{V}^-\left(2\frac{d}{v}\right)\right]$. Substituting and solving for $\mathcal{V}^+\left(2\frac{d}{v}\right)$, which is the wave traveling in the $+x$-direction at the source end after $2\frac{d}{v}$ seconds have elapsed, results in

$$\mathcal{V}^+\left(2\frac{d}{v}\right) = \frac{V_g\left(2\frac{d}{v}\right)R_0}{R_g + R_0} + \mathcal{V}^-\left(2\frac{d}{v}\right)\frac{R_g - R_0}{R_g + R_0}$$

But, $\mathcal{V}^-\left(2\frac{d}{v}\right) = V_g(0)K_L$, so we can write

$$\mathcal{V}^+\left(2\frac{d}{v}\right) = \frac{V_g\left(2\frac{d}{v}\right)R_0}{(R_g + R_0)} + V_g(0)K_LK_g$$

where $K_g = \dfrac{R_g - R_0}{R_g + R_0}$.

As pointed out in the preceding paragraph, the signal reflects at the source end as though it were a load of R_g ohms.

Example 2–2

A numerical example for the more general case will now be considered. The details of the example are shown in Fig. 2–9. The transmission line is RG–8A/U, which has

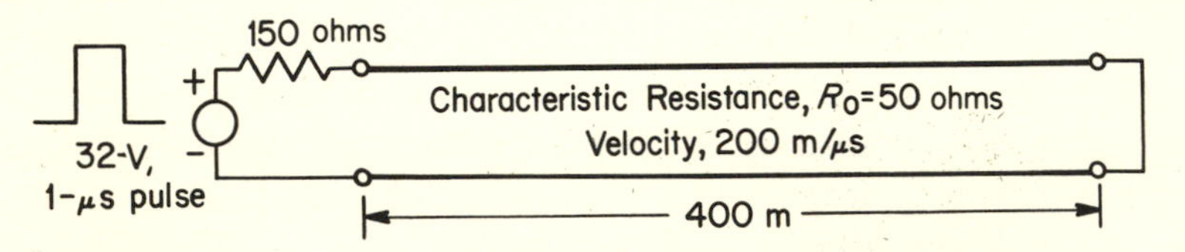

Fig. 2–9. Example of multiple reflections on transmission line.

a characteristic resistance of 50 ohms and a velocity of propagation of 200 m/μs. The steps in the analysis of this problem are as follows:

1. For a +32-V pulse from source, a +8-V signal will propagate toward the load and will reach the load in 2 μs.
2. A −8-V pulse will be reflected and will reach the source in an additional 2 μs.

3. A −4-V pulse will be reflected at the source and will travel to the load in an additional 2 μs.
4. The process repeats itself over and over from this point on, with the signal inverted at the load and halved at the source on each trip.

The voltage and current waveforms at the input end of the transmission line are then as shown in Fig. 2–10. A significant point to note is that the actual voltages and cur-

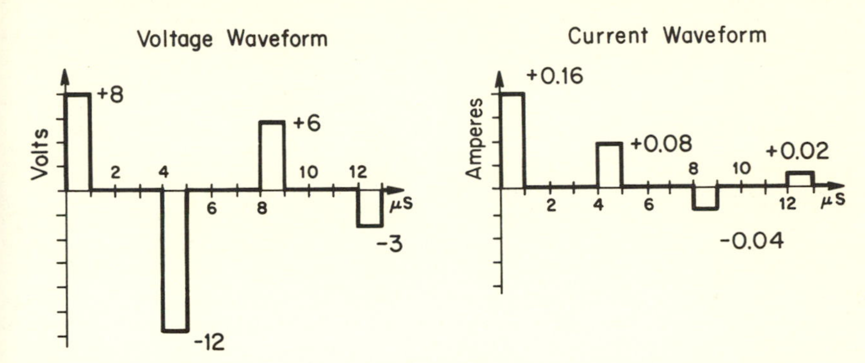

Fig. 2–10. Voltage and current waveforms at input end of line of Fig. 2–9.

rents at a point of reflection are found as the sum of the incident and reflected waves. For example, the −8-V pulse of step 2 must be added to the −4-V pulse of step 3 to find the total or net voltage at the input of the line (see second pulse in Fig. 2–10). It is interesting to note that the voltage pulse at the input end of the line, caused by the first reflection, is larger than the incident pulse from the source, but this result is expected for the particular choice of numbers.

If the input signal had been a pulse long enough so that the reflected signal reached the source before the input pulse vanished, then the problem would be solved exactly the same way. If the input pulse should be 6 μs long, then the voltage and current waveforms would be as in Fig. 2–11. These results are obtained directly from Fig. 2–10 by expanding each pulse to a 6-μs length and adding them where they overlap.

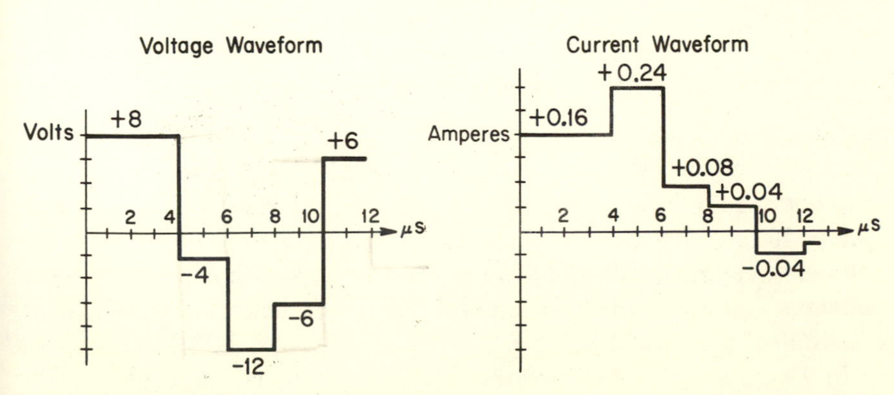

Fig. 2–11. Voltage and current waveforms at input end of line for 6-μs pulse.

In order to simplify the construction of voltage and current waveforms as shown in Fig. 2–10, it is convenient to construct a chart such as that shown in Fig. 2–12, where distance is measured along one axis and time

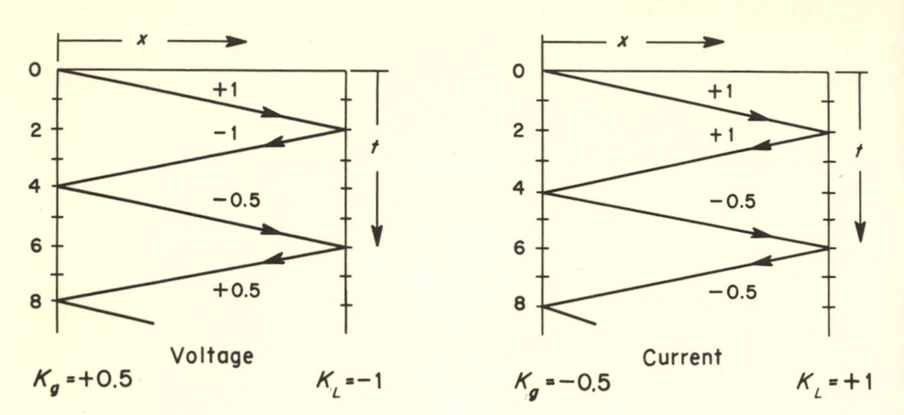

Fig. 2–12. Charts for simplifying construction of voltage and current waveforms.

along the other axis. The signals at various points are labeled in terms of an assumed incident waveform of 1-unit amplitude and negligible width. The actual values of the voltage and currents must be appropriately scaled. For the example given in Fig. 2–9, the incident voltage pulse is +8 V, and so the voltage-scale factor must be 8. Similarly, the current scale factor is 0.16.

Example 2–3

The versatility of the chart form of analysis shown in Fig. 2–12 may be illustrated by using the chart to determine the voltage waveform at the center of the line. This calculation is carried out as shown in Fig. 2–13 for the conditions of Fig. 2–9. A dotted line is drawn in the time direction at the center of the line in the voltage waveform chart of Fig. 2–12, and the voltages and times are read directly.

The results discussed in this article may be illustrated rather emphatically by the photographs shown in Fig. 2–14. These photographs were obtained by using a section of RG–8A/U transmission line approximately 200 m long. In each view of the oscilloscope the top trace is the voltage waveform at the input end of the line, and the bottom trace is the voltage waveform at the load end of the line. The generated pulse in each case is 30 V high and 1 μs long.

In Fig. 2–14(a) the generator internal resistance is 50 ohms. The bottom picture shows the open-circuit voltage of the generator as 30 V

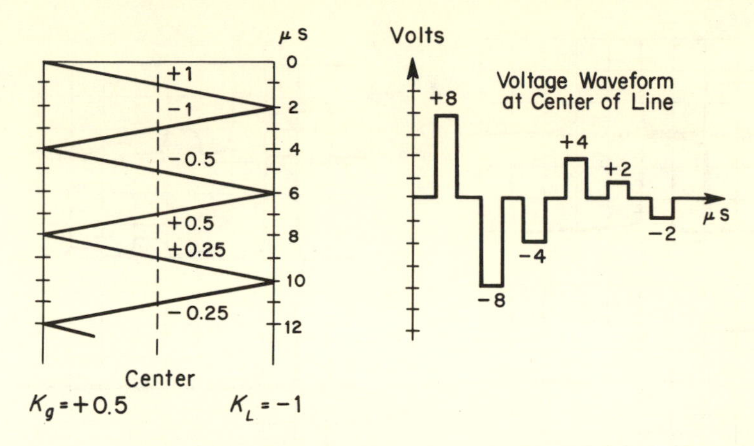

Fig. 2-13. Specific application of line chart.

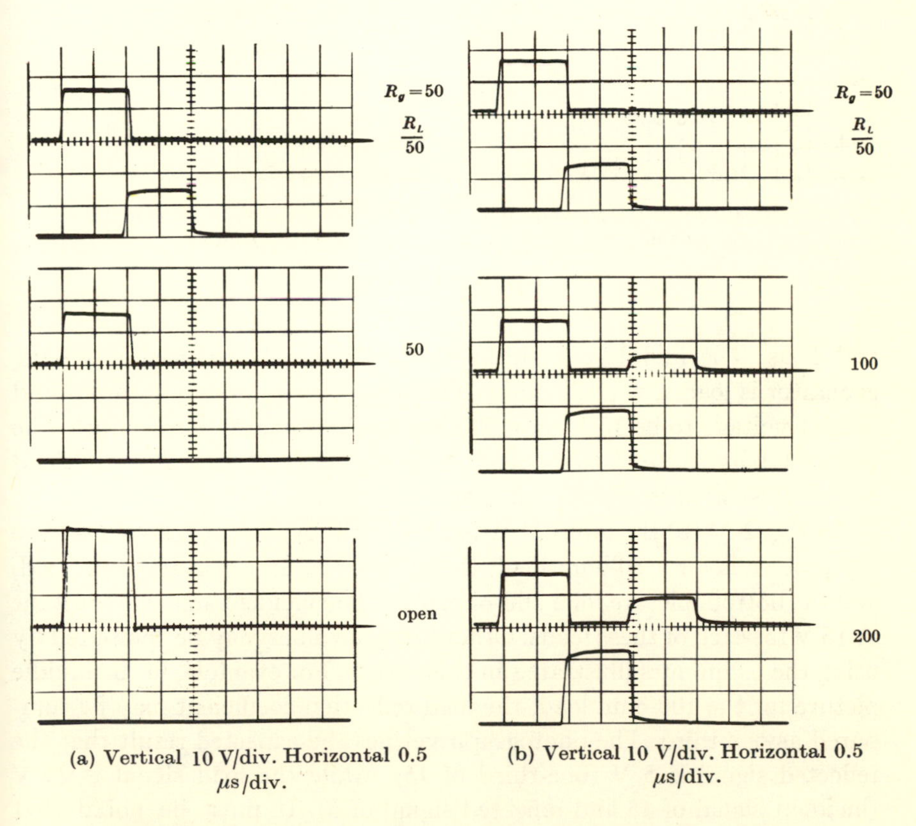

(a) Vertical 10 V/div. Horizontal 0.5 μs/div.

(b) Vertical 10 V/div. Horizontal 0.5 μs/div.

Fig. 2-14. Laboratory demonstration summarizing Art. 2-4.

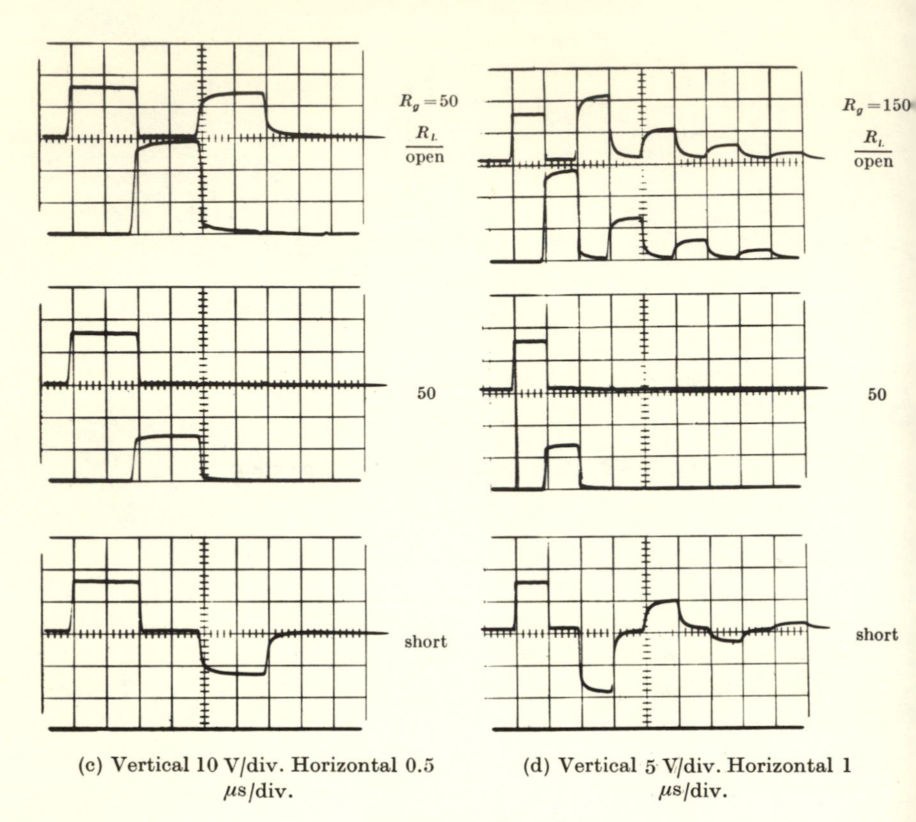

(c) Vertical 10 V/div. Horizontal 0.5 μs/div.

(d) Vertical 5 V/div. Horizontal 1 μs/div.

Fig. 2–14 (*Cont.*)

and 1 μs. The middle one shows the 15-V signal produced when the generator is loaded with 50 ohms. The top picture shows both input and output voltage to be 15 V on a 50-ohm line terminated in 50 ohms. The delay of 1 μs required to travel the 200 m between generator and load is also illustrated in the top picture.

In Fig. 2–14(b) the internal resistance of the generator is 50 ohms. The top picture is for a 50-ohm load, the middle picture for a 100-ohm load, and the bottom picture for a 200-ohm load. The incident signal is constant at 15 V for each of these loads. Other voltage values may be computed by using the techniques illustrated in this article. For example, in the middle picture for the 100-ohm load, the load reflection coefficient may be computed as one-third. The photograph verifies the expected result that the reflected signal is 5 V (one-third of 15), while the load signal is 20 V (incident signal of 15 and reflected signal of 5). It must be noted that the reflected signal can be read on the input waveform because the generator internal resistance is exactly 50 ohms, and so the reflected signal is dissipated there without further reflection. This would not be true for any other generator internal resistance.

In Fig. 2–14(c) the internal resistance of the generator is 50 ohms. The top picture is for an open-circuit load, the middle picture for a 50-ohm load, and the bottom picture for a short-circuit load. By inspection it may be verified that the incident signal from the generator is 15 V for this wide variety of loads. Furthermore the top picture shows that the signal is reflected without sign change for an open circuit, the middle picture shows that no reflection occurs for a 50-ohm load, and the bottom picture shows that the signal is reflected with a sign change for a short circuit. As a consequence of these reflections, the top picture shows that the load voltage on open circuit is twice the incident signal, or 30 V; the middle picture for 50-ohm termination shows the load voltage to be the incident signal of 15 V; and the bottom picture shows the voltage at the short circuit to be zero.

In Fig. 2–14(d) the internal resistance of the generator has been increased to 150 ohms to illustrate multiple reflections. In this case the 4:1 division at the source produces an incident signal of 7.5 V. The top picture is for an open-circuit load, the middle picture for R_0 termination, and the bottom picture for a short-circuit load.

All these photographs show the effect of losses in the system as rounding and attenuation of the signals in the processes of reflection. However, the results are sufficiently close to the predicted values to be very practically used.

2–5. SOLUTION OF THE TRANSMISSION-LINE EQUATIONS FOR SINUSOIDAL SIGNALS

The results given in the previous articles may be applied directly to the sinusoidal analysis of the lossless line. However, the solution obtained in this fashion cannot be easily generalized to include losses. The easiest way to satisfy these conflicting aims of simplicity and generality is to solve the lossless case in the fashion of previous articles, and then to consider the alternate approach, involving phasor notation for the general line equations, which gives the same answers for the lossless line but also works for the general case with losses.

Consider first the case shown in Fig. 2–15, where a generator with

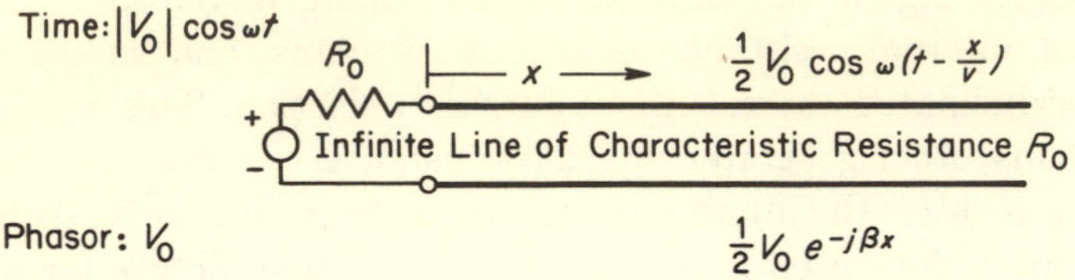

Time: $|V_0| \cos \omega t$ R_0 $x \longrightarrow$ $\frac{1}{2} V_0 \cos \omega (t - \frac{x}{v})$

Infinite Line of Characteristic Resistance R_0

Phasor: V_0 $\frac{1}{2} V_0 e^{-j\beta x}$

Fig. 2–15. Connection between time and phasor notation for transmission line.

internal resistance of R_0 is connected to an infinite-length transmission line. The solution to this problem is known from Art. 2–4 to be

$$\mathcal{V} = \frac{1}{2} \mathcal{V}_g \left(t - \frac{x}{v}\right) \tag{2–17}$$

Only the voltage signal needs to be carried along, since the current tags along like an obedient shadow and may be found at any time by rules formulated in Art. 2–4. If the generator voltage should be the signal

$$\mathcal{V}_g = V_0 \cos \omega t \tag{2–18}$$

then the voltage at any point along the line would be given by

$$\mathcal{V} = \frac{V_0}{2} \cos \omega \left(t - \frac{x}{v}\right) = \frac{V_0}{2} \cos (\omega t - \beta x) \tag{2–19}$$

In these equations V_0 is a real constant, while β has been used to replace (ω/v).

Examination of the voltage expression of Eq. (2–19) shows, as expected, a voltage wave propagating in the $+x$-direction since one could track a constant phase of the voltage by moving in the $+x$-direction along the line with a velocity $v = \omega/\beta$. Considering the voltage distribution along the line at an instant of time, say t_1, shows that adjacent voltage crests are separated by a distance λ such that $\beta\lambda = 2\pi$. The distance λ is called the *wavelength* and is given by the relationship

$$\lambda = \frac{2\pi}{\beta}$$

or

$$\lambda = \frac{v}{f}$$

Now, quite clearly the voltage waveform given in Eq. (2–19) can be written in phasor notation as simply

$$V = \frac{V_0}{2} e^{-j\beta x} \tag{2–20}$$

This expression is of little value in itself, but it should be examined in comparison with Eq. (2–19) to give the connection between the time notation and the phasor notation. Multiple reflections are handled in the fashion of Art. 2–4, but in this article on sinusoidal signals the quantities involved will be written in phasor rather than in the more cumbersome time notation.

Suppose now that the reflection coefficient for a load impedance Z_L is to be found. For this computation it is supposed that a transmission line with characteristic resistance R_0 is connected to the load Z_L, and that, measured at the load, an incident signal[1] of 1, in phasor notation, is forced by a generator connected to the line. Quite clearly, then, a reflection must exist, since the load is different from the characteristic resistance of the line. Suppose that the voltage reflection coefficient is defined to be K_L (note that K_L may now be a complex number to account for possible phase shift of the reflected signal relative to the incident signal). Then, in phasor notation, the condition imposed by the load requires that the voltage reflection coefficient K_L satisfy

$$\frac{\text{Phasor net load voltage}}{\text{Phasor net load current}} = Z_L = \frac{1 + K_L}{(1/R_0) - (K_L/R_0)} = R_0 \frac{1 + K_L}{1 - K_L} \qquad (2\text{–}21)$$

The expression given in Eq. (2–21) may be solved for the value of the voltage reflection coefficient with the result

$$K_L = \frac{Z_L - R_0}{Z_L + R_0} \qquad (2\text{–}22)$$

It should be noted that in the special case when the load is purely resistive, the result given by this formula reduces to that found earlier.

Example 2–4

As an example of the use of Eq. (2–22), suppose that we find the voltage reflection coefficient for a load of $+j50$ ohms on a line with a characteristic resistance of 50 ohms. By direct application of the formula, the value of the voltage reflection coefficient is found to be $j1$. This result is interpreted in phasor notation to mean that the reflected voltage signal is advanced 90 deg relative to the incident signal. Thus, if the incident signal should be $50 + j0$ V, the reflected signal would be $0 + j50$ V. The currents

[1] We may suppose for this analysis that the generator has an internal resistance of R_0, so that any signal reflected from the load is entirely dissipated in the source without re-reflection to the load. If the source were not perfectly matched to the line, multiple reflections would be present, and it would be found that the net incident wave and the net reflected wave would still be related by exactly the same reflection coefficient. This is true because each pair of incident and reflected waves is related by K_L. For signals other than sine waves it would not in general be possible to define a net reflection coefficient in the sense used here.

would evidently be $1 + j0$ A and $0 - j1$ A. The ratio of net load voltage ($50 + j50$ V) to the net load current ($1 - j1$ A) would then quite evidently be $j50$ ohms, as originally stated.

Up to this point it has been possible to show that the excitation of an infinite-length transmission line and the reflection coefficient at a load involve the same basic computations in phasor notation as were considered in the transient analysis of Art. 2–4. We recognize that in using phasor notation the system voltages and currents have reached steady-state values so that the time dependence can be suppressed. One is forced to wonder how much further the transient analysis can carry us toward the solution of the voltages and currents on a terminated transmission line with sinusoidal excitation. The answer is that for all practical purposes we have taken all that we can from the transient analysis and now must return to the differential equations relating the voltage and current on the line to complete the steady-state analysis. Start by considering the equations for the lossless transmission line, Eqs. (2–3) and (2–4), and assume phasor voltages and currents $V(x)$ and $I(x)$ such that

$$\mathcal{V}(x, t) = \text{Re}\,\{V(x)e^{j\omega t}\} \tag{2–23}$$

and

$$\mathcal{I}(x, t) = \text{Re}\,\{I(x)e^{j\omega t}\} \tag{2–24}$$

Substituting into Eqs. (2–3) and (2–4) results in

$$\frac{-dV}{dx} = j\omega LI \tag{2–25}$$

$$\frac{-dI}{dx} = j\omega CV \tag{2–26}$$

Equations (2–25) and (2–26) can be manipulated to yield

$$\frac{d^2V}{dx^2} + \omega^2 LCV = 0 \tag{2–27}$$

and

$$\frac{d^2I}{dx^2} + \omega^2 LCI = 0 \tag{2–28}$$

These equations can now be solved for voltage and current with the results shown in Eqs. (2–29) and (2–30):

$$V(x) = V^+e^{-j\beta x} + V^-e^{+j\beta x} \tag{2–29}$$

$$I(x) = I^+e^{-j\beta x} + I^-e^{+j\beta x} \tag{2–30}$$

where $V^+ = |V^+|e^{j\phi^+}$, $V^- = |V^-|e^{j\phi^-}$, $I^+ = |I^+|e^{j\psi^+}$, $I^- = |I^-|e^{j\psi^-}$ and $\beta = \omega\sqrt{LC}$.

These are general expressions for the voltage and current waveforms on the lossless transmission line in complex phasor form. It is instructive to develop these expressions in real time by application of the phasor transform relationship of Eqs. (2–23) and (2–24):

$$\begin{aligned}\mathcal{V}(x, t) &= \text{Re}\,\{(V^+e^{-j\beta x} + V^-e^{+j\beta x})e^{j\omega t}\} \\ &= \text{Re}\,\{|V^+|e^{j(\omega t-\beta x+\phi^+)} + |V^-|e^{j(\omega t+\beta x+\phi^-)}\} \\ &= |V^+|\cos(\omega t - \beta x + \phi^+) + |V^-|\cos(\omega t + \beta x + \phi^-)\end{aligned}$$

Similarly,

$$\mathcal{I}(x, t) = |I^+|\cos(\omega t - \beta x + \psi^+) + |I^-|\cos(\omega t + \beta x + \psi^-)$$

It is seen that the voltage and current distributions represent the superposition of two waves traveling in opposite directions on the transmission line as we expect. By means of Eqs. (2–25) and (2–26), it is easily verified that

$$I(x) = \frac{V^+}{R_0}e^{-j\beta x} - \frac{V^-}{R_0}e^{+j\beta x} \tag{2–31}$$

and

$$V(x) = I^+R_0e^{-j\beta x} - I^-R_0e^{+j\beta x} \tag{2–32}$$

Ordinarily it is not necessary to solve for the voltages and currents along a line for a given source and internal resistance because, for the special case of sinusoidal signals, the behavior of the system can be much more easily expressed in terms of impedance relationships along the line. In other words, the absolute power level of the signals does not modify the general behavior of the system but serves only to determine the scale factors on the voltage and current. Our problem ordinarily is then to determine the relationship between input impedance, load impedance, and line length. This problem, as illustrated in Fig. 2–16, is very simply solved by the same artifice used in determining the reflection coefficient of the load impedance. Suppose that an incident signal of 1 in phasor notation is incident toward the load (voltage reflection coefficient, K_L) at the input terminals. Then the net input voltage and net input current are found directly, and the input impedance is subsequently found to be (see Fig. 2–16):

$$\frac{\text{Phasor net input voltage}}{\text{Phasor net input current}} = Z_{\text{in}} = R_0\frac{1 + K_Le^{-j2\beta d}}{1 - K_Le^{-j2\beta d}}$$

With reference to the input terminals, the incident voltage is 1, while the reflected voltage is multiplied by the voltage reflection coefficient of K_L and "suffers" a phase delay of $2\beta d$ radians for its round trip of $2d$ meters at a delay constant of β radians/meter. The current is similarly affected

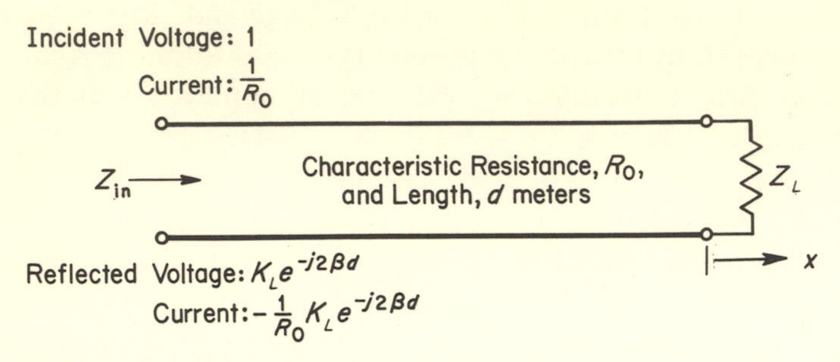

Fig. 2–16. Input impedance of a line.

except for the additional factor of -1 at the load as a result of the positive current convention. Substituting Eq. (2–22) into the equation for Z_{in} above results in an expression for the input impedance in terms of R_0, Z_L, β, and the line length d:

$$Z_{in} = R_0 \frac{Z_L + jR_0 \tan \beta d}{R_0 + jZ_L \tan \beta d} \tag{2–33}$$

Another way of determining the input impedance of the system shown in Fig. 2–16 depends on the simple relationship that connects the reflection coefficient at the load with the value existing at other points along the line. By using the phasor incident voltage of 1 as before, measured at the input of the line, we find that the phasor reflected voltage, measured also at the input of the line, is $K_L e^{-j2\beta d}$. Consequently, since the voltage reflection coefficient is by definition just the ratio of reflected to incident voltage, the voltage reflection coefficient at the input of the line is related to that at the load by the simple equation

$$K_{in} = K_L e^{-j2\beta d} \tag{2–34}$$

Thus the voltage reflection coefficient at some point along the line has the same magnitude as that directly at the load, but it is delayed in phase by $2\beta d$ radians. The connection between the reflection coefficient and impedance methods of relating input and load quantities is summarized in Fig. 2–17.

Example 2–5

As an example of the use of these results, suppose that the input impedance of a 50-ohm line terminated in $+j50$ ohms is desired for a line length such that $\beta d = \pi/2$ radian. Quite clearly the load-voltage reflection coefficient is $(j1 - 1)/(j1 + 1) = +j1$. Then the voltage reflection coefficient at the input end of the line may be found

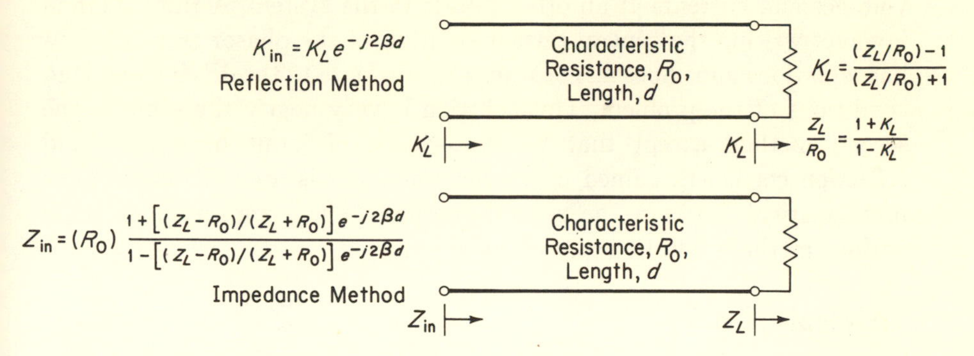

Fig. 2–17. Comparison of reflection coefficient and impedance methods of analysis.

as $(j1)(e^{-j\pi}) = -j1$. And finally from this the input impedance is found to be $50(1 - j1)/(1 + j1) = -j50$ ohms.

The relationship between reflection coefficients shown in Eq. (2–34) will be the basis of a graphical method of relating input impedance, line length, and load impedance in Chapter 3. A very fortunate part of this relationship is that it can be generalized for the lossy transmission line by just having the exponent be $-2\gamma d$, where γ is the complex constant $\alpha + j\beta$. When the losses are zero (i.e., $\alpha = 0$), the result becomes that given above, $-j2\beta d$. The lossy transmission-line problems may also be solved graphically. These graphical techniques are very much faster and less subject to computational error than direct mathematical manipulations. As a consequence, much of the work will be done graphically after this chapter.

The general problem of a generator with an internal impedance of R_0 connected to a line of arbitary length and arbitrary termination can be analyzed by the techniques illustrated thus far. However, if the internal impedance of the generator is other than the characteristic resistance of the line, multiple reflections occur, and then the solution of the problem is less obvious. The problem may be attacked in several different ways. One method is to compute "all" the reflections and re-reflections and sum the results. Fortunately the infinite series involved here is geometric and can be summed by using the familiar result

$$\text{Sum} = \frac{\text{first term}}{1 - \text{ratio}}$$

Another method is to find by solution of simultaneous equations the value of the net incident voltage wave. This result may then be used to predict voltages and currents at all other points in the system. A third method involves writing the general phasor solution of the phasor transmission-line equations and choosing constants appropriate to the driving and load conditions of the problem. This solution is very nearly the same as the second method except that the knowledge of input impedance and reflection coefficient gained in previous analysis is used to simplify the mathematical work in the second method. The comparison of these various methods will be explored by example.

Example 2–6

Consider a source having zero internal resistance driving a $\lambda/4$ section of line terminated in a short circuit, as shown in Fig. 2–18. If the source voltage is $50 + j0$ V

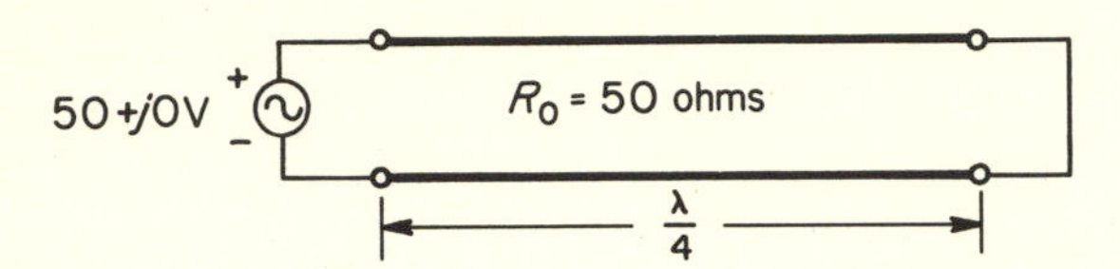

Fig. 2–18. Classical example of a system with multiple reflections.

then the initial incident signal will have a value of $50 + j0$ V. Since the electrical equivalent of one-quarter wavelength is $\pi/2$ electrical radians, the input impedance of the line may be easily found as $+j\infty$ ohms. By using the first method indicated above, it is possible to write down the load current in phasor form as the sum of an infinite series of reflections and re-reflections:

$$I_{\text{load}} = 1e^{(-j\pi)/2} + 1e^{(-j\pi)/2} + 1e^{(-j3\pi)/2} + 1e^{(-j3\pi)/2} + \cdots$$

This expression really contains two geometric series, and the result can be combined and summed as

$$I_{\text{load}} = \frac{2e^{(-j\pi)/2}}{1 - e^{-j\pi}} = -j1$$

Consequently it is seen that a 50-V source produces a load current of 1 A lagging the source by 90 deg.

The second method outlined for solving this problem depends on first finding that the input impedance of this length of line is infinite by Eq. (2–33). The voltage reflection coefficient at the load is found to be -1 by Eq. (2–22). Now suppose that we denote the net phasor incident voltage on the line, as measured at the source, by C^+, which is simply a constant as yet unknown. Using the fact that the reflection coefficient of the load is -1 permits us to write the source voltage in terms of C^+ so that the value of this constant may be found from the condition

$$\text{Source voltage} = 50 + j0 = C^+ + (-1)C^+e^{-j\pi} = 2C^+$$

The value of the net phasor incident voltage is then clearly $25 + j0$. Consequently the load current may be easily written as

$$I_{\text{load}} = \frac{25e^{(-j\pi)/2}}{50} + \frac{25e^{(-j\pi)/2}}{50} = -j1 \text{ A}$$

The third method is to solve Eq. (2–29) subject to the conditions $V(0) = 50 + j0$, $V(\lambda/4) = 0$. Substituting:

$$50 = I^+R_0 - I^-R_0$$
$$0 = I^+R_0e^{-j(\pi/2)} - I^-R_0e^{+j(\pi/2)}$$

Since $R_0 = 50$ ohms, we can write:

$$1 = I^+ - I^-$$
$$0 = I^+ + I^-$$

Thus:

$$I^+ = -I^- = \tfrac{1}{2}$$

Now by means of Eq. (2–30), one can write

$$I(x) = \tfrac{1}{2}e^{-j\beta x} - \tfrac{1}{2}e^{+j\beta x}$$

At $x = \lambda/4$, $\beta x = \pi/2$, and the load current is

$$I_{\text{load}} = I(\lambda/4) = \tfrac{1}{2}e^{-j(\pi/2)} - \tfrac{1}{2}e^{+j(\pi/2)} = -j\,1 \text{ A}$$

Obviously the incident signal on a transmission line driven from a source with an internal impedance equal to the characteristic resistance of the line will be half of the open-circuit source voltage and completely independent of the load. Quite reasonably then but not so obviously, the incident signal on a transmission line driven from a source with an internal impedance value other than that of the characteristic resistance of the line will not be independent of the load. The example given in Fig. 2–18 will serve to emphasize this point. With the load being a short circuit, it is found that the incident signal at the input of the line is $25 + j0$ V. If, however, the load had been 50 ohms, then quite clearly the incident signal would have been $50 + j0$ V. The difference is, of course, very marked. The importance and significance of having an incident wave independent of the load will be discussed further in Art. 2–7 on calculation of power in lossless transmission-line systems.

2–6. STANDING WAVES ON TRANSMISSION-LINE SYSTEMS

An interesting and important aspect of transmission-line analysis involves the standing-wave pattern that exists on a transmission line with a mismatched load. The signal reflected from the mismatched load tends at some points along the line to interfere destructively with the incident signal, while at other points it tends to interfere constructively. The net effect is a standing-wave pattern having crests or troughs every half-wavelength. This phenomenon is similar to that observed on a violin string when it is plucked at some point. The string is vibrating harmonically at any given point along its length; however, the amplitude of the vibration is a function of position along the string. Similarly, a transmission line has a sinusoidal voltage at each point, but the amplitude of the sinusoidal variation is a function of position along the line. Illustrations of typical standing-wave patterns are given in Fig. 2–19 for various values

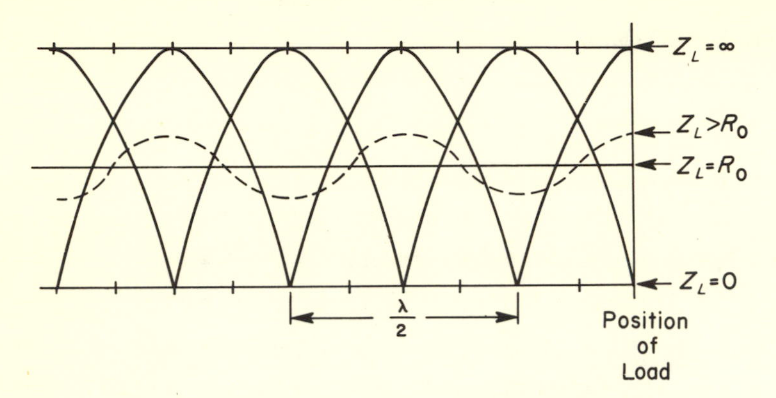

Fig. 2–19. Standing waves on a transmission line.

of load. For the purposes of this figure it has been assumed that the source has an internal resistance equal to the characteristic resistance of the line so that the incident wave is independent of the load.

The standing-wave pattern shown in Fig. 2–19 can be more clearly understood by reference to Fig. 2–20, where the phasor representation of the incident and reflected signals has been shown along with the resultant standing-wave pattern for the particular case when the load is an open

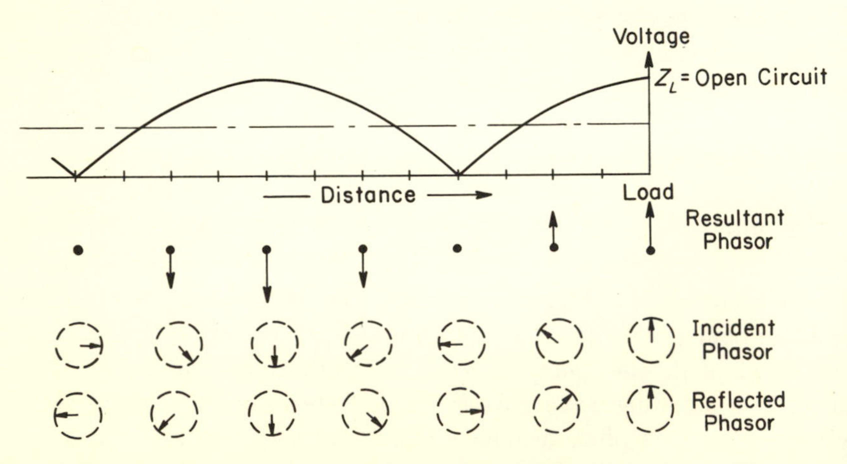

Fig. 2–20. Standing waves in terms of incident and reflected traveling waves.

circuit. It should be noted that the signal lags in phase as it "travels" in the direction of propagation, regardless of whether the discussion involves the incident or the reflected signal. It also should be noted that the destructive interference shows up every half-wavelength along the line. The fact that the spacing is every half-wavelength may be easily explained by noting that a full-wavelength variation occurs in the difference between incident and reflected signals for every half-wavelength movement along the line because the total path (down and back) has been increased by a full wavelength.

An interesting aspect of the results shown in Fig. 2–20 is that the voltage is in time phase all along the line for the special case of reactive termination. In this case the termination is infinite. This result for an open-circuit load makes it seem more reasonable to assume in the antenna chapters of this text that the current varies sinusoidally in amplitude along the antenna but that it is in time phase at each point along the antenna. Comparison of the antenna with the open-circuited transmission line does not prove that the assumption is correct, but it does make the assumption easier to accept.

The interference phenomena described in Fig. 2–20 can be derived mathematically by considering Eq. (2–29). If the line is open-circuited, $V^- = +V^+$. Substituting into Eq. (2–29) results in

$$V(x) = V^+(e^{-j\beta x} + e^{+j\beta x})$$

or

$$V(x) = 2V^+ \cos \beta x \tag{2–35}$$

This assumes that the open circuit exists at $x = 0$ and the line extends back in the $-x$-direction. If the line is short-circuited, the voltage distribution becomes

$$V(x) = -2jV^+ \sin \beta x \tag{2–36}$$

In both of the cases just considered, the voltage distribution is called a *pure standing wave.* The reason for the term "pure standing wave" can be seen by considering the real-time expressions for the voltage distributions (assuming $\phi^+ = 0$):

$$\mathcal{V}(x, t) = 2|V^+| \cos \beta x \cos \omega t \qquad \text{(open-circuit case)} \tag{2–37}$$

$$\mathcal{V}(x, t) = 2|V^+| \sin \beta x \sin \omega t \qquad \text{(short-circuit case)} \tag{2–38}$$

In either case the voltage waveform is fixed (standing) along the line although it is oscillating as a sinusoidal function of time.

The current waveforms for these terminations are easily developed by considering Eq. (2–30). For an open-circuit termination, $I^- = -I^+$ at the load. Substitution results in

$$I(x) = I^+e^{-j\beta x} - I^+e^{+j\beta x}$$

$$I(x) = -2jI^+ \sin \beta x = -2j\frac{V^+}{R_0} \sin \beta x \tag{2–39}$$

In a similar manner we can show that a short-circuit termination results in a current distribution of

$$I(x) = \frac{2V^+}{R_0} \cos \beta x \tag{2–40}$$

In real time these current distributions become

$$\mathcal{I}(x, t) = \frac{2|V^+|}{R_0} \sin \beta x \sin \omega t \qquad \text{(open-circuit case)} \tag{2–41}$$

$$\mathcal{I}(x, t) = \frac{2|V^+|}{R_0} \cos \beta x \cos \omega t \qquad \text{(short-circuit case)} \tag{2–42}$$

Notice that in both cases the voltage and current are 90 degrees out of phase and have maxima and minima along the line that are shifted from one another by one-quarter of a wavelength. That is to say that a maximum in the voltage distribution is coincident along the line with a minimum in the current distribution, and vice versa.

In the case where a unity reflection coefficient termination is not present, a general expression for the voltage distribution along the line can be written as

$$V(x) = V^+e^{-j\beta x} + K_L V^+e^{+j\beta x} \tag{2–43}$$

where K_L is the voltage reflection coefficient at the load. This can be rewritten as

$$V(x) = V^+e^{-j\beta x}(1 - K_L) + 2K_L V^+ \cos \beta x \tag{2–44}$$

Equation (2–44) is seen to be the superposition of a traveling wave and a standing wave.

The magnitude of the voltage distribution for this case is easily seen to be

$$|V(x)| = |V^+|\sqrt{1 + |K_L|^2 + 2|K_L| \cos (2\beta x + \phi_L)} \tag{2–45}$$

where $K_L = |K_L|e^{j\phi_L}$. Notice that for $2\beta x + \phi_L = \pm 2m\pi$; $m = 0, 1, 2, \ldots,$

$$|V(x)| = |V^+|(1 + |K_L|)$$

and when $2\beta x + \phi_L = \pm(2m + 1)\pi$; $m = 0, 1, 2, \ldots,$

$$|V(x)| = |V^+|(1 - |K_L|)$$

These conditions correspond to the maxima and minima shown by the dashed line in Fig. 2–19. That particular curve corresponds to $\phi_L = 0$.

One of the important reasons for considering the standing-wave patterns on a transmission line is that the value of the load impedance and the resultant standing-wave pattern are so closely related that the impedance may be computed from a knowledge of only certain key pieces of information about the pattern: the voltage standing-wave ratio and the distance between the load and a voltage minimum. This information, needed to determine the load impedance, may be found very easily with the aid of a slotted line.

This slotted line is, as its name implies, merely a section of coaxial line with a very small slot along its length. A probe is inserted to a very shallow depth into this slot, and the amplitude of signal coupled to this probe is proportional to the amplitude of the voltage between the conductors. The position of the voltage minimum may be observed with the aid of a suitably calibrated scale along the length of the line. The voltage standing-wave ratio (VSWR, or simply S) may be found from the signal picked up on the probe with the aid of a suitably calibrated detecting system.

An important relationship exists between the magnitude of the reflection coefficient of the load and the value of the VSWR. This relationship, as expressed by the formula

$$S = \frac{1+|K_L|}{1-|K_L|} = \frac{V_{\max}}{V_{\min}} \tag{2–46}$$

follows directly by considering an incident voltage signal with an amplitude of 1 V. The reflected voltage signal would evidently be K_L volts. Thus the maximum voltage on the line would occur at such a position that the incident and reflected values were *in phase*, and the voltage would be $1 + |K_L|$ volts. The minimum voltage would similarly be found at such a position that the incident and reflected values were *out of phase*, and the voltage would be $1 - |K_L|$ volts. Since by definition the VSWR is just the ratio of the maximum voltage to the minimum voltage, the value of S is found as in Eq. (2–46).

Several important features of the voltage standing-wave ratio are: (1) The minimum value of S is 1 and occurs when the load is R_0; (2) the value of S for a resistive load is the larger of R_L/R_0, or R_0/R_L; (3) the value of S for a purely reactive load is infinity. A proof of these results is left to the problems.

For many purposes the value of the VSWR on a system is a sufficient measure of the nearness of match of the load to the line. And quite equally well, the magnitude of the reflection coefficient will give the same information, though presented in a different fashion, and is more conveniently measured. The connection between S and $|K_L|$ may be presented graphically as shown in Fig. 2–21. The value of S is often read more conveniently

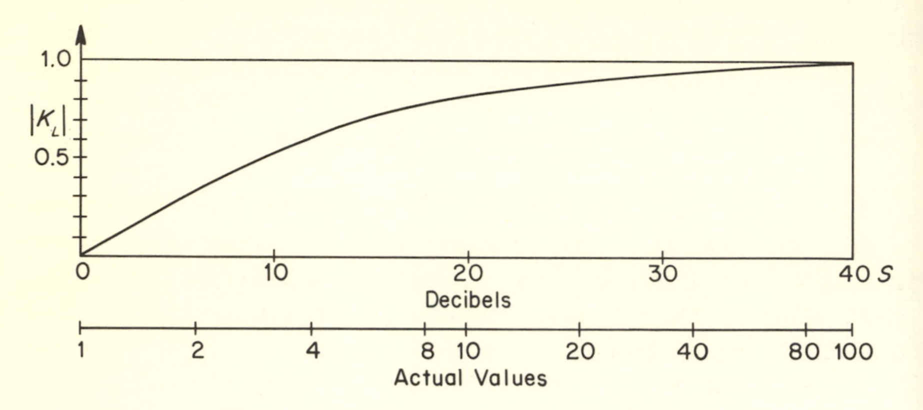

Fig. 2–21. Relationship between S and $|K_L|$.

in terms of decibels. A scale of S in decibels has been included in Fig. 2–21. As a practical matter the decibel rating of S gives a more informative figure of the match or mismatch of a system. For example, the difference in performance of systems having standing-wave ratios of 1 and 4 is very great, whereas the difference for standing-wave ratios of 10 and 14 is quite small. On a decibel scale these results are viewed in a better perspective as a 12-dB change in the first case and a 3-dB change in the second case.

There are many reasons why the load should be reasonably well matched to the line. Among these reasons are the desirability of achieving minimum signal distortion, maximum load power, and minimum line losses. These concepts will be discussed in considerable detail in Chapter 4, which deals with transmission-line matching systems. At this point it is sufficient to point out the nearness of impedance matching necessary to achieve reasonable system performance. For many practical radio-frequency applications, a standing-wave ratio of 1.5 is considered acceptable. In this case the magnitude of the reflection coefficient is 0.2 and results in a reflected power[2] which is only 4 per cent of the incident power. For

[2] A certain amount of caution must be exercised in using the concepts of incident and reflected power, since they have practical meaning only on a system that is driven from a matched source. Another limitation occurs in the consideration of lossy lines, where the concept loses practical significance if the losses are large.

delay-line applications the situation is more critical and the load must be chosen very carefully to minimize signal distortion from reflection. In this case the matching may be commonly done with resistances having tolerances of 5 per cent or less, and the resulting reflection coefficient will be found to be 2.5 per cent or less. For telephone lines the matching may be even more critical to maintain adequate performance over very long distances where many repeating stations are required. Even a 1-dB error in this case amounts to a fantastic 500-dB error when cascaded over 500 repeating links. For power-line applications, matching is possible only in a very general sense, since the load is primarily fixed by the consumer and is not in the control of the power companies except to a very limited degree (as, for example, tap changing on transformers and the addition of capacitors for power factor correction).

2–7. POWER ON LOSSLESS LINES

On lossless or nearly lossless lines it is possible to talk about power flow in a rather worthwhile and informative manner. Basically, one can go to any point on the line and formulate the instantaneous power flowing in the $+x$-direction as the product of the time-dependent voltage and current at that point. That is,

$$p(x, t) = \mathcal{V}(x, t)\mathcal{I}(x, t)$$

Converting Eqs. (2–29) and (2–31) to time-dependent form with $V^- = K_L V^+$ and substituting into the instantaneous power expression results in

$$p(x, t) = \frac{|V^+|^2}{R_0}\left[\cos^2 (wt - \beta x + \phi^+) - |K_L|^2 \cos^2 (wt + \beta x + \phi^+ + \phi_L)\right]$$

The time-averaged power flowing in the $+x$-direction at x is defined as

$$P(x) = \frac{1}{T}\int_0^T p(x, t)\, dt$$

where T is one complete period of the sinusoidal signal. In the case being considered it is easy to see that

$$P(x) = \frac{|V^+|^2}{2R_0}(1 - |K_L|^2)$$

by performing the integration of the instantaneous power expression.

One can arrive at the same result with much less effort by working with the phasor forms of the voltage and current and formulating the product of the phasor voltage and the conjugate of the phasor current at the point x on the line. The result of this operation is

$$V(x)I^*(x) = [V^+e^{+j\beta x}(1 + K_Le^{-j2\beta x})]\left[\frac{V^{+*}}{R_0}e^{+j\beta x}(1 - K_L^*e^{-j2\beta x})\right]$$
$$= \frac{|V^+|^2}{R_0}(1 - |K_L|^2 + K_Le^{j2\beta x} - K_L^*e^{-j2\beta x})$$
$$= \frac{|V^+|^2}{R_0}[1 - |K_L|^2 + j2|K_L| \sin(2\beta x + \phi_L)] \tag{2-47}$$

Since we are interested in the time-averaged power flow on the line, which was determined in the preceding paragraph, we see that

$$P(x) = \tfrac{1}{2}\,\text{Re}\,\{V(x)I^*(x)\}$$
$$= \frac{1}{2}\frac{|V^+|^2}{R_0}(1 - |K_L|^2)$$
$$= \frac{(V_{\text{rms}}^+)^2}{R_0}(1 - |K_L|^2) \tag{2-48}$$

The imaginary part of Eq. (2–47) is a reactive power flow (VAR flow to power engineers) which represents the energy stored in the standing waves on the transmission line and the reactance of the load. If the line is lossless and if the source is perfectly matched to the line (these are very important qualifications), it is possible to discuss power transfer on the system in terms of incident and reflected power.

For a lossless line the relationship between voltage and current is fixed at the characteristic resistance of the transmission line for a wave traveling in either direction along the line. Consequently, if the source is perfectly matched to the line so that no re-reflections may occur at the source, then the power flow on the line may be measured in terms of incident power, given as $(V_{\text{rms}}^+)^2/R_0$, and reflected power, given as $(V_{\text{rms}}^-)^2/R_0$. The rms values of incident and reflected voltages are related by the magnitude of the reflection coefficient as $V_{\text{rms}}^- = |K_L|V_{\text{rms}}^+$. Therefore the ratio of reflected power to incident power is easily found to be

$$\frac{\text{Power reflected}}{\text{Power incident}} = |K_L|^2 \tag{2-49}$$

Although the time-averaged power reflected from a termination is always related to the time-averaged power incident on the termination by the magnitude of the reflection coefficient squared, the time-averaged incident power is independent of the load only for the condition that the source is matched to the line.

Often the incident power under these conditions is termed the available power since it represents the maximum power that could be delivered to a load.

Example 2–7

As an example of the use of the result given in Eq. (2–49), suppose that the ratio of reflected to incident power is desired for a 75-ohm load on a 50-ohm line. In this case the reflection coefficient is found to be $25/125 = 0.2$, and so Eq. (2–49) may be used to find that only 4 per cent of the power is reflected from the load. This means that if the matched source could supply 50 W to a matched load of 50 ohms, it would supply only 48 W to a mismatched load of 75 ohms.

This same conclusion can be deduced by an alternate means. The source voltage required to supply 50 W to a 50-ohm matched load may be very easily found to be 100 V, rms (remember the 50 ohms internal resistance of the source). Now, since the conclusion presented in Eq. (2–49) does not include the line length in any way, we are forced to conclude that this ratio of power is unaffected by line length (only for a matched source as assumed in its derivation). Suppose then that the line length is zero and that the 75-ohm load is connected across the 100-V source with 50 ohms internal resistance. Clearly the load power in the 75-ohm resistor would be 48 W, as given earlier.

For personal satisfaction this result may also be verified for a line length such as $\lambda/4$. The input impedance in this case could be calculated to be 33.3 ohms, and the power dissipated in this resistance, which must be the same as the load power because the line is lossless, is again found to be 48 W.

These results are presented in summary form in Fig. 2–22.

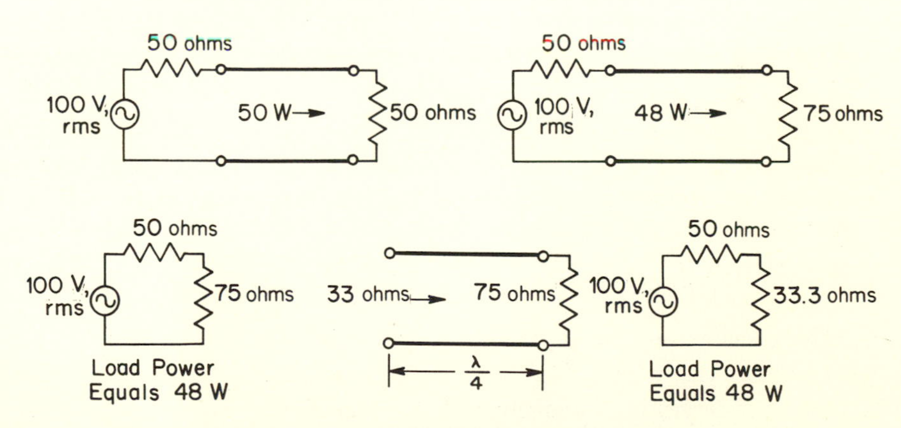

Fig. 2–22. Power calculation for line with a matched source.

A reasonable question at this point is how to make certain that the source is correctly matched to the line. One method would be to use matching techniques as given in Chapter 4. This method has the advantage of being nearly lossless. A method that finds considerable use in the laboratory "pads" the signal source with an attenuator having a characteristic resistance equal to that of the line. This technique produces considerable power loss but has the advantage of simplicity, since it does not require any adjustment such as do the matching schemes. The "matching" produced in this fashion works both ways and hence also minimizes the effect of changing loads on the power output and frequency of the power source. The effectiveness of this method of "matching" may be seen by example.

Example 2–8

The pads used to isolate the power source from the transmission line would commonly produce attenuations in a matched system ranging from 6 dB upward. A typical value is 20 dB; this produces a 100-times power loss in a matched system. The constants of a pi section having 20 dB of attenuation for a 50-ohm characteristic resistance are shown in Fig. 2–23.

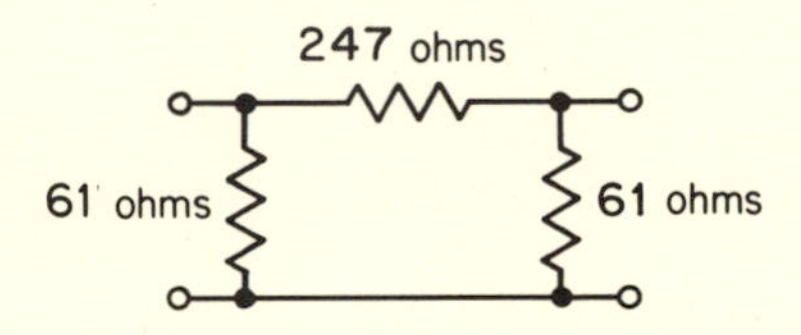

Fig. 2–23. Constants of a 20-dB, 50-ohm attenuator.

The effect of the pad on the system may be seen by consideration of several special cases. The pad is designed so that if the resistance of the generator is 50 ohms, then the impedance looking back into the pad connected to the generator will be exactly 50 ohms. The extreme cases of mismatched generator would be for internal impedances of either zero or infinity. In these extreme cases the impedance looking back into the pad would be either 49 or 51 ohms, and clearly this result is sufficiently close to 50 ohms to produce an exceptionally good match. In fact, a 10-dB pad produces a match which is sufficiently close for most practical laboratory use, although it may be noted that the closer match is produced by pads with the higher values of attenuation.

The measurement of the magnitude of the reflection coefficient is in general more easily done than measurement of VSWR. One of the simpler schemes for determination of the reflection coefficient is illustrated in Fig. 2–24. In this figure a short section of transmission line is placed in proximity to the main line but is loosely coupled to it to prevent disturbing the main line; this auxiliary line is correctly terminated in its characteristic impedance at each end. The coupling from the main line starts a wave traveling on the auxiliary line in the opposite direction as the wave

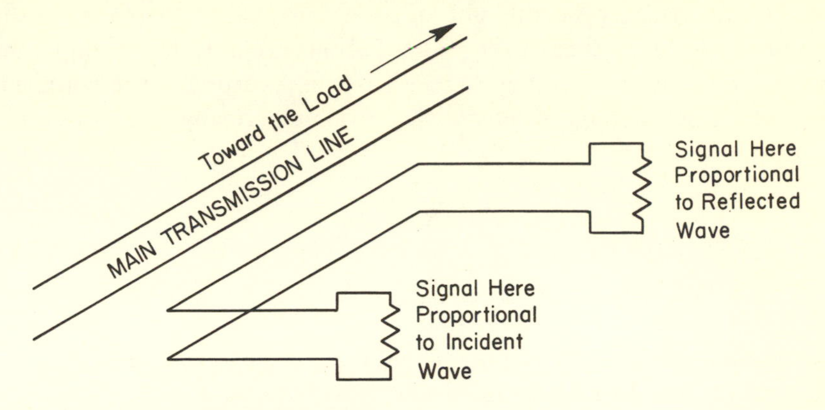

Fig. 2–24. Simple directional coupler.

on the main line, and proper termination of the auxiliary line prevents reflections. Thus the voltage measured across the termination at each end of the auxiliary line is found to be proportional to the main-line incident wave at that end of the coupling region. The ratio of the magnitudes of the incident and reflected components obtained in this fashion may be used to compute the magnitude of the reflection coefficient.

Another device for measurement of the magnitude of the reflection coefficient is shown in Fig. 2–25 and has the distinction of working all the

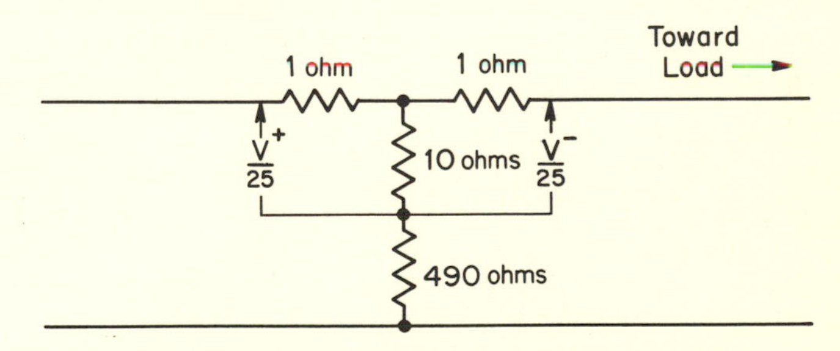

Fig. 2–25. Device to measure magnitude of reflection coefficient on a 50-ohm line.

way down to direct current. Its operation depends on the relationship between the voltage and current at any point on the line as expressed by the equations

$$V = V^+ + V^- \tag{2-50}$$

and

$$I = \frac{V^+}{R_0} - \frac{V^-}{R_0} \tag{2-51}$$

These equations may be thought of as having been written with the reference point for x taken at the point of observation. If, for example, the current is observed as a voltage across a 1-ohm resistor and the voltage is observed across a voltage divider with a ratio numerically equal to R_0, the results will be

$$V_v = \frac{V^+}{R_0} + \frac{V^-}{R_0} \tag{2-52}$$

and

$$V_i = \frac{V^+}{R_0} - \frac{V^-}{R_0} \tag{2-53}$$

Now the sum and differences may be obtained to yield independently the values proportional to the incident and reflected waves:

$$V_v + V_i = \frac{2V^+}{R_0} \tag{2-54}$$

and

$$V_v - V_i = \frac{2V^-}{R_0} \tag{2-55}$$

Consequently the magnitude of the reflection coefficient may be found as the ratio of the magnitudes of each of these signals. The physical construction of a device to do this is shown in Fig. 2–25. The constants shown are not the optimum values but are sufficiently close for practical purposes and illustrate the point more clearly than the exact values. One point of discrepancy is that the total shunt resistance should be of such a value that the characteristic resistance of the inserted device will be exactly 50 ohms; the value of resistance in this case is closer to 1250 ohms than to the 500 ohms used.

Example 2–9

As an example of the use of the device shown in Fig. 2–25, suppose that a generator with an internal resistance of 50 ohms is used to drive a 75-ohm load. If the

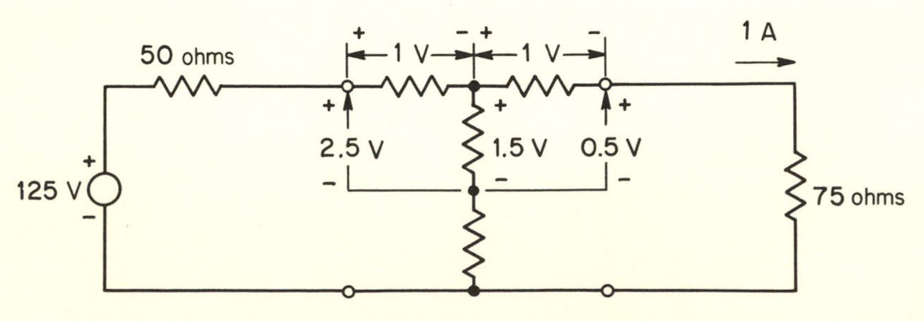

Fig. 2–26. Approximate voltages and currents in Example 2–9.

open-circuit voltage of the source is 125 V, then the voltages and currents will be very nearly as indicated in Fig. 2–26. Very clearly the value proportional to the incident signal is 2.5 V, whereas the value proportional to the reflected signal is 0.5 V. Consequently the reflection coefficient may be computed to have a magnitude of 0.2.

2–8. EFFECT OF LOSSES ON AC SOLUTION

The general ac transmission-line equations are developed from Eqs. (2–1) and (2–2), assuming a sinusoidal time dependence and retaining the line parameters R and G. The result is the pair of equations

$$-\frac{dV}{dx} = ZI \tag{2–56}$$

and

$$-\frac{dI}{dx} = YV \tag{2–57}$$

In these equations $V = V(x)$ and $I = I(x)$ represent the phasor voltage and current along the line, respectively. The quantities Z and Y represent $R + j\omega L$ and $G + j\omega C$, respectively.

Differentiating and combining Eqs. (2–56) and (2–57) give two equations:

$$\frac{d^2V}{dx^2} = ZYV \tag{2–58}$$

and

$$\frac{d^2I}{dx^2} = ZYI \tag{2–59}$$

The solutions of these equations are easily seen to be:

$$V(x) = V^+e^{-\gamma x} + V^-e^{+\gamma x} \tag{2–60}$$

and

$$I(x) = I^+e^{-\gamma x} + I^-e^{+\gamma x} \tag{2–61}$$

where $\gamma = \sqrt{YZ}$ is the propagation constant. In general, γ will be a complex number which can be expressed in terms of its real and imaginary parts as

$$\gamma = \alpha + j\beta = \sqrt{(R + j\omega L)(G + j\omega C)} \tag{2–62}$$

The phasor voltage of Eq. (2–60) and the phasor current of Eq. (2–61) can be written in time-dependent form as

$$\begin{aligned}
\mathcal{V}(x, t) &= \mathrm{Re}\{(V^+e^{-\gamma x} + V^-e^{+\gamma x})e^{j\omega t}\} \\
&= \mathrm{Re}\{|V^+|e^{-\alpha x}e^{j(\omega t - \beta x + \phi^+)} + |V^-|e^{+\alpha x}e^{j(\omega t + \beta x + \phi^-)}\} \\
&= |V^+|e^{-\alpha x} \cos(\omega t - \beta x + \phi^+) + |V^-|e^{+\alpha x} \cos(\omega t + \beta x + \phi^-)
\end{aligned}$$

and

$$\begin{aligned}\mathcal{I}(x, t) &= \mathrm{Re}\{(I^+e^{-\gamma x} + I^-e^{+\gamma x})e^{j\omega t}\} \\ &= |I^+|e^{-\alpha x} \cos (\omega t - \beta x + \psi^+) + |I^-|e^{+\alpha x} \cos (\omega t + \beta x + \psi^-)\end{aligned}$$

It is apparent that the traveling-wave nature of the solution which was evident in the lossless-line analysis has been retained, but in the lossy-line situation the signal is attenuated in an exponential manner as it propagates along the line.

The current function can be related to the voltage by

$$I(x) = \frac{V^+e^{-\gamma x}}{Z_0} - \frac{V^-e^{+\gamma x}}{Z_0} \tag{2-63}$$

The term Z_0 represents the characteristic impedance and is given by the expression

$$Z_0 = \sqrt{Z/Y} = \sqrt{\frac{R + j\omega L}{G + j\omega C}}$$

In general both γ and Z_0 are complex functions of frequency, which means that signals of different frequency will be affected differently as they travel over the transmission line. This effect can degrade a signal made up of many frequency components to an intolerable extent.

In the special case where $L/R = C/G$, the propagation constant reduces to

$$\gamma = \alpha + j\beta = RG + j\omega\sqrt{LC} \tag{2-64}$$

The facts that the attenuation constant is independent of frequency and that the phase shift constant is linearly dependent on frequency assure that the propagation is distortionless. This viewpoint of distortionless transmission can be verified by considering any input signal as being decomposed into frequency components by Fourier analysis. Since each component is attenuated by the same factor and each component is delayed by the same time interval, recombination of the results will clearly be an attenuated and delayed replica of the input signal. The characteristic impedance Z_0 becomes purely real and equal to

$$Z_0 = R_0 = \sqrt{L/C} \qquad \text{(distortionless case only)} \tag{2-65}$$

when the condition for distortionless transmission is satisfied.

Perhaps the greatest advantage of the analysis of distortionless transmission is that it points out the possibility of correcting a transmission system so that it will be distortionless. For example, in a telephone transmission line it is usually found that the shunt conductance is very small, whereas the series resistance of the conductors is very great.

In order to obtain distortionless transmission in this case, it is necessary to increase the series inductance of the transmission system. An increase of the series inductance on a distributed base is not particularly feasible, but it is nevertheless practical to consider adding lumped inductors at suitably close intervals in order to produce essentially the same effect. Since the highest frequency of interest on the telephone lines may be 5 kHz, which has a wavelength in air of some 37 miles, it is possible to space the inductors or "loading coils" at intervals of a few miles and achieve a practical equivalent of distributed loading.

2-9. AC CALCULATIONS FOR SMALL LOSSES

In many cases the transmission-line losses are very small; in this case certain simplifying approximations can be made. In particular we may show that

$$\gamma = \alpha + j\beta = \left(\frac{R}{2R_0} + \frac{G}{2G_0}\right) + j\omega\sqrt{LC} \tag{2-66}$$

and

$$Z_0 = R_0 = \sqrt{\frac{L}{C}} \tag{2-67}$$

if the line losses are sufficiently small.[3]

The attenuation constant α can also be determined in the case of small losses by a perturbation method of analysis. Imagine an ac signal source supplying power to a characteristic impedance load through a length of transmission line. The sinusoidal voltage and current waveforms will be attenuated by a term $e^{-\alpha x}$ as they travel toward the load. If V^+_{rms} represents the rms voltage at $x = 0$, then the power transmitted toward the load at the arbitrary distance x is given by

$$P_T = \frac{(V^+_{\text{rms}})^2}{R_0}\, e^{-2\alpha x} \tag{2-68}$$

[3] For example, the propagation constant γ may be written as

$$\gamma = \sqrt{ZY} = \sqrt{(R + j\omega L)(G + j\omega C)}$$

$$= \sqrt{(j\omega L)(j\omega C)\left(1 + \frac{R}{j\omega L}\right)\left(1 + \frac{G}{j\omega C}\right)}$$

Now using the approximations $(1 + a)(1 + b) = (1 + a + b)$ and $\sqrt{(1 + c)} = 1 + c/2$, which hold for a, b, and c small compared with 1, will permit us to write

$$\gamma = j\omega\sqrt{LC}\left(1 + \frac{R}{j2\omega L} + \frac{G}{j2\omega C}\right) = \left(\frac{R}{2R_0} + \frac{G}{2G_0}\right) + j\omega\sqrt{LC}$$

as stated.

This result is obtained by noting that if the transmission line is severed at the point x, then the input impedance "looking" toward the load will be R_0, while the rms voltage across the line will be $V_{\text{rms}}^{+}e^{-\alpha x}$. The formula of Eq. (2–68) expresses the simple fact that, because of losses, the power passing a given cross-section of the line decays exponentially with distance from the source. If the losses are assumed to be small, then we may note that

$$-\left(\frac{dP_T}{dx}\right) = +2\alpha \frac{(V_{\text{rms}}^{+})^2}{R_0} e^{-2\alpha x} = 2\alpha P_T \tag{2–69}$$

represents the power lost in dissipation per unit length of the line. Consequently, if P_L denotes the power lost per unit length of line, then we may express the attenuation constant as

$$\alpha = \frac{P_L}{2P_T} \tag{2–70}$$

An important point to note about this formula is that it can be applied to any *system* where the power transmission obeys an exponential law of decay. As a result, Eq. (2–70) may be applied practically to attenuation in waveguide structures where an apparently more direct method of calculation would be quite impractical.

In a transmission-line system we note that the power lost per unit length of line in the series resistance and shunt conductance may be formulated at $x = 0$ in the form

$$P_L\,dx = \left(\frac{V_{\text{rms}}^{+}}{R_0}\right)^2 (R\,dx) + (V_{\text{rms}}^{+})^2(G\,dx) \tag{2–71}$$

This result was obtained by considering a dx length of line at $x = 0$ and computing the series and shunt losses. The power transmitted down the line at $x = 0$ is evidently just

$$P_T = \frac{(V_{\text{rms}}^{+})^2}{R_0} \tag{2–72}$$

Now, combining the values of P_L and P_T as indicated by the formula given for the attenuation constant α in Eq. (2–70) gives

$$\alpha = \frac{R}{2R_0} + \frac{G}{2G_0} \tag{2–73}$$

PROBLEMS

2–1. Verify that Eqs. (2–9) and (2–10) are correct, by direct substitution into Eqs. (2–3) and (2–4). At the same time verify that v and R_0 are as given by Eqs. (2–11) and (2–12).

2–2. Discuss the fact that the two traveling waves of Eq. (2–8) are evidence of the same phenomena from several different approaches. *Hint:* Consider that v appears as v^2, that x appears symmetrically in the equations, and that the directional choice of x was arbitrary.

[For Problems 2–3 through 2–5 the transmission-line system is assumed to be a generator, with an internal resistance of 150 ohms, supplying a 400-m transmission line with a load resistance of 30 ohms. The characteristic resistance of the line is assumed to be 50 ohms and the velocity to be 200 m/μs. The generated pulse is 32 V in amplitude.]

2–3. Sketch and label the voltage and current waveforms as a function of time at the input end of the transmission line for a 1-μs pulse. Repeat for a 6-μs pulse.

2–4. Sketch and label the voltage and current waveforms as a function of time at the load end of the line for a 1-μs pulse.

2–5. Sketch and label the voltage and current waveforms as a function of time at a point midway along the line for a 1-μs pulse.

2–6. Suppose now that the 30-ohm load of Problems 2–3 to 2–5 is replaced by 150 m of 30-ohm line terminated in a short circuit. Sketch and label the voltage and current at the junction of the two lines for a 1-μs pulse. *Hint:* Combine two charts for this analysis.

2–7. By analogy with wave propagation of light, justify the fact that the incident wave along the transmission line is independent of the load.

2–8. By reference to Appendix A it may be noted that the parameters L and C for a transmission line involve the dimensions in a reciprocal fashion so that the product LC is independent of the geometrical configuration of the line; the product is in fact $\mu\epsilon$ under certain idealizing assumptions. Show that for air-insulated lines the velocity is that of light, c. Also show that for simple dielectric insulation ($\mu_r = 1$), the velocity of propagation is reduced by the square root of the relative dielectric constant.

2–9. A common dielectric for insulation of transmission lines is polyethylene, which has a relative dielectric constant of about 2.25. Show that the velocity of propagation on such a line is essentially 200 m/μs.

2–10. Explain how to determine the length of a transmission line if only one end of the line is available for measurement but it is known that the line is open at the far end.

2–11. Explain how an oscilloscope, a pulse generator, and a precision-calibrated variable resistor may be used to determine accurately the characteristic resistance of a transmission line.

2–12. For delay lines it is convenient to measure the capacity parameter C and the characteristic resistance R_0. Explain how these data can be used to determine the velocity of propagation along such a line. Determine the length of line required to produce a delay of 1 μs if $C = 1000$ pF/m and $R_0 = 4000$ ohms.

2–13. For RG–8A/U it is known that $R_0 = 50$ ohms and $v = 200$ m/μs. Determine from these data the values of L and C.

2–14. A 6-in. section of polyethylene insulated ($\epsilon_r = 2.25$) coaxial line has a characteristic resistance of 75 ohms and is shorted at the far end. Find the voltage waveform at the input if the input signal is in the form of a 10-μs, 1-A current pulse with a rise time of 0.5 ms and a fall time of 1 μs (assume both rise and fall are linear). Assume a 75-ohm load connected across input of charging line.

2–15. Plot the input voltage of a 100-ft length of RG–8A/U (with $R_0 = 50$ and $\epsilon_r = 2.25$) to a 10-μs, 100-V pulse applied from a generator with an internal resistance of 1000 ohms. The line is open at the far end. What value of capacitor could be used in place of the line to give essentially the same waveform? Compare this capacity with the total capacity of the line (open circuited, and neglecting L, R, and G). Repeat this problem for a generator internal resistance of 50 ohms to show how this process of equivalents degenerates.

2–16. Find the input impedance of a lossless line for the special cases in which the load is (a) an open circuit, (b) a short circuit, and (c) R_0. Comment on the significance of each of these results.

2–17. By using the relationship $v = f\lambda$, where λ stands for wavelength, f for frequency, and v for velocity, show that the "electrical length" βd of a transmission line, in radians, may be expressed equivalently as $2\pi(d/\lambda)$ radians. The quantity (d/λ) is the line length expressed in wavelengths.

2–18. Express the length of a line in wavelengths if it is known that the "electrical length" of the line is $(\pi/2)$ radian. Find the "electrical length" of a line in radians if the line is a three-quarter wavelength. Comment on this statement: The line is 180 electrical degrees long.

2–19. What is the "electrical length," in degrees, of 100 miles of line at 60 hertz? What fraction of a wavelength does this represent? Assume air insulation.

2–20. What is the "electrical length," in degrees, of 100 ft of 300-ohm twin-lead at 60 MHz? How many wavelengths does this represent? Assume the velocity of propagation is 0.82 times that of light.

2–21. Find the connection between input impedance and load impedance for a lossless line for the special cases in which the line is (a) one-quarter wavelength, (b) one-half wavelength, and (c) 1 wavelength.

2–22. Show that the magnitude of the reflection coefficient is exactly 1 for all reactive loads.

2–23. Find the input impedance of a three-eighth wavelength section of line having a characteristic resistance of 50 ohms and a load impedance of $50 + j50$ ohms.

2–24. Starting with Eq. (2–34), find a relationship that expresses the load impedance in terms of the input impedance and line length.

2–25. If the load impedance is known to be a short circuit and the input impedance is found to be $+j50$ ohms, find the line length(s) for which these conditions hold when the characteristic resistance of the line is 50 ohms.

2–26. Suppose that a one-quarter wavelength section of 50-ohm line is driven from a voltage source of 50 V. If the line is shorted at the far end, determine the current supplied by the generator and the current through the short circuit.

2–27. Suppose that a three-quarter wavelength section of 50-ohm line is formed into a closed loop. Find the input impedance at any point on the loop. *Hint:* Use symmetry.

2–28. Show that the current standing-wave ratio is numerically the same as the VSWR for a given load.

2–29. Show that the value of S for a purely resistive load is the larger of the quantities R_L/R_0 and R_0/R_L.

2–30. Find the VSWR for a 20-ohm, 100-ohm, and 200-ohm load if the characteristic resistance of the line is 50 ohms.

2–31. Show that if a transmission line were severed at a position of minimum voltage, the input impedance measured toward the load would be minimum and resistive with the value R_0/S. Repeat for a voltage maximum to show that the impedance measured toward the load would be maximum and resistive with the value SR_0.

2–32. Plot the amplitude of the voltage across the line as a function of position along the line for a 50-ohm line terminated in (a) 25 ohms, (b) 50 ohms, (c) $+j50$ ohms, (d) 0 ohms, and (e) infinite ohms. Assume an incident voltage signal of 1 V. Use a graphical technique for evaluation of the amplitude as shown in Fig. 2–20.

2–33. Compute the phase-shift constant β for RG–8A/U cable and normalize the expression to a frequency base of 100 MHz. That is, the expression for β should be found at 100 MHz and the result for other frequencies expressed in terms of the value at 100 MHz. The only data for this evaluation are that the cable dielectric is polyethylene with $\epsilon_r = 2.25$.

2–34. Find the load impedance on a line if S is known to be 4 and the distance from the load to a voltage minimum on the line is one-quarter wavelength. The value of R_0 is 50 ohms.

2–35. Derive a formula giving the load impedance on an R_0 line in terms of S and the electrical length d between a voltage minimum on the line and the load impedance.

2–36. Express the magnitude of the reflection coefficient in terms of the VSWR by a formula. Find the magnitude of K if the value of S is 10.

2–37. Show that the power dissipated in the load is 48 W for the special case when the line length is $\lambda/8$ in Example 2–7 (Fig. 2–22).

2–38. Show from impedance considerations that the power dissipated in the load is independent of line length when the source is perfectly matched to the line.

2–39. If the "attenuation" produced by a mismatched load on an otherwise matched system is defined by the expression

$$\text{Attenuation} = 10 \log_{10} \left(\frac{\text{power to matched load}}{\text{power to actual load}} \right)$$

show that it may be related to the standing-wave ratio on a line by the expression

$$\text{Attenuation} = 10 \log_{10} \frac{(S+1)^2}{4S}$$

Find the "attenuation," in decibels, produced by a load mismatched to the extent that the VSWR on the line is 1.5.

2–40. Explain why the device shown in Fig. 2–25 for measurement of the magnitude of the reflection coefficient inherently has a small error even though the shunt resistance be chosen to make the section have a characteristic resistance of exactly 50 ohms. Compute this error for the specific example given in Fig. 2–26. Can this error be made zero? If so, redesign the circuit to accomplish this result.

2–41. Design a device similar to that shown in Fig. 2–25 which will work satisfactorily on a 75-ohm line.

2–42. Find the magnitude of the reflection coefficient of a 30-ohm load on a 50-ohm line, using calculations made as in Fig. 2–26. Verify the result by conventional computation. Repeat for a load of $50 + j50$ ohms.

2–43. Write the mathematical expression for the voltage along a three-eighth wavelength section of line driven with a voltage of 50 V on one end and a voltage of $+j50$ V on the other end. Measure x from the 50-V source. Assume a 50-ohm line. What is the standing-wave ratio on the line?

2–44. Show that the input impedance of a lossless line may be expressed in terms of the load impedance as

$$Z_{\text{in}} = R_0 \frac{Z_L \cos \beta d + jR_0 \sin \beta d}{R_0 \cos \beta d + jZ_L \sin \beta d}$$

3

Graphical Aids to Transmission-Line Analysis

3–1. INTRODUCTION

Graphical aids may be conveniently constructed for relating input impedance, line length, and load impedance. These graphical aids simplify the work, minimize the possibility of mistakes, and markedly increase the power of visualization of the user over what can be done with direct use of equations.

The only graphical aid to be discussed in detail in this chapter will be a reflection coefficient chart, more commonly called a Smith chart in honor of its originator.[1] There are many variations of the Smith chart, with rectangular and polar calibration of impedance or admittance, and with common bases of 1 ohm, 50 ohms, 1 mho, and 20 millimhos. These variations of the Smith chart may be summarized in the following tabulation.

[1] P. H. Smith, "Transmission Line Calculator," *Electronics*, vol. 12 (January 1939), pp. 29–31.

Impedance		Admittance	
Rectangular	*Polar*	*Rectangular*	*Polar*
1 ohm	1 ohm	1 mho	1 mho
50 ohms	50 ohms	20 millimhos	20 millimhos

Another chart, the impedance chart, is sometimes found useful. However, in contrast with the Smith chart, it is not readily extensible to analysis of lossy lines, it is more difficult to apply to line matching problems, and because of the finite range of impedance values covered, it cannot be used for all problems. Because of its limited usefulness and our desire to avoid unnecessary detail, discussion of this chart has been reserved for Appendix C. Examples of the per-unit form of the rectangular and polar Smith chart and the rectangular impedance chart are shown in Fig. 3–1.

One of the very important uses of the transmission-line charts is in the reduction of laboratory data. The data may be in a variety of different forms, depending on the instrument used for the measurements. The three common presentations of the data are (1) actual impedance in polar form from an impedance bridge; (2) actual admittance in rectangular form from an admittance bridge; and (3) standing-wave ratio and position of minimum from a slotted line. Certain variations in the techniques of analysis are necessary for each of these different forms of data. These variations will be discussed in a later article of this chapter.

One of the most interesting applications of the transmission-line charts involves the design of transmission-line matching networks using combinations of various-length sections of transmission line. This application is discussed at length in Chapter 4.

3–2. THE SMITH CHART

In the interest of simplicity we shall restrict ourselves to discussion of the 1-ohm rectangular Smith chart, illustrated in Fig. 3–1(a). It should be noted, however, that this restriction to discussion of the 1-ohm rectangular Smith chart is not really limiting in scope since this chart may be used to solve any problem which can be solved on the other charts. The other charts merely have special scales which make particular computations simpler.

The Smith chart is based upon the relationship between reflection coefficients at two different points along a line. For a lossless line this relationship may be written as

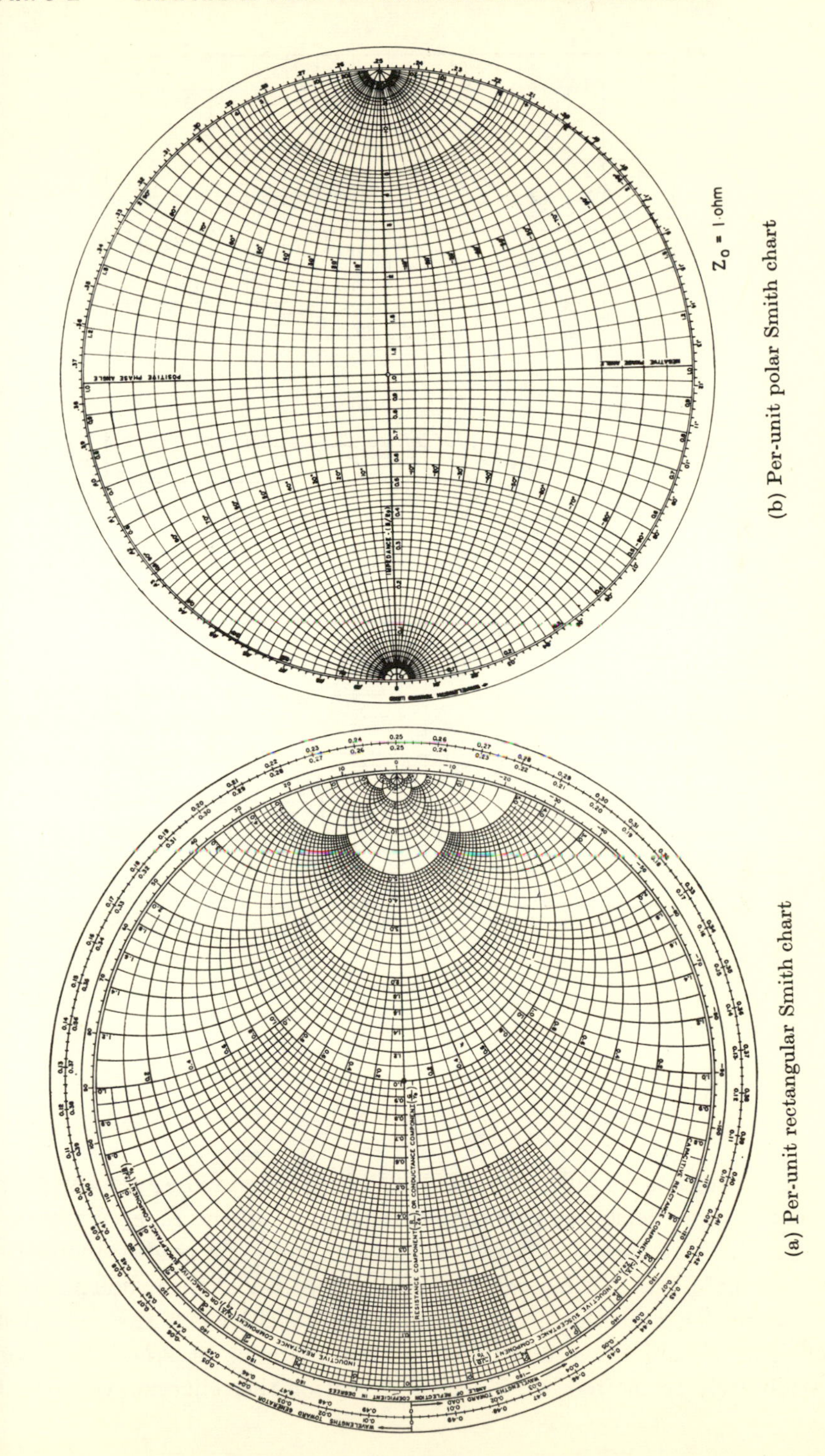

(a) Per-unit rectangular Smith chart

(b) Per-unit polar Smith chart

Fig. 3–1. Examples of various forms of transmission-line graphical aids.

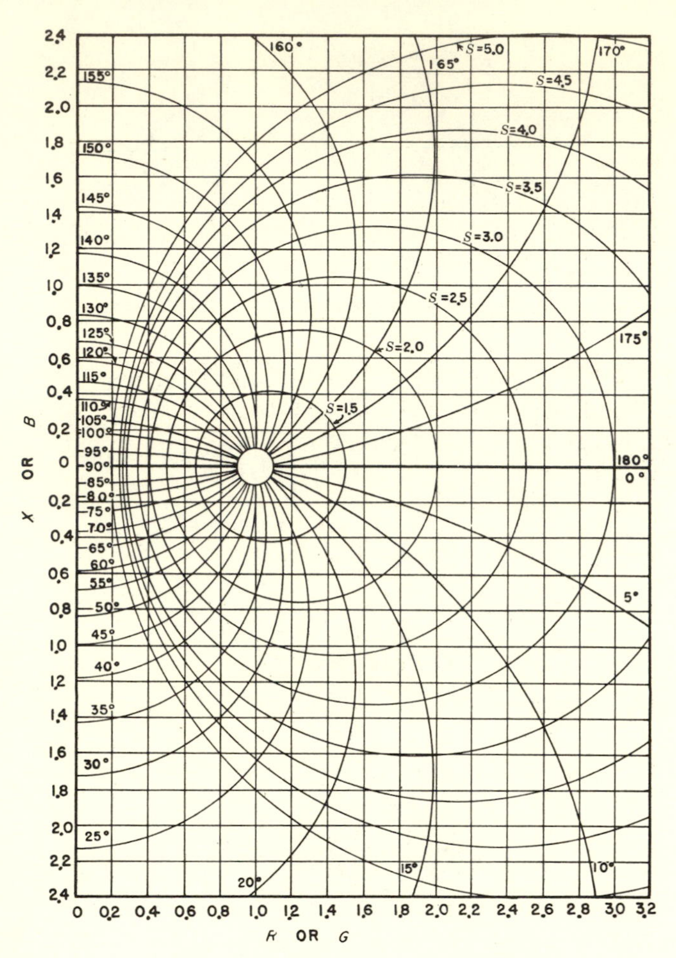

(c) Per-unit rectangular impedance chart

Fig. 3–1 (*Cont.*)

$$K' = Ke^{-j2\beta d} \tag{3–1}$$

In this expression the reflection coefficient K' is taken to be at a position d meters toward the source from the measurement of the reflection coefficient K. This relationship between K and K' may be readily found by inspection of Fig. 3–2. An incident signal of $1 + j0$ V at the prime position will be reflected at the prime position as $Ke^{-j2\beta d}$ when the reflection coefficient K and the round-trip electrical distance of $2\beta d$ have been appropriately accounted for. Since by definition the reflection coefficient

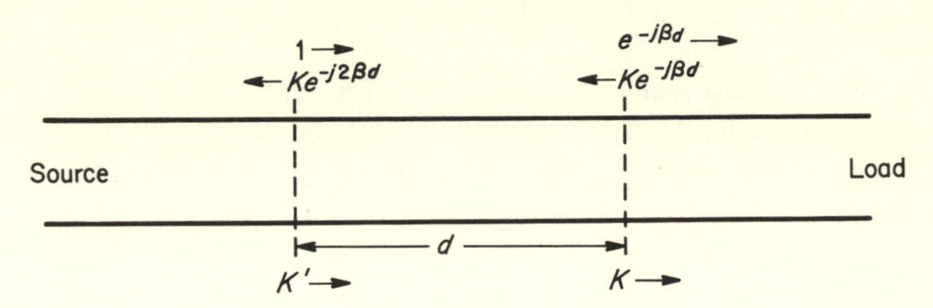

Fig. 3–2. Relationship between reflection coefficients along a lossless line.

is the ratio of reflected to incident signal at a given point on the line, the result in Eq. (3–1) is seen to be correct as given.

In order to make it as versatile as possible, the chart is commonly calibrated in terms of per-unit values of impedance or admittance. This normalization process may be viewed simply as an artifice to use one chart for problems with lines having various values of characteristic resistance. It may be also viewed as means of providing a single chart sufficing when the entire problem is scaled by the value of the characteristic resistance of the line. Regardless of the reasons for normalization, the actual impedance Z (note use of capital letter) will be normalized with respect to the characteristic resistance R_0 to yield the corresponding per-unit impedance z (note use of small letter) as defined by the equation

$$z = \frac{Z}{R_0} \tag{3–2}$$

The relationships between this per-unit impedance z and the corresponding reflection coefficient K are

$$K = \frac{z - 1}{z + 1} \tag{3–3}$$

and

$$z = \frac{1 + K}{1 - K} \tag{3–4}$$

An interesting and important feature of the result given in Eq. (3–3) is that the absolute magnitude of the reflection coefficient K can never exceed 1 for passive load if the characteristic impedance of the line is purely resistive.[2] This simple fact is important because it means that for

[2] If the characteristic impedance is purely resistive and the load is passive, then the per-unit value of z_L may be written as $r + jx$, where r is a non-negative number. Consequently the magnitude of K_L given by the expression

$$|K_L|^2 = \frac{(r - 1)^2 + x^2}{(r + 1)^2 + x^2}$$

is never greater than 1.

all passive loads, the Smith chart may be drawn inside a circle of one unit radius.

The Smith chart will serve equally well for admittance calculations as for impedance calculations. This fact is particularly important in matching systems where stubs are placed in parallel with the lines. The proof of this result is left as an exercise.

The construction of a Smith chart will be discussed at this point in terms of a specific example. Our problem, as summarized in Fig. 3–3, is to

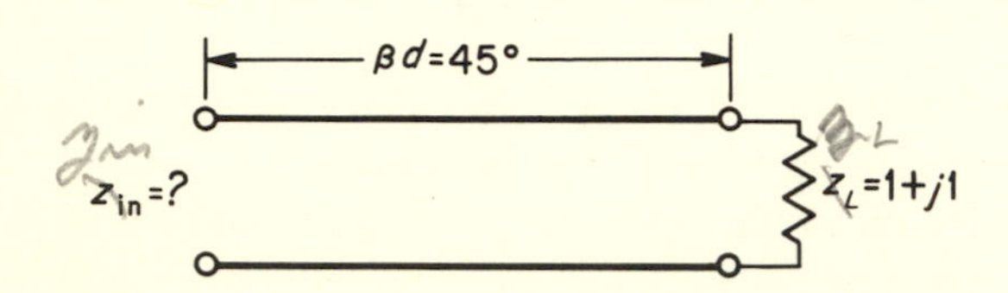

Fig. 3–3. Example problem analyzed in Fig. 3–4.

find the input impedance of a line 45 electrical degrees long which is terminated in a per-unit load impedance of $1 + j1$ per-unit ohms. Details of the analysis process may be presented in three simple steps:

1. Compute the load reflection coefficient from Eq. (3–3) as $0.447 \underline{/ 63.4°}$ and plot the results as shown in Fig. 3–4(a).
2. Compute the input reflection coefficient from Eq. (3–1) and plot the result as shown in Fig. 3–4(b). The exponential factor of $-j2\beta d$ merely contributes a rotation of $-2(45)$ deg or -90 deg.
3. Compute the input impedance from Eq. (3–4) as $2 - j1$ per-unit ohms and plot as shown in Fig. 3–4(c).

Analysis like that of Fig. 3–4 is as yet more difficult than direct substitution into the equations, but it can be refined to a marked degree by the following observations:

1. There is a one-to-one correspondence[3] between values of z and K so that the per-unit values may be labeled directly on the chart to avoid completely the calculations in steps 1 and 3 above.
2. The rotation by the exponential factor of $-j2\beta d$ can be simplified by suitable calibration of a scale around the chart.

An example of a per-unit rectangular Smith chart is shown in Fig. 3–5. Note that the electrical distance scale is calibrated in wavelengths and

[3] See Appendix C.

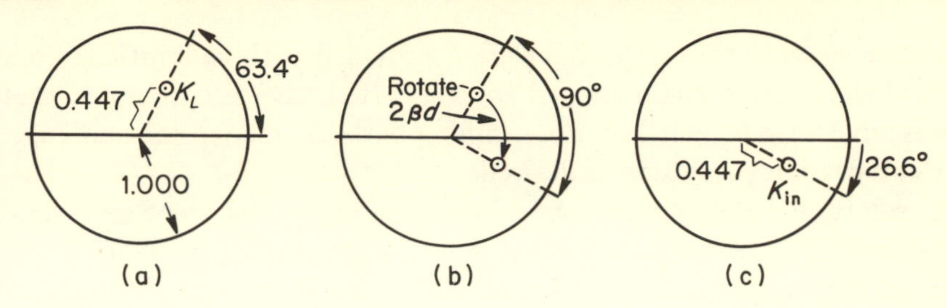

Fig. 3–4. Steps in computing input impedance from load impedance for problem illustrated in Fig. 3–3.

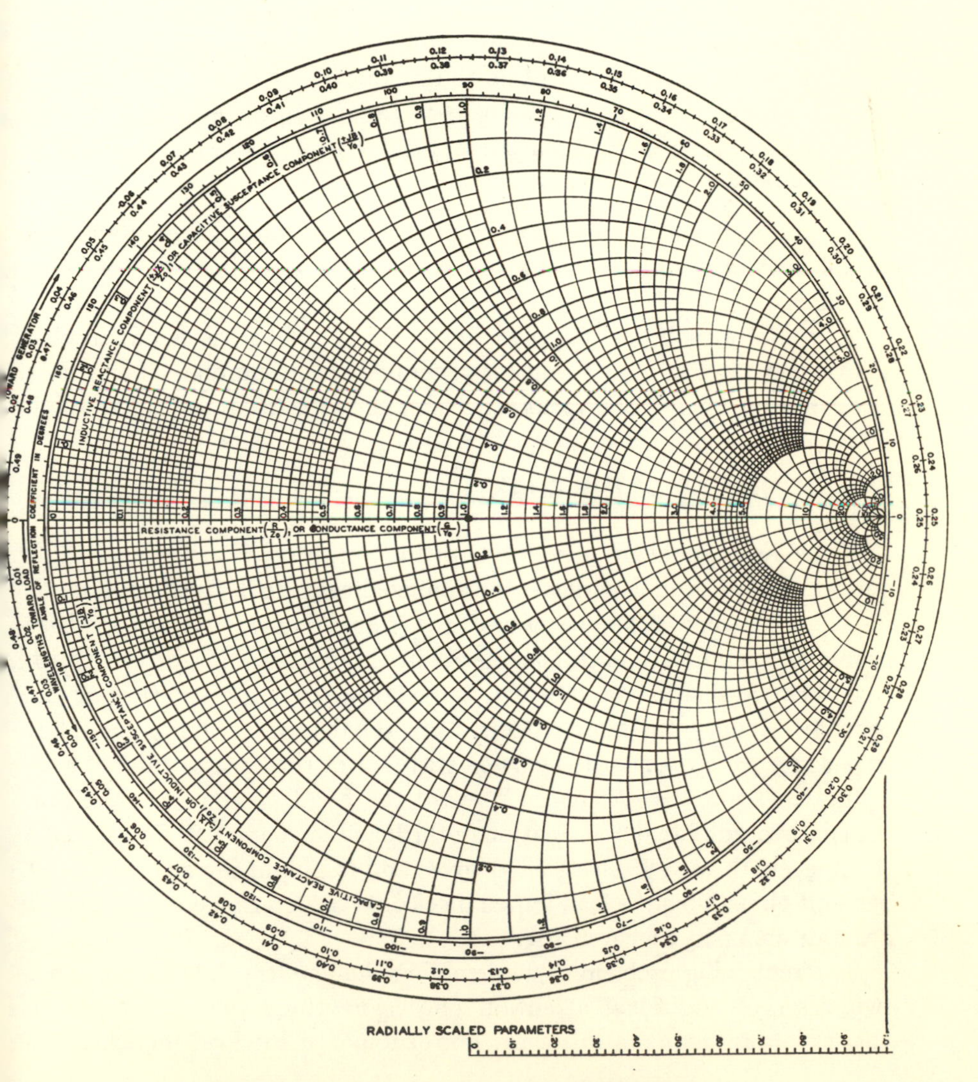

Fig. 3–5. A per-unit rectangular Smith chart.

that, for convenience, scales have been provided with calibrations both toward the generator and toward the load. It should also be noted that scales have been included for determining both the magnitude and angle of the reflection coefficient, if desired.

A solution of the example given in Fig. 3–3 by direct steps on a completed per-unit rectangular Smith chart is shown in Fig. 3–6(a). The various steps required in the solution are:

1. Plot the load of $1 + j1$ per-unit ohms as shown at point A in Fig. 3–6(a).
2. Construct a radial line from the center through point A and note on the "Wavelengths Toward Generator" scale the value 0.162. Add 0.125 wavelength to obtain the value 0.287 and construct a radial line from the center through this value on the scale.
3. Swing an arc about the center from point A on the first radial line to point B on the second radial line. Read the input impedance at point B to be $2 - j1$ per-unit ohms.

The work is thereafter very simple: plot the load impedance, move on a circular path the required distance, and read the resulting value of input impedance.

A graphical solution requires about 25 sec, whereas an analytical solution requires at least 5 min for general choice of numbers; hence it is clear that the chance of error is much smaller with the graphical method.

3–3. USING THE SMITH CHART

Three basic problems solvable with the Smith chart are to determine one of the three quantities—load impedance, line length, and input impedance—from a knowledge of the remaining two. One of these problems has already been considered in the construction of the Smith chart. Determination of the load impedance from specification of the other two follows the same pattern as in the determination of the input impedance except that the chart is traveled in the reverse direction. A calibrated scale, "Wavelengths Toward the Load," has been placed on the chart to simplify determination of the load impedance. If, for example, the line length were 45 electrical degrees and the input impedance $2 - j1$ per-unit ohms, then the load impedance could be determined to be $1 + j1$ per-unit ohms, as shown in Fig. 3–6(b).

The remaining problem of determining the line length from the other two deserves additional attention only from the standpoint that the answer is not unique in this case. For example, if the load impedance is

$1 + j1$ per-unit ohms and the input impedance is known to be $2 - j1$ per-unit ohms, then one may be tempted to say that the line length must be 0.125λ in view of the results of Figs. 3–6(a) and 3–6(b). However, note carefully that lines $0.125 + n0.500\lambda$ long would give exactly the same result for any integer value of n. This means that the line length can be determined only from the given data to within an integral number of half-wavelengths. Such an ambiguity could be expected from the fact that the load impedance and input impedance of a half-wavelength section of line are identical (which follows directly from the fact that a complete rotation about the chart requires 0.5λ).

A very practical problem which may be solved with the aid of the Smith chart is the determination of the VSWR (voltage standing wave ratio) for any given load. The solution of this problem is probably most easily explained by example.

Example 3–1

Suppose that the VSWR on a line with a load impedance of $1 + j1$ per-unit ohms is desired. The problem could be solved by laboriously computing the reflection coefficient to be $0.447 \underline{/63.4^\circ}$ and then finding the VSWR from the magnitude of the reflection coefficient as $S = 1.447/0.553 = 2.62$. Of course one way to simplify the work would be to read the value for the reflection coefficient directly from the Smith chart, as shown in Fig. 3–6(c), and then to compute S as before. However, there is a much simpler way, as shown in Fig. 3–6(d). The scheme here is simply to find the input impedance of the line for such a line length that it will be purely resistive and at a maximum. This per-unit value of resistance may then be shown[4] to have exactly the same numerical value as the VSWR (S). In this case, the value is found to be 2.62 per-unit ohms, and the value of S is thus 2.62.

As indicated in the example, the VSWR for any load may be found by the simple expedient of plotting the value on a normalized Smith chart, rotating this point on a constant-radius circle to the resistive axis, and reading the maximum per-unit value of the resistance as the corresponding VSWR.

The work in Example 3–1 also shows how the Smith chart may be used to advantage for conversion between the magnitude of the reflection coefficient $|K|$ and the VSWR (S). In this case the process is merely one of reading between two scales; if the scale for the magnitude of the reflection coefficient is placed from center to edge of the Smith chart

[4] Suppose that this resistive value is r_{max} per-unit ohms. The reflection coefficient will be real and will have a magnitude $|K| = (r_{max} - 1)/(r_{max} + 1)$. The corresponding value of VSWR found from $S = (1 + |K|)/(1 - |K|)$ then easily works out to be simply $S = r_{max}$.

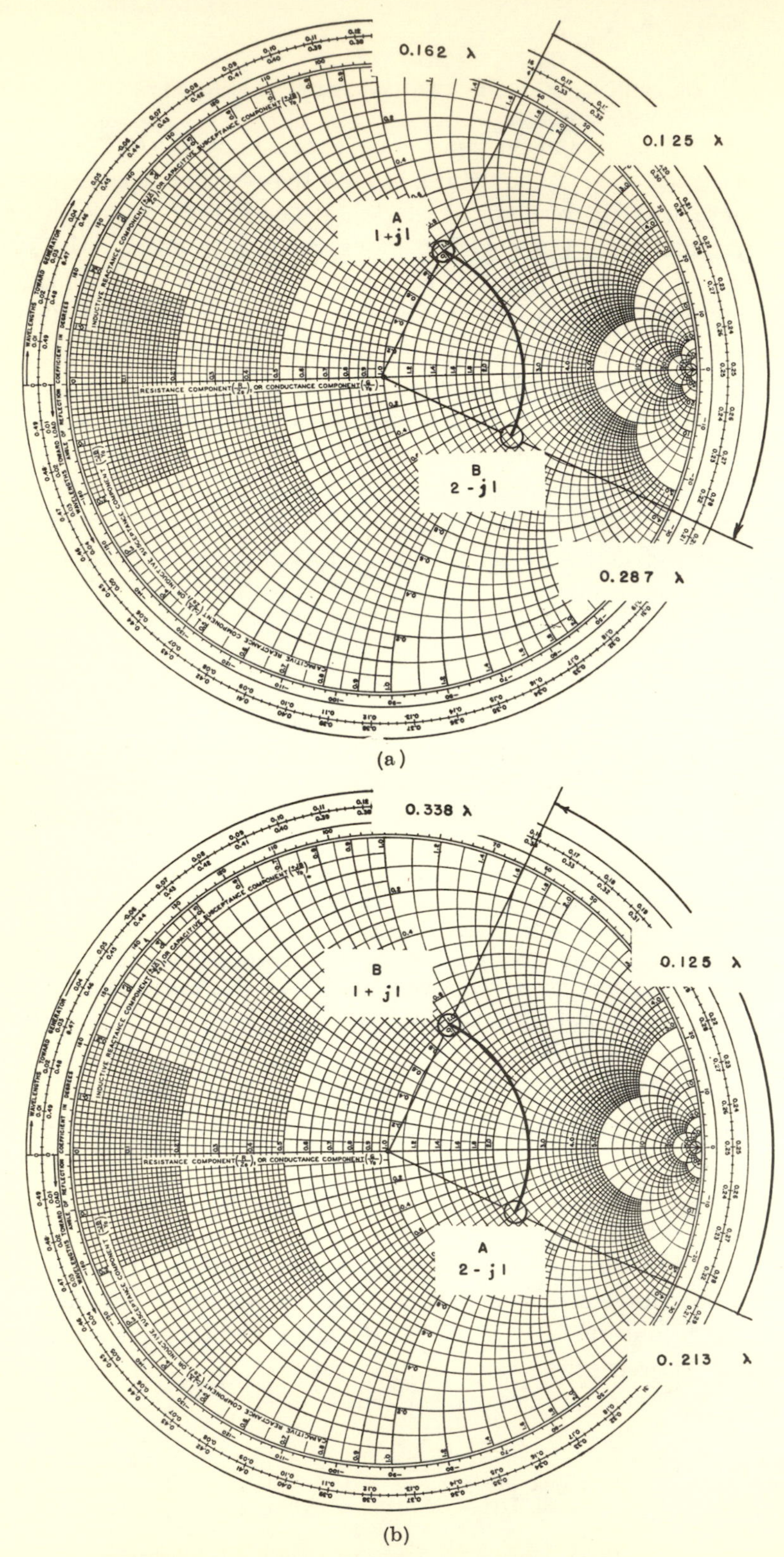

Fig. 3–6. Examples of various uses of the Smith chart.

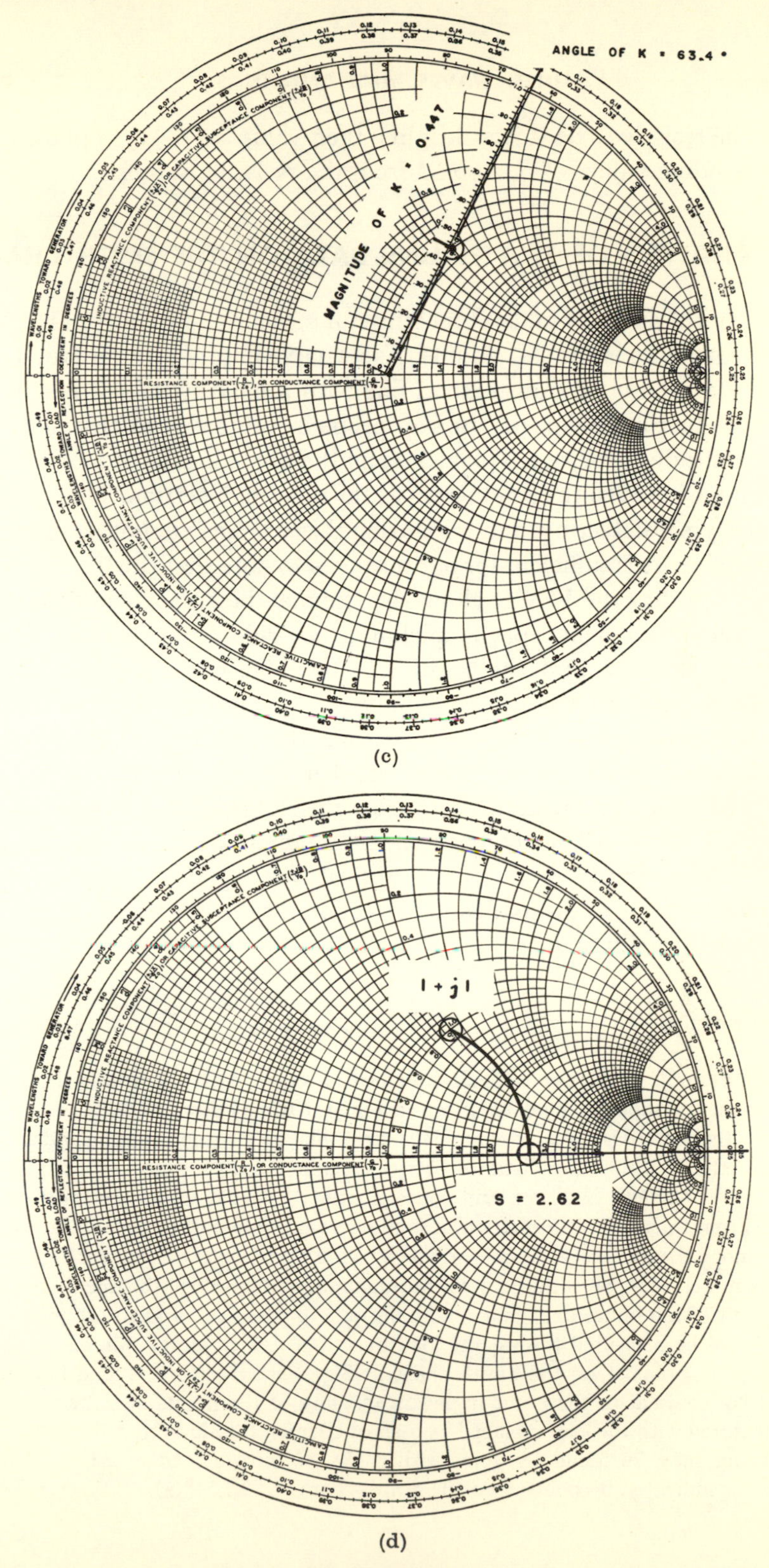

Fig. 3–6 (Cont.)

along the resistive axis (toward right), then values of $|K|$ are opposite the corresponding values of S, as read from the resistive values.

3–4. ANALYSIS OF SLOTTED-LINE DATA ON THE SMITH CHART

One of the simplest and yet most useful measuring instruments at very high frequencies is the slotted line. As its name implies, this instrument is actually a section of transmission line with a lengthwise slot cut in the outer conductor. A capacity pickup probe is inserted into this slot, and the magnitude of the signal picked up on the probe is thus proportional to the voltage between the conductors at the point of insertion. The methods of amplifying, detecting, and measuring this voltage will not be discussed at this point. It is sufficient for our purposes to note that the instrumentation is such that the VSWR along the line may be either read directly or easily computed. The capacity pickup probe rides on a carriage along a calibrated scale so that the positions of voltage minima can be found. Voltage minima are used in the analysis, since positions of the minima may be more accurately determined than the positions of the maxima.

Correction must be made for line length on all transmission-line measurements, and the data taken from the slotted line are no exception. The correction is ordinarily determined by temporarily replacing the load with a short circuit. This reference measurement then determines the electrical position of the load, and correction for the line length to the load can then be easily made.

Example 3–2

As an example of the use of the Smith chart to reduce data taken from slotted line measurements, suppose the data shown in Fig. 3–7 to be known.

Since the slotted section is commonly air-insulated (so the probe may be easily moved), the fact that the distance between minima is 40 cm may be used to find that the wavelength is 80 cm, and consequently the signal frequency must be 375 MHz. From the $S = 5$ information, it is known that values of impedance at various points along the line are on a circle on the per-unit Smith chart passing through $r = 5$. Furthermore, it is known that a point of minimum voltage along the line corresponds to a point of minimum impedance (actually purely resistive), and so the impedance that would be measured toward the load at either of the voltage minima would be as shown at point A on the Smith chart in Fig. 3–8(a). Since the load impedance would be identical to the input impedance at any of the points of voltage minima (as measured with the load replaced by a short circuit), it follows that the actual load impedance may be found by going either $15\lambda/80$ toward load or $25\lambda/80$ toward source, as indicated by point B on the Smith chart in Fig. 3–8(a).

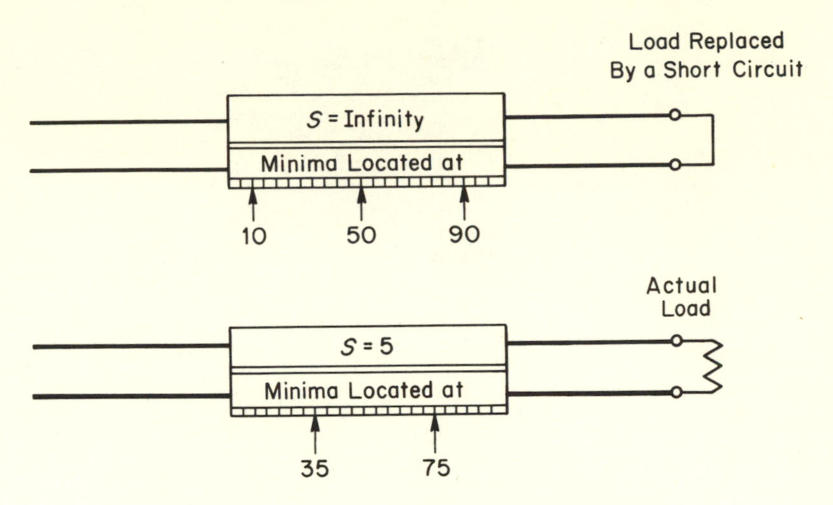

Fig. 3–7. Data from slotted-line measurements.

The value of the load impedance may then be read directly as $1.13 - j1.89$ per-unit ohms. If the characteristic resistance of the lines should be 50 ohms, then the actual value of load impedance would clearly be $56.5 - j94.5$ ohms.

TEST I to here

3–5. ANALYSIS OF IMPEDANCE AND ADMITTANCE BRIDGE DATA ON SMITH CHART

The data read on an impedance or admittance bridge generally never represent the actual value of the load because there must, of necessity, be a transmission line connecting the load to the measuring device. Variable-length sections of coaxial line having constant characteristic resistance are available and can be inserted into the measuring system to make the bridge direct reading, but these are very expensive. The scheme here is to adjust the line length between the load and the point of measurement within the bridge to be exactly some multiple of a half-wavelength so that the load impedance and the measured impedance are exactly the same. This critical length adjustment is usually made by replacing the unknown load by some known load (commonly a short circuit) and adjusting the line length so that the value measured on the bridge agrees with the value of the known load. The usefulness of this technique is limited severely if the lines are even slightly lossy, since correction must be made for the losses if the results are to be reasonably accurate, and the further correction for line length requires very little added time and effort.

The correction for losses will be discussed in the next article. The

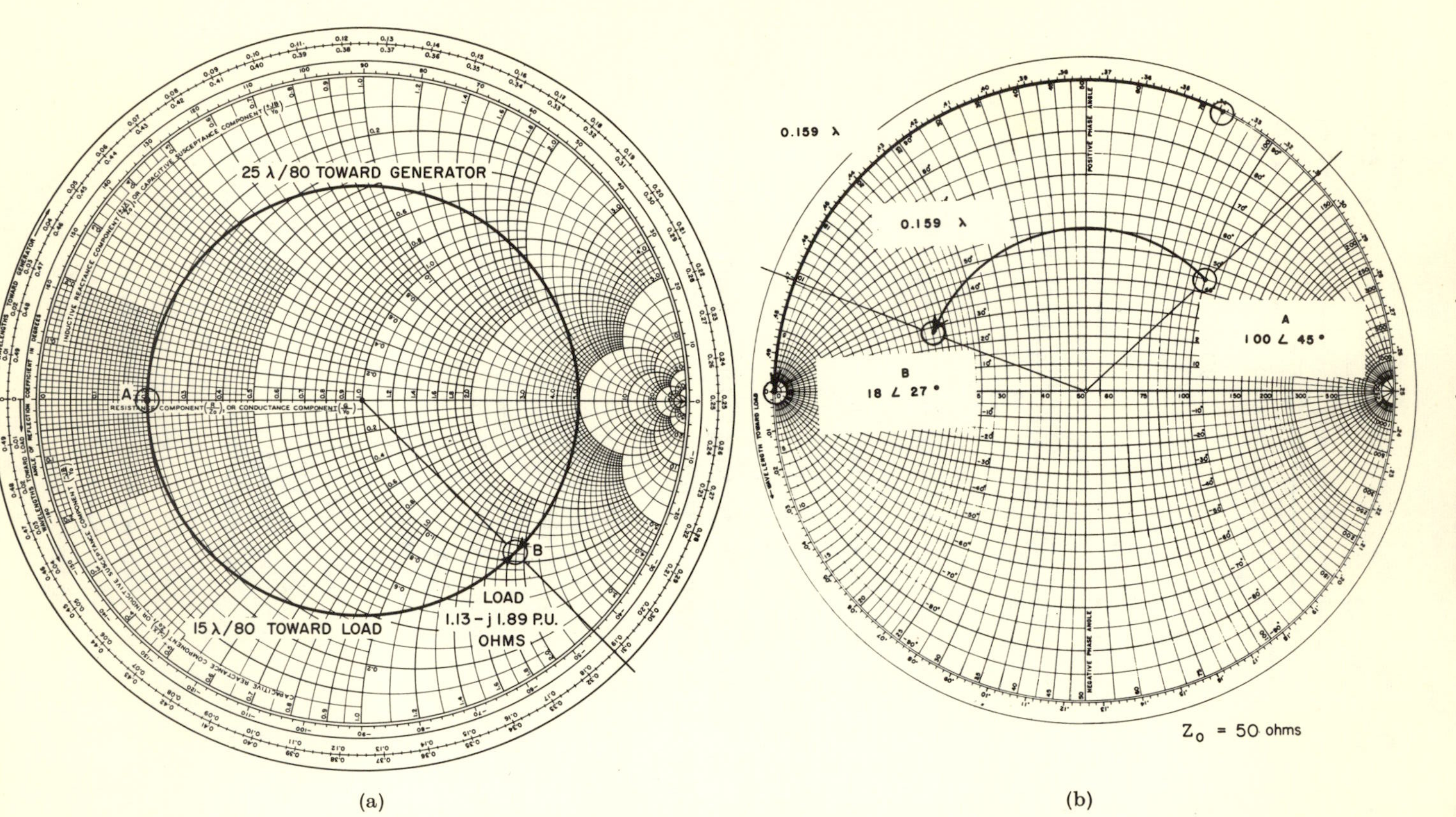

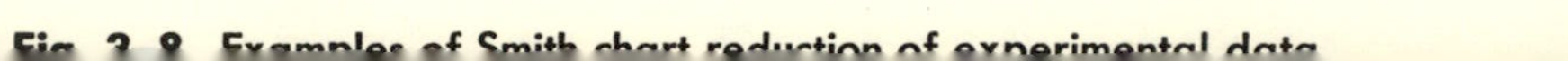

Fig. 3.8 Examples of Smith chart reduction of experimental data

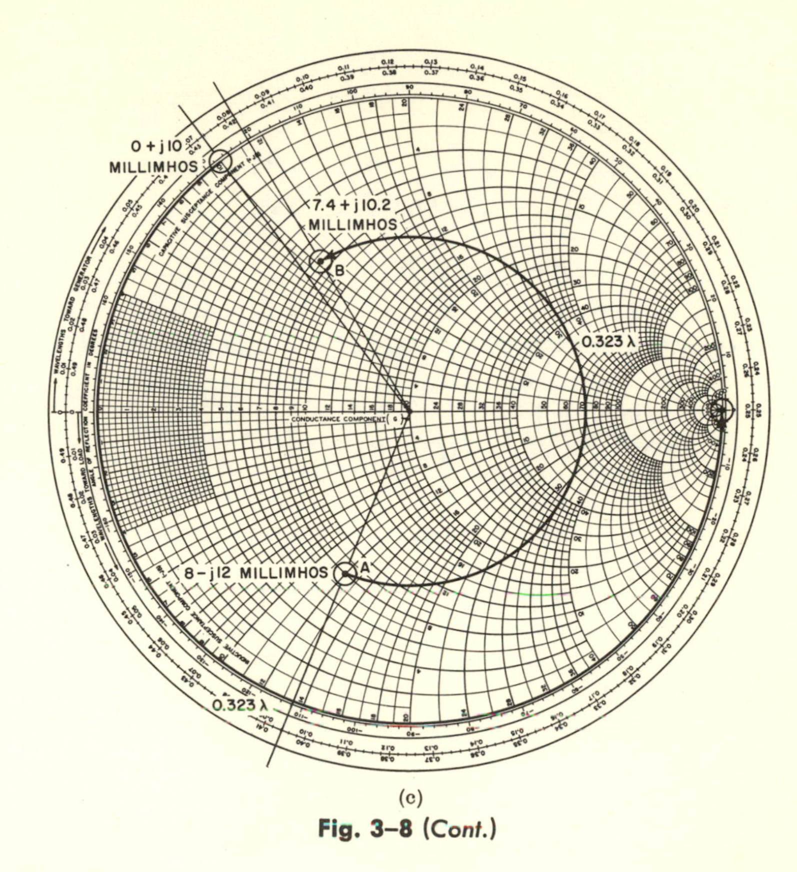

(c)

Fig. 3–8 (*Cont.*)

correction for line length is easily handled, as shown in the following examples.

Example 3–3

Suppose a load on a 50-ohm line reads as a measured value of 100 $\underline{/45^\circ}$ ohms on an impedance bridge. The value for the load replaced by a short circuit is found to be 80 $\underline{/90^\circ}$ ohms. The second set of data is used first in finding the line length (within the $n\lambda/2$ ambiguity) to be 0.159λ, as shown in Fig. 3–8(b). Then the load impedance is found in straightforward steps between points A and B in Fig. 3–8(b) to be 18 $\underline{/29^\circ}$ ohms.

Example 3–4

Suppose a load on a 50-ohm (20-millimho) line reads as a measured value of 8 − j12 millimhos on an admittance bridge. The value for the load replaced by a short circuit is $+j10$ millimhos. Again the second set of data is used first in finding the line length (within the 0.5λ ambiguity) to be 0.323λ, as shown in Fig. 3–8(c). Then the load admittance is found in straightforward steps between points A and B in Fig. 3–8(c) to be $7.4 + j10.2$ millimhos. One point of difference here is that the admittance of a short circuit is infinite.

3–6. EFFECTS OF LOSSES ON SMITH CHART ANALYSIS

The basic effect of losses on Smith chart analysis is to make the path between reflection coefficients (impedances) an exponential spiral rather than the circle obtained when the losses are zero. For an assumed incident signal of V^+ at the input of a lossy line of length d, the reflected signal measured at the input will be $V^+K_Le^{-2\gamma d}$ because of the reflection coefficient K_L at the load and a round trip path of $2d$ at a propagation constant γ. The reflection coefficient K_{in} at the input of the line is therefore

$$K = K_{in}e^{-2\gamma d} = K_Le^{-2\alpha d}e^{-j2\beta d} \tag{3–5}$$

The effect of losses on the Smith chart method of analysis may best be understood by considering specific examples. Suppose that we analyze the problem shown in Fig. 3–9.

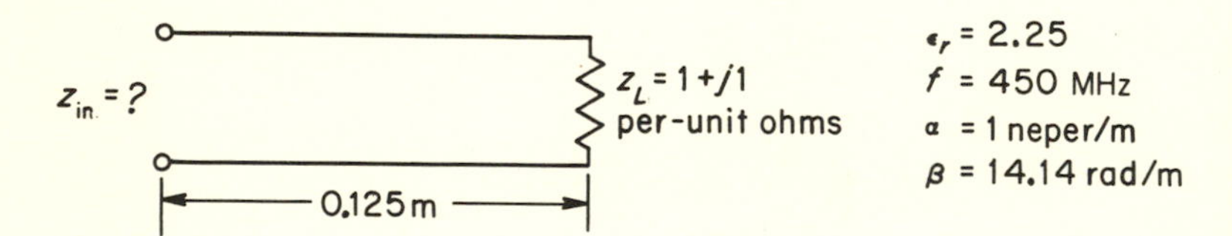

Fig. 3–9. A lossy line example.

Example 3–5

By using the data given in Fig. 3–9, it is possible to compute $2\gamma d$ as $2\alpha d + j2\beta d =$ 0.25 neper $+ j3.54$ radians. Consequently, on the Smith chart, the input reflection coefficient is related to the load reflection coefficient by a scale factor $e^{-0.25} = 0.779$ and a rotation of $e^{-j3.54}$, which amounts to a clockwise rotation of 3.54 radians or 202 deg. Starting at the load of $1 + j1$ per-unit ohms [point A in Fig. 3–10(a)] and following the preceding steps will show that the input impedance is $0.55 - j0.27$ per-unit ohms [point B in Fig. 3–10(a)]. The rotation of 202 degrees clockwise is rather obvious, but it should be noted how the scaling by 0.779 is handled. A *reflection coefficient scale* calibrated from 0 to 1 has been supplied with the charts, and by using it, the magnitude of the load reflection coefficient is found to be about 0.44. Multiplication of this value by 0.779 yields a value of about 0.34, which has been used in Fig. 3–10(a) in determining the input impedance.

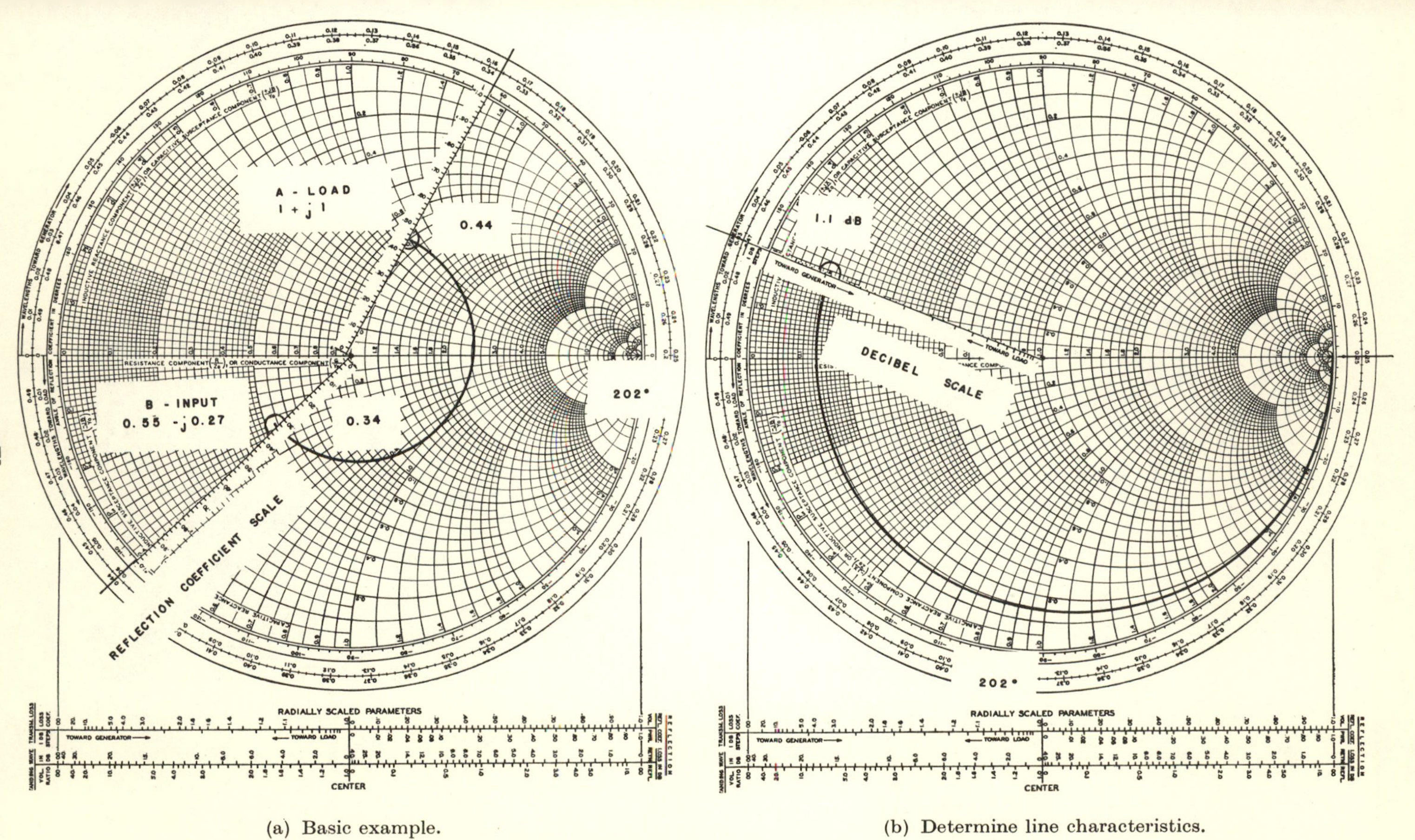

(a) Basic example.

(b) Determine line characteristics.

Fig. 3–10. Smith chart analysis of lossy lines.

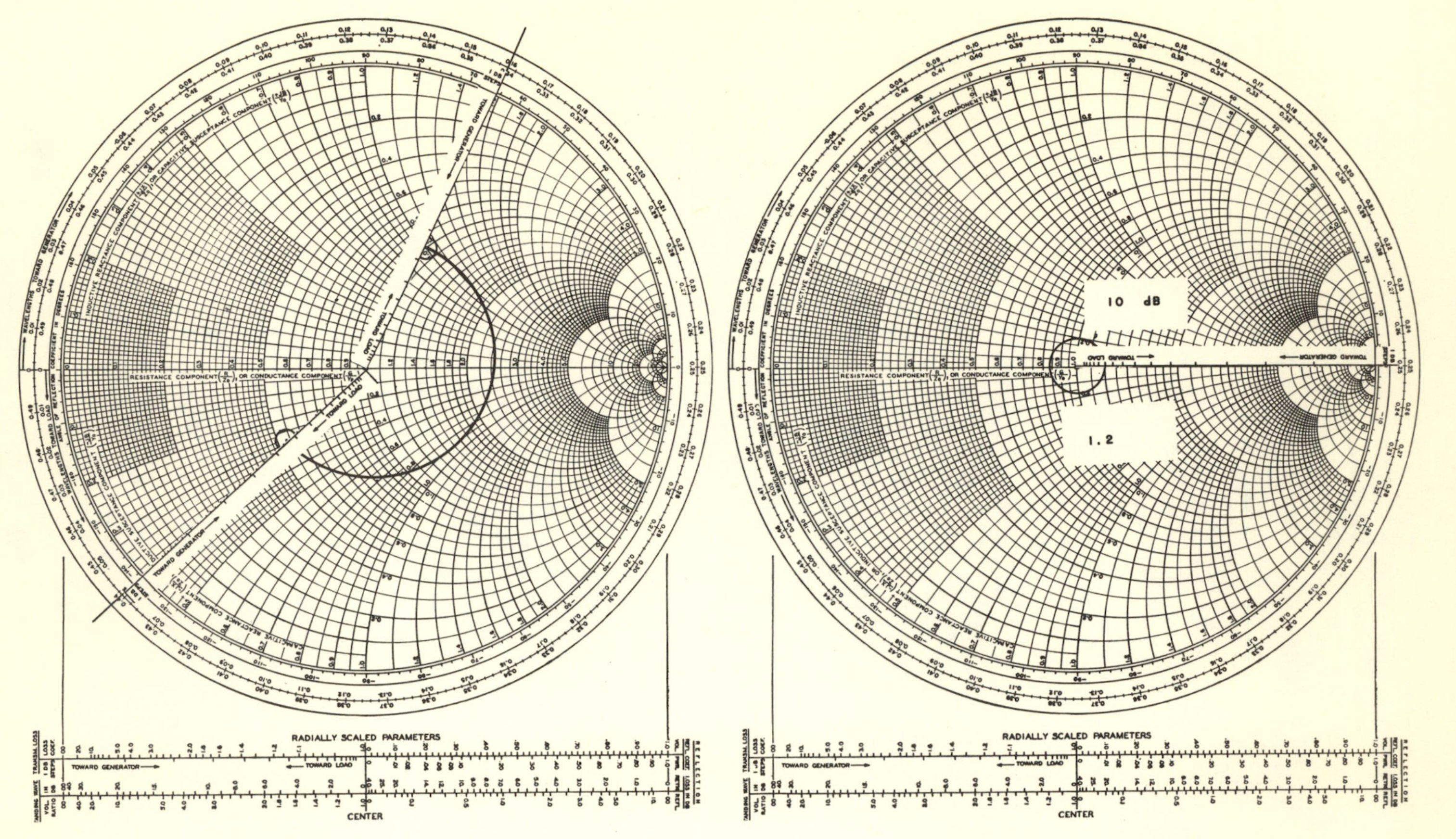

(c) Direct impedance calculation.

(d) Attenuator example.

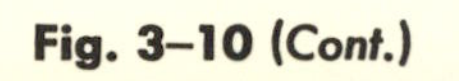

Fig. 3–10 (Cont.)

This problem was worked in terms of α in nepers/meter and β in terms of radians/meter. However, the more common situation is that the quantities αd and βd are known from measurements made on the line with a known load, such as a short circuit. Since for convenience the charts have been calibrated in terms of αd in decibels and βd in fractional wavelengths, it is worthwhile to learn how to work with these scales.

Example 3–6

Consider again the example of Fig. 3–9. Suppose now, however, that the length is not known to be 0.125 m, but rather suppose that the input impedance with an open-circuit load is measured to be $0.13 + j0.19$ per-unit ohms. By using the *transmission loss* scale, which is calibrated in 1-dB steps, the loss in the line is found to be 1.1 dB. [See Fig. 3–10(b).] The line length may be found by conventional means to be 0.280λ (using the information that the physical length is *about* 0.125 m long). These values may be compared with Example 3–5 for consistency. The factor $\alpha d =$ 0.125 neper converts by the standard conversion factor of 8.686 to be 1.08 dB, which compares very favorably with the graphical value of 1.1 dB. The factor of $\beta d = 1.77$ radians converts by division by 2π to 0.282λ, which compares very favorably with the graphical value of 0.280λ.

The next step in the example is to use the data determined from measurement on the shorted line to determine the input impedance for the load of $1 + j1$ per-unit ohms. The process is illustrated in Fig. 3–10(c).

Starting at the load of $1 + j1$ per-unit ohms, find the input impedance by going *toward the source* 0.280λ on the *wavelength* scale around the edge of chart and *toward the source* 1.1 dB on the *transmission loss* scale. Notice that there are deliberately no calibrations on the *transmission loss* scale; rather, distance between each pair of marks represents a 1-dB change. Interpolation between marks is used for increments of less than 1 dB.

It is important to note that the factor of 2 which shows up in Eq. (3–5) has been calibrated into the scales on the chart so that no consideration need be taken on this point.

The chart can be used for any of the common calculations for lossy lines by applying the techniques illustrated in this chapter. Corrections for line losses ordinarily must be made in the laboratory if high accuracies are to be obtained. The important feature is that very little more effort is required to account for losses on the Smith chart.

The Smith chart may also be used to considerable advantage in working with general symmetrical networks, as shown in the following example.

Example 3–7

Suppose that it is desired to find the input impedance of a 10-dB, 50-ohm pad for various loads. Using the *transmission loss* scale, it is possible to show, as in Fig. 3–10(d), that the input impedance for all loads must be within a VSWR circle of 1.2. For an open-circuit termination then, the input impedance will be 60 ohms, while for a short-circuit termination, the input impedance will be 41.5 ohms. In a similar fashion the input impedance for any load may be found. Note that in this case $\beta d = 0$ and no circular rotation takes place.

PROBLEMS

3–1. Convert 70 electrical degrees to wavelength and 0.15λ to electrical degrees.

3–2. Construct a simplified Smith chart with r and x contours of 0, ± 0.5, ± 1.0, and ± 2.0; check against an actual chart.

3–3. Find the reflection coefficient for a load of $2 + j2$ per-unit ohms. Also find the standing-wave ratio.

3–4. Find the magnitude of the reflection coefficient for a standing-wave ratio of 10 by direct use of the Smith chart. Illustrate with a sketch of the Smith chart.

3–5. (a) Find the input impedance of a 0.375λ line terminated in $2 + j3$ per-unit ohms.

(b) Find the load impedance of a 0.225λ line if the input impedance is measured to be $1 - j1$ per-unit ohms.

(c) Find the length of a line within an integral number of half-wavelengths if the load impedance is $0.5 + j0.7$ per-unit ohms and the input impedance is $1.4 - j1.4$ per-unit ohms.

3–6. Repeat Problem 3–5, but with the numerical values as per-unit mhos instead of per-unit ohms.

3–7. Find the actual value of the input impedance of a 3.4-m length of RG–8A/U cable terminated in a 100-ohm resistor for a frequency of 200 MHz. The value of characteristic resistance is 50 ohms, and the dielectric of the RG–8A/U is polyethylene with $\epsilon_r = 2.25$.

3–8. Suppose it is known that the standing-wave ratio, S, is 3 and a point of minimum input impedance occurs at a point 0.225λ from the load. Determine the per-unit value of the load impedance.

3–9. Show that a per-unit Smith chart is equally as applicable for use as a 1-mho chart as for a 1-ohm chart.

3–10. With the load in place on a slotted line, the value of S is 4, and the minima occur at 15, 25, 35 cm, etc., on a scale calibrated toward the load. With the load temporarily replaced by a short circuit, the minima are found at 12, 22, 32 cm, etc. Find the load impedance if the line is 50 ohms. Also determine the frequency.

3–11. Show that the input impedance of a lossless line terminated in a short circuit is $+jR_0 \tan \beta d$, whereas it is $-jR_0 \cot \beta d$ for an open-circuit termination.

From this show that the value of R_0 may be found as the square root of the product of the open- and short-circuit values of impedance. If the open- and short-circuit impedance of a cable are measured to be $-j88.5$ and $+j28.2$ ohms, find the length and characteristic resistance of the line.

3–12. Determine the length of 50-ohm line, short-circuited at the end, required to produce reactances of $+j100$, $-j100$, 0, and infinite ohms.

3–13. What desirable features does coaxial cable have for laboratory use as compared to parallel wire line? Explain.

3–14. Find the input impedance of a distortionless lossy line with a $2 + j0$ per-unit ohm load if the input impedance with a short circuit on the end is $0.4 - j0.2$ per-unit ohms. How many decibels of attenuation is there in this line? If the line length is about 0.9 m, the value of ϵ_r is 2.25, and the frequency is 450 MHz, determine the attenuation of the line per 100 ft.

3–15. Find the load impedance of a lossy line with a measured input impedance of $0.5 - j0.4$ per-unit ohms. Assume the line to be the same as in Problem 3–14.

3–16. Find the input impedance of a lossy line with $1 + j0$ per-unit ohm load impedance. Assume the line is the same as in Problem 3–14.

4

Transmission-Line Matching

4–1. INTRODUCTION

The basic concern of this chapter will be matching of radio-frequency lines, using transmission-line elements for matching. Matching on other systems consists in general of very specialized detail and can as a practical matter be treated only in general terms such as in Art. 4–2.

There are basically three different and important considerations involved in the connection of a source to a load through a transmission line: (1) choosing a transmission line; (2) matching the load to line; and (3) matching between line and source. Each of these will be discussed in general terms in Art. 4–3. In addition, some of the practical details of impedance matching will be discussed in this chapter.

Some of the matching methods that have been found useful are discussed in detail in the remaining articles of this chapter. These various methods represent only a few of the many possible methods of impedance matching. The only "stray" in this group is the balun (*bal*anced to *un*balanced transformer). As its name implies, this device is used to connect between balanced and unbalanced lines without disturbing the equilibrium conditions on either. The various matching schemes to be discussed are: (1) balun; (2) tapered line and quarter-wavelength transformer; and (3) single-, double-, and triple-stub tuners. The stub, balun, and the various stub tuners find wide application in the laboratory because of their ease of adjustment to different frequencies and load conditions. The others, while not easy to adjust, may be economical in construction for a fixed application and thus still be quite practical.

4-2. MATCHING OTHER THAN RADIO-FREQUENCY LINES

The matching of lines other than radio frequency presents a problem of considerable scope. We shall only outline some of the practical considerations involved for some particular situations.

On power lines, for example, matching can only be done in a rather rudimentary manner because the load is set by the consumer and is constantly changing. However, the problem is not so hopeless as it may at first seem because the loads on the large distribution lines may be reasonably constant. In this case transformers can be used to adjust impedance levels to optimum values for the system. In addition, matching in the form of power-factor correction may be imposed on the system by the use of physical capacitors or by use of a synchronous machine with a suitably excited field. The power line is ordinarily not analyzed by the transmission-line methods discussed thus far, primarily because the line is electrically short and because line losses are appreciable. Instead, it is usually analyzed in terms of equivalent electrical networks. In any case, matching in the sense discussed with regard to reflection is of secondary importance. This is primarily true because the "signal" does not carry any information of value and because practical lines are usually small fractions of a wavelength long. The basic importance of the rudimentary matching is in reducing losses to a minimum. This goal is attained by matching in the sense discussed with regard to radio-frequency lines. As a practical consideration it may be noted that the characteristic impedance of conventional power lines is of the order of 400 ohms.

On telephone lines, matching is extremely important because faulty matching can produce echos which prove annoying to the subscriber. Furthermore, if many links are involved, the effect of a small mismatch in each link may be compounded to proportions that make the desirable circuits unusable. Note, for example, that if the mismatch should be such as to cause a 1-dB error per link, the error would be compounded to the fantastic figure of 500 dB over 500 links. Matching on telephone lines can thus be very critical and is ordinarily accomplished by a load in the form of a reasonably complex electrical network. This network will match the characteristic impedance of the line with considerable precision at all frequencies of interest. The situation is complicated in comparison with radio-frequency lines in that losses are not negligible in most cases (because the frequency is low, and it is the ratio of ωL to R which is important). Consequently the characteristic impedance changes considerably with frequency rather than being a constant resistance independent of frequency.

On audio cables, matching is usually of no importance. Again, the lines are usually only a small fraction of a wavelength long at the highest frequency of interest, and the "transmission-line effects" are of no importance. For most audio purposes the cable may be simply treated in effect as a shunt capacitance on the driving source. A practical example will illustrate the point.

Example 4–1

Suppose that a shielded cable is used to connect between a tuner and a remote amplifier over a distance of 50 ft (15.24 m). For a common cable with a capacitance of 20 pF/ft, this would give an equivalent capacitance of 1000 pF. As a check on the cable, it may be noted that at 15 kHz the wavelength in free space is 20,000 m, and clearly the cable is electrically short. Suppose that the source is assumed to have an internal resistance of 10,000 ohms and the load is assumed to be 1 megohm. The frequency response from source to load would be such that the output would be down 3 dB at an upper frequency of about 16 kHz. As we shall see in Chapter 6, this capacitive loading may be viewed in terms of multiple reflections on the line.

Let us see how proper matching would affect the situation. Assume that the characteristic resistance is 100 ohms[1] and that the load is 100 ohms rather than the previously assumed 1 megohm. In this case the band width is virtually unlimited by the cable, but the signal capabilities and gain have been reduced. For an original signal level of 1 V, the output is now only about 0.01 V. The amplitude cannot be conveniently raised back to 1 V because of possible overload in the driving amplifier.

On video cables, matching is usually of great importance. In this case, band widths of 5 to 10 MHz are required, and for reasonably long cable runs (over a few meters) the penalty illustrated in the previous example must be paid to obtain the band width. In this case it is usual to match only at the load, since matching at the source increases the driving requirements by a factor of two. Alternately, the source may be very well matched and the load left as an open circuit. In this case the reflected signal is of no importance because it is dissipated completely in the

[1] The characteristic resistance must be in the range 20 to 200 ohms because of the logarithmic connection to physical dimensional ratio.

matched source. It is interesting to note that the driving power of each is equally efficient.

Example 4-2

Suppose that the source is a 100-mA pulse current source, as from a pentode or the collector of a transistor. For a 50-ohm line terminated in 50 ohms, the load voltage will evidently be a 5-V pulse. If the 50 ohms is placed at the input, on the other hand, the current pulse will only initiate a 2.5-V pulse toward the load. However, since the load is open-circuited in this case, a unit reflection takes place at the load to give the same load-voltage pulse amplitude of 5 V, as in the first case. Either will work, but the properly terminated load is usually preferred.

The cables used for video work are usually chosen so that the losses are negligible and proper termination is in the form of a fixed resistor. However, sometimes this is not possible (as with delay cables where losses are usually appreciable), and it is necessary to terminate the cable in an impedance that matches the characteristic impedance of the line at all frequencies. The actual circuit for this purpose usually involves a resistor for the predominant part of the impedance and a supplementary inductor, capacitor, or both for reasonably precise matching.

4-3. MATCHING ON RADIO-FREQUENCY LINES

Although the techniques that will be illustrated for matching are based on the assumption that the lines are lossless, the lines are known to be slightly lossy in the actual case. Our justification for neglecting losses, as before, is that the results derived by this assumption are close enough to the practical case to be useful and can be derived with a significant saving of labor.

The choice of a transmission line for a given application usually depends on the availability of manufactured products. However, several basic, important points should be noted about the choice. First, the characteristic resistance of the lines is fixed in a limited range, about 30 to 500 ohms, by the physical geometry of the line. This situation is graphically illustrated in Fig. 4-1, which gives the relationship between characteristic resistance and geometry for coaxial and parallel-wire lines with air insulation.

There is an optimum physical geometry to handle maximum power in the case of the coaxial line. If the inner conductor is large, then the electric

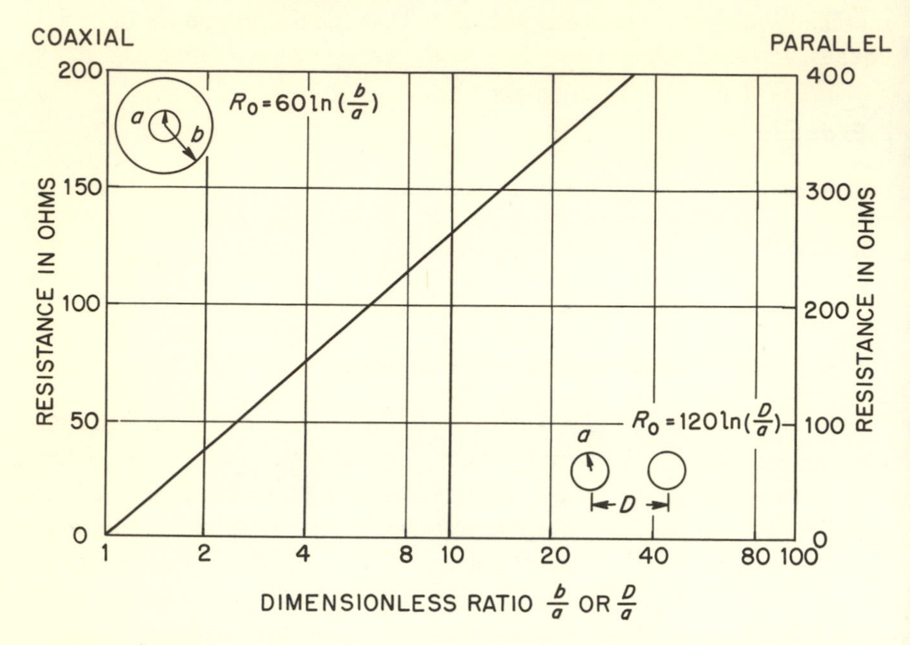

Fig. 4–1. Relationship between characteristic resistance and geometry.

field intensity between the conductors is great because of the small spacing. On the other hand, if the inner conductor is small, the electric field intensity near the center conductor is great because of the small radius of curvature. The optimum radius may be found by calculus to be such that $b/a = e = 2.718$, which corresponds to characteristic resistance of 60 ohms for air-insulated lines. Normally, however, the line is insulated with a dielectric like polyethylene to provide mechanical support of the inner conductor. In this case the characteristic resistance for the optimum power-handling capability is reduced by $\sqrt{\epsilon_r} = \sqrt{2.25} = 1.5$ to a value of 40 ohms. In actual practice the commercial coaxial lines have a characteristic resistance of either 50 or 75 ohms. Normal values of characteristic resistance for parallel-wire lines are in the range of 200 to 450 ohms.

In recent years two transmission-line configurations, known as *strip transmission line* and *microstrip*, have come into widespread usage. There are two reasons for this popularity; (1) these lines are planar in nature, thus lending themselves to automatic production-line techniques, and (2) these lines are readily adapted to hybrid integrated circuit technology at microwave frequencies. The characteristic impedance of strip transmission line as a function of the line parameters is shown in Fig. 4-2. Note that this characteristic impedance is normalized so that the effect of the dielectric is removed from the curves.

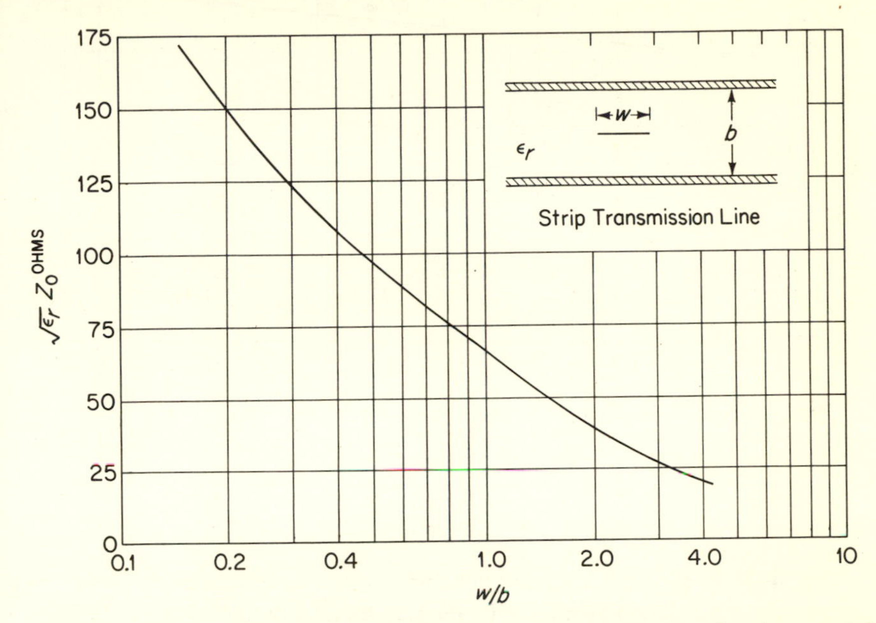

Fig. 4-2. Characteristic impedance of strip transmission line for the special case of a very thin center strip. (After S. B. Cohn, "Problems in Strip Transmission Lines," *IRE Transactions on Microwave Theory and Techniques,* vol. 3, No. 2, March 1955.)

The situation for microstrip is much more complicated because the line does not support a true TEM[2] mode in a non-static situation. This is due to the fact that the fields are in both the dielectric supporting the strip and the free-space region above the strip. A plot of the ratio of free-space wavelength (λ_0) to microstrip wavelength (λ_m) and the characteristic impedance of microstrip versus the strip width (W) to dielectric height (H) ratio for the case of a relative dielectric constant of 15 is shown in Fig. 4-3. Both strip transmission line and microstrip are generally used in low-power applications.

Matching of the load to the characteristic impedance of the transmission line is accomplished by methods to be discussed in detail in Arts.

[2] The acronym TEM (Transverse Electric and Magnetic) denotes a field pattern having its electric and magnetic fields contained in a plane transverse to the direction of propagation. This subject will be discussed in greater detail in Chapter 9.

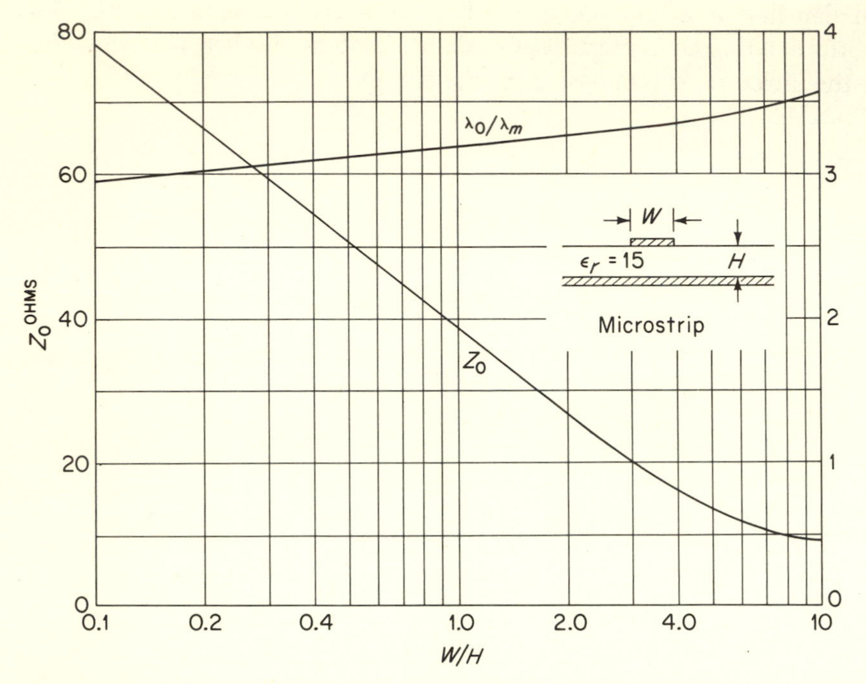

Fig. 4–3. Characteristic impedance and λ_0/λ_m ratio for microstrip transmission line. (From the 1969 *Microwave Engineers' Handbook and Buyers' Guide,* as presented by Burke, Gelnovatch, and Chase, based on Wheeler's work.)

4–5 and 4–6. Note carefully in this later discussion the effect of frequency deviations on matching.

Matching between line and source is the remaining problem of interest. Consider, for example, the situation where the power source has an internal impedance of $75 + j0$ ohms over the frequency band of interest. The problem is how to match between a 50-ohm line (which is assumed to have an input impedance of $50 + j0$ ohms over the band because of matching at the load) and a $75 + j0$ ohm source impedance.

First, it would be worthwhile to check and see how much power loss results from this degree of mismatch. If it is imagined, as a "crutch" in the analysis, that the source is connected through a section of 75-ohm line, then a 50-ohm load would have a voltage reflection coefficient of $+0.2$ on this line. Consequently the reflected power would only be 4 per cent of the incident power, and matching could not be expected to give much improvement in power transfer.

If, however, for reasons other than maximum power transfer, it seems desirable to match the system, then a matching network is needed to transform the 50-ohm load up to the 75-ohm internal impedance of the source. This matching, while only increasing the load power by 4 per cent, would have the desirable feature of making matching at the load less critical. This feature depends on two factors; multiple reflections die out as the product of the reflection coefficients "looking outward" at each end of the line, and the matching from line to source tends to produce matching from source to line. In the particular case when the load, as fixed by the characteristic resistance of the line, is purely resistive at $50 + j0$ ohms, it may be shown that matching to the source for maximum power transfer happens to make the impedance looking back toward the source exactly $50 + j0$ ohms.

The actual relationship between matching conditions at the two ends of a lossless network is summarized in Fig. 4-4. If a load Z_2 is placed at

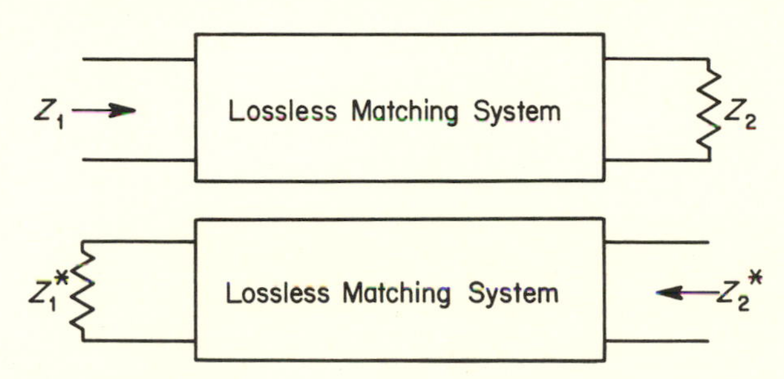

Fig. 4-4. Matching conditions for a lossless network.

one end of the network and the impedance measured at the other end is Z_1, and then if the load Z_1^* is placed at the first end, it will be found that the impedance measured at the opposite end will be Z_2^* (the superscript asterisk means conjugate). This result may be established by the reciprocity theorem, and the proof is left as an exercise. For the special case when Z_2 is purely resistive, then Z_2^* will be purely resistive, and perfect matching toward the source is produced by making the matching-network conjugate match the internal impedance of the source for maximum power transfer.

4-4. BALUNS

A balun, as explained in Art. 4-1, is a device used to connect between balanced transmission lines (parallel wire) and unbalanced transmission lines (coaxial). It might at first be thought that the connection would be trivial, since each consists of two conductors and no mathematical distinction has yet been made between the balanced and unbalanced lines. However, if the balanced (parallel wire) transmission line is to behave according to theory, it is necessary that the currents in the two lines be equal and opposite and that the voltages to ground of each of the two lines be equal and opposite. These conditions on the voltage and current are actually forced on us by the assumptions made in deriving the values of L and C for the transmission line parameters.

If the balanced transmission line is not both driven antisymmetrically and loaded symmetrically, then any mathematical analysis made by using the methods outlined thus far will be incorrect. The results obtained from an unbalanced mode of operation of a balanced line are predictable, but with considerably less ease than if the line is driven in a balanced fashion. As a practical matter, the balanced line is never driven "exactly" balanced. However, if the unbalance is small, then the deviation of results from the balanced case will also be small.

While many baluns exist, only one form using transmission-line elements will be discussed here in detail. The basic form of the balun may be indicated in transformer form as shown in Fig. 4–5(a). Baluns are actually

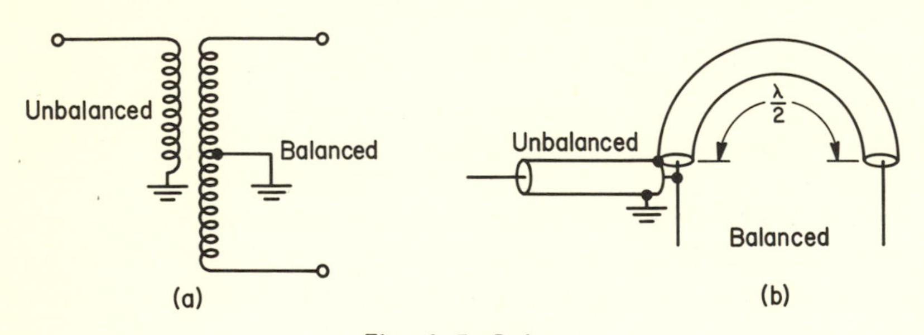

Fig. 4–5. Baluns.

constructed in this fashion at frequencies below about 200 MHz, but the precision of the device is not sufficient to permit accurate laboratory measurements. Instead, laboratory measurements are usually made by using a balun constructed as shown in Fig. 4–5(b). The balun in this latter case depends upon the fact that voltages and currents are preserved in magnitude over the precise distance of $\lambda/2$ and that the phase shift is 180 degrees. This 180-degree phase shift in the voltage amounts to a reversal of polarity, which is needed to drive the other wire of a parallel-wire line. It should be noted that a 4:1 impedance transformation is produced in the balun of Fig. 4–5(b), as illustrated in Fig. 4–6. If the

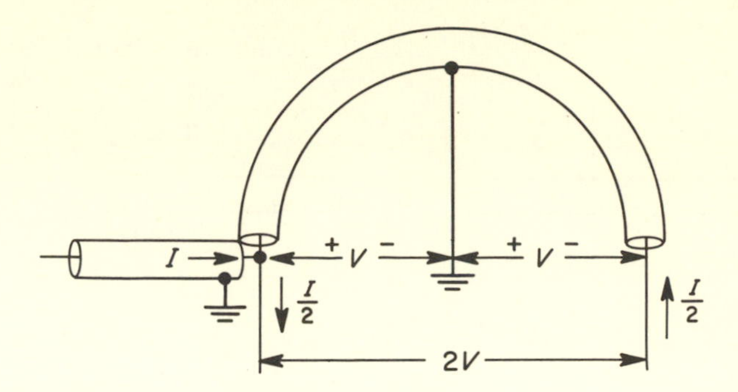

Fig. 4–6. Four-to-one impedance transformation produced by this balun.

voltage and current on the unbalanced line are V and I, respectively, then the voltage and current on the balanced line will be $2V$ and $I/2$, respectively. As a consequence, any load at the balanced terminals of the balun will be reflected as a fourth of this value at the unbalanced terminals of the balun. Note, however, that conventional corrections for line length are required for measurements other than at the immediate terminals of the balun. The lone exception to this correction occurs in the special case when the reflected load on the unbalanced coaxial line is identical to its characteristic resistance. For example, if the characteristic resistance of the coaxial line is 50 ohms and the characteristic resistance of the parallel-wire line is 200 ohms, then a 200-ohm load anywhere on the balanced system will be reflected as a 50-ohm load anywhere on the unbalanced system, and vice versa.

The balun described in Fig. 4–5(b) is usually constructed for laboratory use in a slightly different fashion than illustrated. Ordinarily it is quite difficult to make an adjustable-length line having constant characteristic resistance. However, fixed-length sections of line and variable-length shorted stubs are reasonably easy and economical to construct. As a consequence of these practical considerations, the section of line one-half wavelength long is usually replaced by an equivalent network, as shown in Fig. 4–7. It consists of two fixed-length sections of line and two adjustable, shorted stubs. For practical reasons of construction, one stub is placed in the center and the other stub is placed at one end.

Example 4–3

As an example of the use of the theory developed in the preceding chapter, consider the following balun analysis.

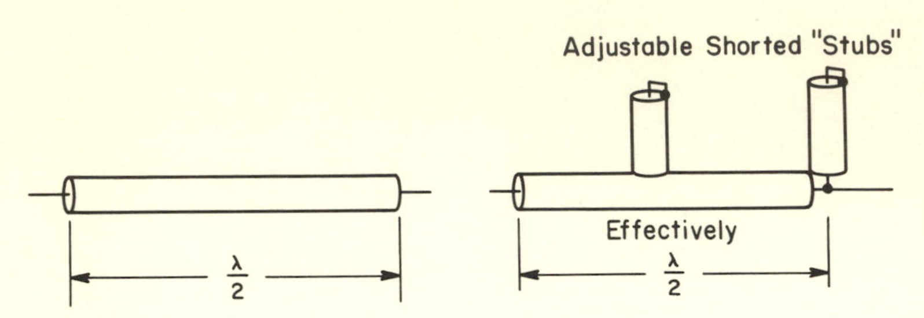

Fig. 4–7. Line equivalent of a half-wavelength obtained by proper adjustment.

For any given frequency and choice of fixed line lengths, there corresponds settings of the two shorted stubs which will make the system behave as a one-half wavelength section of line. It will be stated without proof that the two systems may be made electrically equal at a given frequency if their input–output characteristics are matched for two different sets of conditions. The two sets of conditions, chosen primarily for simplicity, are that the input impedance must be the same as the load impedance for the two special cases of open- and short-circuit loads. The notation for this example is given in Fig. 4–8 for the assumed conditions that the two fixed

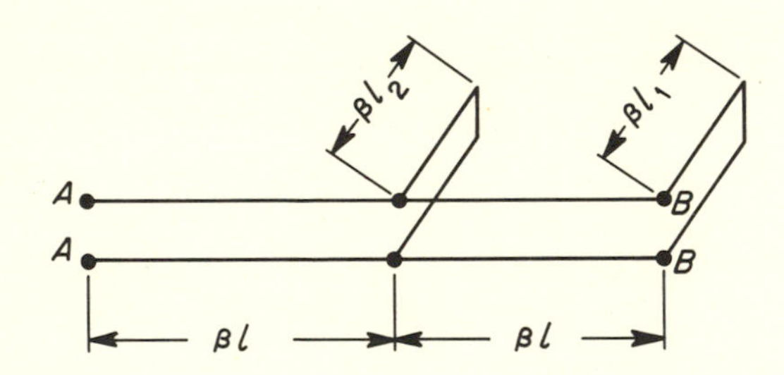

Fig. 4–8. Notation for line equivalent to one a half-wavelength long.

lengths of line are identical lengths. It is also assumed that all lines have a common characteristic resistance equal to R_0. The system is most easily analyzed by considering it broken at the junction with the line βl_2, the βl length of line in one part and other lines in the other part.

With a short at BB, then the impedance looking into AA must also be a short circuit. Since the admittance of a short-circuited line βl long is $-jG_0 \cot \beta l$, while the admittance of an open-circuited line βl long is $+jG_0 \tan \beta l$, it follows that the admittance conditions required for analysis of this problem are as illustrated in Fig. 4–9. With the network broken into the two parts as shown, it is possible to consider the actual parallel combination of lines on the right as equivalent to a shorted section $(\pi - \beta l)$ long for Fig. 4–9(a) and as an open section $(\pi - \beta l)$ long for Fig.

4–9(b). The equations which must thus be satisfied are

$$\cot(\pi - \beta l) = \cot \beta l_2 + \cot \beta l \tag{a}$$

and

$$-\tan(\pi - \beta l) = \cot \beta l_2 + \cot(\beta l + \beta l_1) \tag{b}$$

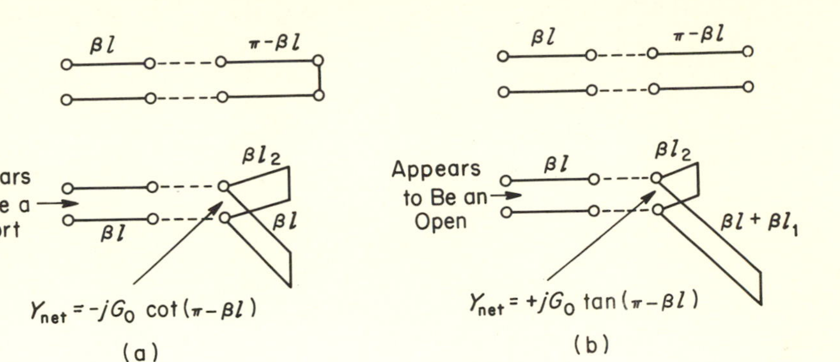

Fig. 4–9. Admittance relationships.

By considerable manipulation of these equations it may be shown that the two stubs are the same length, with the common length given by the equation

$$\cot \beta l_1 = \cot \beta l_2 = -2 \cot \beta l \tag{c}$$

The proof of this expression is left as an exercise, since it merely involves routine trigonometric manipulations of Eqs. (a) and (b). Alternately, the equation may be written in terms of tangents as

$$\tan \beta l_1 = \tan \beta l_2 = \frac{-\tan \beta l}{2} \tag{d}$$

This connection between stub spacing and stub length may be conveniently expressed in graphical form, as shown in Fig. 4–10. The scales in this figure have been calibrated in terms of wavelengths for convenience.

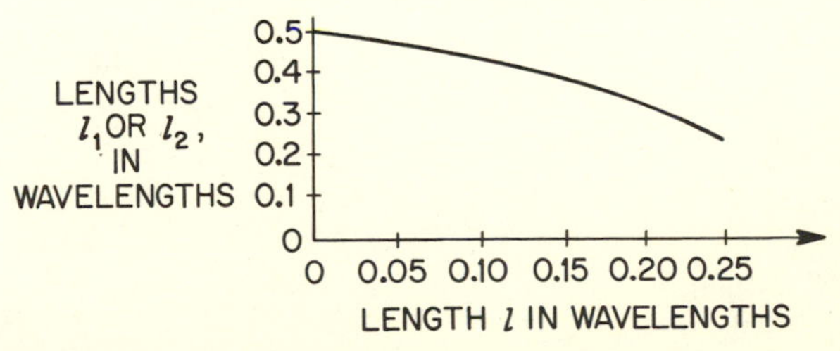

Fig. 4–10. Relationship between stub spacing and length for balun.

4–5. MULTIPLE-STUB TUNERS

In this article predominant attention will be directed toward analysis of single-stub and double-stub tuners. Triple-stub tuners may be analyzed exactly in the fashion of double-stub tuners and so will not be treated separately.

The basic matching arrangement for the single-stub tuner is shown in Fig. 4–11. For convenience it will be assumed that all the lines have the

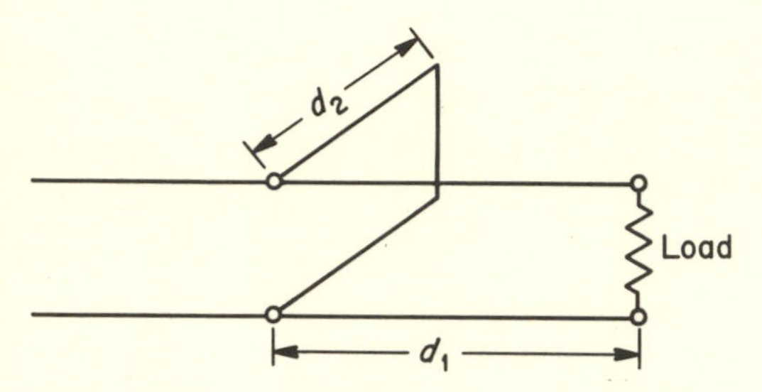

Fig. 4–11. Basic single-stub tuner system.

same characteristic resistance. The distance d_1 from the load to the stub, as well as the length d_2 of the shorted stub, must be adjustable. Our problem is to determine these two lengths to match a given load to the characteristic resistance of the line. This problem can be handled most efficiently on a transmission-line chart. Since the stub is connected in parallel with the main transmission line, it follows that an admittance form of the Smith chart would be most practical to use in the analysis. On a per-unit admittance form of Smith chart, the principle of the single-stub tuner may be summarized as follows:

Principle of the Single-Stub Tuner

> The length d_1 is chosen such that the load is transformed to the unit conductance circle on the Smith chart, and this choice permits the length d_2 to be chosen such that the pure susceptance it presents will exactly cancel that reflected by the load at its terminals, with the resulting admittance being $1 + j0$ per-unit mhos, which is the condition of perfect match.

Consider now the analysis of the particular problem where the load is $100 + j100$ ohms and the lines all have a characteristic resistance of 50 ohms.

Example 4–4

The first step in the analysis is to convert $100 + j100$ ohms to $2 + j2$ per-unit ohms and then by inversion to $0.25 - j0.25$ per-unit mhos. This value of load admittance is plotted on a per-unit admittance form of Smith chart as shown in Fig. 4–12(a). The permissible locus of admittances which may be converted to $1 + j0$ ohms by the pure susceptance of the shorted stub is shown as the unit conductance contour in Fig. 4–12(b). Now it is possible, as illustrated in Figs. 4–12(c) and 4–12(d), to choose the distance d_1 to be either 0.219λ or 0.362λ and attain a point on this locus. In case I the stub length d_2 must be chosen so that the stub presents a susceptance of $-j1.6$ per-unit mhos. The Smith chart may be used to obtain this length as 0.089λ. In case II the stub length d_2 must be chosen so that the stub presents a susceptance of $+j1.6$ per-unit mhos, and again the Smith chart may be used to obtain this length as 0.411λ. Note carefully in using the chart to find the stub lengths that the admittance of a short circuit is infinite.

The two possible solutions of this single-stub matching example are given in Fig. 4–13. Since losses increase directly with line length and standing-wave ratio, the single-stub matching solution shown in Fig. 4–13(a), where the lines are the shortest, is preferred to that of Fig. 4–13(b). However, the difference in performance between the two solutions is small enough so that other factors may determine the selection. For example, if the shorted stub could not be adjusted as short as 0.089λ, then the other solution might be selected. Note, however, that this line might be increased by 0.5λ to 0.589λ to avoid adjustment difficulties without affecting the matching in any way.

Before proceeding to the double-stub tuner, let us consider why it is more desirable than the single-stub tuner. The primary disadvantage of the single-stub tuner is that it requires an adjustable-length line of constant characteristic resistance. Such a line is quite expensive and difficult to build, and so its use is avoided whenever possible. The construction of an adjustable shorted stub is illustrated in Fig. 4–14, and the construction of an adjustable-length line of constant characteristic resistance is shown in Fig. 4–15. Before looking at Fig. 4–15 closely, you may find it instructive to try designing an adjustable-length line of constant characteristic resistance. It may be harder than first imagined. Very evidently, from inspection of these two figures, the adjustable shorted stub is much the easiest to construct. Consequently it is desirable to develop a matching system which uses only adjustable-length shorted stubs and, of course, fixed sections of line. Such a scheme is illustrated in Fig. 4–16 and is called a *double-stub tuner*. The degree of freedom offered

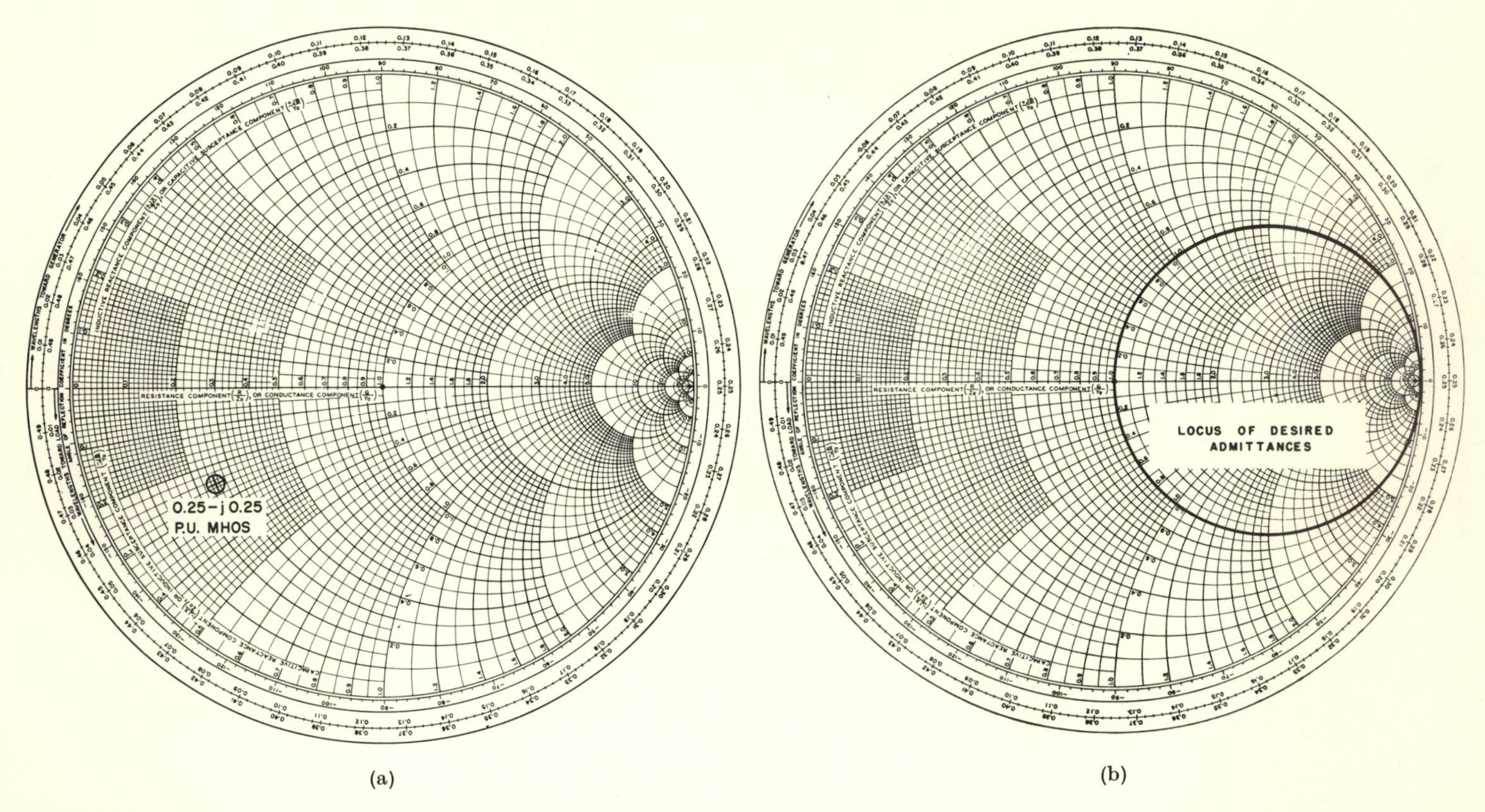

(a)

(b)

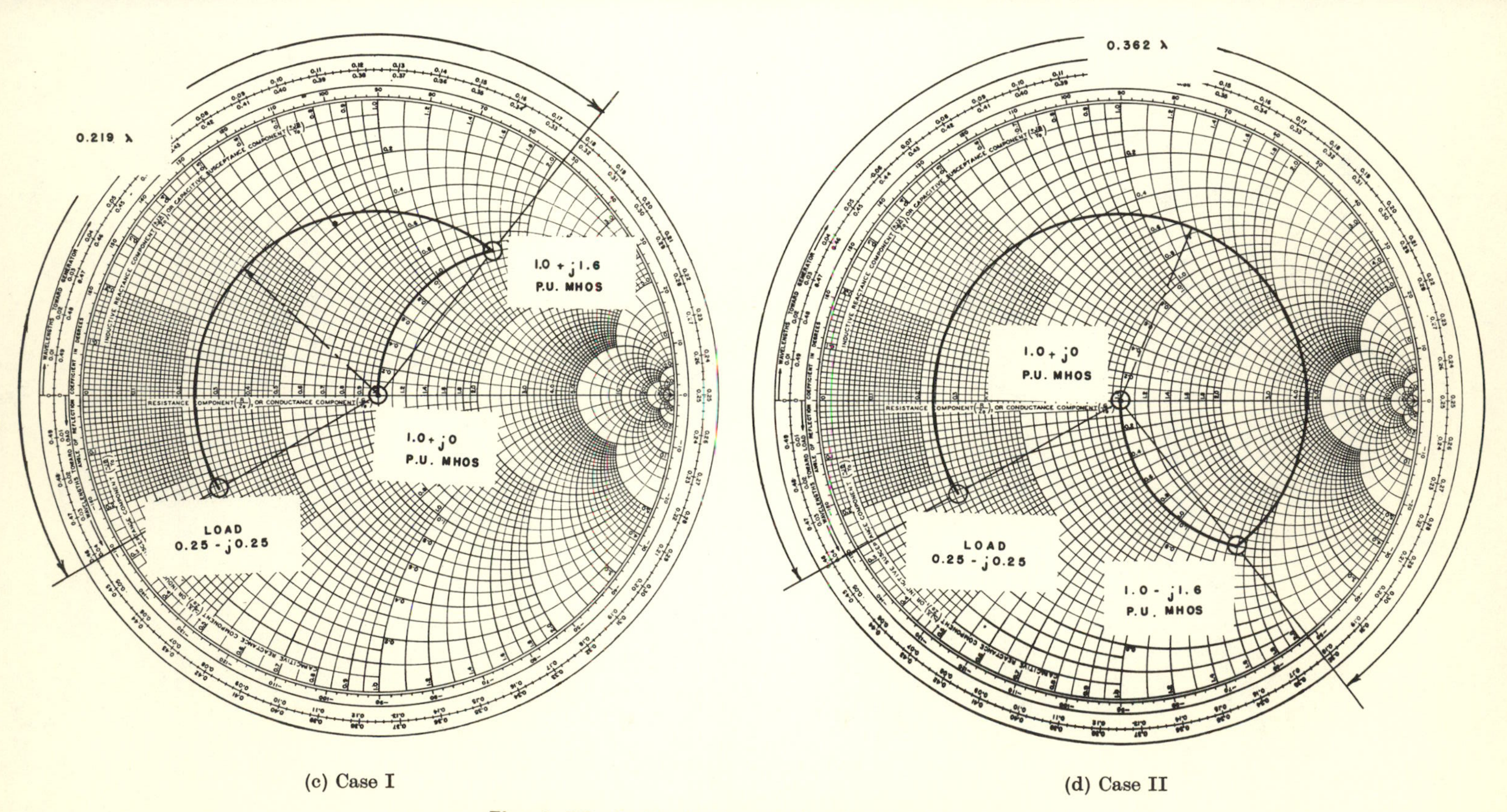

(c) Case I

(d) Case II

Fig. 4–12. Smith chart analysis of single-stub tuner.

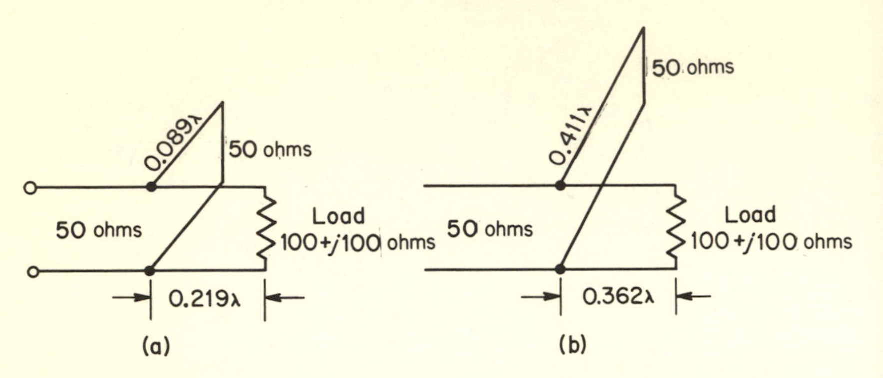

Fig. 4–13. The two solutions of the single-stub matching problem.

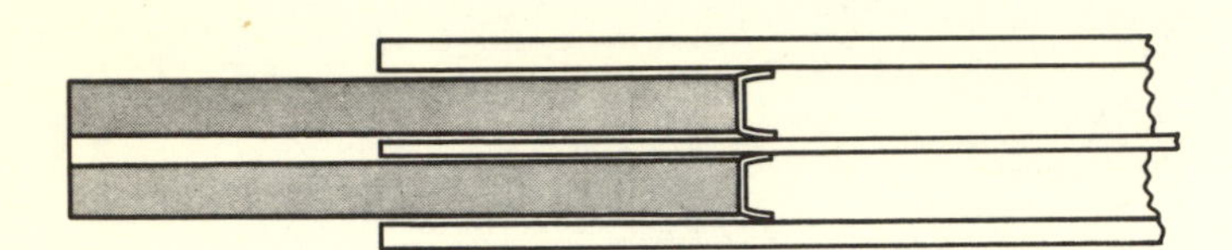

Fig. 4–14. Adjustable shorted coaxial stub.

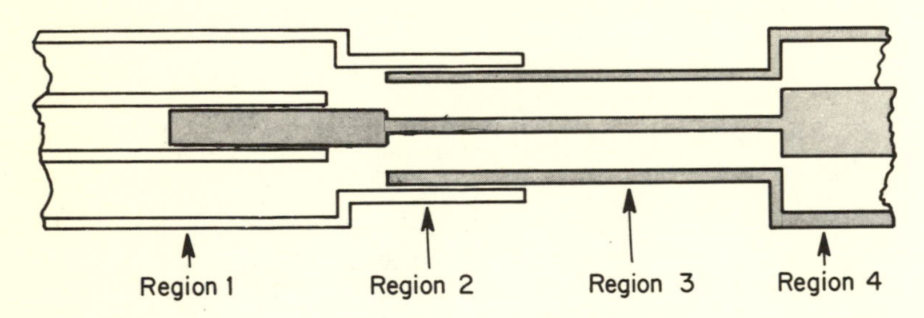

Fig. 4–15. Sectional view of adjustable-length coaxial line having constant R_0.

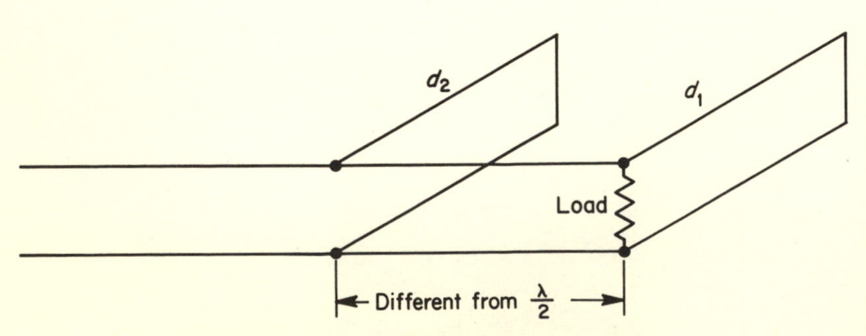

Fig. 4–16. Basic double-stub tuner system.

by the adjustable-length section of line has been replaced by an adjustable shorted stub and a fixed length of line. Unfortunately a double-stub tuner will not match an arbitrary load, as would the single-stub tuner. The addition of another adjustable shorted stub and fixed-length line may be used to correct this situation. However, the additional complication of the triple-stub tuner is usually not warranted, since the double-stub tuner may be made to work satisfactorily and is not nearly so difficult or critical to adjust. The techniques of analysis for the triple-stub tuner are virtually the same as for the double-stub tuner, and hence they will not be discussed again in this section.

The double-stub tuner system of Fig. 4-16 has two adjustable-length shorted stubs with the lengths d_1 and d_2 and a fixed-length section of line between them which must be other than a half-wavelength long. Common values of this fixed length are $\lambda/8$ and $3\lambda/8$ for reasons which will be discussed after a system has been analyzed in detail. Our problem is to determine the lengths d_1 and d_2 of the two shorted stubs to match a given load to the characteristic resistance of the line. This problem may be handled most efficiently on a transmission-line chart. On a per-unit admittance form of Smith chart the principle of the double-stub tuner may be summarized as follows:

Principle of the Double-Stub Tuner

> The length d_1 must be chosen such that the load and stub in parallel is transformed to the unit conductance circle by transferral down the fixed-length section of line. This choice permits the length d_2 to be chosen such that the pure susceptance which it presents will exactly cancel that reflected by the load at its terminals, with the resulting admittance being $1 + j0$ per-unit mhos, which is the condition of perfect match.

The most difficult step in analyzing the double-stub tuner initially comes in choosing the length d_1 such that the load and stub in parallel is transformed to the unit conductance circle by transferral down the fixed-length section of line. Once this step is accomplished, the process reduces exactly to that of the single-stub tuner when choosing the length d_2. Although the length d_1 can be chosen by trial and error, such a time-consuming task is not necessary. The secret of success lies in finding the locus of points on the chart such that transferral down the fixed-length line toward the source will yield the locus of points represented by the unit conductance circle. Such a locus is really quite simple to find because by going backwards, from source to load, it is easily seen to be the unit conductance circle rotated toward the load by the length of the fixed section of line. In the special case when the fixed length of line is $\lambda/8$ long, the locus appears as shown in Fig. 4–17(a). Transferral of this locus of points toward the source by $\lambda/8$ yields the unit conductance circle, as illustrated in Fig. 4–17(b).

All that is necessary, then, is to add susceptance by the stub at the load so that the combination is on the required locus [shown for the special case of $\lambda/8$ stub spacing in Fig. 4–17(a)]. This admittance may then be transferred toward the source by the length of the fixed line, with complete assurance that the resulting admittance will lie on the unit conductance circle so that the remaining stub may be used to cancel the susceptive portion and yield a perfect match.

Example 4–5

Suppose the load is $100 + j100$ ohms, the lines have a characteristic resistance of 50 ohms, and the spacing between stubs is $\lambda/8$. The per-unit load admittance is easily computed as $0.25 - j0.25$ per-unit mhos. There are two solutions, just as for the single-stub tuner. The steps have been summarized in Figs. 4–17(c) and 4–17(d). In verbal summary form the steps are:

1. Add $+j0.59$ per-unit mhos ($+j1.89$ in case II) with 0.335λ (0.422 in case II) of line to reach the admittance of $0.25 + j0.34$ per-unit mhos ($0.25 + j1.64$ in case II), which is on locus of 4–17(a) for $\lambda/8$ stub spacing.
2. Travel toward the source $\lambda/8$ to reach the admittance of $1.00 + j1.63$ per-unit mhos ($1.00 - j3.50$ in case II).
3. Add $-j1.63$ per-unit mhos ($+j3.50$ in case II) with 0.088λ (0.455 in case II) of line to reach the admittance $1 + j0$ per-unit mhos ($1 + j0$ in case II).

A summary of these results is presented in Fig. 4–18 for the various lengths of stubs required in the two solutions. In this case there is no clear distinction between the two solutions with regard to losses in the lines. Very similar results may be obtained for $3\lambda/8$ spacing of the stubs.

With the background of Example 4–5 it is now possible to discuss the choice of spacing between the stubs. By reference to Fig. 4–17(a) it may be noted that load admittances inside the $g = 2$ circle may never intersect the desired locus of admittances at stub 1 for $\lambda/8$ stub spacing. Consequently these loads may not be matched with such a double-stub tuner. In a similar fashion, it is possible to find the loads which may or may not be matched for other spacings between the stubs. A summary of these results is presented in Fig. 4–19, where the shaded areas represent the loads that cannot be matched. It is interesting to note that, except for the special case when the distance between stubs is *precisely* a half-wave-

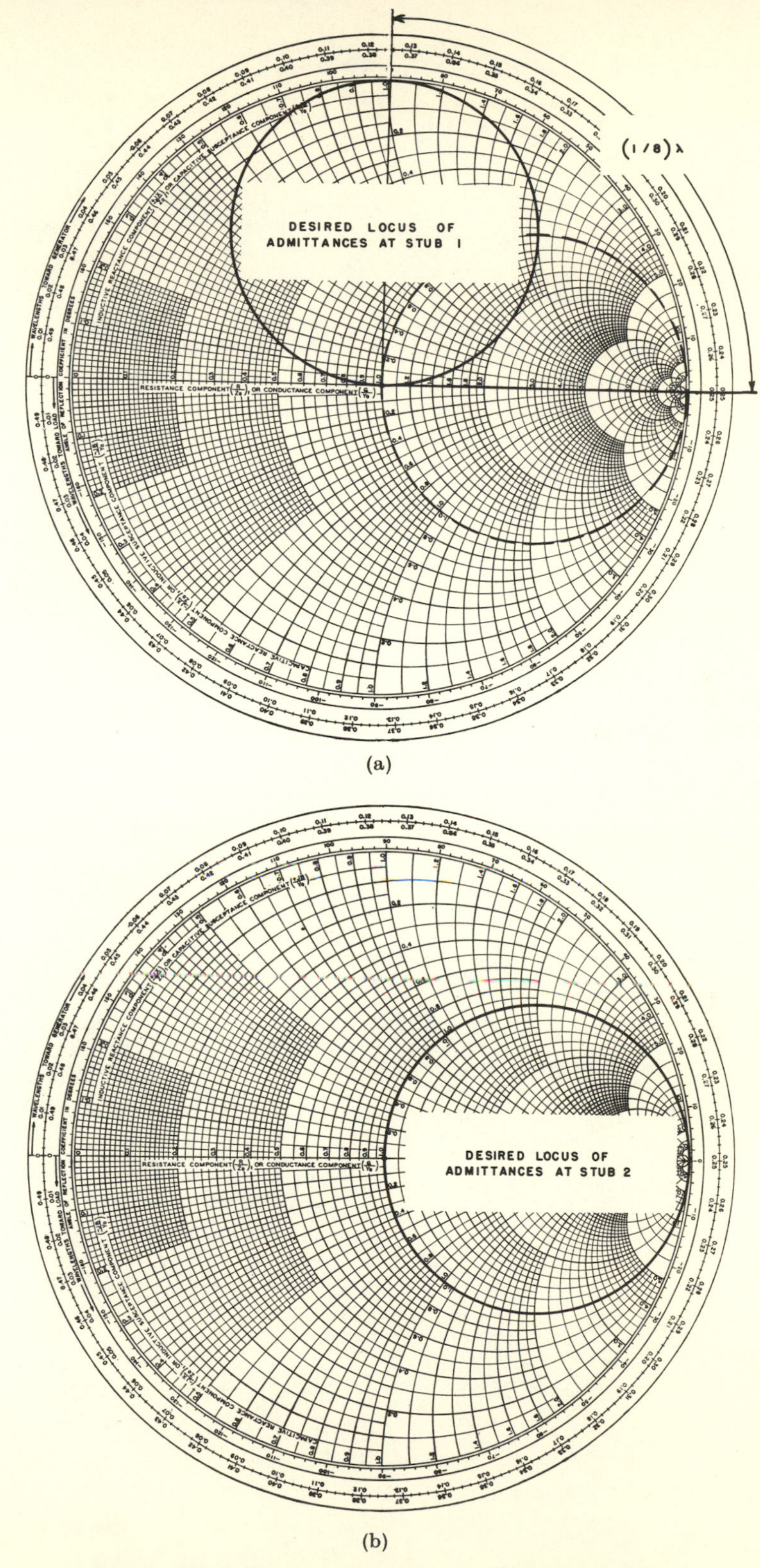

Fig. 4–17. Smith chart analysis of double-stub tuner.

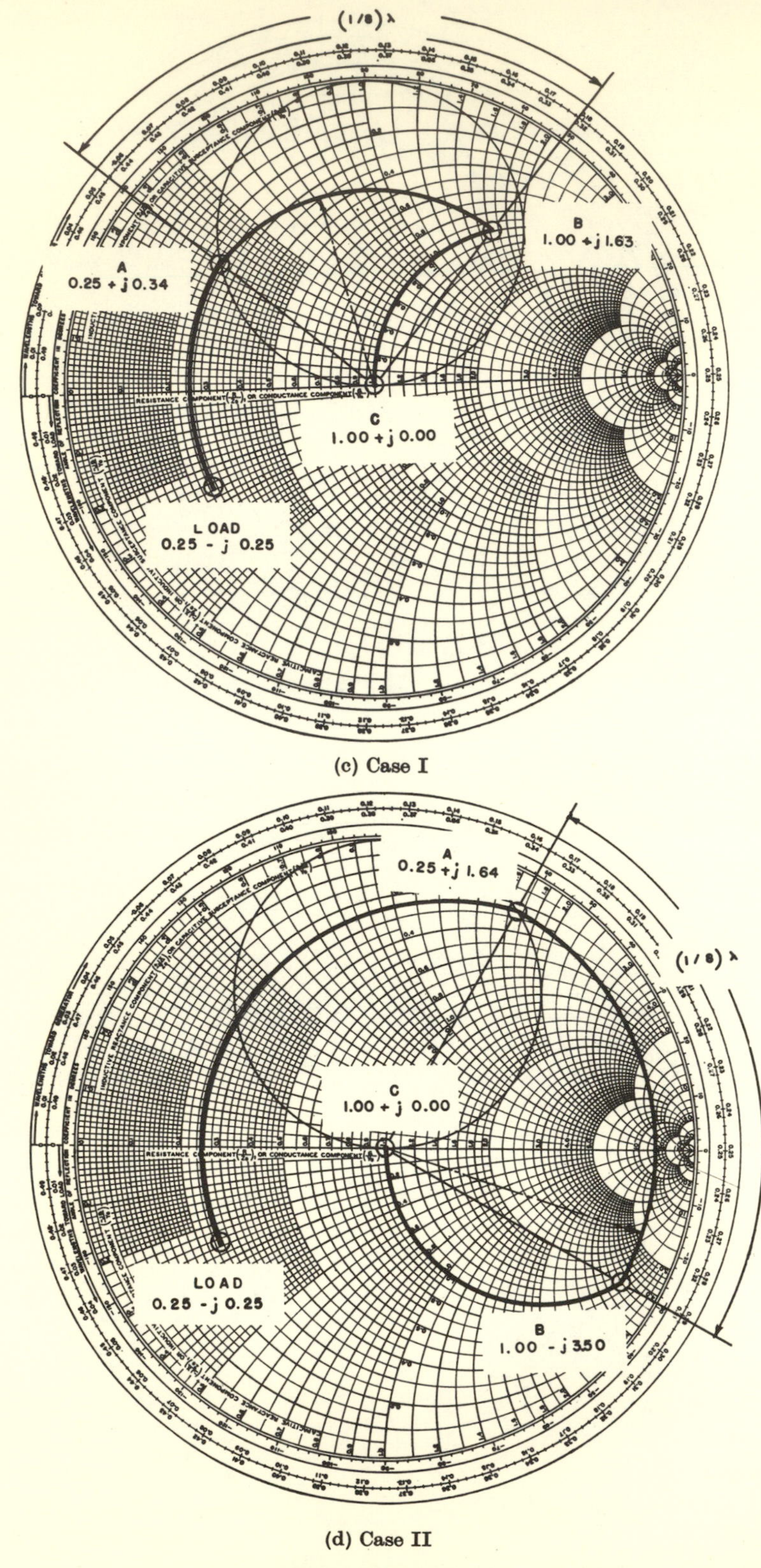

(c) Case I

(d) Case II

Fig. 4–17 (*Cont.*)

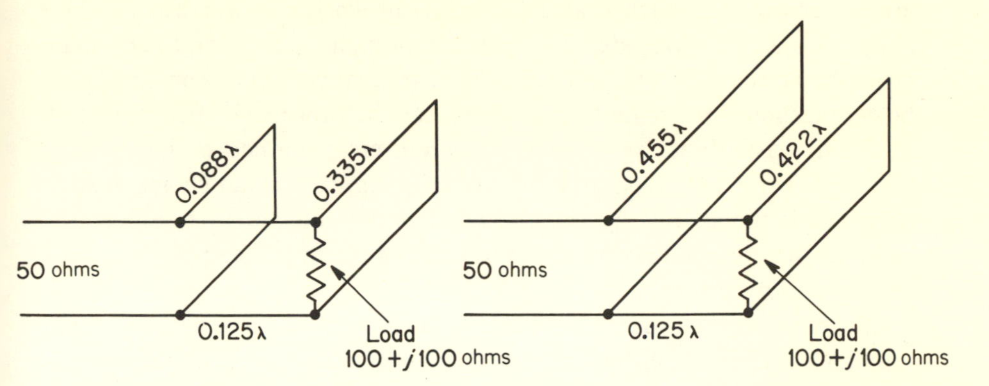

Fig. 4–18. The two solutions of the double-stub matching problem.

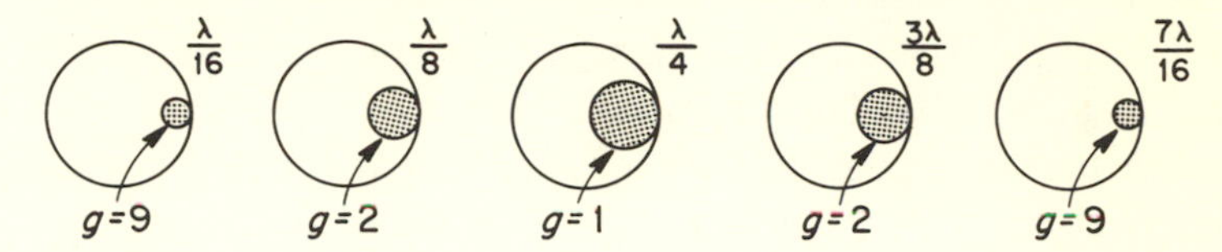

Fig. 4–19. Spacing between stubs. The ranges of load which cannot be matched are shaded.

length, the widest possible ranges of loads can be matched for a distance between stubs *approaching* a half-wavelength. The smallest range of load can be matched for a stub spacing of a quarter-wavelength. However, the adjustments become exceedingly critical as the distance between stubs approaches $\lambda/2$, and for this reason a choice of spacing like $\lambda/8$ or $3\lambda/8$ represents a reasonable compromise between matching range and critical adjustment.

If the load should happen to fall on such a portion of the chart that matching is not directly possible, then it is merely necessary to add a fixed length of line between the load and first stub to adjust the load to a value that can be matched. The analysis then proceeds in a conventional manner.

An important laboratory consideration is that (unless the advantage of not requiring an adjustable length line is to be lost) the distance between stubs will not be a precise distance like $\lambda/8$. Instead, with fixed sections of line, the distance will be some odd value like 0.147λ and the analysis will have to be made on this basis.

Because of slight inaccuracies in the system, you will almost invariably find that a small amount of "touching up" is required on the tuners. You might then logically ask if all the analytical work was worthwhile. The answer would be a very definite "yes." Although each of the tuners may be set by trial and error, the process is in general very time consuming. A better method is to compute the values, set the tuner, and then "touch it up." The basic difficulty with the trial-and-error method is that it may fail to converge to a good match if an unfortunate original choice of stub positions is made. But by calculation it is possible to start sufficiently close to the correct lengths that final adjustments to the proper values can be rapidly made.

Matching of the load to the characteristic impedance of the transmission line is accomplished by methods discussed in detail in the previous articles. The underlying reasons for the matching, as well as practical considerations, need some further discussion at this point. As a matter of orientation, we are talking about radio-frequency lines. Hence the discussion of matching basically depends upon matching at a single frequency or over a narrow band of frequencies, since in general it is virtually impossible to match an arbitrary load over the entire frequency spectrum. Three variations with frequency are important here:

1. Variation of the load impedance with frequency.
2. Variation of the matching network with frequency.
3. Variation of the characteristic impedance of the line with frequency.

For the usual band widths of interest (10 per cent or less of center frequency) the variations of the load impedance (1) and variation of the characteristic impedance of the line (3) are usually small with respect to the variation of the matching network with frequency (2). For example, this variation for the single-stub and double-stub tuners is shown in Fig. 4–20 for the special case when the load is assumed to be 100 ohms (inde-

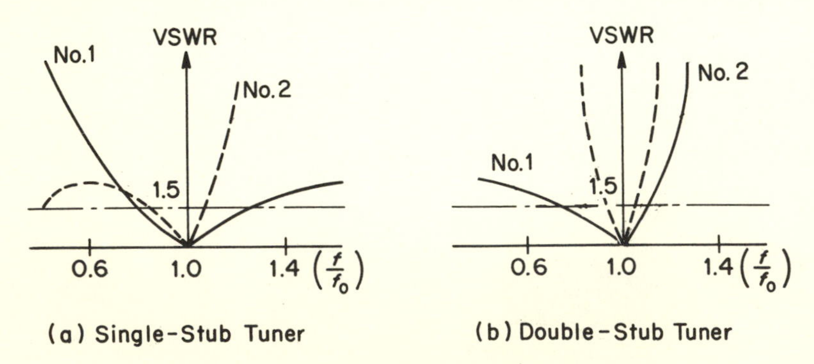

Fig. 4–20. Variation in matching networks with frequency.

pendent of frequency) and the line is assumed to be 50 ohms (independent of frequency).[3] There are two solutions in each case and the results for both solutions are shown on the curves. The mismatch at frequencies different from the design frequency is shown by finding the VSWR on the line to the left of the matching section. Commonly a VSWR of 1.5 (corresponding to 4 per cent reflected power) is accepted as the permissible upper limit for satisfactory system operation. In this case the band widths are 16 and 40 per cent for the single-stub tuners and 10 per cent and 4 per cent for the double-stub tuners.

4–6. TAPERED LINE AND QUARTER-WAVELENGTH TRANSFORMER

The tapered line involves a gradual transition in geometry from one characteristic resistance to another [see Fig. 4–21(a)]. If this transition

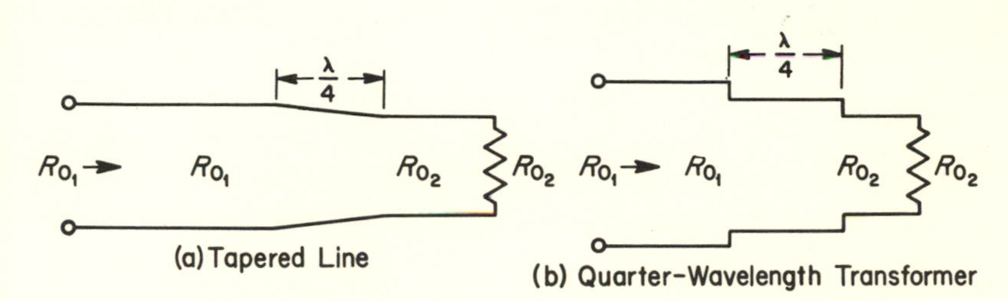

Fig. 4–21. Tapered line and quarter-wavelength transformer.

takes place over an electrical distance of one quarter-wavelength or greater, then the device behaves in the fashion of a transformer. Various tapers are used for this application, common ones being linear taper, exponential taper, and hyperbolic taper. Although each of these could be analyzed in detail, for our purposes it is sufficient to note that the transition must take place over a distance of one quarter-wavelength or more. As a consequence of this specification of taper in terms of electrical length, it follows that the matching characteristic of the tapered line is high-pass in form. For above the design frequency the tapered section will be more than one quarter-wavelength long and correct matching will take place, while below the design frequency the tapered section will be less than one quarter-wavelength and it will fail in its matching function.

[3] The double-stub tuner is assumed to have $3\lambda/8$ spacing between stubs.

The quarter-wavelength transformer is similar to the tapered line, except in this case the section must be exactly a quarter-wavelength long and the geometry constant at a value intermediate between that at either end [see Fig. 4–21(b)]. Analysis of the quarter-wavelength transformer is exceptionally simple and follows directly from the general lossless impedance formula

$$Z_{\text{in}} = R_0 \frac{Z_L \cos \beta l + jR_0 \sin \beta l}{R_0 \cos \beta l + jZ_L \sin \beta l} \tag{4–1}$$

In this formula Z_{in} and Z_L are the input and load impedances, respectively, while R_0 is the characteristic resistance of the line and βl is the electrical length of the line. For the special case when the line is a quarter-wavelength long, electrically equivalent to $\beta l = 90$ degrees, then Eq. (4–1) reduces to

$$Z_{\text{in}} Z_L = R_0^2 \tag{4–2}$$

This remarkably simple equation expresses the fact that perfect matching between lines with characteristic resistances R_{0_1} and R_{0_2} may be obtained by using a quarter-wavelength section of line having a characteristic resistance which is the geometric mean of the two as given by

$$R_0 = \sqrt{R_{0_1} R_{0_2}} \tag{4–3}$$

Example 4–6

Suppose that it is desired to connect two antennas, each with an internal impedance of 300 ohms, to a 300-ohm section of transmission line. Very clearly, the problem here

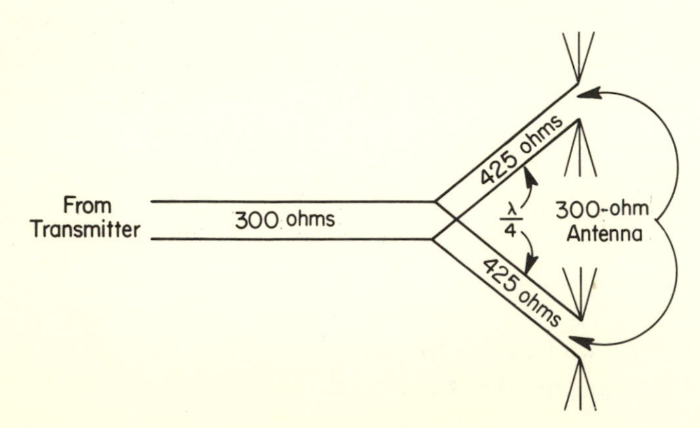

Fig. 4–22. Quarter-wavelength transformers for impedance-matching antennas.

may be solved by building two quarter-wavelength transformers to work between 300 ohms and 600 ohms, as shown in Fig. 4–22. The characteristic resistance of each of the transformers in this case must be about 425 ohms.

At this point, one may assume that the matching is not dependent on frequency. Actually, of course, at a given frequency a specific physical length of line is required for a quarter-wavelength transformer, and this physical length does not change with frequency. Consequently the matching as illustrated in Fig. 4–22 is correct at only one frequency.

An interesting aspect of matching is illustrated in Fig. 4–22 with regard to whether the system is matched both in receiving and transmitting. From antenna theory it is known that the internal impedance of an antenna does not depend on whether it is used for transmitting or receiving. The matching network very clearly is correct for the transmitting direction because it was designed that way. However, the fact that the matching is also correct in the receiving direction is much less obvious. The system of Fig. 4–22 has been redrawn in Fig. 4–23 to aid in the

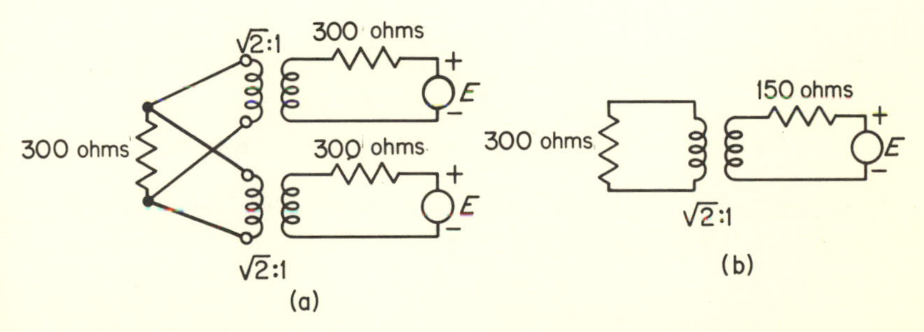

Fig. 4–23. System of Fig. 4–22 redrawn in circuit notation.

discussion. The assumed condition that both antennas receive the same signal has been subtly included in Fig. 4–23(a) by showing both sources as E with the same relative polarity. Transformers with the appropriate turns ratio of $\sqrt{2}:1$ have been used to replace the quarter-wavelength sections of line. Because of the symmetry in the circuit of Fig. 4–23(a), it is possible to reduce it further as shown in Fig. 4–23(b), and in this reduced form the system is obviously matched for receiving. Note carefully the important role played by both antennas receiving the same signal.

4–7. THE TWO-SECTION QUARTER-WAVE TRANSFORMER

One of the disadvantages of a single quarter-wave transformer is the sharp rise in the reflection coefficient at the input as the frequency is shifted away from the design frequency f_0. The frequency response of a single quarter-wave transformer designed to match a 100-ohm load to a 50-ohm transmission line is shown in Fig. 4–26. The obviously limited bandwidth of the single-section transformer can be extended by connecting two or more quarter-wave transformers in series and making the proper choice for the characteristic impedance of the transformer sections. By analyzing the two-section quarter-wave transformer, the essential features of series-connected transformers can be illustrated while keeping the mathematics relatively simple.

The two-section quarter-wave transformer to be considered is shown in Fig. 4–24.

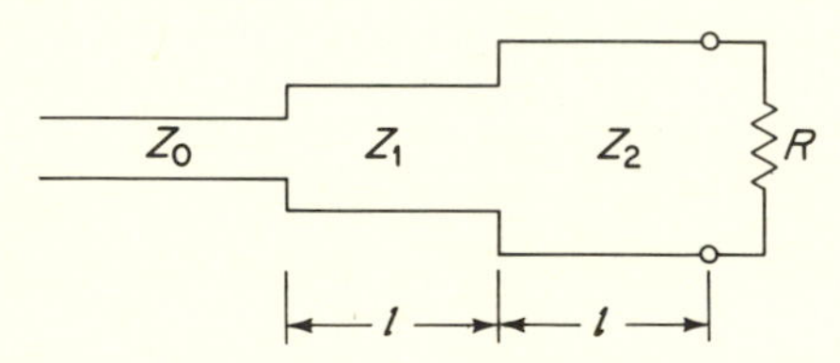

Fig. 4–24. The two-section quarter-wave transformer.

To facilitate the analysis of this structure it will be helpful to define the parameters as shown in Fig. 4–25.

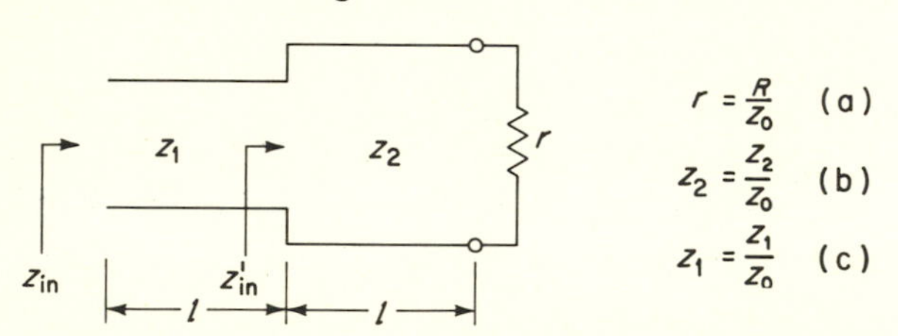

Fig. 4–25. Notation for analysis of two-section transformer.

The normalized impedance z'_{in} is easily found by application of Eq. (2–33) to a transmission line of normalized characteristic impedance z_2 and of length l, and terminated in a normalized impedance of r:

$$z'_{in} = z_2 \frac{r + jz_2t}{z_2 + jrt} \tag{4–4}$$

In Eq. (4–4), z'_{in} is normalized to Z_0 and $t = \tan \beta l$.

Note also that $\beta l = \frac{2\pi}{\lambda}\frac{\lambda_0}{4} = \frac{\pi}{2}\frac{\lambda_0}{\lambda} = \frac{\pi}{2}\frac{f}{f_0}$. The normalized impedance z_{in} serves as the load for the line of characteristic impedance z_1 and length l. The normalized input impedance of the composite structure can now be written as

$$z_{in} = z_1 \frac{z'_{in} + jz_1 t}{z_1 + jz'_{in} t}$$

or

$$z_{in} = z_1 \frac{(z_2 r - z_1 r t^2) + j(z_1 z_2 + z_2^2)t}{(z_1 z_2 - z_2^2 t^2) + j(z_1 r + z_2 r)t}. \tag{4-5}$$

Since the reflection coefficient K_{in} is related to z_{in} by $K_{in} = (z_{in} - 1)/(z_{in} + 1)$, we can look for points of zero reflection ($K_{in} = 0$) as solutions of $z_{in} - 1 = 0$, or

$$(z_1 z_2 - z_2^2 t^2) + j(rz_1 + rz_2)t = (z_1 z_2 r - z_1^2 r t^2) + j(z_1^2 z_2 + z_1 z_2^2)t$$

Equating the real and imaginary parts results in two equations:

$$z_1 z_2 - z_2^2 t^2 = z_1 z_2 r - z_1^2 r t^2$$

$$(z_1 + z_2) r t = z_1 z_2 (z_1 + z_2) t$$

These can be rewritten as

$$(z_1^2 r - z_2^2) t^2 = z_1 z_2 (r - 1) \tag{4-6}$$

and

$$r = z_1 z_2 \tag{4-7}$$

We can consider r, the normalized load resistance, and t, the tangent function at which the input reflection coefficient (K_{in}) goes to zero, to be independent variables, or known quantities, so Eqs. (4-6) and (4-7) can be solved for z_1 and z_2.

Example 4-7(a)

$$K_{in} = 0 \text{ at } f = f_0$$

In this case $t \to \infty$ at $f = f_0$. This requires that

$$z_1^2 r - z_2^2 = 0$$

and

$$r = z_1 z_2$$

in order to satisfy the conditions imposed by Eqs. (4-6) and (4-7). Solving for z_1 and z_2 results in $z_1 = r^{1/4}$ and $z_2 = r^{3/4}$. Substituting these values back into z_{in} yields the following expressions for z_{in} and K_{in}

$$z_{in} = \frac{(r - r^{1/2} t^2) + j(r^{1/4} + r^{3/4})t}{(1 - r^{1/2} t^2) + j(r^{1/4} + r^{3/4})t} \tag{4-8}$$

and

$$K_{in} = \frac{(r-1)}{(r+1) - 2r^{1/2}t^2 + 2j(r^{1/4} + r^{3/4})t} \tag{4-9}$$

Now let us consider the variation in K_{in} with frequency. Differentiating K_{in} with respect to f results in

$$\frac{dK_{in}}{df} = \frac{dK_{in}}{dt}\frac{dt}{df} = -\frac{(r-1)\left[-4r^{1/2}t + 2j(r^{1/4} + r^{3/4})\right]}{[(r+1) - 2r^{1/2}t^2 + 2j(r^{1/4} + r^{3/4})t]^2}\left(\sec^2\frac{\pi}{2}f/f_0\right)\left(\frac{\pi}{2f_0}\right)$$

As $f \to f_0$, $t \to \infty$ and $\sec \pi f/2f_0 \to \infty$, but the denominator involves t^4 and controls the function and forces dK_{in}/df to be zero at the frequency f_0. Thus, for this case, the reflection coefficient is not only zero at f_0, it also has zero slope. This means that the bandwidth will be wider than was the case for the single-section transformer.

Example 4–7(b)

$$K_{in} = 0 \text{ at } f = f_1 \neq f_0$$

Under these conditions, Eqs. (4–6) and (4–7) require that

$$(z_1^2 r - z_2^2)t_1^2 = z_1 z_2 (r-1)$$

and

$$r = z_1 z_2$$

In this case r and t_1 are assumed to be known, and it is necessary to determine z_1 and z_2. Since the tangent function t has odd symmetry about f_0, there are two values of f_1 (symmetrical about f_0) which satisfy the equations. Forcing the reflection coefficient to zero at two points symmetrical about f_0 results in a non-zero reflection coefficient at f_0. That is:

$$z_{in} = \frac{z_1^2 r}{z_2^2} \tag{4-10}$$

and

$$K_0 = \frac{z_1^2 r - z_2^2}{z_1^2 r + z_2^2}, \tag{4-11}$$

at $f = f_0$. The value of K_0 is usually chosen as the criterion to define the bandwidth of the transformer and the passband is considered to be the frequency band for which $|K_{in}| \leq K_0$. For this case, z_{in} and K_{in} can be written as

$$z_{in} = r\frac{(r - z_1^2 t^2) + j(z_1 + z_2)t}{(r - z_2^2 t^2) + jr(z_1 + z_2)t} \tag{4-12}$$

$$K_{in} = \frac{r(r-1) + (z_2^2 - z_1^2 r)t^2}{r(r+1) - (z_2^2 + z_1^2 r)t^2 + j2(z_1 r + z_2 r)t} \tag{4-13}$$

Figure 4–26 shows a plot of the magnitude of the reflection coefficient, $|K_{in}|$, for four transformers, each designed to match a 100-ohm load to a 50-ohm transmission line. The four cases are: (1) a single-section quarter-wave transformer, (2) a two-section quarter-wave transformer with $|K_{in}| = 0$ at $f = f_0$, and (3) a two-section quarter-wave transformer with $|K_{in}| = 0.1$ at $f = f_0$ and (4) a two-section quarter-wave transformer with $|K_{in}| = 0.05$ at $f = f_0$. The increased bandwidth of the two-

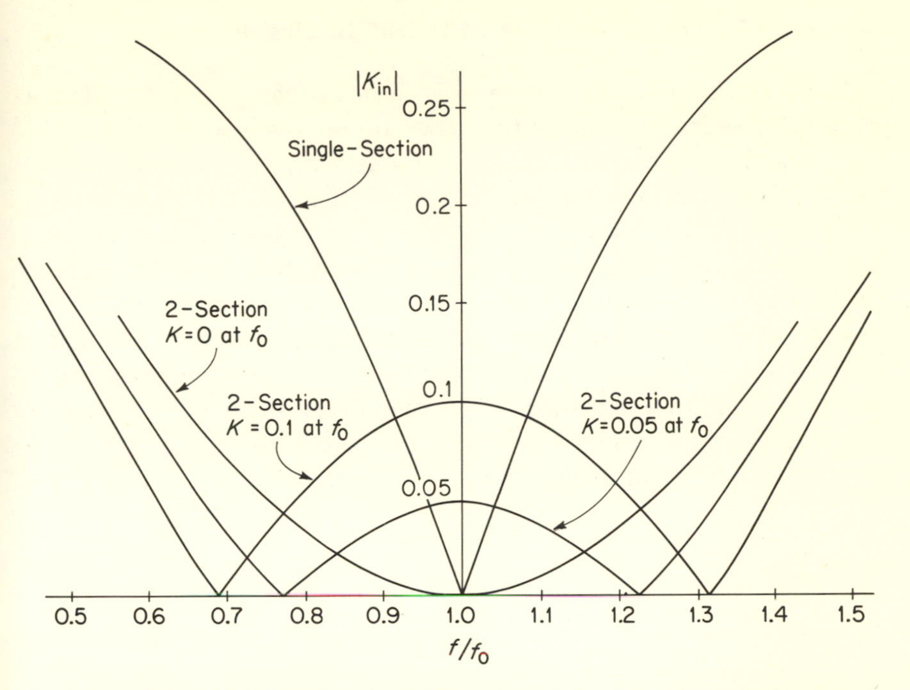

Fig. 4–26. A comparison of single-section and two-section transformers.

section transformers is easily seen. If $|K_{in}| = 0.1$ is used as a criterion for determining bandwidth, we see that the fractional bandwidth of the single-section transformer is 0.19 whereas the two-section transformers show a fractional bandwidth of 0.72 for case 2 and a fractional bandwidth of 0.92 for case 3. The increased bandwidth in case 3 is gained as a result of the finite reflection coefficient at $f = f_0$. Case 4 shows that a reduction in the reflection coefficient at f_0 results in a decrease of the bandwidth. The characteristic impedances for the quarter-wave sections are easily calculated. For the examples cited above, the following results were obtained:

Case 1: $Z_1 = 70.7$ ohms
Case 2: $Z_1 = 59.46$ ohms
$Z_2 = 84.09$ ohms
Case 3: $Z_1 = 62.52$ ohms
$Z_2 = 79.97$ ohms
Case 4: $Z_1 = 60.96$ ohms
$Z_2 = 82.01$ ohms

More than two quarter-wave transformers can be connected in series and the transmission-line impedances selected to obtain even better reflection coefficient characteristics in the passband. The analysis of such transformers is complicated and will not be discussed here.[4]

[4] R. E. Collin, "Theory and Design of Wide Band Multisection Quarter-wave Transformers," *Proc. IRE*, vol. 43, pp. 179–185, February 1955.

4–8. EFFECTS OF LOSSES ON MATCHING CIRCUITS

Although losses do affect the precision of matching, the effect is not particularly serious as long as the losses are relatively small; this is the usual practical case. Furthermore the losses are so small in normal transmission lines that the behavior of the matching circuits is more nearly ideal than is usually ever obtained in lumped-constant circuits. For example, the quality factor Q of inductive reactances used in lumped-constant circuits is ordinarily less than 100, whereas it is conveniently possible with transmission-line stubs to obtain inductive reactances having Q's of 1000 or more. In addition, if the final tuning of the matching system is made experimentally, then the effects of small losses can be tuned out by slight readjustment of stub lengths from the values theoretically calculated for lossless lines.

PROBLEMS

4–1. Verify Eqs. (a) and (b) in Example 4–3 (see pp. 86–87). Also establish Eq. (c) from the previous two equations.

4–2. Prove the matching conditions of Fig. 4–4. Specialize to the case when Z_2 is purely resistive.

4–3. Explain why matching both at source and load tends to make matching less critical at either.

4–4. Find the two solutions of matching a $100 + j0$-ohm load using 50-ohm line and a single-stub tuner.

4–5. Find the two solutions of matching a $100 + j0$-ohm load using 50-ohm line and a double-stub tuner with $3\lambda/8$ spacing.

4–6. Find the two solutions of matching a $50 + j50$-ohm load using 50-ohm line and a double-stub tuner with $3\lambda/8$ spacing.

4–7. Compute and plot the standing-wave ratio S on line from source versus frequency for (a) a single-stub tuner and (b) a double-stub tuner under the assumption that the load is 0.5 per-unit mho. Plot the frequency scale as a percentage above and below the matched frequency, from $0.60f_0$ below to $1.5f_0$ above. Compare with Fig. 4–20 and discuss. Assume $3\lambda/8$ stub spacing at f_0 for double-stub tuner.

4–8. Design a single-stub matching system to match a $10 + j10$-ohm load to a 50-ohm line. Find the physical lengths of air line required at a frequency of 500 MHz.

4–9. Design a double-stub matching system to match a $10 + j10$-ohm load to a 50-ohm line. Assume all lines are 50 ohms and that the stubs are shorted. You will note that the per-unit load admittance is $2.5 - j2.5$ mhos and cannot be matched with $3\lambda/8$ spacing between stubs. Assume for this problem that the spacing between stubs is $\lambda/16$.

4–10. Ordinarily you will not be able to have exactly $3\lambda/8$ or any other given length of line between the stubs, since the basic advantage of the double-stub tuner is that you do not *need* an adjustable-length line of constant characteristic impedance. In the laboratory you will have sections of line 10 cm, 20 cm, and 30 cm long; tee sections to connect in the stubs are shorted adjustable stubs. Assume that there is a fixed 3-cm length on either side of the tee and that the minimum length of the stubs (but measured to the center of the tee) is 6 cm. With these restrictions, match a 200-ohm load to the system, assuming all lines are 50 ohms. Also assume that the frequency is 500 MHz. *Caution:* Do not forget that there is a fixed 3 cm on both sides of the tee junction and that this means the load will actually have to be transferred to the first stub in order to apply directly the results in the text. Assume an air dielectric.

4–11. Assume that you are facing the same equipment limitations as listed in Problem 4–10. Find the solution to Problem 4–9 under these limitations.

4–12. Assume that you are facing the same equipment limitations as listed in Problem 4–10. Find the solution of Problem 4–9 by picking the distance between stubs to be as near $3\lambda/8$ as possible and by using a fixed length of line between the load and the first stub. There are many possible solutions to this problem. Do one.

4–13. Find the characteristic resistance of a quarter-wavelength transformer to match between 200 ohms and 800 ohms.

4–14. Design a stub and quarter-wavelength transformer to match a $100 + j100$-ohm load to a 50-ohm line. Assume the stub is in parallel with load.

4–15. Design a two-section quarter-wave transformer to match a 200-ohm load to a 50-ohm transmission line with zero reflection at f_0.

4–16. Design a two-section quarter-wave transformer to match a 200-ohm load to a 50-ohm transmission line with a reflection coefficient $|K_{\text{in}}| = 0.1$ at $f = f_0$.

5

Matrix Representation of Transmission-Line Circuits

5–1. INTRODUCTION

One of the most useful means of characterizing an electrical component or device is in the formalism of matrix notation. One reason for this utility lies in the fact that a matrix equation can usually be solved quite rapidly on a digital computer. The matrix representation is also very helpful in analyzing electrical structures. For these reasons we will now develop several matrix representations for transmission-line sections and use these representations in developing an understanding of specific transmission-line circuits.

5–2. IMPEDANCE, ADMITTANCE, AND *ABCD* MATRIX REPRESENTATION OF A SECTION OF TRANSMISSION LINE

Several matrix representations of a given device are usually available to the engineer, with the most useful choice to be made on the basis of the application. For example, consider the simple two-port shown in Fig. 5–1.

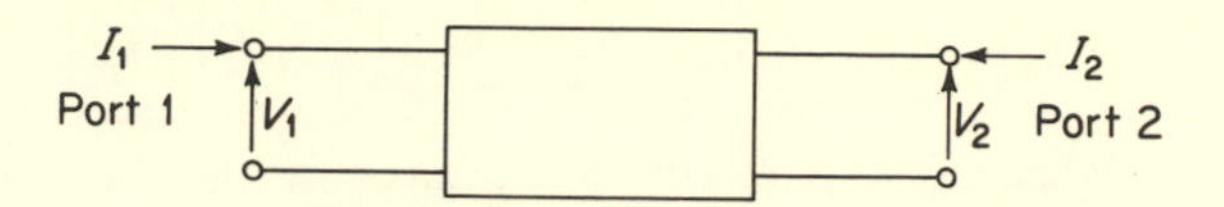

Fig. 5-1. Basic two-port.

Two common matrix descriptions of this circuit are the impedance matrix and the admittance matrix:

$$\begin{matrix} V_1 = Z_{11}I_1 + Z_{12}I_2 \\ V_2 = Z_{21}I_1 + Z_{22}I_2 \end{matrix} \Rightarrow \begin{bmatrix} V_1 \\ V_2 \end{bmatrix} = \begin{bmatrix} Z_{11} & Z_{12} \\ Z_{21} & Z_{22} \end{bmatrix} \begin{bmatrix} I_1 \\ I_2 \end{bmatrix} \Rightarrow [V] = [Z][I] \quad (5\text{-}1)$$

$$\begin{matrix} I_1 = Y_{11}V_1 + Y_{12}V_2 \\ I_2 = Y_{21}V_1 + Y_{22}V_2 \end{matrix} \Rightarrow \begin{bmatrix} I_1 \\ I_2 \end{bmatrix} = \begin{bmatrix} Y_{11} & Y_{12} \\ Y_{21} & Y_{22} \end{bmatrix} \begin{bmatrix} V_1 \\ V_2 \end{bmatrix} \Rightarrow [I] = [Y][V] \quad (5\text{-}2)$$

where the variables V and I are as defined in Fig. 5-1.

The main advantage of either matrix representation lies in the ability to evaluate the elements of the matrix by a series of short circuit–open circuit tests. For a reciprocal structure it can be shown that $Z_{12} = Z_{21}$ and $Y_{12} = Y_{21}$. It is often useful to represent the two-ports by equivalent T or π networks. Two such representations for reciprocal two-ports are shown in Fig. 5-2.

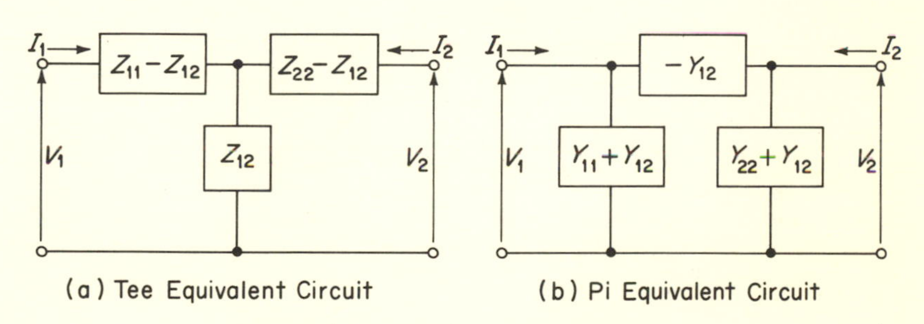

Fig. 5-2. Two-port equivalences.

Another useful representation is the general circuit parameter or *ABCD* matrix. This matrix is defined by Eq. (5-3) with the variables positive as defined in Fig. 5-3.

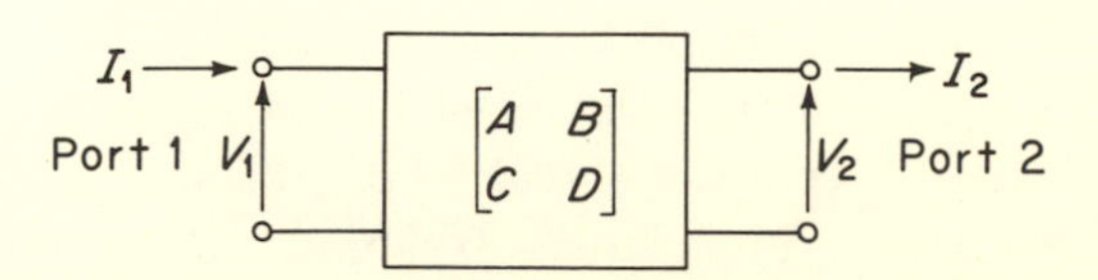

Fig. 5-3. *ABCD* matrix model.

$$V_1 = AV_2 + BI_2 \qquad I_1 = CV_2 + DI_2 \qquad \begin{bmatrix} V_1 \\ I_1 \end{bmatrix} = \begin{bmatrix} A & B \\ C & D \end{bmatrix} \begin{bmatrix} V_2 \\ I_2 \end{bmatrix} \tag{5-3}$$

The utility of the general circuit parameter matrix lies in the ease in which cascaded two-ports can be represented. For example, consider a pair of two-ports connected in series as shown in Fig. 5–4.

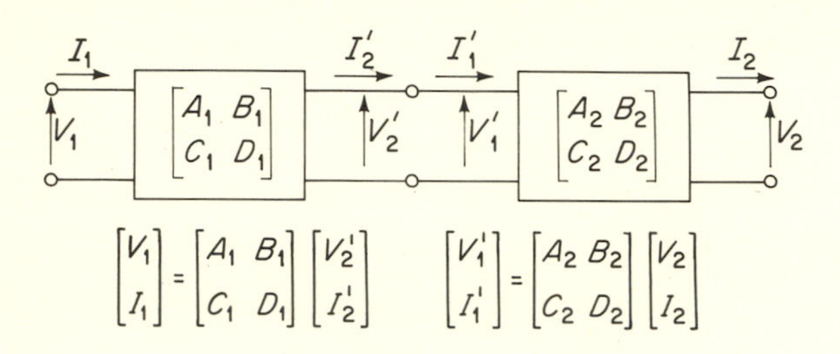

Fig. 5–4. Cascaded two-ports.

It is seen that $V_2' = V_1'$ and $I_2' = I_1'$. Thus,

$$\begin{bmatrix} V_1 \\ I_1 \end{bmatrix} = \begin{bmatrix} A_1 & B_1 \\ C_1 & D_1 \end{bmatrix} \begin{bmatrix} A_2 & B_2 \\ C_2 & D_2 \end{bmatrix} \begin{bmatrix} V_2 \\ I_2 \end{bmatrix} \tag{5-4}$$

The general circuit parameter matrix representation of cascaded two-ports is the matrix product of the general circuit parameter matrices of the individual two-ports.

One can easily derive relationships between the impedance or admittance matrix and the general circuit parameter matrix. For example,

$$\text{(a)} \quad A = \frac{Z_{11}}{Z_{21}} \qquad \text{(b)} \quad B = \frac{Z_{11}Z_{22}}{Z_{21}} - Z_{12}$$
$$\text{(c)} \quad C = \frac{1}{Z_{21}} \qquad \text{(d)} \quad D = \frac{Z_{22}}{Z_{21}} \tag{5-5}$$

Equation 5–5 relates the general circuit parameter matrix elements to the impedance matrix elements. (Note that one must set I_2 of the Z or Y matrix equal to $-I_2$ of the $ABCD$ matrix.)

Other useful lumped-element two-port circuits and their matrix representations are summarized in Figs. 5–5(a) and 5–5(b).

At this point let us consider a section of transmission line of length l having a characteristic impedance Z_0 ohms and a propagation $\gamma = (\alpha + j\beta)$ as shown in Fig. 5–6.

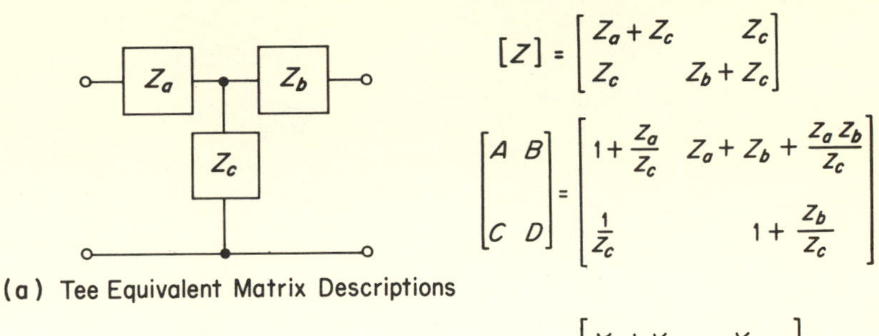

(a) Tee Equivalent Matrix Descriptions

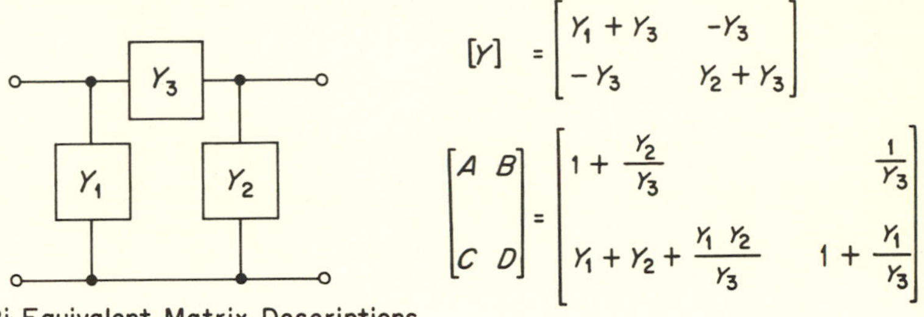

(b) Pi Equivalent Matrix Descriptions

Fig. 5–5. Matrix equivalences for lumped-element two-ports.

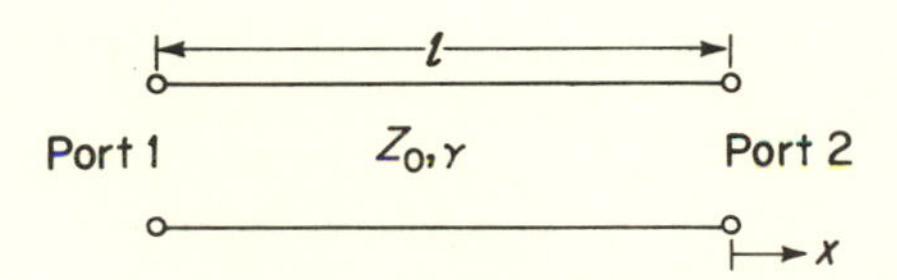

Fig. 5–6. Section of transmission line.

The voltage and current distributions on the line are given by the expressions

$$V(x) = V^+e^{-\gamma x} + V^-e^{+\gamma x}$$

$$I(x) = \frac{V^+}{Z_0}\,e^{-\gamma x} - \frac{V^-}{Z_0}\,e^{+\gamma x}$$

If we want to determine the $ABCD$ matrix representation of this length of transmission line, we can see by comparing Figs. 5–3 and 5–6 that we must force $V(0) = V_2$ and $I(0) = I_2$, where positive current flows out of port 2 as required in the general circuit parameter representation. That is,

$$V^+ + V^- = V_2$$
$$V^+ - V^- = I_2 Z_0$$

Solving for V^+ and V^- yields $V^+ = \frac{1}{2}(V_2 + I_2 Z_0)$ and $V^- = \frac{1}{2}(V_2 - I_2 Z_0)$. Substituting these values into the voltage and current distribution expressions results in

$$\begin{aligned} V(x) &= V_2 \left(\frac{e^{+\gamma x} + e^{-\gamma x}}{2}\right) + I_2 Z_0 \left(\frac{e^{-\gamma x} - e^{+\gamma x}}{2}\right) \\ &= V_2 \cosh \gamma x - I_2 Z_0 \sinh \gamma x \\ I(x) &= \frac{V_2}{Z_0}\left(\frac{e^{-\gamma x} - e^{+\gamma x}}{2}\right) + I_2 \left(\frac{e^{-\gamma x} + e^{+\gamma x}}{2}\right) \\ &= -\frac{V_2}{Z_0} \sinh \gamma x + I_2 \cosh \gamma x \end{aligned}$$

The correspondence between Figs. 5–3 and 5–6 can be completed by evaluating $V(x)$ and $I(x)$ at $x = -l$ since it is apparent that $V(-l) = V_1$ and $I(-l) = I_1$.

$$\begin{aligned} V(-l) &= V_1 = V_2 \cosh \gamma l + I_2 Z_0 \sinh \gamma l \\ I(-l) &= I_1 = \frac{V_2}{Z_0} \sinh \gamma l + I_2 \cosh \gamma l \end{aligned}$$

We can now write the general circuit parameter matrix representation of the transmission-line section as shown in Fig. 5–7.

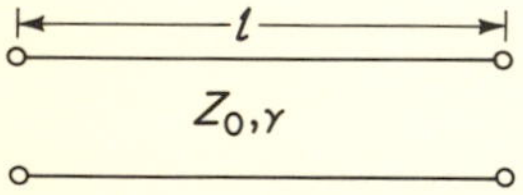

l

Z_0, γ

$$\begin{bmatrix} A & B \\ C & D \end{bmatrix} = \begin{bmatrix} \cosh \gamma l & Z_0 \sinh \gamma l \\ \frac{1}{Z_0} \sinh \gamma l & \cosh \gamma l \end{bmatrix}$$

Fig. 5–7. ***ABCD*** **matrix representation of a transmission-line section.**

The impedance matrix and admittance matrix representations of this section of line are readily determined, with the results given in Eq. (5–6).

$$\text{(a)}\ [Z] = \begin{bmatrix} Z_0 \coth \gamma l & \dfrac{Z_0}{\sinh \gamma l} \\ \dfrac{Z_0}{\sinh \gamma l} & Z_0 \coth \gamma l \end{bmatrix} \qquad \text{(b)}\ [Y] = \begin{bmatrix} Y_0 \coth \gamma l & \dfrac{-Y_0}{\sinh \gamma l} \\ \dfrac{-Y_0}{\sinh \gamma l} & Y_0 \coth \gamma l \end{bmatrix} \tag{5–6}$$

5–3. SCATTERING MATRIX REPRESENTATION OF TWO-PORTS

A matrix representation that is particularly useful in characterizing structures at microwave frequencies is the scattering matrix. For simplicity, let us consider a two-port again (Fig. 5–8).

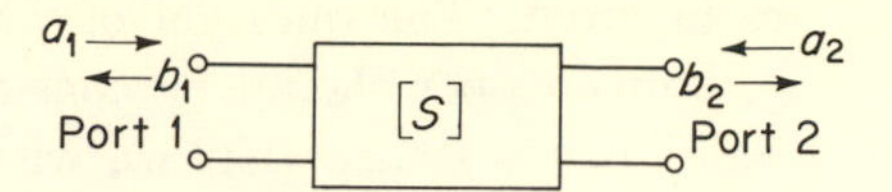

Fig. 5–8. Scattering matrix model.

The variables are designated a_1, a_2, b_1, and b_2. These variables are defined as follows:

$$a_1a_1^* = \text{time-averaged power flowing toward port 1}$$
$$b_1b_1^* = \text{time-averaged power flowing away from port 1}$$
$$a_2a_2^* = \text{time-averaged power flowing toward port 2}$$
$$b_2b_2^* = \text{time-averaged power flowing away from port 2}$$

The variables are related as follows:

$$\begin{aligned} b_1 &= S_{11}a_1 + S_{12}a_2 \\ b_2 &= S_{21}a_1 + S_{22}a_2 \end{aligned} \Rightarrow \begin{bmatrix} b_1 \\ b_2 \end{bmatrix} = \begin{bmatrix} S_{11} & S_{12} \\ S_{21} & S_{22} \end{bmatrix} \begin{bmatrix} a_1 \\ a_2 \end{bmatrix} \Rightarrow [b] = [S][a] \quad (5\text{–}7)$$

where the matrix $[S]$ is called the scattering matrix.

There are several advantages to the scattering-matrix representation of a structure at microwave frequencies. One advantage is the ease of experimentally determining the elements of the scattering matrix. For example, connecting a matched termination at port 2 means that all the energy flowing away from port 2 is dissipated in the termination so no energy is returned to the structure. This means that $a_2 = 0$. When this fact is used to simplify the scattering matrix equations (Eq. 5–7), the results are similar to the simplification realized when performing short-circuit–open-circuit tests with the impedance or admittance matrix representations. An important consequence of using matched or reference terminations in determining the elements of the scattering matrix becomes apparent when one tries to characterize a structure containing an active element such as a transistor. The matched termination condition would most likely be the normal operating state for the circuit and if one were to try to use short-circuit–open-circuit tests the circuit could become unstable or inoperative.

Another advantage lies in the fact that most microwave sources are constant-power sources. This is in contrast to most low-frequency sources, which are either constant-voltage or constant-current sources or approximately so. We know that the voltage and current distributions on a transmission line are anything but constant. Thus a variable that maintains a constant magnitude (such as square root of time-averaged power) is useful in characterizing the two-port.

A few comments are in order. The question of what constitutes a matched termination at a given port should be considered. We are at liberty to choose any value of resistance that we wish to serve as the matched termination. This resistance is often called the *reference resistance* or *reference termination.* The elements of the scattering matrix will be different for different values of reference termination. Thus the reference termination must be specified along with the scattering matrix in order that the matrix representation be useful in designing circuits.

The variables a and b are complex quantities so we must be more specific when referring to power flow into and out of the ports. We measure the elements of the scattering matrix at points along the transmission lines connected to the structure being tested. These points of measurement are called "reference planes." The phase of the variables a and b depends upon the position of the reference planes. Thus, a given scattering matrix characterizes a structure which encompasses everything between the reference planes.

The relationship between the scattering matrix and the impedance matrix or admittance matrix can be developed by considering the two-port shown in Fig. 5–9.

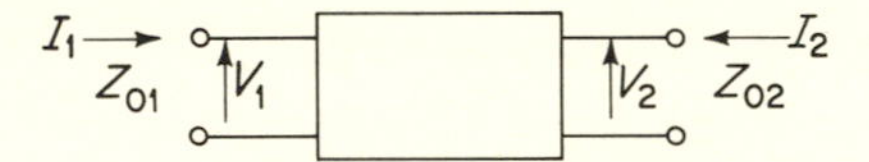

Fig. 5–9. Basic two-port.

Z_{01} and Z_{02} can be considered the characteristic impedances of transmission lines connected to the two-ports. (They could also be considered to be the reference impedances at the reference planes.)

On any given transmission line, we can write

$$V(x) = V^+e^{-\gamma x} + V^-e^{+\gamma x} = Z_0I^+e^{-\gamma x} - Z_0I^-e^{+\gamma x}$$

$$I(x) = I^+e^{-\gamma x} + I^-e^{+\gamma x} = \frac{V^+}{Z_0}e^{-\gamma x} - \frac{V^-}{Z_0}e^{+\gamma x}$$

We can also write expressions for the time-averaged power flow in the line as (assuming Z_0 to be real)

$$P_{+x} = \text{time-averaged power flowing in the } +z\text{-direction}$$

$$P_{+x} = \frac{|V^+|^2}{2Z_0}e^{-2\alpha x} = \frac{|I^+|^2}{2}Z_0e^{-2\alpha x}$$

$$P_{-x} = \text{time-averaged power flowing in the } -x\text{-direction}$$

$$P_{-x} = \frac{|V^-|^2}{2Z_0}e^{+2\alpha x} = \frac{|I^-|^2}{2}Z_0e^{+2\alpha x}$$

It is helpful to define normalized voltages and currents such that

$$P_{+x} = |v^+|^2 e^{-2\alpha x} = |i^+|^2 e^{-2\alpha x}$$

and

$$P_{-x} = |v^-|^2 e^{+2\alpha x} = |i^-|^2 e^{+2\alpha x}$$

Comparing expressions for P_{+x} and P_{-x}, we see that

$$v^+ = \frac{V^+}{\sqrt{2Z_0}} \qquad i^+ = \frac{\sqrt{Z_0}I^+}{\sqrt{2}}$$

$$v^- = \frac{V^-}{\sqrt{2Z_0}} \qquad i^- = \frac{\sqrt{Z_0}I^-}{\sqrt{2}}$$

Normalizing the voltage and current expressions on the transmission line results in the relationships

$$\frac{V(x)}{\sqrt{2Z_0}} = \frac{V^+}{\sqrt{2Z_0}}e^{-\gamma x} + \frac{V^-}{\sqrt{2Z_0}}e^{+\gamma x} = \frac{\sqrt{Z_0}I^+}{\sqrt{2}}e^{-\gamma x} - \frac{\sqrt{Z_0}I^-}{\sqrt{2}}e^{+\gamma x}$$

or

$$v(x) = v^+e^{-\gamma x} + v^-e^{+\gamma x} = i^+e^{-\gamma x} - i^-e^{+\gamma x}$$

and

$$\frac{\sqrt{Z_0}I}{\sqrt{2}}(x) = \frac{\sqrt{Z_0}I^+}{\sqrt{2}}e^{-\gamma x} + \sqrt{Z_0}I^-e^{+\gamma x} = \frac{V^+}{\sqrt{2Z_0}}e^{-\gamma x} - \frac{V^-}{\sqrt{2Z_0}}e^{+\gamma x}$$

or

$$i(x) = i^+e^{-\gamma x} + i^-e^{+\gamma x} = v^+e^{-\gamma x} - v^-e^{+\gamma x}$$

At this point we recall the impedance matrix representation of Eq. 5–1:

$$V_1 = Z_{11}I_1 + Z_{12}I_2$$
$$V_2 = Z_{21}I_1 + Z_{22}I_2$$

These equations can be manipulated as shown with the results given in Eqs. (5–8):

$$\text{(a)} \quad \frac{V_1}{\sqrt{Z_{01}}} = \frac{Z_{11}I_1\sqrt{Z_{01}}}{Z_{01}} + \frac{Z_{12}I_2\sqrt{Z_{02}}}{\sqrt{Z_{01}Z_{02}}} \Rightarrow v_1 = z_{11}i_1 + z_{12}i_2$$

$$\text{(b)} \quad \frac{V_2}{\sqrt{Z_{02}}} = \frac{Z_{21}I_1\sqrt{Z_{01}}}{\sqrt{Z_{01}Z_{02}}} + \frac{Z_{22}I_2\sqrt{Z_{02}}}{Z_{02}} \Rightarrow v_2 = z_{21}i_1 + z_{22}i_2 \qquad (5\text{–}8)$$

Here we have defined

$$z_{ij} = \frac{Z_{ij}}{\sqrt{Z_{0i}Z_{0j}}} \qquad i, j = 1, 2 \qquad (5\text{–}9)$$

We can now define a normalized impedance matrix $[z]$ such that

$$[v] = [z][i] \qquad (5\text{–}10)$$

In a similar manner we can write

$$[i] = [y][v] \tag{5–11}$$

where

$$y_{ij} = \frac{Y_{ij}}{\sqrt{Y_{0i}Y_{0j}}} \qquad i, j = 1, 2 \tag{5–12}$$

If we now define

$$v_i^+ = a_i$$

and

$$v_i^- = b_i$$

it can be seen that (assuming the reference plane to be the plane $x = 0$)

$$v_i = v_i^+ + v_i^- = a_i + b_i$$
$$i_i = v_i^+ - v_i^- = a_i - b_i$$

Solving for a_i and b_i results in

$$a_i = \tfrac{1}{2}(v_i + i_i)$$

and

$$b_i = \tfrac{1}{2}(v_i - i_i)$$

In terms of the column matrices $[a]$, $[b]$, $[v]$, and $[i]$,

$$\begin{aligned} &\text{(a)} \qquad [a] = \tfrac{1}{2}\{[v] + [i]\} \\ &\text{(b)} \qquad [b] = \tfrac{1}{2}\{[v] - [i]\} \end{aligned} \tag{5–13}$$

Equation (5–13) can be manipulated to show that

$$[v] = [z][i] = [a] + [S][a] = [I + S][a]$$
$$[i] = [y][v] = [a] - [S][a] = [I - S][a]$$

Combining, we see that

$$[z][i] = [z][I - S][a] = [I + S][a]$$

or

$$[z] = [I + S][I - S]^{-1} \tag{5–14}$$

where $[I]$ is the identity matrix. In a similar manner, we can develop three additional very useful relationships:

$$[S] = [z - I][z + I]^{-1} \tag{5–15}$$

$$[y] = [I - S][I + S]^{-1} \tag{5–16}$$

and

$$[S] = [I - y][I + y]^{-1} \tag{5–17}$$

There are two interesting properties of the scattering matrix that we will find useful. These are:

1. If the structure is reciprocal, the scattering matrix is symmetrical.
2. If the structure is lossless, the scattering matrix is unitary. That is, $[S^*]_t[S] = [I]$. (The subscript t means "transposed.")

The first statement can be proved by showing that the scattering matrix $[S]$ is equal to its transpose. Starting with $[v] = [z][i]$, where $[v] = [v^+] + [v^-] = [a] + [b]$ and $[i] = [v^+] - [v^-] = [a] - [b]$, we can write

$$[z + I][b] = [z - I][a]$$

or

$$[S] = [z + I]^{-1}[z - I]$$

If we take the transpose of this,

$$[S]_t = [z - I]_t[z + I]_t^{-1} = [z - I][z + I]^{-1}$$

which, by Eq. (5–15), is equal to $[S]$, so the scattering matrix must be symmetrical.

The second statement can be proved by forcing the net power flowing into the structure to be zero, as it must be if there is no loss within the structure. Mathematically stated, we will require that

$$\sum_{i=1}^{N} b_i^* b_i - \sum_{j=1}^{N} a_j^* a_j = 0$$

In matrix notation this can be written as

$$[b^*]_t[b] - [a^*]_t[a] = [0]$$

Substituting Eq. (5–7):

$$([S][a])_t^*[S][a] - [a^*]_t[a] = 0$$

The transpose of the matrix product can be rewritten as

$$[a^*]_t[S^*]_t[S][a] - [a^*]_t[a] = 0$$

This reduces to

$$[a^*]_t([S^*]_t[S] - [I])[a] = 0$$

which will be satisfied if

$$[S^*]_t[S] = [I]$$

Now let us consider a simple two-port characterized by a scattering matrix $[S]$ (Fig. 5–10). We will terminate port 2 with a matched or reference termination and drive port 1 with unit power.

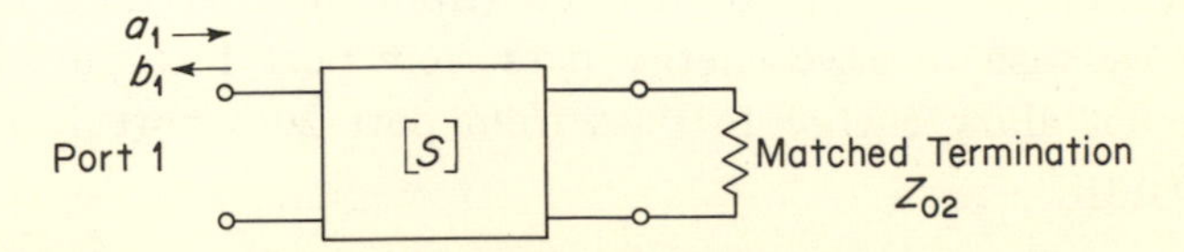

Fig. 5–10. Terminated two-port.

Our statement of the terminal conditions means that $a_1 = 1$, $a_2 = 0$ (there is no energy reflected from a matched termination). Substituting into Eq. (5–7) results in

$$\begin{bmatrix} b_1 \\ b_2 \end{bmatrix} = \begin{bmatrix} S_{11} & S_{12} \\ S_{21} & S_{12} \end{bmatrix} \begin{bmatrix} a_1 \\ 0 \end{bmatrix} \Rightarrow \begin{matrix} b_1 = S_{11}a_1 \\ b_2 = S_{21}a_1 \end{matrix}$$

We recall that $b_1 = v^-$ and $a_1 = v^+$, the normalized reflected and incident voltages on the transmission line attached to port 1. Thus,

$$S_{11} = \frac{v^-}{v^+}$$

which is the voltage reflection coefficient at reference plane 1 (port 1). Letting $a_1 = 1$, we have

$$b_1 = S_{11}$$
$$b_2 = S_{21}$$

From the definition of the scattering matrix variables, we determine the outward flow of power through reference planes 1 and 2:

$$b_1b_1^* = \text{reflected power at port } 1 = S_{11}S_{11}^* = |S_{11}|^2$$
$$b_2b_2^* = \text{power transmitted to load } Z_{02} = S_{21}S_{21}^* = |S_{21}|^2$$

Thus we have the physical interpretation of S_{11} as a voltage reflection coefficient at reference plane 1, and S_{21} is a voltage transmission coefficient at reference plane 2.

Up to this point only two-port structures have been considered. It should be recognized, however, that the foregoing discussion could easily be extended to multiport structures. In the next section examples of three-port and four-port devices will be discussed.

Scattering Matrix Representation of Microwave Devices. Most practical microwave devices are easily characterized by their scattering matrix representation. One of the simplest examples is the isolator (Fig. 5–11). The isolator is a two-port structure which passes energy in only one direction. The device presents a matched termination to a generator having an internal impedance equal to the reference impedance of the driven port when the other port is terminated in its reference impedance. The scattering matrix can be deduced from the preceding statements. Suppose we wish to have energy pass from port 1 to port 2 without loss but not allow energy to pass from port 2 to port 1. Expanding $[b] = [S][a]$,

$$b_1 = S_{11}a_1 + S_{12}a_2$$
$$b_2 = S_{21}a_1 + S_{22}a_2$$

Passing energy from port 1 to port 2 without loss means that $|b_2| = |a_1|$ or $S_{21} = e^{j\varphi}$. If no energy is allowed to pass from port 2 to port 1, $b_1 = 0$ when $a_2 \neq 0$. This means that $S_{12} = 0$. A matched condition at the driven port when the undriven port is match-terminated means that $S_{11} = S_{22} = 0$. The scattering matrix for the isolator is now seen to be of the form shown in Fig. 5–11. This is a valid representation of an isolator and will

$$[S]_{\text{isolator}} = \begin{bmatrix} 0 & 0 \\ e^{j\varphi} & 0 \end{bmatrix}$$

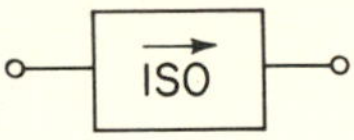

ISO

Fig. 5–11. Scattering matrix and symbol for an isolator.

serve to point out another useful property of the scattering matrix. Let us ask ourselves whether or not we could realize this device as a lossless structure. We can if the scattering matrix is unitary. Is $[S^*]_t[S] = [I]$?

$$\begin{bmatrix} 0 & e^{-j\varphi} \\ 0 & 0 \end{bmatrix} \begin{bmatrix} 0 & 0 \\ e^{j\varphi} & 0 \end{bmatrix} = \begin{bmatrix} 0 & 0 \\ 0 & 0 \end{bmatrix} \neq [I]$$

Thus we see that a lossless structure having the desired scattering matrix cannot be realized. Some provision to dissipate the energy entering port 2 must be made. This, in fact, is done in commercial isolators.

Now let us consider a three-port with the following properties. Each port will present a matched load to a source with internal impedance equal to the reference impedance when driven individually while the undriven ports are terminated in reference impedances. Further, we will require that energy into port 1 pass without attenuation out port 2; energy into port 2 will pass without attenuation out port 3; and energy into port 3 will pass without attenuation out port 1. We can apply the arguments of the preceding example to show that the scattering matrix must be of the form shown in Fig. 5–12.

$$[S] = \begin{bmatrix} 0 & 0 & e^{j\alpha} \\ e^{j\beta} & 0 & 0 \\ 0 & e^{j\delta} & 0 \end{bmatrix}$$

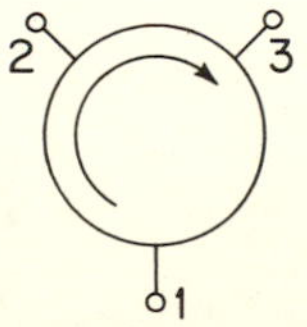

2 3 1

Fig. 5–12. Scattering matrix and symbol for a three-port circulator.

Such a device is known as a *circulator*. Now let us consider the question whether or not this structure can be lossless. Is $[S^*]_t[S] = [I]$?

$$\begin{bmatrix} 0 & e^{-j\beta} & 0 \\ 0 & 0 & e^{-j\delta} \\ e^{-j\alpha} & 0 & 0 \end{bmatrix}\begin{bmatrix} 0 & 0 & e^{j\alpha} \\ e^{j\beta} & 0 & 0 \\ 0 & e^{j\delta} & 0 \end{bmatrix} = \begin{bmatrix} 1 & 0 & 0 \\ 0 & 1 & 0 \\ 0 & 0 & 1 \end{bmatrix}$$

We see that $[S^*]_t[S] = [I]$, so we conclude that a circulator can be realized as a lossless non-reciprocal structure.

As a final example of a scattering matrix let us consider a four-port device known as a *directional coupler*. The directional coupler is symbolically represented as shown in Fig. 5–13.

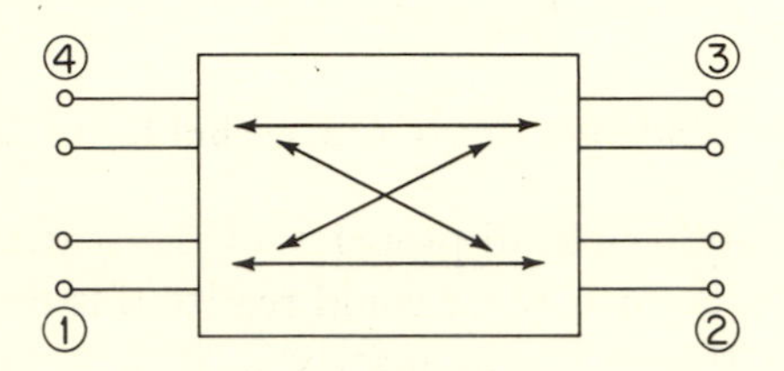

Fig. 5–13. Symbol for a four-port directional coupler.

We will begin by specifying that any port be matched when driven by a generator of internal impedance equal to the reference impedance while all other ports are terminated in reference impedances. This means that $S_{11} = S_{22} = S_{33} = S_{44} = 0$. We will further require that energy does not couple from port 1 to port 4 nor from port 2 to port 3. This means that $S_{14} = S_{41} = S_{23} = S_{32} = 0$. Now the scattering matrix takes the form (assuming a reciprocal structure)

$$[S] = \begin{bmatrix} 0 & S_{12} & S_{13} & 0 \\ S_{12} & 0 & 0 & S_{24} \\ S_{13} & 0 & 0 & S_{34} \\ 0 & S_{24} & S_{34} & 0 \end{bmatrix}$$

Applying the unitary property $[S^*]_t[S] = [I]$ results in six equations that must be satisfied if the structure is to be lossless.

$$\begin{aligned} |S_{12}|^2 + |S_{13}|^2 &= 1 \qquad & S_{12}S_{24}^* + S_{13}S_{34}^* &= 0 \\ |S_{12}|^2 + |S_{24}|^2 &= 1 \qquad & S_{13}^*S_{12} + S_{34}^*S_{24} &= 0 \\ |S_{13}|^2 + |S_{34}|^2 &= 1 \qquad & |S_{24}|^2 + |S_{34}|^2 &= 1 \end{aligned}$$

We see that $|S_{13}| = |S_{24}|$ and $|S_{12}| = |S_{34}|$. If we choose $S_{12} = S_{34} = C_1$ and $S_{13} = S_{24} = jC_2$ by proper choice of reference planes (C_1 and C_2 real), we can satisfy the equations that ensure a lossless structure as long as $C_1^2 + C_2^2 = 1$. The resultant scattering matrix is

$$[S] = \begin{bmatrix} 0 & C_1 & jC_2 & 0 \\ C_1 & 0 & 0 & jC_2 \\ jC_2 & 0 & 0 & C_1 \\ 0 & jC_2 & C_1 & 0 \end{bmatrix} \tag{5-18}$$

To see how the directional coupler works, let us put a source of unit power on port 1 and matched terminations on ports 2, 3, and 4. This means we can write

$$[b] = [S][a] \Rightarrow \begin{bmatrix} b_1 \\ b_2 \\ b_3 \\ b_4 \end{bmatrix} = \begin{bmatrix} 0 & C_1 & jC_2 & 0 \\ C_1 & 0 & 0 & jC_2 \\ jC_2 & 0 & 0 & C_1 \\ 0 & jC_2 & C_1 & 0 \end{bmatrix} \begin{bmatrix} 1 \\ 0 \\ 0 \\ 0 \end{bmatrix}$$

Expansion results in the set of equations

$$\begin{aligned} b_1 &= 0 \\ b_2 &= C_1 \\ b_3 &= jC_2 \\ b_4 &= 0 \end{aligned}$$

The reflected power at port 1 is $b_1b_1^* = 0$; the transmitted power to port 2 is $b_2b_2^* = |C_1|^2$; the coupled power to port 3 is $b_3b_3^* = |C_2|^2$; and the power flow out of port 4 is $b_4b_4^* = 0$. The coupling in decibels is defined as

$$\text{Coupling } (C) = 10 \log_{10} \frac{P_1}{P_3} = 10 \log_{10} \frac{1}{|C_2|^2}$$

We will discuss some specific directional couplers in greater detail later in the chapter.

5-4. SHIFTING REFERENCE PLANES IN THE SCATTERING MATRIX DESCRIPTION

In the last example the comment was made that the elements of the scattering matrix could be selected to be real or imaginary by an appropriate choice of reference planes. This is a very useful concept in the application of the scattering matrix to practical structures so we will now consider the effects of moving the reference planes in greater detail. Consider a two-port with attached transmission lines as shown in Fig. 5-14.

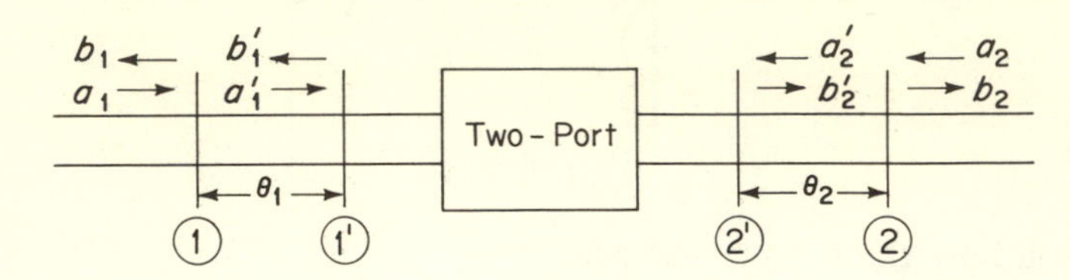

Fig. 5–14. Model for shifting reference planes.

In terms of reference planes at (1) and (2)

$$b_1 = S_{11}a_1 + S_{12}a_2$$
$$b_2 = S_{21}a_1 + S_{22}a_2$$

Now let us shift the reference planes to (1′) and (2′). The scattering matrix representation referred to planes (1′) and (2′) is

$$b_1' = S_{11}'a_1' + S_{12}'a_2'$$
$$b_2' = S_{21}'a_1' + S_{22}'a_2'$$

Our knowledge of the behavior of the waves on the transmission lines allows us to write

$$b_1' = b_1e^{+j\theta_1} \quad a_1' = a_1e^{-j\theta_1} \quad b_2' = b_2e^{+j\theta_2} \quad a_2' = a_2e^{-j\theta_2}$$

Substituting in the equations for the primed reference planes results in

$$b_1e^{j\theta_1} = S_{11}'a_1e^{-j\theta_1} + S_{12}'a_2e^{-j\theta_2}$$
$$b_2e^{j\theta_2} = S_{21}'a_1e^{-j\theta_1} + S_{22}'a_2e^{-j\theta_2}$$

or

$$b_1 = S_{11}'e^{-j2\theta_1}a_1 + S_{12}'e^{-j(\theta_1+\theta_2)}a_2$$
$$b_2 = S_{21}'e^{-j(\theta_1+\theta_2)}a_1 + S_{22}'e^{-j2\theta_2}a_2$$

Comparing coefficients of a_1 and a_2,

$$S_{11}'e^{-j2\theta_1} = S_{11} \qquad S_{12}'e^{-j(\theta_1+\theta_2)} = S_{12}$$
$$S_{21}'e^{-j(\theta_1+\theta_2)} = S_{21} \qquad S_{22}'e^{-j2\theta_2} = S_{22}$$

We want the elements of the scattering matrix at the primed reference planes. That is,

$$S_{11}' = S_{11}e^{+j2\theta_1} \qquad S_{12}' = S_{12}e^{j(\theta_1+\theta_2)}$$
$$S_{21}' = S_{21}e^{j(\theta_2+\theta_1)} \qquad S_{22}' = S_{22}e^{j2\theta_2}$$

In general, it can be shown that

$$S_{ij}' = S_{ij}e^{j(\theta_i+\theta_j)} \tag{5–19}$$

where positive θ represents shifting the reference plane toward the junction.

5–5. THE TRANSMISSION MATRIX REPRESENTATION

The last matrix representation we will consider is the transmission matrix. The transmission matrix is defined as shown in Fig. 5–15.

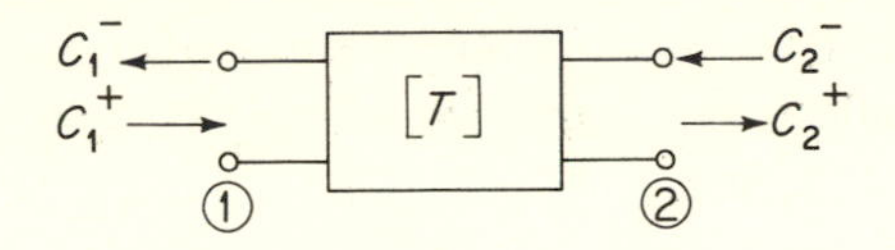

Fig. 5–15. Model for transmission matrix.

Here the variables are C_1^+, C_1^-, C_2^+, C_2^-, which are defined as follows:

$C_1^+C_1^{+*}$ = time-averaged power flow toward port 1
$C_1^-C_1^{-*}$ = time-averaged power flow away from port 1
$C_2^+C_2^{+*}$ = time-averaged power flow away from port 2
$C_2^-C_2^{-*}$ = time-averaged power flow toward port 2

The variables are related by the transmission coefficients:

$$\begin{aligned} C_1^+ &= T_{11}C_2^+ + T_{12}C_2^- \\ C_1^- &= T_{21}C_2^+ + T_{22}C_2^- \end{aligned} \qquad \begin{bmatrix} C_1^+ \\ C_1^- \end{bmatrix} = \begin{bmatrix} T_{11} & T_{12} \\ T_{21} & T_{22} \end{bmatrix} \begin{bmatrix} C_2^+ \\ C_2^- \end{bmatrix} \tag{5–20}$$

It is clear from the definition of the variables that $C_1^+ = a_1$, $C_1^- = b_1$, $C_2^+ = b_2$, and $C_2^- = a_2$. The advantage of the transmission matrix is the ease with which cascaded devices or structures can be characterized. For example, the transmission matrix of a pair of two-ports connected in series (Fig. 5–16) is simply the matrix product of the transmission matrices of the constituent two-ports.

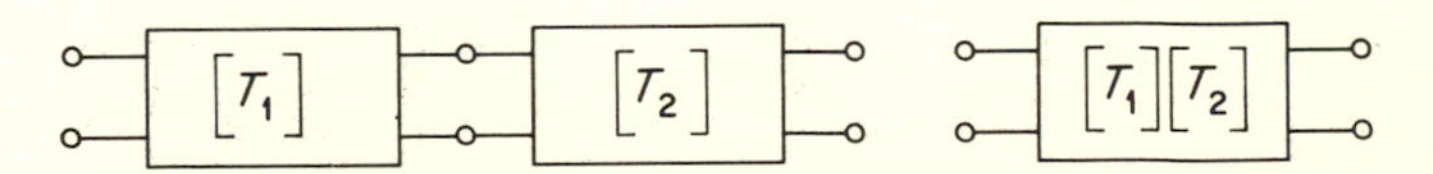

Fig. 5–16. Transmission matrix for cascaded two-ports.

Since the transmission matrix variables are related to the scattering matrix variables, we should expect some relationship between the elements of the scattering matrix representation of a two-port and its transmission matrix representation. This relationship can easily be shown to be

$$\begin{bmatrix} T_{11} & T_{12} \\ T_{21} & T_{22} \end{bmatrix} = \begin{bmatrix} \dfrac{1}{S_{21}} & \dfrac{-S_{22}}{S_{21}} \\ \dfrac{S_{11}}{S_{21}} & S_{12} - \dfrac{S_{11}S_{22}}{S_{21}} \end{bmatrix} \tag{5-21}$$

Remember that the elements of the scattering matrix are relatively easy to measure experimentally or to deduce on the basis of the properties of the scattering matrix. Knowledge of the scattering matrix elements enables us to formulate the transmission matrix, which is useful when connecting devices together.

5–6. MICROWAVE DEVICES

In this article we will consider the application of matrix techniques to the analysis of several structures that are commonly encountered by the microwave engineer.

We will first consider two related 3-dB directional couplers. These are the branch-line hybrid and the ring hybrid or TEM transmission-line magic tee. The structures are shown in Fig. 5–17.

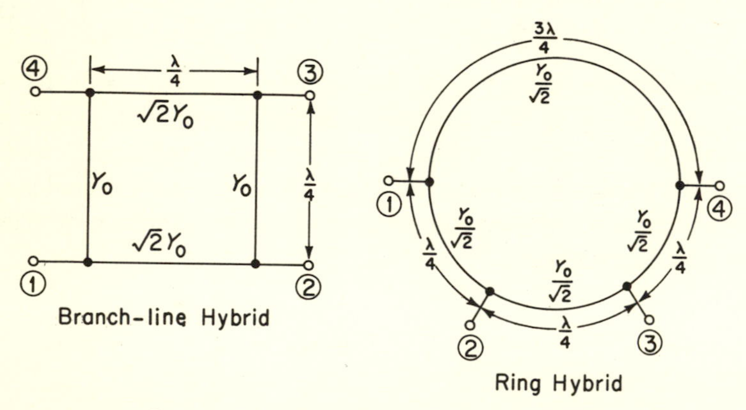

Fig. 5–17. Two 3-dB directional couplers.

These structures will be analyzed by the useful technique of considering models for even and odd excitation independently and then applying the principle of linear superposition to deduce the behavior of the composite structure. The branch-line hybrid will serve as an illustrative example.

Let us begin by applying a source of 2 volts with source impedance Z_0 to port 1. Ports 2, 3, and 4 will be terminated in reference impedances (Z_0). The circuit being analyzed now looks like that shown in Fig. 5–18.

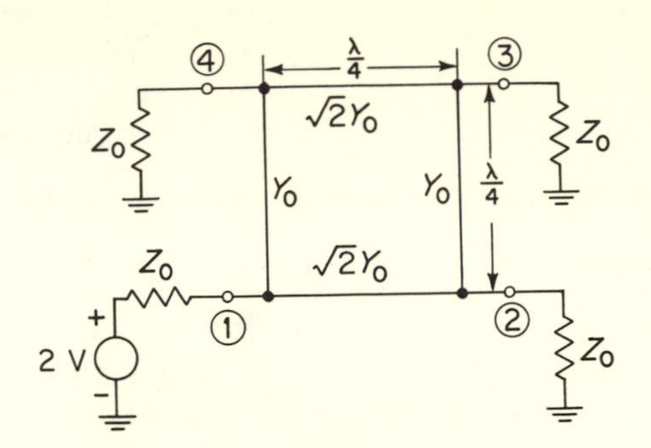

Fig. 5–18. Terminated branch-line hybrid coupler.

This can be represented as the superposition of two excitation situations: an even excitation model, as shown in Fig. 5–19(a), and an odd excitation model, as shown in Fig. 5–19(b).

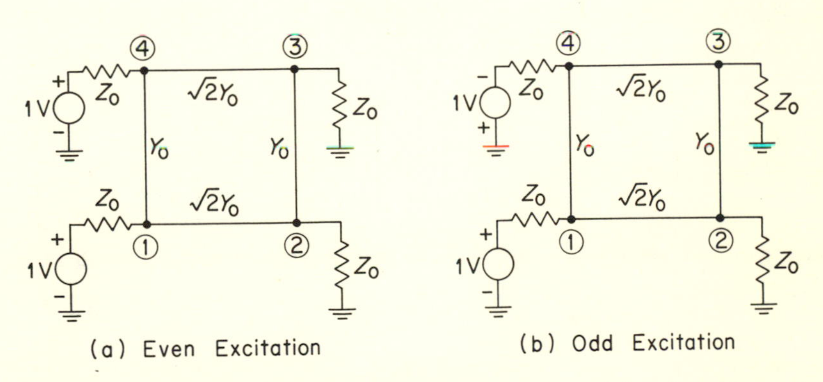

Fig. 5–19. Even-odd excitation models for the branch-line hybrid coupler.

We recognize that planes of electrical symmetry exist in the models of Fig. 5–19(a) and (b) that will allow us to reduce the four-port problem to equivalent two-port problems (Fig. 5–20).

These circuits can be solved for V_{1e}, V_{1o}, I_{1e}, I_{1o}, V_{2e}, V_{2o}, I_{2e}, and I_{2o} by determining the $ABCD$ matrix representation of the two-ports for each excitation. After these voltages and currents have been determined, the

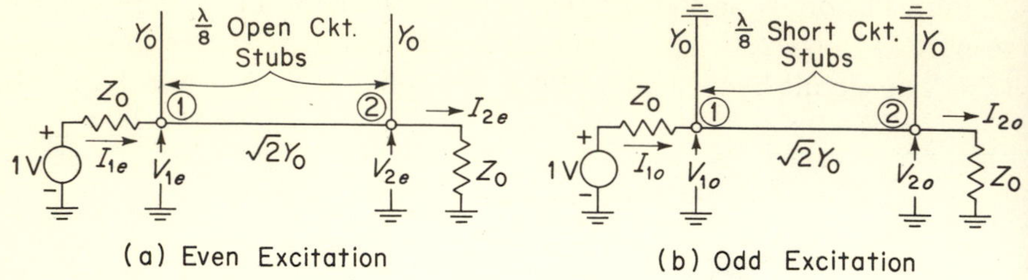

Fig. 5–20. Modified even-odd excitation models.

voltages and currents of the original four-port can be determined from the following relationships:

$$\begin{aligned} V_1 &= V_{1e} + V_{1o} & I_1 &= I_{1e} + I_{1o} \\ V_2 &= V_{2e} + V_{2o} & I_2 &= I_{2e} + I_{2o} \\ V_3 &= V_{2e} - V_{2o} & I_3 &= I_{2e} - I_{2o} \\ V_4 &= V_{1e} - V_{1o} & I_4 &= I_{1e} - I_{1o} \end{aligned}$$

The *ABCD* matrix representations for the even and odd excitation two-ports can be written as follows:

$$\begin{bmatrix} A & B \\ C & D \end{bmatrix}_{\text{even}} = \begin{bmatrix} 1 & 0 \\ jY_0 & 1 \end{bmatrix} \begin{bmatrix} 0 & \frac{jZ_0}{\sqrt{2}} \\ j\sqrt{2}Y_0 & 0 \end{bmatrix} \begin{bmatrix} 1 & 0 \\ jY_0 & 1 \end{bmatrix} = \begin{bmatrix} -\frac{1}{\sqrt{2}} & \frac{jZ_0}{\sqrt{2}} \\ j\frac{Y_0}{\sqrt{2}} & -\frac{1}{\sqrt{2}} \end{bmatrix}$$

$$\begin{bmatrix} A & B \\ C & D \end{bmatrix}_{\text{odd}} = \begin{bmatrix} 1 & 0 \\ -jY_0 & 1 \end{bmatrix} \begin{bmatrix} 0 & j\frac{Z_0}{\sqrt{2}} \\ j\sqrt{2}Y_0 & 0 \end{bmatrix} \begin{bmatrix} 1 & 0 \\ -jY_0 & 1 \end{bmatrix} = \begin{bmatrix} \frac{1}{\sqrt{2}} & j\frac{Z_0}{\sqrt{2}} \\ j\frac{Y_0}{\sqrt{2}} & \frac{1}{\sqrt{2}} \end{bmatrix}$$

where we have recognized the *ABCD* matrix relationships shown in Fig. 5–21.

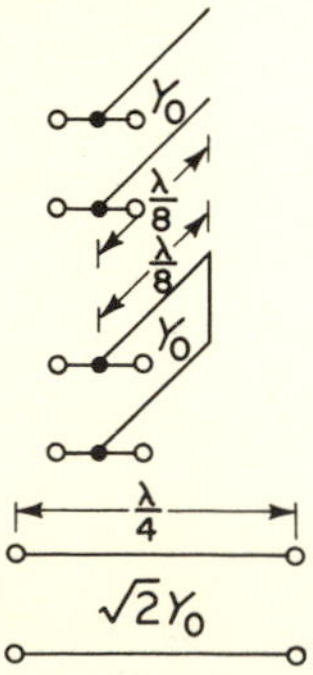

$$\begin{bmatrix} A & B \\ C & D \end{bmatrix} = \begin{bmatrix} 1 & 0 \\ jY_0 \tan\frac{\pi}{4} & 1 \end{bmatrix} = \begin{bmatrix} 1 & 0 \\ jY_0 & 1 \end{bmatrix}$$

$$\begin{bmatrix} A & B \\ C & D \end{bmatrix} = \begin{bmatrix} 1 & 0 \\ -jY_0 \cot\frac{\pi}{4} & 1 \end{bmatrix} = \begin{bmatrix} 1 & 0 \\ -jY_0 & 1 \end{bmatrix}$$

$$\begin{bmatrix} A & B \\ C & D \end{bmatrix} = \begin{bmatrix} \cos\frac{\pi}{2} & j\frac{Z_0}{\sqrt{2}}\sin\frac{\pi}{2} \\ j\sqrt{2}Y_0\sin\frac{\pi}{2} & \cos\frac{\pi}{2} \end{bmatrix} = \begin{bmatrix} 0 & j\frac{Z_0}{\sqrt{2}} \\ j\sqrt{2}Y_0 & 0 \end{bmatrix}$$

Fig. 5–21. Useful ABCD matrix relationships.

Now we can write

$$V_{1e} = -\frac{1}{\sqrt{2}} V_{2e} + j\frac{Z_0}{\sqrt{2}} I_{2e} \qquad V_{1o} = \frac{1}{\sqrt{2}} V_{2o} + j\frac{Z_0}{\sqrt{2}} I_{2o}$$

$$I_{1e} = j\frac{Y_0}{\sqrt{2}} V_{2e} - \frac{1}{\sqrt{2}} I_{2e} \qquad I_{1o} = j\frac{Y_0}{\sqrt{2}} V_{2o} + \frac{1}{\sqrt{2}} I_{2o}$$

Port 2 is terminated in an impedance Z_0 and port 1 is driven by a source of 1 volt with an internal impedance Z_0. Thus, we can apply the terminal conditions: $V_2 = I_2 Z_0$ and $V_1 + I_1 Z_0 = 1$. This results in the voltage and current relationships

$$V_{1e} = \frac{Z_0}{\sqrt{2}}(j - 1)I_{2e} \qquad V_{1o} = \frac{Z_0}{\sqrt{2}}(j + 1)I_{2o}$$

$$I_{1e} = \frac{1}{\sqrt{2}}(j - 1)I_{2e} \qquad I_{1o} = \frac{1}{\sqrt{2}}(j + 1)I_{2o}$$

$$I_{2e} = \frac{1}{\sqrt{2}Z_0(j - 1)} \qquad I_{2o} = \frac{1}{\sqrt{2}Z_0(j + 1)}$$

The voltages and currents for the two-port models can now be written, with the result

$$V_{1e} = \frac{1}{2} \qquad V_{1o} = \frac{1}{2}$$

$$I_{1e} = \frac{1}{2Z_0} \qquad I_{1o} = \frac{1}{2Z_0}$$

$$V_{2e} = \frac{1}{\sqrt{2}(j - 1)} \qquad V_{2o} = \frac{1}{\sqrt{2}(j + 1)}$$

$$I_{2e} = \frac{1}{\sqrt{2}Z_0(j - 1)} \qquad I_{2o} = \frac{1}{\sqrt{2}Z_0(j + 1)}$$

These in turn can be used to determine the voltages and currents at the ports of the original four-port structure, as demonstrated in Eq. 5–21:

(a) $$V_1 = \frac{1}{2} + \frac{1}{2} = 1$$

(b) $$V_2 = \frac{1}{\sqrt{2}(j - 1)} + \frac{1}{\sqrt{2}(j + 1)} = \frac{-j}{\sqrt{2}}$$

(c) $$V_3 = \frac{1}{\sqrt{2}(j - 1)} - \frac{1}{\sqrt{2}(j + 1)} = -\frac{1}{\sqrt{2}}$$

$$\text{(d)} \qquad V_4 = \frac{1}{2} - \frac{1}{2} = 0$$

$$\text{(e)} \qquad I_1 = \frac{1}{2Z_0} + \frac{1}{2Z_0} = \frac{1}{Z_0} \tag{5–21}$$

$$\text{(f)} \qquad I_2 = \frac{1}{\sqrt{2}Z_0(j-1)} + \frac{1}{\sqrt{2}Z_0(j+1)} = \frac{-j}{\sqrt{2}Z_0}$$

$$\text{(g)} \qquad I_3 = \frac{1}{\sqrt{2}Z_0(j-1)} - \frac{1}{\sqrt{2}Z_0(j+1)} = -\frac{1}{\sqrt{2}Z_0}$$

$$\text{(h)} \qquad I_4 = \frac{1}{2Z_0} - \frac{1}{2Z_0} = 0$$

Thus it is seen that port 4 is isolated and energy is coupled in equal measure to ports 2 and 3 with a -90-degree phase shift between port 2 and port 1 and a 180-degree phase shift between ports 3 and 1. We recognize from the physical symmetry of the structure that driving port 2 would couple to ports 1 and 4; driving port 3 would couple to ports 4 and 1; and driving port 4 would couple to ports 3 and 2. It should be recognized that the input impedance at port 1 is equal to Z_0. We conclude that the structure is matched at the driven port when all other ports are match-terminated.

We can deduce scattering matrix representation for the branch-line hybrid on the basis of the foregoing statements. Clearly it is a 4×4 matrix with $S_{11} = S_{22} = S_{33} = S_{44} = 0$. In addition, $S_{21} = -j/\sqrt{2}$, $S_{31} = -1/\sqrt{2}$, $S_{41} = 0$, $S_{32} = 0$, $S_{42} = -1/\sqrt{2}$, and $S_{43} = -j/\sqrt{2}$. These values plus the fact that the structure is reciprocal allow us to complete the scattering matrix representation of the 90-degree hybrid:

$$[S]_{\text{branch-line hybrid}} = \begin{bmatrix} 0 & -j/\sqrt{2} & -1/\sqrt{2} & 0 \\ -j/\sqrt{2} & 0 & 0 & -1/\sqrt{2} \\ -1/\sqrt{2} & 0 & 0 & -j/\sqrt{2} \\ 0 & -1/\sqrt{2} & -j/\sqrt{2} & 0 \end{bmatrix}$$

Now let us turn our attention to the ring hybrid. A little reflection shows us that the even and odd excitation two-port models for the case of the source applied to port 1 are as shown in Fig. 5–22.

The *ABCD* matrix representations for even and odd excitation are

$$\begin{bmatrix} A & B \\ C & D \end{bmatrix}_{\text{even}} = \begin{bmatrix} 1 & 0 \\ -j\dfrac{Y_0}{\sqrt{2}} & 1 \end{bmatrix} \begin{bmatrix} 0 & jZ_0\sqrt{2} \\ j\dfrac{Y_0}{\sqrt{2}} & 0 \end{bmatrix} \begin{bmatrix} 1 & 0 \\ j\dfrac{Y_0}{\sqrt{2}} & 1 \end{bmatrix} = \begin{bmatrix} -1 & j\sqrt{2}Z_0 \\ j\sqrt{2}Y_0 & 1 \end{bmatrix}$$

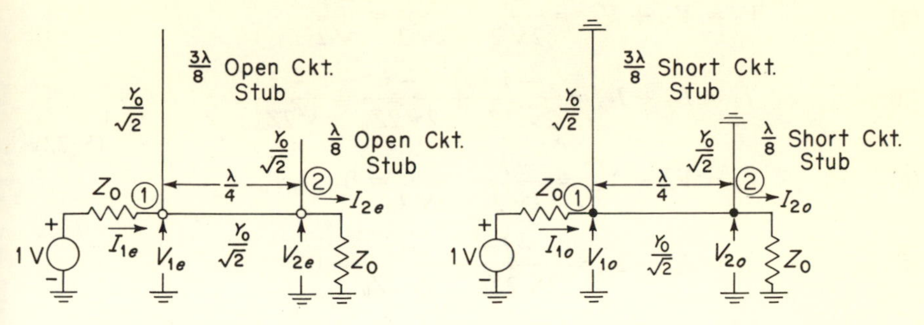

Fig. 5–22. Modified even-odd excitation models for the ring hybrid.

and

$$\begin{bmatrix} A & B \\ C & D \end{bmatrix}_{\text{odd}} = \begin{bmatrix} 1 & 0 \\ j\dfrac{Y_0}{\sqrt{2}} & 1 \end{bmatrix} \begin{bmatrix} 0 & jZ_0\sqrt{2} \\ j\dfrac{Y_0}{\sqrt{2}} & 0 \end{bmatrix} \begin{bmatrix} 1 & 0 \\ -j\dfrac{Y_0}{\sqrt{2}} & 1 \end{bmatrix} = \begin{bmatrix} 1 & j\sqrt{2}Z_0 \\ j\sqrt{2}Y_0 & -1 \end{bmatrix}$$

The voltages and currents on the models must satisfy the relationships

$$V_{1e} = -V_{2e} + j\sqrt{2}Z_0I_{2e} \qquad V_{1o} = V_{2o} + j\sqrt{2}Z_0I_{2o}$$
$$I_{1e} = j\sqrt{2}Y_0V_{2e} + I_{2e} \qquad I_{1o} = j\sqrt{2}Y_0V_{2o} - I_{2o}$$

Applying the terminal-condition equations $V_2 = I_2Z_0$ and $V_1 + I_1Z_0 = 1$ results in the terminal voltage and current expressions

$$V_{1e} = \frac{\sqrt{2}+j}{2\sqrt{2}} \qquad V_{1o} = \frac{\sqrt{2}-j}{2\sqrt{2}}$$

$$I_{1e} = \frac{\sqrt{2}-j}{2\sqrt{2}Z_0} \qquad I_{1o} = \frac{\sqrt{2}+j}{2\sqrt{2}Z_0}$$

$$V_{2e} = \frac{-j}{2\sqrt{2}} \qquad V_{2o} = \frac{-j}{2\sqrt{2}}$$

$$I_{2e} = \frac{-j}{2\sqrt{2}Z_0} \qquad I_{2o} = \frac{-j}{2\sqrt{2}Z_0}$$

Now the voltages and currents for the four-port when driven at port 1 can be formulated with the result tabulated in Eq. (5–22):

(a) $$V_1 = V_{1e} + V_{1o} = \frac{\sqrt{2}+j}{2\sqrt{2}} + \frac{\sqrt{2}-j}{2\sqrt{2}} = 1$$

(b) $$I_1 = I_{1e} + I_{1o} = \frac{\sqrt{2}-j}{2\sqrt{2}Z_0} + \frac{\sqrt{2}+j}{2\sqrt{2}Z_0} = \frac{1}{Z_0}$$

$$\text{(c)} \quad V_2 = V_{2e} + V_{2o} = \frac{-j}{2\sqrt{2}} + \frac{-j}{2\sqrt{2}} = \frac{-j}{\sqrt{2}}$$

$$\text{(d)} \quad I_2 = I_{2e} + I_{2o} = \frac{-j}{2\sqrt{2}Z_0} + \frac{-j}{2\sqrt{2}Z_0} = \frac{-j}{\sqrt{2}Z_0}$$

$$\text{(e)} \quad V_3 = V_{2e} - V_{2o} = \frac{-j}{2\sqrt{2}} - \frac{-j}{2\sqrt{2}} = 0 \qquad (5\text{–}22)$$

$$\text{(f)} \quad I_3 = I_{2e} - I_{2o} = \frac{-j}{2\sqrt{2}Z_0} - \frac{-j}{2\sqrt{2}Z_0} = 0$$

$$\text{(g)} \quad V_4 = V_{1e} - V_{1o} = \frac{\sqrt{2}+j}{2\sqrt{2}} - \frac{\sqrt{2}-j}{2\sqrt{2}} = \frac{j}{\sqrt{2}}$$

$$\text{(h)} \quad I_4 = I_{1e} - I_{1o} = \frac{\sqrt{2}+j}{2\sqrt{2}Z_0} - \frac{\sqrt{2}-j}{2\sqrt{2}Z_0} = \frac{j}{\sqrt{2}Z_0}$$

In this case port 3 is isolated and energy splits equally between ports 2 and 4. Note that the energies out of ports 2 and 4 are 180 degrees out of phase. One (port 4) is leading the input by 90 degrees and the other (port 2) is lagging the input by 90 degrees. Due to symmetry considerations, we would expect similar operation if we were to drive port 4; that is, energy would divide between ports 3 and 1 with port 2 isolated.

To determine the output conditions when ports 2 or 3 are driven we must go back to our even and odd excitation models and connect the source to port 2 and the load to port 1. If we think about this a little bit we recognize that a $\lambda/8$ open stub looks like a $3\lambda/8$ shorted stub and a $\lambda/8$ shorted stub looks like a $3\lambda/8$ open stub. Thus odd excitation at the port where the $\lambda/8$ stub is connected produces the same voltages and currents as even excitation at the port where the $3\lambda/8$ stub is connected, and vice versa. All we have to do then is to interchange V_{1o} and I_{1o} with V_{2e} and I_{2e}, and V_{2o} and I_{2o} with V_{1e} and I_{1e}. The resultant equations applicable to driving port 2 are

$$V_{1e} = \frac{-j}{2\sqrt{2}} \qquad V_{1o} = \frac{-j}{2\sqrt{2}}$$

$$I_{1e} = \frac{-j}{2\sqrt{2}Z_0} \qquad I_{1o} = \frac{-j}{2\sqrt{2}Z_0}$$

$$V_{2e} = \frac{\sqrt{2}-j}{2\sqrt{2}} \qquad V_{2o} = \frac{\sqrt{2}+j}{2\sqrt{2}}$$

$$I_{2e} = \frac{\sqrt{2}+j}{2\sqrt{2}Z_0} \qquad I_{2o} = \frac{\sqrt{2}-j}{2\sqrt{2}Z_0}$$

The voltages and currents for the four-port when driven at port 2 can now be determined in a straightforward manner with the results tabulated in Eq. 5-23:

$$
\begin{aligned}
&\text{(a)} \quad V_1 = V_{1e} + V_{1o} = \frac{-j}{2\sqrt{2}} + \frac{-j}{2\sqrt{2}} = \frac{-j}{\sqrt{2}} \\
&\text{(b)} \quad I_1 = I_{1e} + I_{1o} = \frac{-j}{2\sqrt{2}Z_0} + \frac{-j}{2\sqrt{2}Z_0} = \frac{-j}{\sqrt{2}Z_0} \\
&\text{(c)} \quad V_2 = V_{2e} + V_{2o} = \frac{\sqrt{2}-j}{2\sqrt{2}} + \frac{\sqrt{2}+j}{2\sqrt{2}} = 1 \\
&\text{(d)} \quad I_2 = I_{2e} + I_{2o} = \frac{\sqrt{2}-j}{2\sqrt{2}Z_0} + \frac{\sqrt{2}+j}{2\sqrt{2}Z_0} = \frac{1}{Z_0} \\
&\text{(e)} \quad V_3 = V_{2e} - V_{2o} = \frac{\sqrt{2}-j}{2\sqrt{2}} - \frac{\sqrt{2}+j}{2\sqrt{2}} = \frac{-j}{\sqrt{2}} \\
&\text{(f)} \quad I_3 = I_{2e} - I_{2o} = \frac{\sqrt{2}-j}{2\sqrt{2}Z_0} - \frac{\sqrt{2}+j}{2\sqrt{2}Z_0} = \frac{-j}{\sqrt{2}Z_0} \\
&\text{(g)} \quad V_4 = V_{1e} - V_{1o} = \frac{-j}{2\sqrt{2}} - \frac{-j}{2\sqrt{2}} = 0 \\
&\text{(h)} \quad I_4 = I_{1e} - I_{1o} = \frac{-j}{2\sqrt{2}Z_0} - \frac{-j}{2\sqrt{2}Z_0} = 0
\end{aligned}
\tag{5-23}
$$

Now we see the interesting result that driving port 2 results in isolation at port 4 and the input energy splits equally between ports 1 and 3. The outputs at ports 1 and 3 are in phase and lag the input by 90 degrees.

Again, it will be instructive to deduce the scattering matrix representation of the ring hybrid from the terminal voltage and current relationships of Eqs. (5-22) and (5-23). We can begin by recognizing that it will be a 4×4 matrix with $S_{11} = S_{22} = S_{33} = S_{44} = 0$. From Eq. (5-22) we can write $S_{21} = -j/\sqrt{2}$, $S_{31} = 0$, and $S_{41} = j/2$. From Eq. (5-23) we can write $S_{32} = -j/\sqrt{2}$ and $S_{42} = 0$. By symmetry, $S_{43} = S_{12} = -j/\sqrt{2}$. These statements plus the reciprocal nature of the structure allow us to write

$$
[S]_{\text{ring hybrid}} = \begin{bmatrix} 0 & -j/\sqrt{2} & 0 & j/\sqrt{2} \\ -j/\sqrt{2} & 0 & -j/\sqrt{2} & 0 \\ 0 & -j/\sqrt{2} & 0 & -j/\sqrt{2} \\ j/\sqrt{2} & 0 & -j/\sqrt{2} & 0 \end{bmatrix}
$$

Some of these statements about the behavior of the voltages and currents at the ports make good physical sense and can be deduced with a little reasoning. For example, looking at the branch-line hybrid driven at port 1, we see that energy can get to port 3 via two paths, each $\lambda/2$ long. Energy from port 1 to port 4 or port 2 goes via two paths, one of which is $\lambda/2$ longer than the other. Trying to explain how these two paths produce additive signals at port 2 and a canceling signal at port 4 breaks down our intuition. This behavior is due to the nature of the reflections at the transmission-line junctions. When considering the ring hybrid driven at port 1, it is seen that signals passing to port 4 traverse equal-length paths ($3\lambda/4$ in length), signals to port 3 traverse paths that differ in length by $\lambda/2$, and signals to port 2 travel paths that differ in length by one wavelength. Thus port 3 is isolated and the net path length difference for signals at ports 2 and 4 is $\lambda/2$. If the source is switched to port 2, we see that the two paths via which energy can pass to port 4 differ in length by $\lambda/2$, thus isolating the port. The two paths to port 1 and the two paths to port 3 differ by one wavelength, so energy should add at these ports and the two ports should be in phase. It is difficult to know what line impedances to use to produce matched conditions and proper terminal conditions without a detailed analysis, however.

An interesting application of the ring-type hybrid junction is in the realization of a balanced microwave mixer. This circuit is diagramed in Fig. 5–23. The signal input at port 4 splits equally between ports 1 and 3

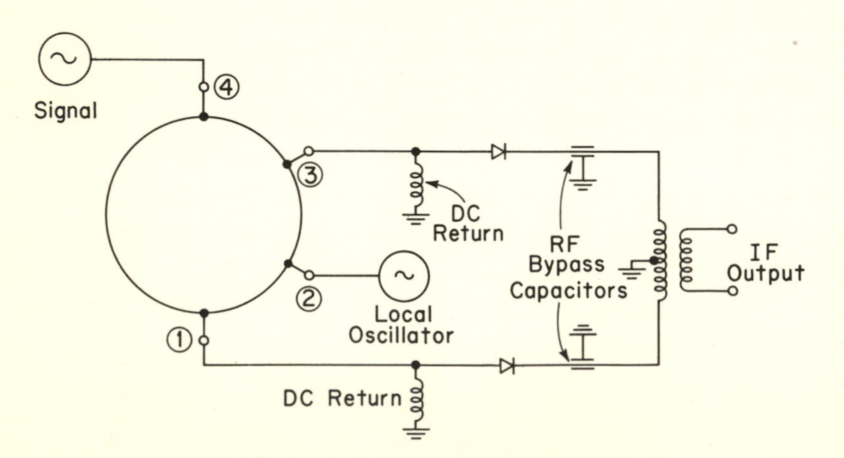

Fig. 5–23. The balanced mixer using a ring hybrid.

with a 180-degree phase difference between the signals emerging from ports 1 and 3. The local oscillator signal is injected at port 2. This energy also splits between ports 1 and 3 but the local oscillator signals emerge from ports 1 and 3 in phase. These signals are rectified by the diodes but only the 180-degree out-of-phase component (that due to the signal) passes through the transformer to the IF amplifier. Oscillator noise appears in phase across the transformer and will not appear in the output. Another advantage of this circuit is the isolation between the signal line and the local oscillator line. If the signal is coming from a receiving antenna, this isolation prevents energy from the local oscillator getting to the antenna and possibly radiating into space.

Another interesting device is the binary power divider. This is a three-port structure that presents a matched termination at the driven port when the other ports are match-terminated. When driven at one port (called the sum port) the input power splits equally between the remaining two ports. When driven at one of the other ports, with the sum port and the remaining port match terminated, power is delivered (with some loss) to the sum port but the other port is isolated. This structure is shown in Fig. 5–24.

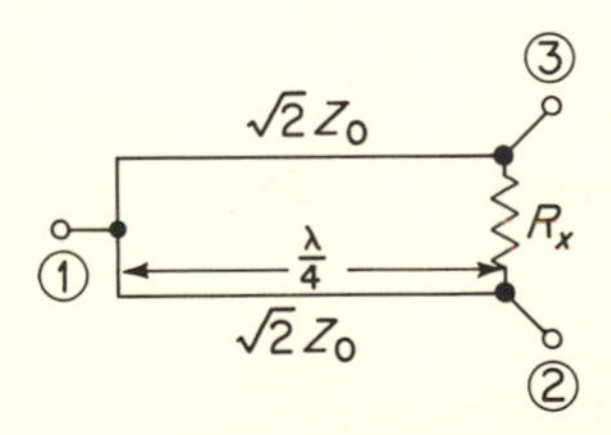

Fig. 5–24. The binary power divider.

It is easy to see that if Z_0 terminations are placed at ports 2 and 3, input energy at port 1 splits equally and no power is dissipated in R_x, called the difference resistor, since each end is at the same potential. We also can see that the input impedance at port 1 is Z_0. Now suppose that the source is placed on port 2 with matched loads (Z_0) on ports 1 and 3. We can apply even and odd excitation analysis to this problem, as shown in Fig. 5–25.

The even excitation–odd excitation models of Fig. 5–25 reduce to the two-port equivalent models shown in Fig. 5–26.

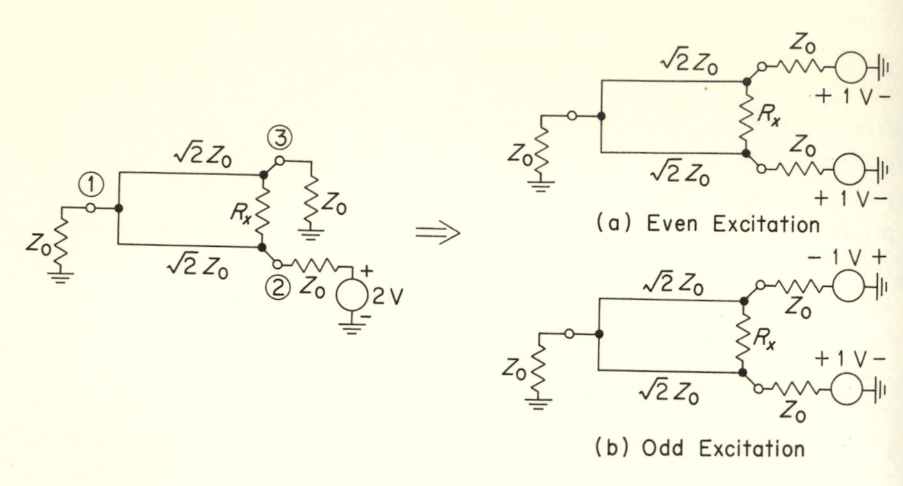

Fig. 5–25. Even-odd excitation models for the binary power divider.

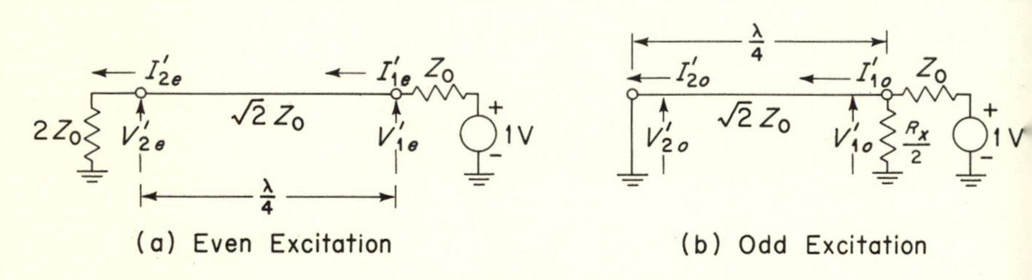

Fig. 5–26. Modified even-odd excitation models for the binary two-port.

We can write directly

$$\begin{bmatrix} A & B \\ C & D \end{bmatrix}_{\text{even}} = \begin{bmatrix} 0 & j\sqrt{2}Z_0 \\ \dfrac{j}{\sqrt{2}Z_0} & 0 \end{bmatrix}$$

$$\begin{bmatrix} A & B \\ C & D \end{bmatrix}_{\text{odd}} = \begin{bmatrix} 1 & 0 \\ \dfrac{2}{R_x} & 1 \end{bmatrix} \begin{bmatrix} 0 & j\sqrt{2}Z_0 \\ \dfrac{j}{\sqrt{2}Z_0} & 0 \end{bmatrix} = \begin{bmatrix} 0 & j\sqrt{2}Z_0 \\ \dfrac{j}{\sqrt{2}Z_0} & \dfrac{j2\sqrt{2}Z_0}{R_x} \end{bmatrix}$$

and

$$V'_{1e} = j\sqrt{2}Z_0 I'_{2e} \qquad V'_{1o} = j\sqrt{2}Z_0 I'_{2o}$$

$$I'_{1e} = \frac{j}{\sqrt{2}Z_0} V'_{2e} \qquad I'_{1o} = \frac{j}{\sqrt{2}Z_0} V'_{2o} + \frac{j2\sqrt{2}Z_0}{R_x} I'_{2o}$$

Applying the terminal conditions $V'_{2e} = 2Z_0 I'_{2e}$, $V'_{1e} + Z_0 I'_{1e} = 1$, $V'_{2o} = 0$, and $V'_{1o} + Z_0 I'_{1o} = 1$ results in the following equations:

$$V'_{1e} = \frac{1}{2} \qquad V'_{1o} = \frac{1}{\left(1 + \frac{2Z_0}{R_x}\right)}$$

$$I'_{1e} = \frac{1}{2Z_0} \qquad I'_{1o} = \frac{2}{R_x\left(1 + \frac{2Z_0}{R_x}\right)}$$

$$V'_{2e} = \frac{-j}{\sqrt{2}} \qquad V'_{2o} = 0$$

$$I'_{2e} = \frac{-j}{2\sqrt{2}Z_0} \qquad I'_{2o} = \frac{-j}{\sqrt{2}Z_0\left(1 + \frac{2Z_0}{R_x}\right)}$$

When port 2 is driven and ports 1 and 3 terminated in Z_0,

$$V_1 = V'_{2e} + V'_{2o} = \frac{-j}{\sqrt{2}}$$

$$V_2 = V'_{1e} + V'_{1o} = \frac{1}{2} + \frac{1}{\left(1 + \frac{2Z_0}{R_x}\right)}$$

$$V_3 = V'_{1e} - V'_{1o} = \frac{1}{2} - \frac{1}{\left(1 + \frac{2Z_0}{R_x}\right)}$$

If we wish to isolate port 3, it is merely necessary to force $V_3 = 0$, which means that $1 + \frac{2Z_0}{R_x} = 2$ or $R_x = 2Z_0$. In this case, the voltages and currents at the terminals become

$$\text{(a)} \quad V_1 = \frac{-j}{\sqrt{2}} \qquad \text{(d)} \quad I_1 = \frac{-j}{\sqrt{2}Z_0}$$

$$\text{(b)} \quad V_2 = 1 \qquad \text{(e)} \quad I_2 = \frac{1}{Z_0} \qquad (5\text{–}24)$$

$$\text{(c)} \quad V_3 = 0 \qquad \text{(f)} \quad I_3 = 0$$

Note that one-half of the input power is delivered to the load (port 1); the rest is dissipated in the difference resistor. This is the price you pay for isolating port 3. If signals are applied to ports 2 and 3, $1/\sqrt{2}$ times the input signal at each port will appear at port 1. That is why port 1 is known as the "sum port."

5–7. COUPLED TRANSMISSION LINES

One of the most useful structures in the realization of microwave directional couplers and filters is the four-port formed by physically locating two transmission lines close enough to each other so that they are coupled via the electric and magnetic fields associated with the voltages and currents on the lines. The type of structure that we are considering is shown diagrammatically in Fig. 5–27. The coupled pair of

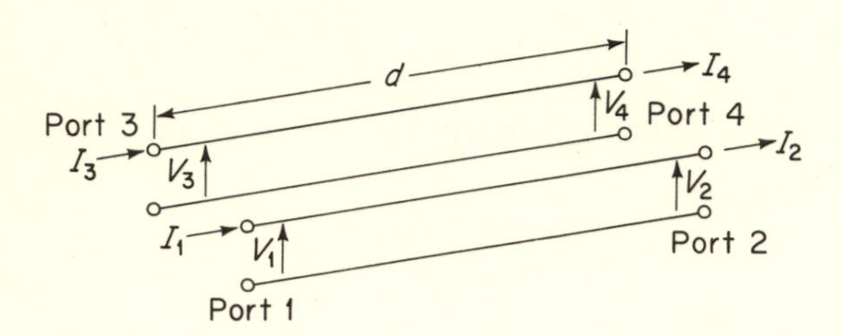

Fig. 5–27. Coupled parallel transmission lines.

transmission lines is easily analyzed by connecting reference impedances ($Z_L = Z_0$) at ports 2, 3, and 4 and driving the structure with a 2-volt source having a source impedance Z_0 (Fig. 5–28).

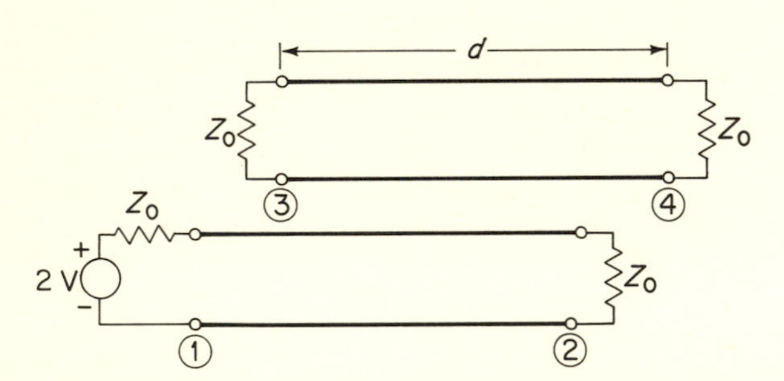

Fig. 5–28. Terminated coupled transmission lines.

We will analyze this circuit by using even excitation–odd excitation analysis, and begin by considering the two excitation conditions shown in Fig. 5–29. Here Z_{0e} is the characteristic impedance of the coupled lines for even excitation, Z_{0o} is the characteristic impedance of the coupled lines for odd excitation, and γ is the propagation constant. It should be noted here that Z_{0e} is not equal to Z_{0o} because the partial capacitances between the lines are different for even and odd excitation. If losses in the

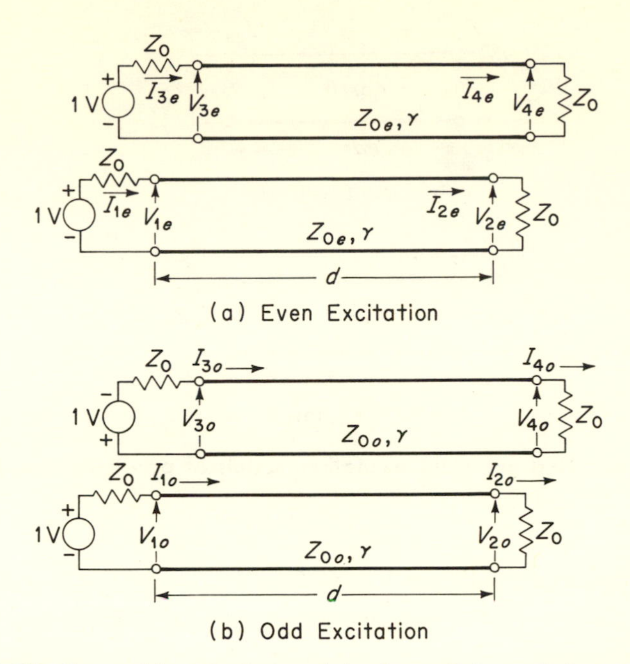

Fig. 5–29. Even-odd excitation models of parallel-coupled transmission lines.

lines are ignored, the propagation constant $\gamma = j\beta$. For TEM-mode lines, we know from field analysis that the velocity of propagation depends on the medium in which the lines are imbedded. This means that the propagation constant is independent of the excitation.

The problem has now been reduced to solving the two transmission-line problems shown in Fig. 5–30. The voltages and currents on the original structure can then be written as a superposition of the even- and odd-mode solutions as follows:

$$\begin{aligned} V_1 &= V_{1e} + V_{1o} & I_1 &= I_{1e} + I_{1o} \\ V_2 &= V_{2e} + V_{2o} & I_2 &= I_{2e} + I_{2o} \\ V_3 &= V_{1e} - V_{1o} & I_3 &= I_{1e} - I_{1o} \\ V_4 &= V_{2e} - V_{2o} & I_4 &= I_{2e} - I_{2o} \end{aligned}$$

We can use the *ABCD* matrix representation of the transmission-line section to write the following relationships between the voltages and currents on the models of Fig. 5–30:

$$\begin{bmatrix} V_{1e} \\ I_{1e} \end{bmatrix} = \begin{bmatrix} \cos\theta & jZ_{0e}\sin\theta \\ jY_{0e}\sin\theta & \cos\theta \end{bmatrix} \begin{bmatrix} V_{2e} \\ I_{2e} \end{bmatrix}$$

$$\begin{bmatrix} V_{1o} \\ I_{1o} \end{bmatrix} = \begin{bmatrix} \cos\theta & jZ_{0o}\sin\theta \\ jY_{0o}\sin\theta & \cos\theta \end{bmatrix} \begin{bmatrix} V_{2o} \\ I_{2o} \end{bmatrix}$$

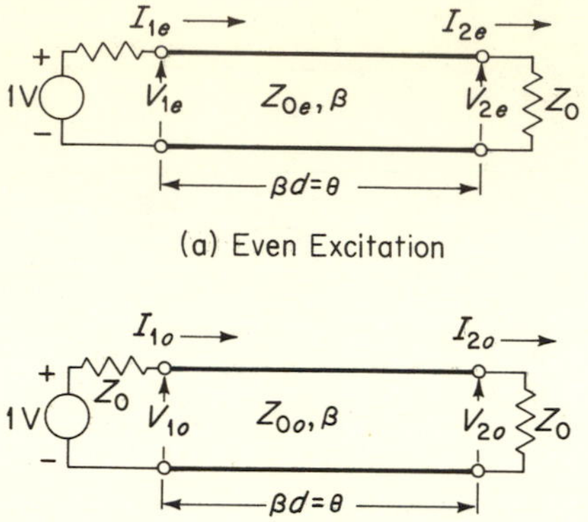

I_{1e} → I_{2e} →
+ 1V − V_{1e} Z_{0e}, β V_{2e} Z_0
$\beta d = \theta$

(a) Even Excitation

I_{1o} → I_{2o} →
+ 1V − Z_0 V_{1o} Z_{0o}, β V_{2o} Z_0
$\beta d = \theta$

(b) Odd Excitation

Fig. 5–30. Modified even-odd excitation models of parallel-coupled transmission lines.

Applying the terminal conditions $V_{2e} = Z_0 I_{2e}$, $V_{2o} = Z_0 I_{2o}$, $V_{1e} + I_{1e}Z_o = 1$, and $V_{1o} + I_{1o}Z_0 = 1$ results in the voltage and current expressions

$$V_{1e} = \frac{Z_{0e}Z_0 \cos\theta + jZ_{0e}^2 \sin\theta}{2Z_{0e}Z_0 \cos\theta + j(Z_{0e}^2 + Z_0^2)\sin\theta}$$

$$I_{1e} = \frac{Z_{0e}\cos\theta + jZ_0 \sin\theta}{2Z_{0e}Z_0 \cos\theta + j(Z_{0e}^2 + Z_0^2)\sin\theta}$$

$$V_{2e} = \frac{Z_0 Z_{0e}}{2Z_{0e}Z_0 \cos\theta + j(Z_{0e}^2 + Z_0^2)\sin\theta}$$

$$I_{2e} = \frac{Z_{0e}}{2Z_{0e}Z_0 \cos\theta + j(Z_{0e}^2 + Z_0^2)\sin\theta}$$

$$V_{1o} = \frac{Z_{0o}Z_0 \cos\theta + jZ_{0o}^2 \sin\theta}{2Z_{0o}Z_0 \cos\theta + j(Z_{0o}^2 + Z_0^2)\sin\theta}$$

$$I_{1o} = \frac{Z_{0o}\cos\theta + jZ_0 \sin\theta}{2Z_{0o}Z_0 \cos\theta + j(Z_{0o}^2 + Z_0^2)\sin\theta}$$

$$V_{2o} = \frac{Z_0 Z_{0o}}{2Z_{0o}Z_0 \cos\theta + j(Z_{0o}^2 + Z_0^2)\sin\theta}$$

$$I_{2o} = \frac{Z_{0o}}{2Z_{0o}Z_0 \cos\theta + j(Z_{0o}^2 + Z_0^2)\sin\theta}$$

To simplify things, let us force the input impedance of the composite structure to be equal to Z_0. That is,

$$Z_0 = \frac{V_1}{I_1} = \frac{V_{1e} + V_{1o}}{I_{1e} + I_{1o}}$$

or

$$Z_0 = \frac{Z_0(Z_{0o}D_e + Z_{0e}D_o)\cos\theta + j(Z_{0o}^2 D_e + Z_{0e}^2 D_o)\sin\theta}{(Z_{0o}D_e + Z_{0e}D_o)\cos\theta + jZ_0(D_e + D_o)\sin\theta}$$

where

$$D_e = 2Z_{0e}Z_0\cos\theta + j(Z_{0e}^2 + Z_0^2)\sin\theta$$

and

$$D_o = 2Z_{0o}Z_0\cos\theta + j(Z_{0o}^2 + Z_0^2)\sin\theta$$

The equality will be established if

$$Z_0^2(D_e + D_o) = Z_{0o}^2 D_e + Z_{0e}^2 D_o$$

Substituting for D_e and D_o results in

$$2Z_0^3(Z_{0e} + Z_{0o})\cos\theta + j(Z_{0e}^2Z_0^2 + Z_{0o}^2Z_0^2 + 2Z_0^4)\sin\theta$$
$$= 2Z_{0e}Z_{0o}Z_0(Z_{0e} + Z_{0o})\cos\theta + j(Z_{0e}^2Z_0^2 + Z_{0o}^2Z_0^2 + 2Z_{0e}^2Z_{0o}^2)\sin\theta$$

This equality will be satisfied if

$$Z_0^2 = Z_{0e}Z_{0o} \tag{5-25}$$

Inserting this relationship into the expressions for the even and odd excitation voltages and currents results in

$$V_{1e} = \frac{Z_0\cos\theta + jZ_{0e}\sin\theta}{2Z_0\cos\theta + j(Z_{0e} + Z_{0o})\sin\theta}$$

$$I_{1e} = \frac{\cos\theta + jZ_{0o}Y_0\sin\theta}{2Z_0\cos\theta + j(Z_{0e} + Z_{0o})\sin\theta}$$

$$V_{2e} = \frac{Z_0}{2Z_0\cos\theta + j(Z_{0e} + Z_{0o})\sin\theta}$$

$$I_{2e} = \frac{1}{2Z_0\cos\theta + j(Z_{0e} + Z_{0o})\sin\theta}$$

$$V_{1o} = \frac{Z_0\cos\theta + jZ_{0o}\sin\theta}{2Z_0\cos\theta + j(Z_{0e} + Z_{0o})\sin\theta}$$

$$I_{1o} = \frac{\cos\theta + jZ_{0e}Y_0\sin\theta}{2Z_0\cos\theta + j(Z_{0e} + Z_{0o})\sin\theta}$$

$$V_{2o} = \frac{Z_0}{2Z_0\cos\theta + j(Z_{0e} + Z_{0o})\sin\theta}$$

$$I_{2o} = \frac{1}{2Z_0\cos\theta + j(Z_{0e} + Z_{0o})\sin\theta}$$

Now we can write expressions for the voltages and currents of the composite structure as tabulated in Eq. (5-26):

(a) $$V_1 = V_{1e} + V_{1o} = \frac{2Z_0\cos\theta + j(Z_{0e} + Z_{0o})\sin\theta}{2Z_0\cos\theta + j(Z_{0e} + Z_{0o})\sin\theta} = 1$$

$$\begin{aligned}
&\text{(b)} \qquad I_1 = I_{1e} + I_{1o} = \frac{2\cos\theta + jY_0(Z_{0e} + Z_{0o})\sin\theta}{2Z_0\cos\theta + j(Z_{0e} + Z_{0o})\sin\theta} = \frac{1}{Z_0} \\
&\text{(c)} \qquad V_2 = V_{2e} + V_{2o} = \frac{2Z_0}{2Z_0\cos\theta + j(Z_{0e} + Z_{0o})\sin\theta} \\
&\text{(d)} \qquad I_2 = I_{2e} + I_{2o} = \frac{2}{2Z_0\cos\theta + j(Z_{0e} + Z_{0o})\sin\theta} \\
&\text{(e)} \qquad V_3 = V_{1e} - V_{1o} = \frac{j(Z_{0e} - Z_{0o})\sin\theta}{2Z_0\cos\theta + j(Z_{0e} + Z_{0o})\sin\theta} \\
&\text{(f)} \qquad I_3 = I_{1e} - I_{1o} = \frac{jY_0(Z_{0o} - Z_{0e})\sin\theta}{2Z_0\cos\theta + j(Z_{0e} + Z_{0o})\sin\theta} \\
&\text{(g)} \qquad V_4 = V_{2e} - V_{2o} = \frac{0}{2Z_0\cos\theta + j(Z_{0e} + Z_{0o})\sin\theta} = 0 \\
&\text{(h)} \qquad I_4 = I_{2e} - I_{2o} = \frac{0}{2Z_0\cos\theta + j(Z_{0e} - Z_{0o})\sin\theta} = 0
\end{aligned} \tag{5-26}$$

These voltages and currents can be further simplified by defining a coupling parameter C such that

$$C = \frac{Z_{0e} - Z_{0o}}{Z_{0e} + Z_{0o}} \tag{5-27}$$

This definition plus the relationship $Z_0^2 = Z_{0e}Z_{0o}$ allows us to write

$$\begin{aligned}
&\text{(a)} \quad V_1 = 1 && \text{(c)} \quad V_3 = \frac{jC\sin\theta}{\sqrt{1-C^2}\cos\theta + j\sin\theta} \\
&\text{(b)} \quad V_2 = \frac{\sqrt{1-C^2}}{\sqrt{1-C^2}\cos\theta + j\sin\theta} && \text{(d)} \quad V_4 = 0
\end{aligned} \tag{5-28}$$

If we pause now and reflect on the structure that we have analyzed, we see that one port is isolated and all ports are matched. This is clearly a directional coupler, but one that is significantly different from the other couplers we have considered. The primary difference lies in the fact that the coupling is continuous along the entire length of the lines. This is in contrast to branch-line hybrid and hybrid ring couplers, where signals traveled different paths and summed in a constructive manner at the coupled port and in a destructive manner at the isolated port. Another feature of this structure is the nature of the wave on the coupled line. If we look at the original structure again, we see that a signal propagating in one direction on one transmission line produces a coupled signal propagating in the opposite direction on the other transmission line. For this reason, this structure is called a *contradirectional coupler*. The maximum coupling occurs when the length of the structure is $\lambda/4$. For this case, $\theta = \pi/2$ and the terminal voltages are

$$V_1 = 1 \qquad V_3 = C$$
$$V_2 = -j\sqrt{1 - C^2} \qquad V_4 = 0$$

The coupling C in dB is:

$$C_{dB} = 10 \log_{10} \frac{1}{|C|^2}$$

Specifying C and Z_0 establishes Z_{0e} and Z_{0o}, since

$$C = \frac{Z_{0e} - Z_{0o}}{Z_{0e} + Z_{0o}}$$

and

$$Z_0^2 = Z_{0e}Z_{0o}$$

Nomograms which enable the engineer to specify the physical dimensions of shielded coupled strip lines, given values of Z_{0e}, Z_{0o}, and ϵ_r (the relative dielectric constant of the medium in which the strip line conductors are located), have been prepared by Cohn and are presented in Figs. 5–31 and 5–32.

Example 5–1

Calculate the strip width, spacing between conductors, and length of the coupling region for a 10-dB directional coupler with a design center frequency of 1 GHz. This coupler is to be constructed of shielded strip line using a dielectric filling with $\epsilon_r = 2.25$ and a thickness between the ground planes of 0.25 inch. The reference impedance is to be 50 ohms.

Solution: For a 10-dB coupler,

$$10 \text{ dB} = 10 \log_{10} \frac{1}{|C^2|} \text{ dB}$$

Solving for the coupling coefficient C results in $C = 0.316$. The coupling coefficient can now be used to relate Z_{0e} and Z_{0o} since

$$0.316 = \frac{Z_{0e} - Z_{0o}}{Z_{0e} + Z_{0o}}$$

or

$$Z_{0e} = 1.925 Z_{0o}$$

Now, using

$$Z_0^2 = Z_{0e}Z_{0o}$$

and the relationship between Z_{0e} and Z_{0o} results in

$$2500 = 1.925\, Z_{0o}^2$$

Solving for Z_{0o} yields

$$Z_{0o} = 36.1 \text{ ohms}$$

Solving for $Z_{0e} = \dfrac{Z_0^2}{Z_{0o}}$ results in

$$Z_{0e} = 69.5 \text{ ohms}$$

Now formulate $\sqrt{\epsilon_r}Z_{0o} = 54.15$ ohms and $\sqrt{\epsilon_r}Z_{0e} = 104.25$ ohms. Referring to the nomograms (Figs. 5–31 and 5–32), we see that

$$\frac{w}{b} = 0.66 \text{ and } \frac{s}{b} = .05$$

Substituting $b = 0.25$ to determine w and s results in

$$w = (0.66)(0.25) = .165 \text{ in.}$$
$$s = (0.05)(0.25) = .0125 \text{ in.}$$

At the center or design frequency the coupling length should be $\lambda/4$ in length, thus

$$d = \frac{3 \times 10^8}{\sqrt{\epsilon_r}(10^9)}\frac{1}{4} = \frac{2 \times 10^8}{4 \times 10^9} = 0.05 \text{ meter}$$

The length of the coupling region will be 5 centimeters.

At this point it is interesting to see how the coupling changes with frequency. If we include the frequency dependence, the coupling in dB is

$$C_{\text{dB}} = 10 \log_{10} \frac{1}{|V_3^2|}$$

where

$$V_3 = \frac{jC \sin\theta}{\sqrt{1 - C^2}\cos\theta + j\sin\theta}$$

In our example, $C = 0.316$. The frequency dependence is contained in the term $\theta = \dfrac{\pi}{2}\dfrac{f}{f_0}$. A plot of the coupling versus f/f_0 is shown in Fig. 5–33. Note the wide band of frequencies over which the coupling stays relatively constant (less than 1 dB variation in the frequency range from 0.7 to 1.3 GHz). This performance is much better than would be possible with the couplers considered up to this point. However, in defense of the others such as the 3-dB branch-line hybrid, it should be noted that a 3-dB contradirectional coupler using the edge coupling described in the shielded strip line example would require an intolerably narrow gap between the conductors. One way to partially circumvent this problem is to couple via the broad sides. That is, locate one strip conductor a short distance above the other.

The scattering matrix description of the coupled strip transmission line directional coupler can easily be deduced by consideration of the general definition of the scattering matrix of a directional coupler, Eq.

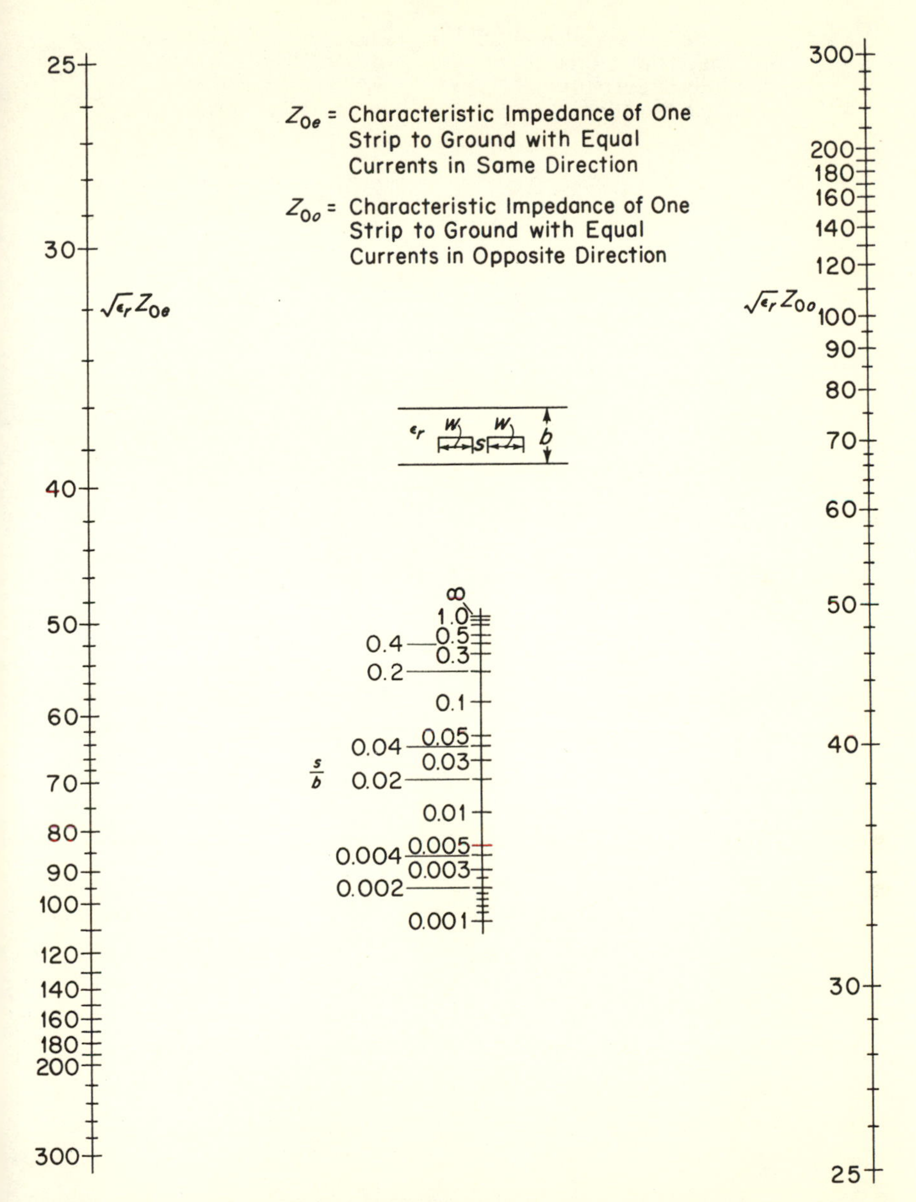

Fig. 5-31. Nomogram, given s/b as a function of Z_{0e} and Z_{0o}, in coupled strip line. (After S. B. Cohn, "Shielded Coupled-Strip Transmission Line," *IRE Transactions on Microwave Theory and Techniques*, vol. 3, p. 33, October 1955.)

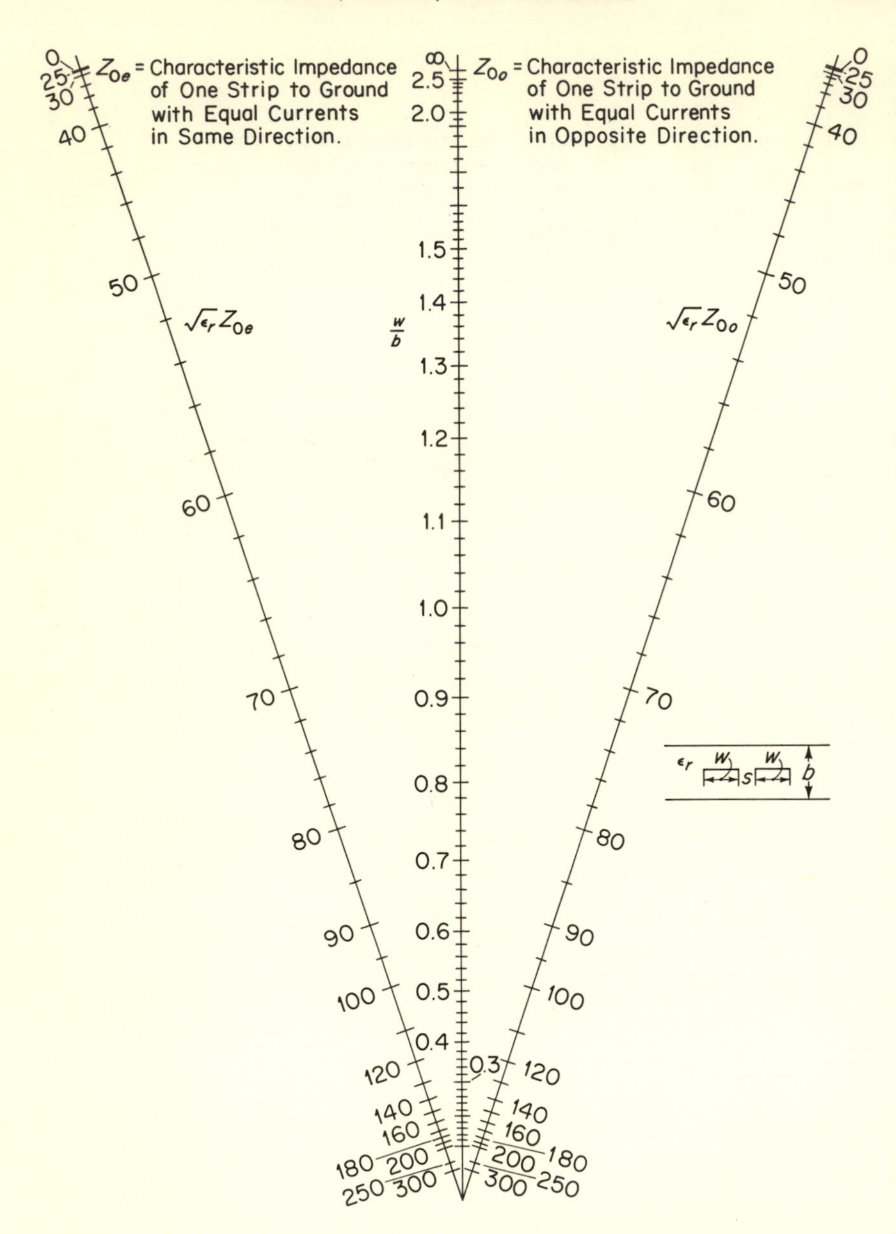

Fig. 5–32. Nomogram, given w/b as a function of Z_{0e} and Z_{0o}, in coupled strip line. (After S. B. Cohn, "Shielded Coupled-Strip Transmission Line," *IRE Transactions on Microwave Theory and Techniques,* vol. 3, p. 32, October 1955.)

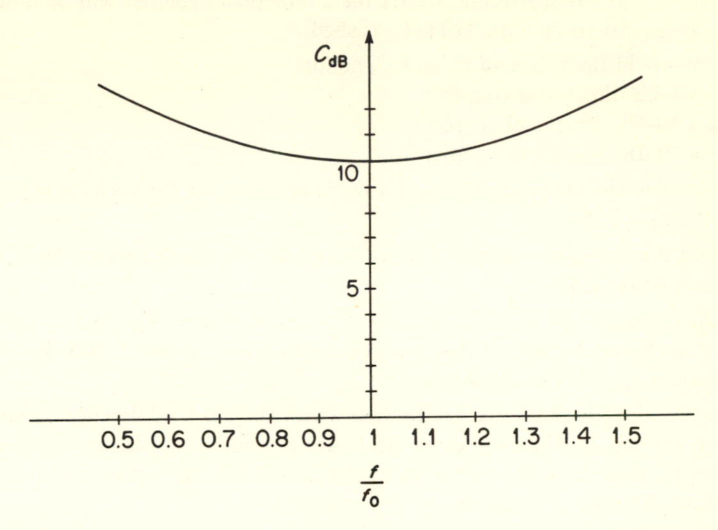

Fig. 5–33. Coupled wave amplitude versus frequency for the coupled strip line directional coupler of Example 5–1.

(5–18), the terminal voltages of the coupled transmission-line directional coupler, Eq. (5–28), and the symmetry of the coupled strip lines. Examination of Eq. (5–18) and Eq. (5–28) indicates that

$$C_1 = \frac{\sqrt{1 - C^2}}{\sqrt{1 - C^2} \cos \theta + j \sin \theta}$$

and

$$C_2 = \frac{C \sin \theta}{\sqrt{1 - C^2} \cos \theta + j \sin \theta}$$

Substitution of these values into Eq. 5–18 results in the scattering matrix description of the coupled strip transmission-line directional coupler.

PROBLEMS

5–1. Derive expressions for the elements of the general circuit parameter matrix representation of a two-port in terms of the elements of the admittance matrix representation.

5–2. Given the measured value of the elements of the scattering-matrix representation of a reciprocal two-port:

$$S_{11} = \frac{1}{3}(1 + j2); \quad S_{12} = j\frac{2}{3}; \quad S_{22} = \frac{1}{3}(1 - j2)$$

Determine the impedance matrix representation of the two-port.

5–3. Confirm that the scattering matrix for a directional coupler will be unitary if the six equations on page 120 are satisfied.

5–4. What would the values of C_1 and C_2 be for:
(a) a 3-dB directional coupler?
(b) a 10-dB directional coupler?
(c) a 20-dB directional coupler?

5–5. Determine the transmission matrix representation for the reciprocal two-port of Problem 5–2.

5–6. Using the terminal voltages of Eq. (5–24), deduce the scattering matrix for a binary power divider.

5–7. Apply a signal of V_s volts to port 1 and a signal of V_o volts to port 4 of a branch-line hybrid. What are the voltages at ports 2 and 3? Assume matched conditions at all ports.

5–8. Analyze the power divider for the case where there is no difference resistor and port 3 is driven while the other ports are terminated in Z_0 impedances. Specifically, determine the reflection coefficient at the drive port and the power delivered to the terminations on port 1 and 2.

5–9. A one-quarter wavelength section of transmission line is connected to port 3 of a branch-line hybrid. Derive the scattering matrix of the composite structure by shifting the reference plane at port 3 of the branch-line hybrid $\pi/2$ radians away from the junction.

5–10. Show that a 3-dB directional coupler could be fabricated by a series connection of two 8.34-dB directional couplers. Use the general directional coupler of Fig. 5–13 for your proof.

5–11. Apply a signal of V_s volts to port 4 and a signal of V_o volts to port 2 of a ring hybrid. Determine the voltages at ports 1 and 3. Assume matched conditions at all ports.

6

The Transmission Line as a Circuit Element

6–1. INTRODUCTION

A transmission line has several properties in common with lumped-constant circuit elements, but at the same time it has several properties very different from them. For example, the transmission line may be used either singly or in combination with inductors and capacitors to produce the equivalent of a series- or parallel-tuned circuit. In this sense it is similar in behavior to lumped-constant circuits. However, contrary to lumped-constant circuit elements, the transmission line so used exhibits not just one or even several resonant frequencies but an *infinite* number of resonant frequencies.

6–2. LOW-FREQUENCY OPERATION OF LINES

Consider the operation of the transmission line in the circuit of Fig. 6–1. Consider this as an application where audio signals are transferred from point A to point B by a coaxial, shielded lead. The length d is assumed to be a very small fraction of a wavelength at the highest frequency involved. It may be recalled that the input impedance of an

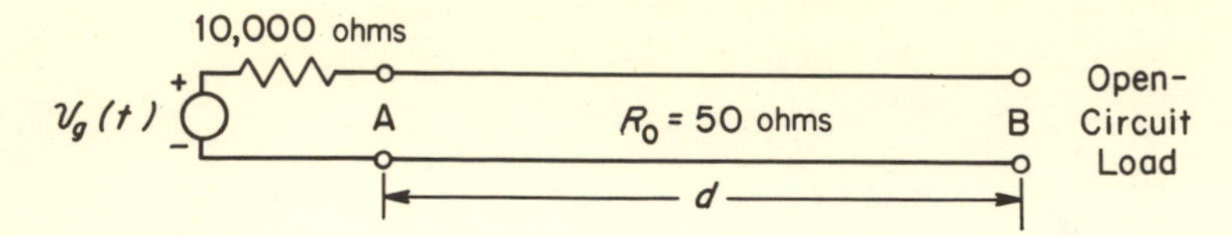

Fig. 6–1. Low-frequency transmission-line operation.

open-circuited section of lossless line of length d may be expressed as

$$Z_i = \frac{-jR_0}{\tan \beta d} \tag{6–1}$$

In this expression $R_0 = \sqrt{L/C}$ and $\beta = \omega\sqrt{LC}$. For sufficiently low frequencies the quantity βd will be so small that the tan βd can be approximated sufficiently well by simply βd, and in this case Eq. (6–1) reduces to

$$Z_i = -j\frac{R_0}{\beta d} = +\frac{\sqrt{L/C}}{j\omega\sqrt{LC}d} = \frac{1}{j\omega(Cd)} \tag{6–2}$$

Equation (6–2), which holds approximately for all ω below some value, clearly shows that at sufficiently low frequencies, the transmission line behaves as a lumped capacitance of Cd farads in an application such as shown in Fig. 6–1. If you had never studied transmission-line theory, you would expect this result, since from a lumped-constant viewpoint the capacitance per-unit length C, times the length d of the line, would give the shunting capacity at audio frequencies. Transmission-line theory gives us the same answer. At the same time, however, it warns us that it cannot be applied outside the assumption that βd is small enough so that tan βd can be approximated satisfactorily by βd. This restriction can also be expressed by saying that the frequency must be sufficiently low that the line is only a small fraction of a wavelength long at the highest frequency.

The fact that the line shown in Fig. 6–1 behaves much like a capacitor can also be shown by considering the voltage waveform at A (or B) for an input which is the unit-step function $u(t)$. Assuming that the source resistance R is very much greater than the R_0 of the line, we may conclude that the incident voltage pulse down the line is about R_0/R volts. This is reflected by a factor of 1 at the open-circuited load and by a factor of *slightly less than* 1 back at the source. Consequently each voltage step at the source caused by a reflected pulse is nearly $2R_0/R$ times the incident pulse step. Since, however, the reflection factor at source is less than 1, the composite voltage waveform at point A versus time will look somewhat as shown in Fig. 6–2. Actually, for the numbers given in Fig. 6–1, there

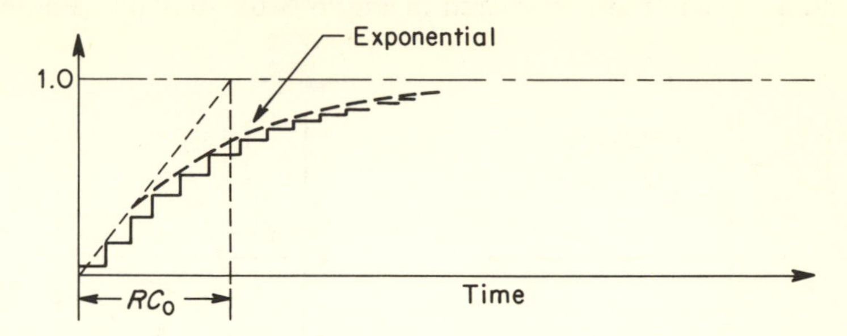

Fig. 6–2. Voltage waveform at point A in Fig. 6–1 for unit step input.

would be many more steps much more closely spaced than those shown in Fig. 6–2. The size of the steps in Fig. 6–2 has been purposely exaggerated for clarity.

A curve drawn through each of the steps would obviously be exponential in form. We can find the time constant of the exponential by computing the length of time required for the steps of voltage to reach 1 V, *presuming* that they continued at the initial rate. Clearly, the round-trip time is $2d/v$ seconds, and exclusive of the first step, the assumed steps would be $2R_0/R$ volts high. Combining these results will show that the time constant T may be found to be

$$\frac{\text{Time per step}}{\text{Volts per step}} = T = \frac{2d}{v}\frac{R}{2R_0} = \frac{Rd}{vR_0} = R(Cd) = RC_0 \qquad (6\text{–}3)$$

Consequently the stepped curve in Fig. 6–2 could be approximated quite well by the exponential curve

$$1 - e^{-t/T} = 1 - e^{-(t/RC_0)} \qquad (6\text{–}4)$$

which again verifies that the line behaves much like a capacitance of $C_0 = Cd$ farads.

6–3. THE LINE AS A TUNED CIRCUIT

Since the input impedance of a line terminated in a reactance (including the special cases of open and short circuits) is a reactance, we might expect that this reactance would behave much like that produced by a lumped-constant element. The only difference between the reactance produced by a line and that by lumped-constant elements is in the variation of that reactance versus frequency. The frequency variations of the input reactance of a shorted line and an inductor are compared in Fig. 6–3(a). The frequency variation of the input reactance of an open line and a capacitor are compared in Fig. 6–3(b). Actually, Fig. 6–3(a)

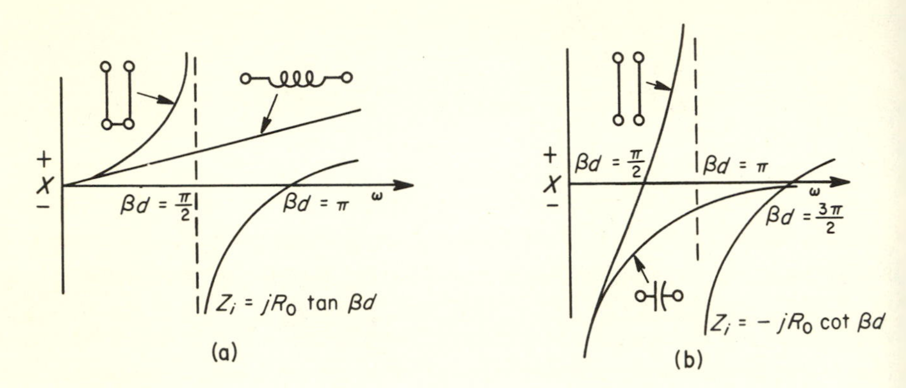

Fig. 6–3. Reactance variations of lumped and distributed elements versus frequency.

would have sufficed for both comparisons, providing the latter comparison was made on a susceptance scale. At sufficiently low frequencies the shorted line and the inductor behave essentially the same, and at sufficiently low frequencies the open line and the capacitor behave essentially the same. However, the reactances of the shorted line and the open line exhibit periodic repetition of the reactance curve versus frequency.

By inspection of the reactance curves for the shorted and open lines shown in Fig. 6–3, it may be seen that they exhibit the properties of series and parallel circuits at certain frequencies. This can be shown analytically by considering the driving-point characteristics of the open-circuited and short-circuited lines in the neighborhood of $\beta d = \pi/2$. The input admittance of a short-circuited stub can be written as ($\omega = \omega_0 + \Delta\omega$)

$$Y_i = -jY_0 \cot (\omega_0 + \Delta\omega) \frac{d}{v} = -jY_0 \cot \left(\frac{\pi}{2} + \frac{\pi}{2}\frac{\Delta\omega}{\omega_0}\right) \cong +jY_0 \frac{\pi}{2}\left(\frac{\omega - \omega_0}{\omega_0}\right) \tag{6–5}$$

Similarly, the input impedance of an open-circuited stub can be shown to be

$$Z_i \approx jZ_0 \frac{\pi}{2}\left(\frac{\omega - \omega_0}{\omega_0}\right) \tag{6–6}$$

The input admittance of a parallel-resonant circuit operating in the neighborhood of resonance is

$$Y_i = j\sqrt{\frac{C}{L}}\left(\frac{\omega^2 - \omega_0^2}{\omega\omega_0}\right) \approx j2\sqrt{\frac{C}{L}}\left(\frac{\omega - \omega_0}{\omega_0}\right)$$

while the input impedance of a series-resonant circuit operating near resonance is

$$Z_i = j\sqrt{\frac{L}{C}}\left(\frac{\omega^2 - \omega_0^2}{\omega\omega_0}\right) \approx j2\sqrt{\frac{L}{C}}\left(\frac{\omega - \omega_0}{\omega_0}\right)$$

It is apparent that the frequency dependence near resonance of the transmission-line circuits is identical to that of the lumped-constant circuits.

The basic reactance curves for both series and parallel circuits are shown in Fig. 6–4 for ready reference. In particular we note that a quarter-

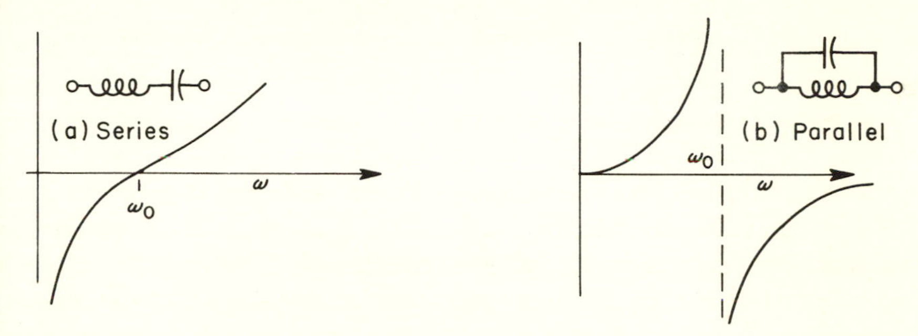

Fig. 6–4. Comparison of reactance curves of series- and parallel-tuned circuits.

wavelength shorted stub and a half-wavelength open stub both behave like parallel-resonant circuits, whereas a half-wavelength shorted stub and a quarter-wavelength open stub both behave like series resonant circuits. Of course, addition of a multiple of a half wavelength to any of these lengths will not materially affect the results. It should also be noted that the line may be made slightly less than or slightly more than the distance required to be resonant (either series or parallel) at a given frequency, and small variable inductors or capacitors, as the case requires, may be used to "fine tune" the resonant circuit to the correct frequency.

Tuned transmission lines are frequently used at very high frequencies (say, above 300 MHz) in the construction of tunable oscillators because of the temperature stability and the extremely high Q which is possible with the method of construction.

6–4. EFFECTS OF LOSSES ON LINES USED AS CIRCUIT ELEMENTS

The effect of losses on stubs used to obtain capacitive and inductive reactances is to add a small resistive or conductive component to the reactance or susceptance. Ordinarily the dissipative part of the impedance will be so small that it can be neglected.

However, in the case of stubs used as tuned circuits, the dissipation becomes the sole determining element of the impedance at resonance and so cannot be neglected. Although the Q quality factor of a line can be determined by inspection of the input impedance curve near resonance, a more generally useful approach is to consider Q defined as

$$Q = 2\pi \frac{\text{maximum energy stored}}{\text{energy dissipated per cycle}} = \frac{\text{maximum energy stored}}{\text{energy dissipated per radian}} \tag{6–7}$$

This result is in agreement with the more common definitions. For example, the Q of a coil is known to be $\omega_0 L/R$, and this value can be computed from the formula by supposing that the inductor is parallel with a lossless capacitor and that the combination tunes to ω_0. If the maximum instantaneous current is $\mathcal{I}_{\text{max}}$, then evidently

$$Q_L = 2\pi \frac{(1/2)L\mathcal{I}_{\text{max}}^2}{(1/2)\mathcal{I}_{\text{max}}^2 r(1/f_0)} = \frac{\omega_0 L}{R} \tag{6–8}$$

Another very useful and interesting way to look at the Q factor involves observation of the shock-excited response of the circuit. All tuned circuits exhibit a shock-excited response in the form of an exponentially damped oscillation, as given by

$$e^{-\sigma t} \sin \omega_0 t \tag{6–9}$$

which is shown in Fig. 6–5. This sort of response is to be expected, since the energy extracted during each cycle is proportional to the amplitude of the oscillation, and this condition is characterized in an exponential form of decay (consider, for example, an RC circuit where the current discharging the capacitor is proportional to the voltage of the capacitor). Now it may be recognized that the maximum energy stored is proportional to the square of the maximum voltage or current, which is proportional to $e^{-2\sigma t}$, while the energy dissipated per cycle is the negative derivative of the maximum energy stored with respect to time multiplied by the time of one cycle. Consequently, the value of Q as defined by Eq.

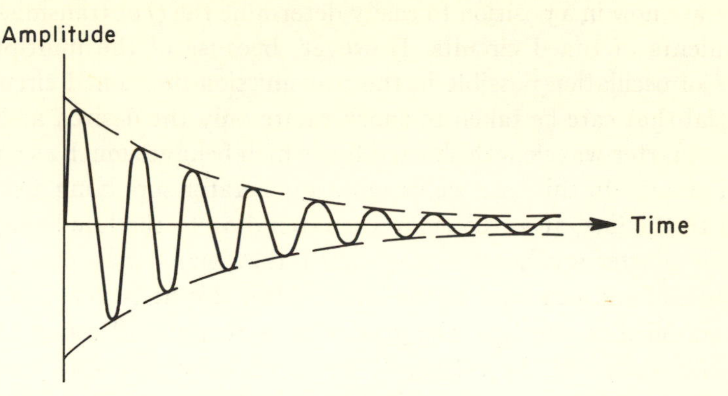

Fig. 6–5. Shock-excited response of a tuned circuit.

(6–8) may be expressed quite simply as

$$Q = 2\pi \frac{(W_{\max} e^{-2\sigma t})}{(-d/dt)(W_{\max} e^{-2\sigma t})(1/f_0)} = 2\pi \frac{1}{2\sigma(1/f_0)} = \frac{\omega_0}{2\sigma} \qquad (6\text{–}10)$$

Substitution of this relationship for Q back into Eq. (6–9) will yield an expression which can be manipulated to show that in a period of Q cycles, the amplitude of oscillation will decay to $e^{-\pi}$, or about 4.32 per cent of its initial value. This permits us to readily estimate the Q of a system by observation of its shock-excited response. For example, a clock pendulum might be displaced and the motion observed. If 40 cycles were required before the amplitude of the swing decayed to 4.32 per cent of its initial value, then we would know that the Q was about 40. This Q-measurement technique is graphically illustrated in Fig. 6–6.

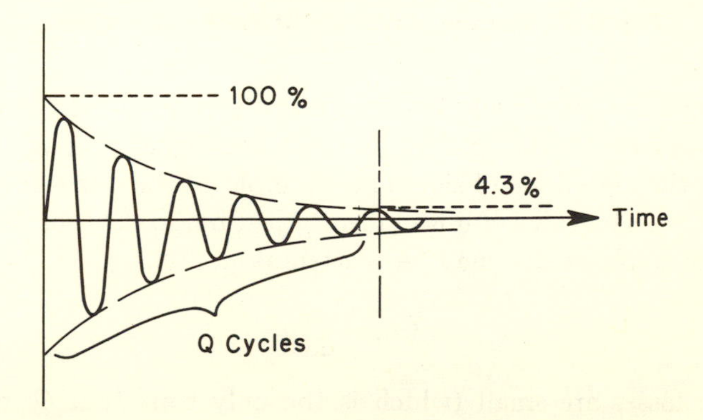

Fig. 6–6. Shock-excited response related to circuit Q.

We are now in a position to easily determine the Q of transmission-line equivalents of tuned circuits. However, because of the multiplicity of modes of oscillation possible in the transmission-line tuned circuits, it is essential that care be taken to shock-excite only the desired mode. Consider a quarter-wavelength shorted line which behaves much as a parallel-tuned circuit. In this case we imagine a generator supplying a sinusoidal signal to the line, at a frequency corresponding to the line being electrically a quarter-wavelength long, until that signal has passed to the shorted end and just back to the input. After this half-cycle of signal has been supplied to the line, the generator is removed (leaving the input open-circuited) and the voltage at any point along the line is observed. The initial conditions that have been supplied to the line do a cat-chasing-its-tail routine and produce a voltage waveform (after removal of the source) at the input end of the line, which appears as shown in Fig. 6–7.

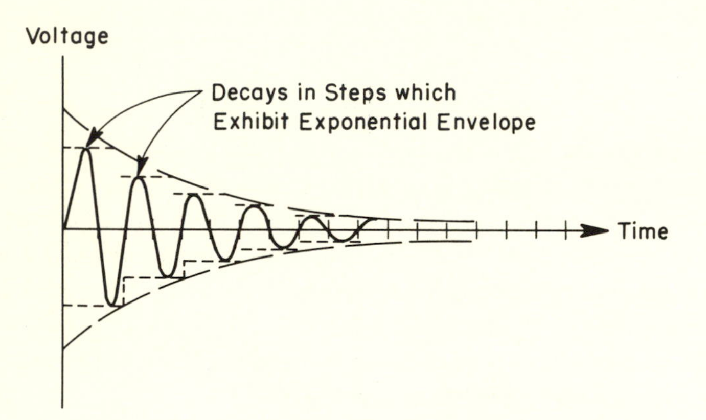

Fig. 6–7. Voltage waveform of shock-excited line.

The value of ω_0 is already known for this problem, and the value of σ can be found by observing that, because the signal propagates with a velocity v, the amplitude of signal at the input of the line will be proportional to $e^{-\alpha v t}$. The value of σ is evidently just αv, so the Q of the quarter-wavelength resonant line may be written as

$$Q_{1/4} = \frac{\omega_0}{2\alpha v} \tag{6–11}$$

When the losses are small (which is the only time that Q makes any particular sense), α may be expressed as $(R/2R_0) + (G/2G_0)$ and the value of $Q_{1/4}$ may be rewritten as

$$Q_{1/4} = \frac{\omega_0\sqrt{LC}}{(R/R_0) + (G/G_0)} \tag{6–12}$$

It is quite interesting to note that if the losses are either wholly in the series resistance or in the shunt conductance, the expression for Q reduces to the simple forms

$$Q_R = \frac{\omega_0 L}{R} \tag{6–13}$$

and

$$Q_G = \frac{\omega_0 C}{G} \tag{6–14}$$

These results strongly resemble those for lumped-constant circuits; the point of difference is that L, R, C, and G are the transmission-line parameters, given in values per unit length of line. By returning to the definition of Q given in Eq. (6–8), the combination of an L and C having different Q's may be made in the manner of paralleling resistors,[1] so that

$$\frac{1}{Q} = \frac{1}{Q_R} + \frac{1}{Q_G} \tag{6–15}$$

By substitution and manipulation it may be shown that Eq. (6–15) does indeed correspond with Eq. (6–12).

6–5. THE LINE AS A DELAY NETWORK

By special construction techniques which involve increasing both the series inductance and capacitance of the line, it is possible to produce substantial time delays in rather short physical lengths of line. Ordinary RG–58A/U has a velocity of propagation of 200 m/μs, and so 200 m would be required to produce 1-μs delay. However, the specially constructed delay lines may have delays on the order of 1 μs/ft. Almost invariably the band width of the cable decreases as the delay is increased. For example, RG–58A/U has a band width of several thousands of megahertz, whereas the special delay cables to produce delays of 1 μs/ft may have band widths of only several megahertz. We have in essence exchanged band width for time delay.

The special construction techniques for high-delay cables include the following:

[1] Note carefully that the same maximum energy storage must be present in L and C if the formula is to hold.

1. Coiling the center conductor to increase the inductance. In this case the outer conductor must not be solid or it will act by transformer action to reduce the inductance of the coiled center conductor. The outer shield usually consists of many strands of insulated wire placed parallel with the length of the cable and connected together only at one end.
2. Winding the center conductor on a high-permeability material such as one of the ferrites.
3. Placing the outer conductor as close as possible to the coiled center conductor in an attempt to increase the capacitance.
4. Taking corrective measures to improve the band width and pulse response of the cable. This involves aluminum foil strips in special geometries.

The characteristic impedance of the high-delay cables is usually greater than for conventional coaxial cables because the predominant increase is made in the inductance parameter of the line. While 50 ohms is common for the ordinary coaxial cables, values as high as 10,000 ohms are not uncommon in high-delay, low-band-width cables.

There is a very close tie between the construction of these delay cables and the construction of delay networks that use lumped-constant elements. In fact the external appearance as well as the electrical circuits resemble each other. An electrical comparison of the two is shown in Fig. 6–8. It is to be understood that distributed capacitance exists between the

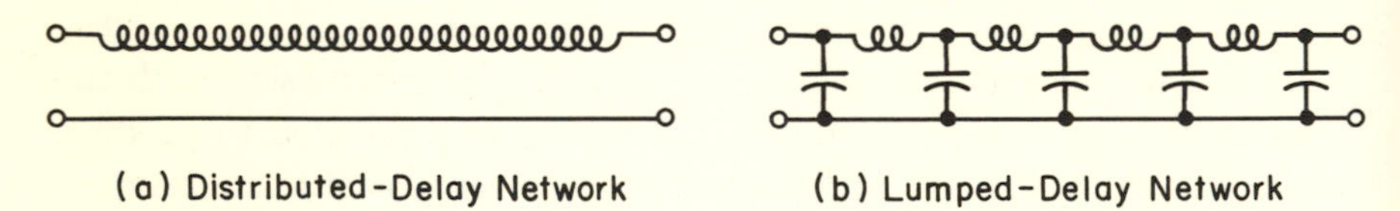

Fig. 6–8. Electrical construction of distributed- and lumped-delay networks.

coiled conductor and the solid conductor in the distributed delay network of Fig. 5–5(a). The physical construction of the two is often very similar because the coils required in the lumped-constant delay network may be wound on a common rod, with some spacing between coils and with taps brought out for connection to the various fixed capacitors. In the distributed delay line with the coiled center conductor, one obtains the required capacitance by physical placement of another shielding-type conductor around the coiled center condutor, as noted earlier.

We may think of making the mathematical transition between the distributed and lumped-constant delay line by merely making each section of the tee network equivalent to a length of the distributed-constant system which is short with respect to a wavelength at the highest frequency component of interest. Or we may start by analysis of the amplitude and phase characteristics of the tee section. This latter viewpoint offers an interesting comparison between the distributed and lumped-constant delay networks, as is illustrated in Fig. 6–9.

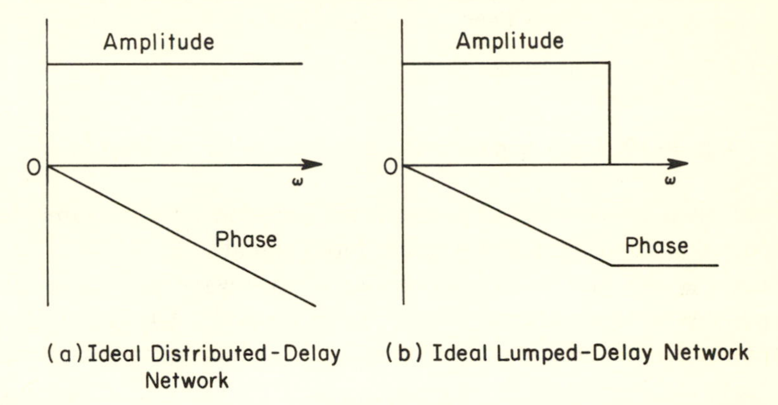

Fig. 6–9. Comparison of amplitude and phase curves for delay devices.

The ideal distributed-constant delay network would exhibit a perfectly constant amplitude characteristic (assuming here that the network is between a voltage source and an R_0 load) and a linear phase characteristic as shown in Fig. 6–9(a). This condition is approximated very well by conventional coaxial cables, but the amplitude characteristic is usually similar to that of the lumped-constant circuit for high-delay cables. The ideal lumped-constant delay network has an amplitude characteristic (again assuming a termination of R_0) which is essentially flat out to some frequency and then approaches zero rapidly for higher frequencies. The phase characteristic for the lumped-constant delay network is essentially linear out to the cutoff frequency and then levels off at some constant values for higher frequencies. The degree of perfection of the lumped-constant delay network is usually measured in terms of how linear the phase characteristic is within the pass band.

A comparison of the responses of the lumped-constant and distributed-delay networks to a step function input is illustrated in Fig. 6–10. The relationship between the time delay τ_d and the phase characteristic of the delay is a simple but very important one. Imagine for the moment that we have an input signal which is a low-frequency square wave rather than a step input. If the period of the square wave is sufficiently long (say, a day

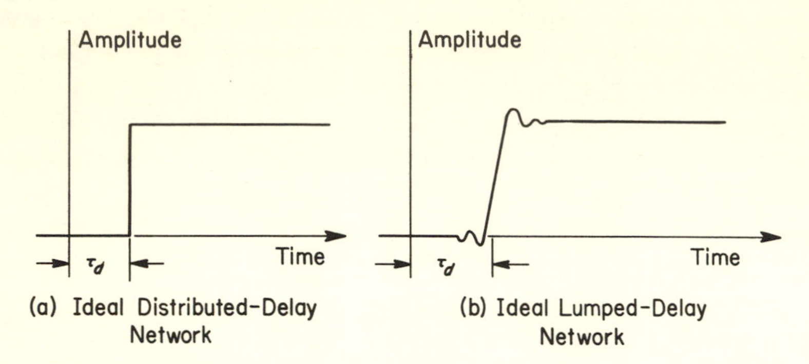

Fig. 6–10. Ideal responses of delay devices for unit step input.

or two), then the response of a circuit will be essentially the same during our period of measurement for either input signal.

Suppose now that we decompose our square wave into its frequency components by conventional Fourier series analysis. A typical frequency component for a fundamental frequency of ω_1 radians per second would be

$$b_n \sin (n\omega_1 t) \tag{6–16}$$

The response of the time delay network to this component would be

$$b_n \sin (n\omega_1 t - \tau n\omega_1) = b_n \sin n\omega_1(t - \tau) \tag{6–17}$$

In writing Eq. (6–17) we have assumed that the signal component is within the pass band of the delay network (which has been assumed to have a unity gain in the pass band) and that the linear phase characteristic of the delay network can be expressed by the equation $\theta = -\omega\tau$. The quantity θ represents the phase shift of the network, ω represents the angular frequency, and τ is a positive proportionality constant which has the dimensions of time and is, in fact, shown by Eq. (6–17) to be the time delay of the network.

The result of Eq. (6–17) shows that every component in the output signal, regardless of frequency is just a duplicate of the input component except for the fixed time delay of τ seconds. Hence, when the output signal components are added back together, the result will be a square wave differing from the input square wave only by the time delay τ. If the pass band of the filter is finite, as shown in Fig. 6–9(b) for the ideal lumped-constant time-delay network, then the response will appear much as shown in Fig. 6–10(b). The response will have a finite "rise time," and depending on the actual shape of the amplitude and phase curves, may or may not have "overshoot" and "ringing" as shown in Fig. 6–10(b). The time delay τ for the network is simply found as the negative of the slope of the phase angle versus angular frequency; that is, simply $-\theta/\omega$ for any θ and the corresponding ω.

The delay per section of a lumped-constant delay network is about the same as the rise time of the system. Since the rise time of the system can be shown to be inversely proportional to the band width, we must conclude that long delays, for a given number of sections, may be had only at the expense of band width. Essentially the same statement may be made for the high-delay distributed constant lines.

The delays provided by transmission lines may be used very advantageously in electronic circuit design. A common application is for producing short delays of pulses. For pulse delays longer than about 10 μs, it is generally found more economical to generate the delays with electronic circuits, such as with a one-shot multivibrator or a phantastron. These electronic circuits can only produce an output pulse delayed with respect to the input pulse, and the size and shape of the input pulse is entirely lost in the delay circuitry. We shall concern ourselves with delays of less than 10 μs, where the transmission line plays a very important role in the circuit design.

Suppose, for example, that you had a series of pulses separated by at least 10 μs with most of them 1 μs and a certain few 2 μs wide. How would you design an electronic circuit that would give an output only after passage of the 2-μs pulses? With a delay line the problem almost becomes trivial, as may be seen from inspection of Fig. 6–11. The coinci-

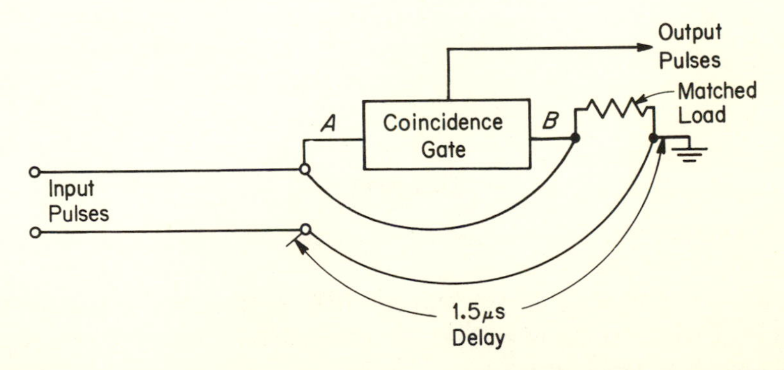

Fig. 6–11. A scheme for selection of 2-μs pulses from a train of pulses both 1 and 2 μs.

dence gate shown in the figure is simply a device which gives an output when signals appear at both A and B but at no other time. It may be made from a single multigrid vacuum tube, a pair of transistors, or even a couple of diodes. It can be seen that the only time voltages will appear simultaneously at points A and B will be for pulses greater than 1.5 μs. Hence the device will give an output for 2-μs pulses but not for 1-μs pulses. One can think of the circuit in Fig. 6–11 as being a filter that separates with regard to time in much the same way as LC filters are used to separate with regard to frequency.

The effect of losses on a transmission line used as a delay network is almost self-evident. If a generator $\mathcal{V}_g(t)$ supplied a characteristic impedance load through a length d of lossy line, then the load voltage will simply be

$$e^{-\alpha d}\mathcal{V}_g\left(t - \frac{d}{v}\right) \tag{6–18}$$

6–6. TEE AND PI AC EQUIVALENT CIRCUITS OF THE TRANSMISSION LINE

At any given frequency it is possible to make either a tee or pi equivalent circuit for a section of transmission line. We can apply the results of the matrix representations discussed in the last chapter to determine the elements of the tee and pi equivalent circuits of a section of transmission line (Fig. 6–12).

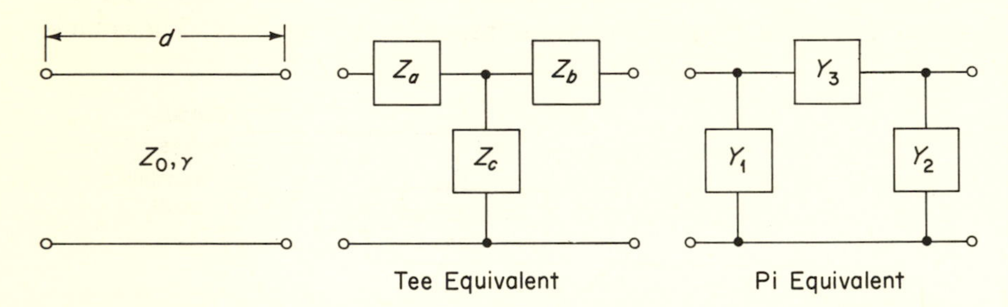

Fig. 6–12. Tee and pi equivalent circuits for a transmission line.

In the last chapter it was shown that the impedance matrix for a section of transmission line of length d, characteristic impedance Z_0, and propagation constant γ is

$$Z = \begin{bmatrix} Z_a + Z_c & Z_c \\ Z_c & Z_b + Z_c \end{bmatrix} = \begin{bmatrix} Z_0 \coth \gamma d & \dfrac{Z_0}{\sinh \gamma d} \\ \dfrac{Z_0}{\sinh \gamma d} & Z_0 \coth \gamma d \end{bmatrix} \tag{6–19}$$

for a tee equivalent circuit and

$$Y = \begin{bmatrix} Y_1 + Y_3 & -Y_3 \\ -Y_3 & Y_2 + Y_3 \end{bmatrix} = \begin{bmatrix} Y_0 \coth \gamma d & \dfrac{-Y_0}{\sinh \gamma d} \\ \dfrac{-Y_0}{\sinh \gamma d} & Y_0 \coth \gamma d \end{bmatrix} \tag{6-20}$$

for the pi equivalent circuit. From this point it is easy to solve for the elements of the equivalent circuit representations. Doing so results in

$$Z_a = Z_b = Z_0 \tanh \frac{\gamma d}{2} \tag{6-21}$$

$$Z_c = Z_0 \operatorname{csch} \gamma d \tag{6-22}$$

and

$$Y_1 = Y_2 = Y_0 \tanh \frac{\gamma d}{2} \tag{6-23}$$

$$Y_3 = Y_0 \operatorname{csch} \gamma d \tag{6-24}$$

Now we consider the specific situation where the transmission line is lossless. In this case the tee and pi equivalent circuits become those shown in Fig. 6-13.

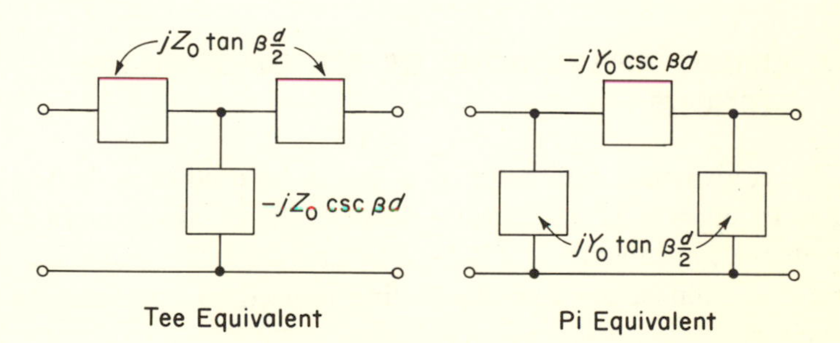

Fig. 6-13. Equivalent circuits.

If the length of the section is a very small fraction of a wavelength for the highest frequency of interest, then approximate equivalent circuits may be made for all frequencies up to the highest frequency of interest by replacing tangents and sines of angles by the angles themselves, with the results shown in Fig. 6-14. These results are as one might have intuitively expected, since for differential lengths of lines the results are exact. A typical result, such as the series inductor in the tee circuit, may be mathematically obtained from the results given in Fig. 6-13 by the steps

$$+jR_0 \tan \frac{\beta d}{2} \approx jR_0 \frac{\beta d}{2} = j\omega \left(L \frac{d}{2}\right) \tag{6-25}$$

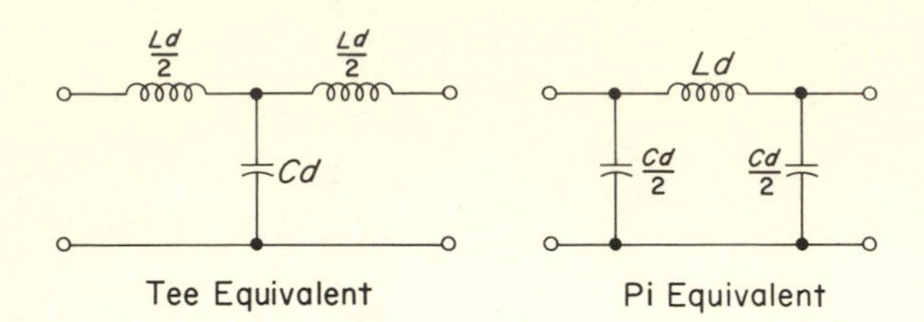

Fig. 6–14. Approximate equivalent circuits for an electrically short line.

By comparison of the result given in Eq. (6–25) with the reactance of an inductor, we conclude that an inductor of $Ld/2$ would produce exactly the same results.

The results given in Figs. 6–13 and 6–14 point out the similarities between distributed-constant and lumped-constant circuits, but at the same time they warn that an all-frequency exact equivalent circuit for a section of transmission line can never be obtained with lumped-constant elements because we cannot, with a finite number of elements, generate a reactance that is a transcendental function of ω, as required of all the elements in the circuit of Fig. 6–13.

6–7. CIRCUIT APPLICATIONS OF TRANSMISSION-LINE SECTIONS

The equivalent circuits of Fig. 6–14 can be used in designing low-pass filters to operate at microwave frequencies. A short section of high-impedance transmission line is used to approximate a series inductor and a short section of low-impedance line is used to approximate a shunt capacitor.

Example 6–1

We will construct a low-pass filter which exhibits a maximally flat or Butterworth response with cutoff frequency of 500 MHz. The cutoff frequency, or the edge of the pass band as it is often referred to, is the frequency where the insertion loss is 3 dB. Insertion loss is defined as the ratio of the actual power delivered to a load to the available power (the power delivered to a load matched to the source). This is readily shown to be

$$\frac{P_{\text{delivered}}}{P_{\text{available}}} = 1 - |K|^2$$

so the insertion loss in decibels is written as

$$\text{Insertion loss} = -10 \log_{10} (1 - |K|^2)$$

The filter will be designed for a reference or matched termination impedance of 50 ohms and will be fabricated from transmission-line sections of 10-ohm characteristic impedance and 150-ohm characteristic impedance. The lumped-constant prototype filter configuration and element values are as shown in Fig. 6–15.

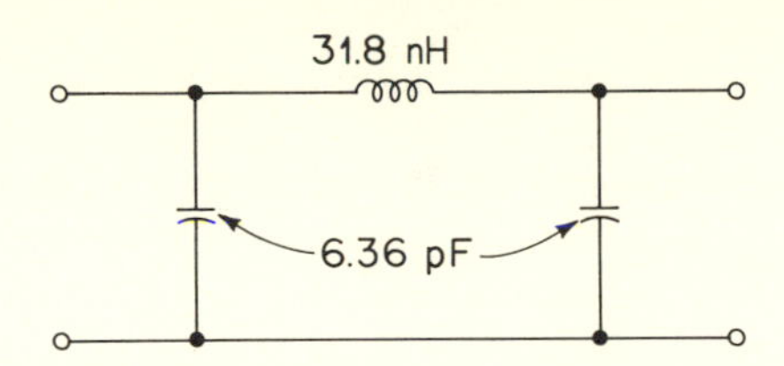

Fig. 6–15. Lumped-constant prototype filter.

We will begin by using a pi equivalent circuit for the section of 150-ohm line that will approximate the series inductor. By forcing the series inductance of the pi equivalent of a section of 150-ohm line to be 31.8 nH, we find that $\sin \beta d = 0.667$ or $\beta d = 41.9°$. This leads to a line length of 0.1163λ. The other elements of the pi equivalent, the two shunt capacitors, have values of 0.76 pF, as determined by solving $\omega C = (1/150) \tan \beta d/2$ at $\omega = (2\pi)(5 \times 10^8)$ rad/sec. Thus, a 0.1163λ length of 150-ohm line has a pi equivalent circuit, as shown in Fig. 6–16.

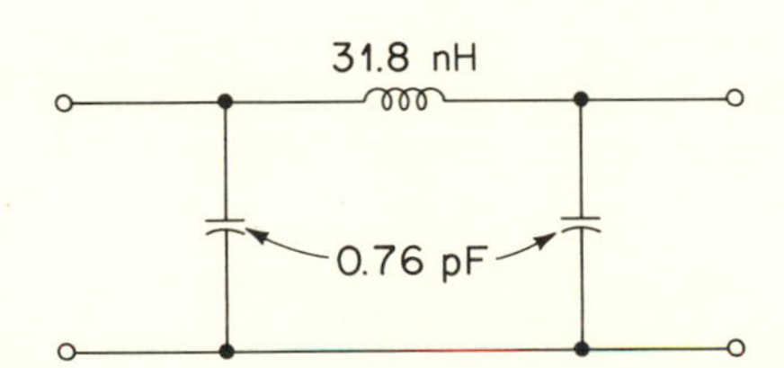

Fig. 6–16. Pi equivalent of 0.1163λ length of 150-ohm transmission line.

To realize the low-pass filter design, we need to add capacitance at each end of the section of 150-ohm line. We can use sections of 10-ohm line to do this. We will consider the tee equivalent of a 10-ohm section and force the length to be such that the shunt capacitance of the tee equivalent is equal to (6.36 − 0.76) or 5.6 pF at 500 MHz. Thus, we write: $(5.6 \times 10^{-12})(2\pi)(5 \times 10^8) = 0.1 \sin \beta d$ or $\sin \beta d = 0.176$ and $\beta d = 10.1°$. This means the length of 10-ohm line required is 0.028λ. The series inductors of the tee equivalent are readily calculated to be $L = .283$ nH. Thus we see that the tee equivalent circuit for a 0.028λ length of 10 ohm is as shown in Fig. 6–17. Note that the wavelength referred to is the wavelength at 500 MHz. Now we can examine the structure formed by connecting two 0.028λ sections of 10-ohm line at each end of a 0.1163λ section of 150-ohm line. This structure (Fig. 6–18) can be considered to be a close approximation to the original lumped-element low-pass prototype that we were trying to realize. Plots of the reflection coefficient versus frequency for the transmission-line structure and the prototype circuit are shown in Fig. 6–19.

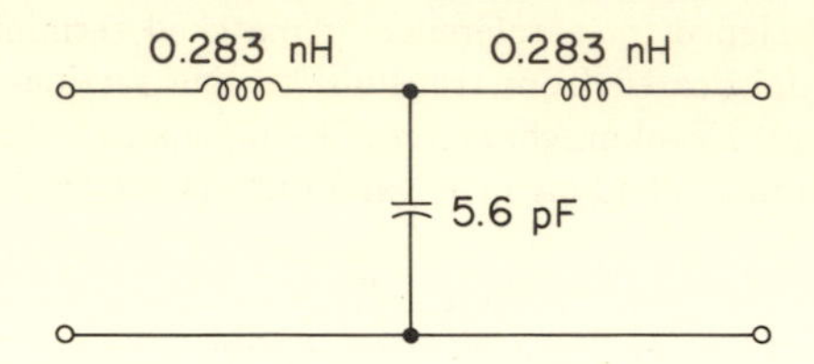

Fig. 6–17. Tee equivalent of 0.028λ length of 10-ohm transmission line.

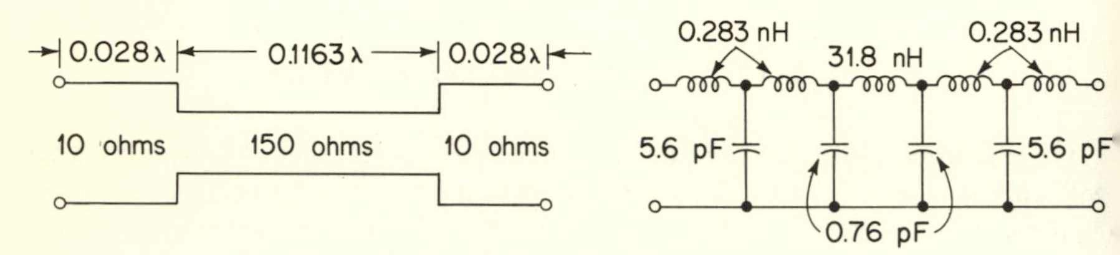

Fig. 6–18. Transmission-line equivalent circuit.

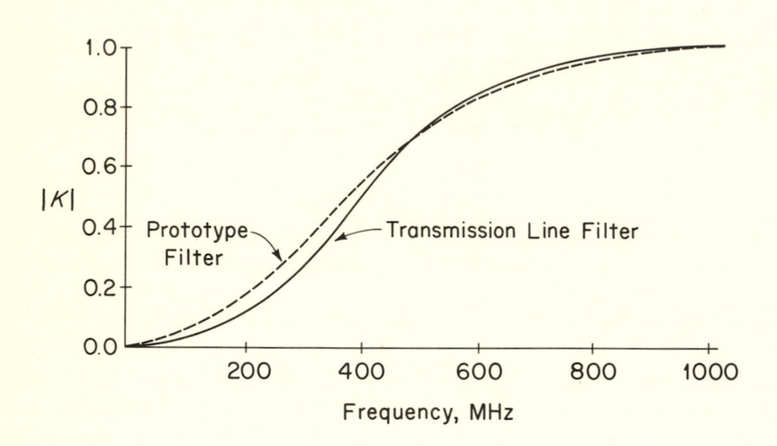

Fig. 6–19. Frequency response of low-pass filter.

We will now consider situations where power is dissipated within the line, as would be the case with a lossy transmission line, although the formulas given in Eqs. (6–21) through (6–24) may also be used to calculate circuit constants for devices not ordinarily considered transmission lines. For example, consider the design of a pad to give 20 dB of attenuation in a matched 50-ohm transmission line. By taking $Z_0 = R_0 =$ 50 ohms and noting that 20 dB corresponds to $\alpha d = 2.30$ nepers, the results of Eqs. (6-23) and (6-24) (using $\beta = 0$) may be applied to show that a pi section would have constants as given in Fig. 6-20.

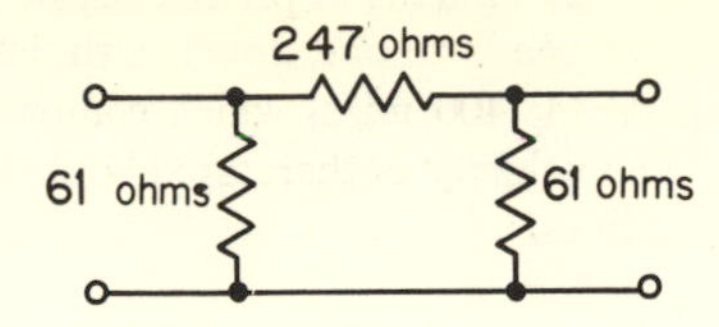

Fig. 6-20. A 20-dB, 50-ohm pi-section attenuator.

Example 6-2

Consider the case of a dissipative line connecting a voltage source and various loads. Suppose that the constants of the line, similar to a telephone line, are given as

$$\begin{aligned} L &= 4.6 \text{ mH/mile} \\ R &= 14 \text{ ohms/mile} \\ C &= 0.01\ \mu\text{F/mile} \\ G &= 0.3\ \mu\text{mho/mile} \end{aligned}$$

The characteristic impedance and propagation constants are given by the formulas

$$Z_0 = \sqrt{\frac{Z}{Y}} \qquad \text{and} \qquad \gamma = \sqrt{ZY}$$

where the values of Z and Y are given by the relations

$$Z = R + j\omega L \qquad \text{and} \qquad Y = G + j\omega C$$

Note carefully that both the characteristic impedance and propagation constants are functions of frequency. Suppose for the sake of definiteness that the analysis is made at a frequency of 2000 Hz. The values of Z_0 and γ are then found quite simply as

$$Z_0 = \frac{59.5 \,\underline{/76.4^\circ}}{126 \times 10^{-6} \,\underline{/89.9^\circ}} = 686 \,\underline{/-6.7^\circ} \text{ ohms}$$

$$\gamma = (59.5 \,\underline{/76.4^\circ})(126 \times 10^{-6} \,\underline{/89.9^\circ}) = 0.0868 \,\underline{/83.2^\circ} \text{ mile}$$

From the value of the propagation constant it is possible to determine the attenuation constant α and the phase constant β as

$$\begin{aligned} \alpha &= 0.0102 \text{ neper/mile} \\ \beta &= 0.0861 \text{ radian/mile} \end{aligned}$$

If the load is the characteristic impedance $686 \,\underline{/-6.7^\circ}$, then the signal propagates from the source to the load, obeying the law

$$E = E_s e^{-\gamma d}$$

Then for a line of, say, 5-mile length, the signal attenuation will be 5(0.0102) neper = 0.051 neper. This value, expressed on a decibel scale, is (0.051)(8.686) = 0.443 dB.

For a 45-mile length of line, the attenuation under the matched-load condition is easily found as 4.0 dB.

The velocity of propagation of the line at this frequency may be determined by noting from the value of β that 2π radians of phase shift occurs over the distance of 72.7 miles. This therefore corresponds to one wavelength, and by using $v = f\lambda$, it is found that the velocity equals 145,400 mi/s, which compares with that of light in free space as 186,000 mi/s. The velocity is therefore about $0.78c$, which is somewhat typical of practical telephone lines.

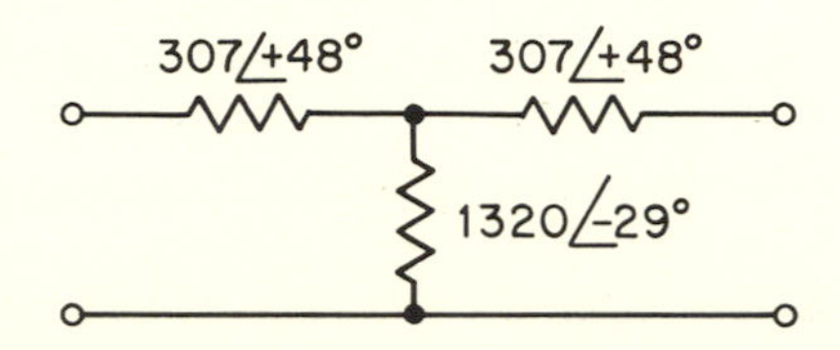

Fig. 6–21. Constants for Example 6-2.

The tee section equivalent of a 45-mile length of the line is shown in Fig. 6–21. From this equivalent circuit it is possible to determine the behavior of the system "viewed from the terminals" for any combination of source and load conditions.

As a final example of the use of the transmission line as a circuit element, we will reconsider the relationship between a section of transmission line and an equivalent lumped-constant element. In Art. 6–2 it was shown that at low frequencies a short length of transmission line terminated in a short circuit exhibits driving point properties equivalent to an inductor while a short length of open-circuited transmission has the driving point characteristics of a capacitor. Of course we recall that the lines must be less than one-quarter wave length long for the equivalence to hold.

This equivalence serves as the basis for defining a new frequency transformation, $S = j \tan \beta d$, called Richards transformation.[2] In the transformed plane (S-plane) the short-circuited section of transmission line becomes an inductor of Z_0 henrys while an open-circuited section of transmission line becomes a capacitor of Y_0 farads. These relationships are summarized in Fig. 6–22.

The S-plane elements can be used in the realization of a prototype filter as though they were lumped-constant elements. There are some fundamental differences, however, due to the nature of the transformation. For example, we see that S is a periodic function of βd which ranges from $-j\infty$ to $+j\infty$ as βd ranges from $-\pi/2$ to $+\pi/2$. Since $\beta d = \pi f/2f_0$, where f_0 is the frequency for which $d = \lambda/4$, it is apparent that S varies

[2] P. I. Richards, "Resistor Transmission-Line Circuits," *Proc. IRE*, vol. 36, pp. 217–220, February 1948.

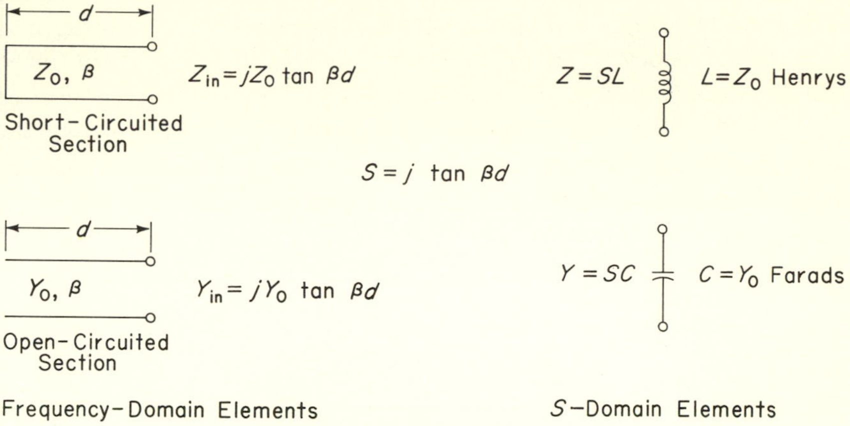

d

Z_0, β $Z_{in} = jZ_0 \tan \beta d$

Short-Circuited Section

$Z = SL$ $L = Z_0$ Henrys

$S = j \tan \beta d$

d

Y_0, β $Y_{in} = jY_0 \tan \beta d$

Open-Circuited Section

$Y = SC$ $C = Y_0$ Farads

Frequency-Domain Elements

S–Domain Elements

Fig. 6–22. Frequency-domain–S-domain equivalences.

with frequency such that S ranges from 0 to $j\infty$ as f ranges from 0 to f_0.

Now suppose we wish to design a filter based on conventional low-pass prototype design techniques. The filter is to be fabricated using open-circuited and short-circuited sections of transmission line of equal length. For example, a three-element low-pass prototype having a maximally flat frequency response will have S-plane elements substituted for the lumped-constant elements and the actual filter will be constructed from transmission-line sections as shown in Fig. 6–23.

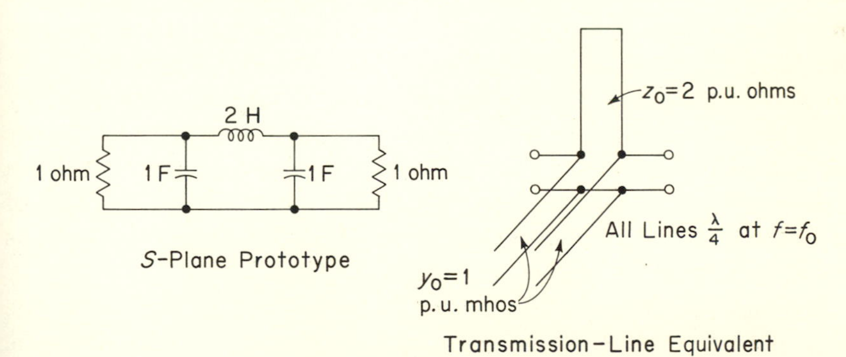

Fig. 6–23. Filter design using S-plane concepts.

Note that a 2-henry S-plane inductor represents a short-circuited stub of normalized characteristic impedance of 2 per-unit ohms, while a 1-farad S-plane capacitor represents an open-circuited stub of normalized characteristic admittance of 1 per-unit mho. The transmission-line elements are of commensurate length; that is, all S-plane elements are $\lambda/4$ in length at the center frequency f_0. The frequency response of the structure in the S-plane has been defined to be maximally flat with a 3-dB insertion loss at $S = j1$. The frequency response in the f-plane can be deduced from the definition of the Richards transformation. A portion of the real frequency response and the corresponding S-plane response is depicted in Fig. 6–24.

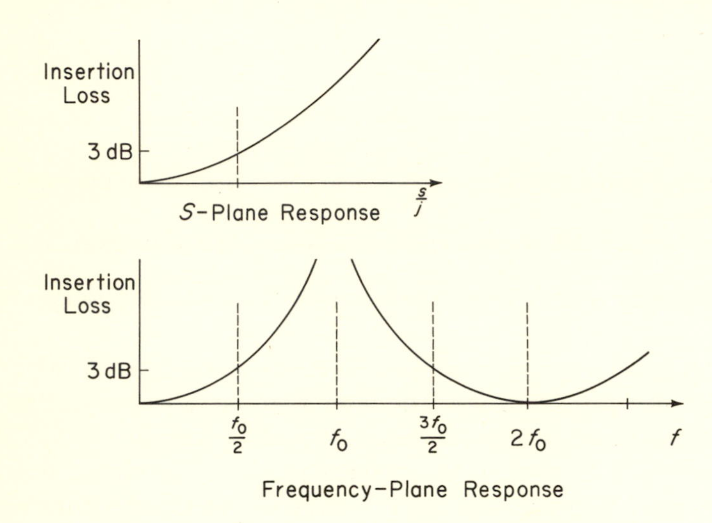

Fig. 6–24. ***S*-plane vs. frequency relationships.**

Note that the response is an even function with period $2f_0$, and the S-plane response from $S = 0$ to $S = j\infty$ is compressed into the frequency range $f = 0$ to $f = f_0$. For the frequency range $f_0 \leq f \leq 2f_0$, the input impedance of the shorted section looks capacitive and the input impedance of the open section looks inductive. Thus, the response in this frequency range is the same as an S-plane high-pass prototype.

The band edge in most prototype designs is set at $S = j$, which forces the band edge to occur at $f = f_0/2$ in the frequency domain. To shift the band edge it is merely necessary to define a new frequency variable $S' = aS$ such that $S' = j$ when $f = f_1$. That is:

$$1 = a \tan \frac{\pi}{2} \frac{f_1}{f_0}$$

or

$$a = \frac{1}{\tan \frac{\pi}{2} \frac{f_1}{f_0}},$$

Thus, we can write

$$S' = j\frac{\tan\dfrac{\pi}{2}\dfrac{f}{f_0}}{\tan\dfrac{\pi}{2}\dfrac{f_1}{f_0}} \tag{6-26}$$

The coefficient a is usually incorporated into the values of the S-plane capacitors and inductors, as indicated in Fig. 6-25.

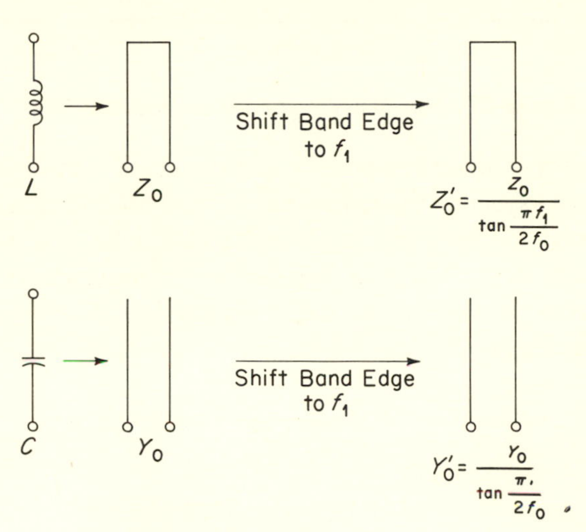

Fig. 6-25. Frequency scaling.

For example, shifting the band edge of the three-element prototype from $0.5f_0$ to $0.9f_0$ results in the transmission-line circuit shown in Fig. 6-26.

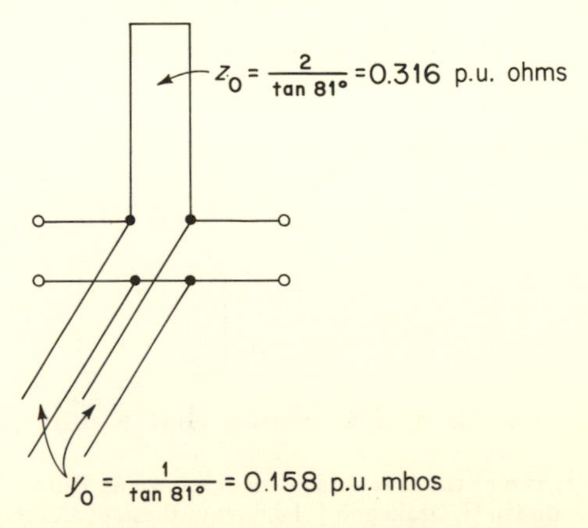

Fig. 6-26. Modified transmission-line filter.

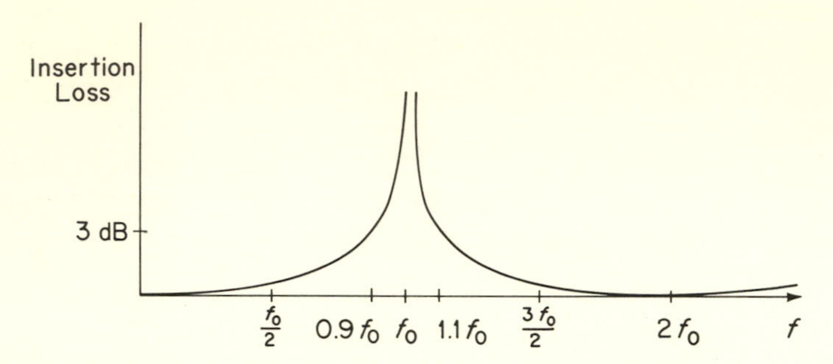

Fig. 6–27. Frequency response of modified filter.

The frequency-domain response of this structure is shown in Fig. 6–27. An unwieldy aspect of the realization of this filter design is the series-connected stub. In most practical transmission lines it is difficult to connect stubs in series. This problem can be circumvented by employing the equivalences known as Kuroda's identities. These identities are summarized in Fig. 6–28.[3]

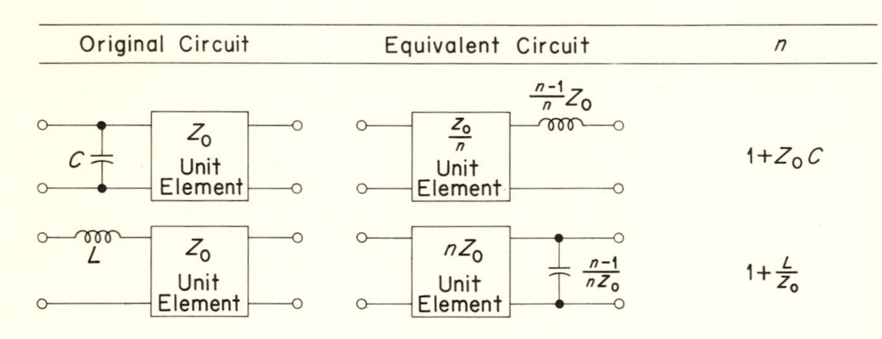

Fig. 6–28. Kuroda's identities.

The circuit element designated by the symbol [Z_0 unit element] is called a *unit element* and is defined by the *ABCD* matrix:

$$\begin{bmatrix} A & B \\ C & D \end{bmatrix} = \frac{1}{\sqrt{1 - S^2}} \begin{bmatrix} 1 & Z_0 S \\ \dfrac{S}{Z_0} & 1 \end{bmatrix}$$

Substitution of $S = j \tan \pi f/2 f_0$ shows that a unit element is just a

[3] Figure 6–28 depicts two of the four equivalences known as Kuroda's identities. A more complete table can be found in H. Ozaki and J. Ishii, "Synthesis of a Class of Stripline Filters," *IRE Trans. on Circuit Theory*, vol. CT-5, pp. 104–109, June 1958.

transmission line of characteristic impedance Z_0 which is one quarter-wavelength long at f_0.

In applying Kuroda's identities to a filter design such as that shown in Fig. 6–23, one must recognize that the insertion loss of an S-plane prototype filter as shown in Fig. 6–29 will be unaffected by the addition

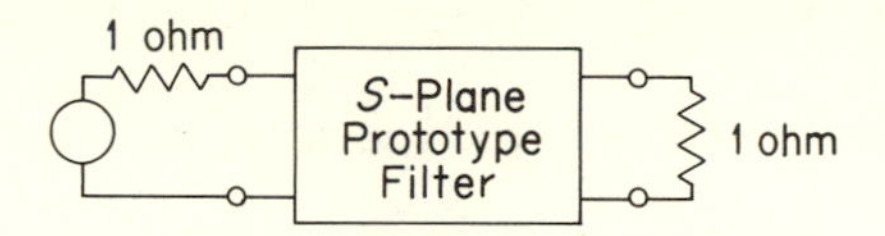

Fig. 6–29. **Matched filter.**

of an arbitrary number of sections of transmission line of normalized characteristic impedance equal to unity inserted between the load and the filter and/or between the source and the filter. Of course, the phase of the transfer function would be affected, but the insertion loss would not be affected. To see how the foregoing can be applied to the realization of the S-plane prototype, consider the steps shown in Fig. 6–30(a) and Fig.

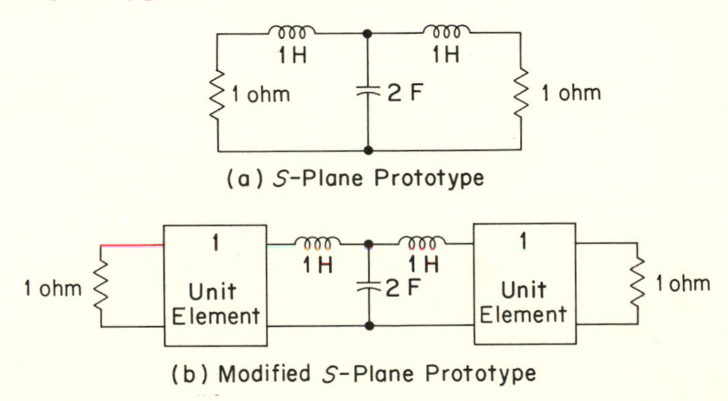

Fig. 6–30. **Insertion of unit elements in an S-plane prototype filter.**

6–30(b), where a unit element of $Z_0 = 1$ is added at each end of the prototype.

The end sections can be changed by means of the Kuroda identity shown in Fig. 6–31.

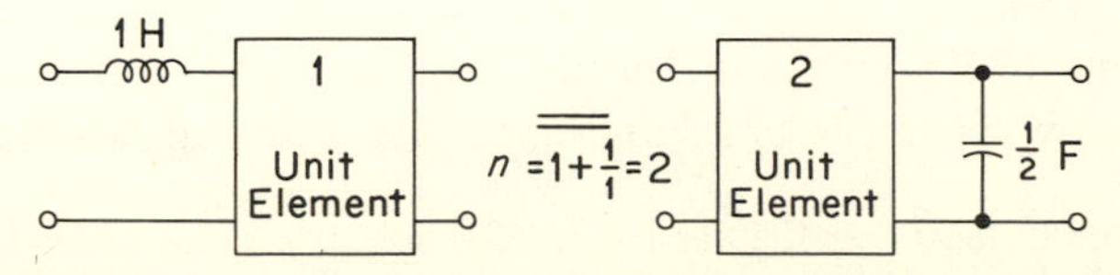

Fig. 6–31. **Kuroda's identity, replacing series L with shunt C.**

Applying this identity to the end sections results in an S-plane prototype as shown in Fig. 6–32.

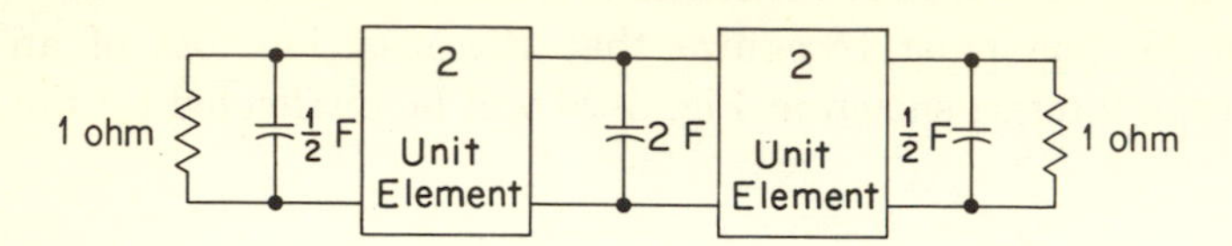

Fig. 6–32. Modified S-plane prototype after the application of Kuroda's identity.

This can be realized in transmission line with open-circuited stubs separated by $\lambda_0/4$ lengths of transmission line, as shown in Fig. 6–33.

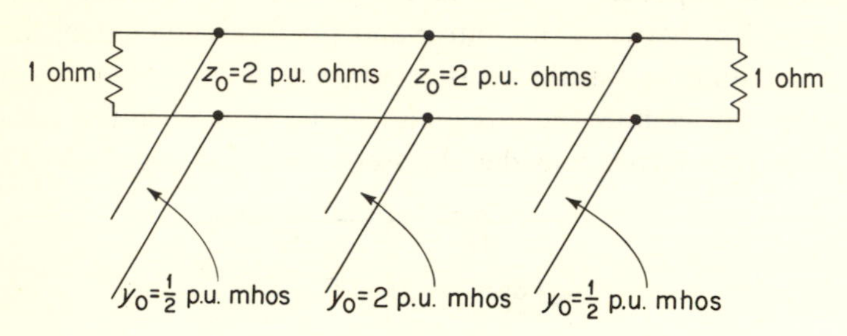

Fig. 6–33. Transmission-line circuit.

All line lengths are one quarter-wavelength long at f_0. Notice that the series-connected stub has been eliminated. This filter design could be easily realized in coaxial line, balanced strip line, microstrip, or twin-lead transmission line.

Many filters designed for use at microwave frequencies are constructed using sections of coupled-strip transmission line. The basic elements for such filters are shown in the equivalences of Fig. 6–34.[4]

The first equivalence is used in the realization of a low-pass filter, while the other two are frequently used in realizing band-pass filters using coupled lines. Sketches depicting typical filters, their S-plane equivalence, and frequency-domain response are shown in Fig. 6–35.

The behavior of these S-plane circuits is readily deduced when one realizes that an S-plane inductor represents a parallel-resonant circuit at the frequency where the coupling region is $\lambda/4$ in length, while an S-plane

[4] See E. M. T. Jones and J. T. Bolljahn, "Coupled-Strip-Transmission Line Filters and Directional Couplers," *IRE Transactions on Microwave Theory and Techniques*, vol. 19, pp. 75–81, April 1956.

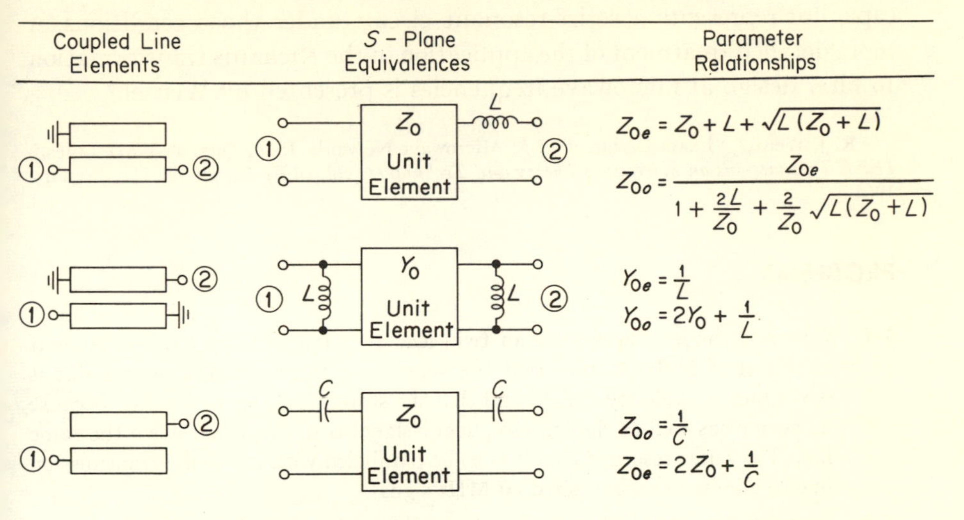

Fig. 6–34. Coupled transmission lines and their *S*-plane equivalences.

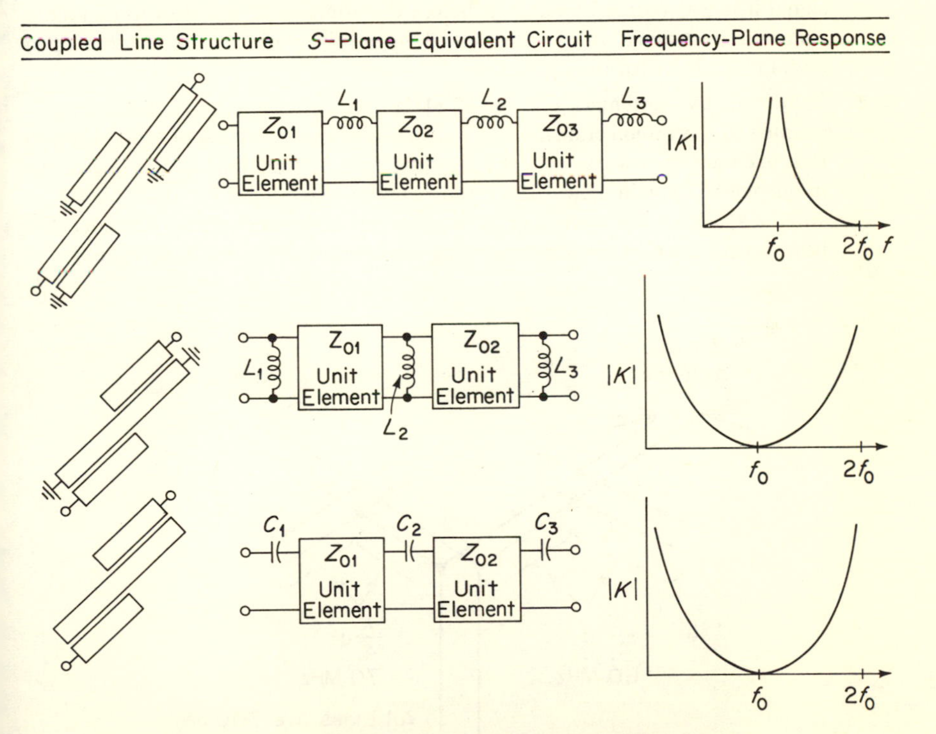

Fig. 6–35. Coupled-transmission-line relationships.

capacitor represents a series-resonant circuit under those conditions. A more detailed treatment of the application of the Richards transformation to filter design at microwave frequencies is presented by Wenzel.[5]

[5] R. J. Wenzel, "Exact Design of TEM Microwave Networks Using Quarter-Wave Lines," *IEEE Transactions on Microwave Theory and Techniques*, vol. MTT-12, pp. 88–111, January 1964.

PROBLEMS

6–1. Compute the length of 300-ohm twin lead required to form a series-resonant circuit at 60 MHz. Assume that the velocity of propagation on such a line is 0.82 times the velocity of light and that the stub is to be open-circuited because dc current as well as the high-frequency signal is to be passed down the same line. The series-resonant circuit is to be paralleled with the signal transmission line to eliminate an undesired 60-MHz signal.

6–2. Suppose the desired signal in Problem 6–1 is at 70 MHz. Explain how another stub can be used so that the system will be correctly matched for the desired signal if it was correctly matched before the addition of the stub to eliminate the undesired signal at 60 MHz. Consider only the addition of another open stub in parallel with the signal line.

6–3. Show how two antennas, one for 60 MHz and the other for 70 MHz, can be coupled to a common transmission line of 300 ohms by means of tuned lines so that each antenna works into the required load of 300 ohms at the respective frequencies. Explain how the system can be extended to a multiplicity of antennas, with each operating at a different frequency. *Hint:* Consider the basic form shown in the accompanying illustration. Also, see Problem 6–4.

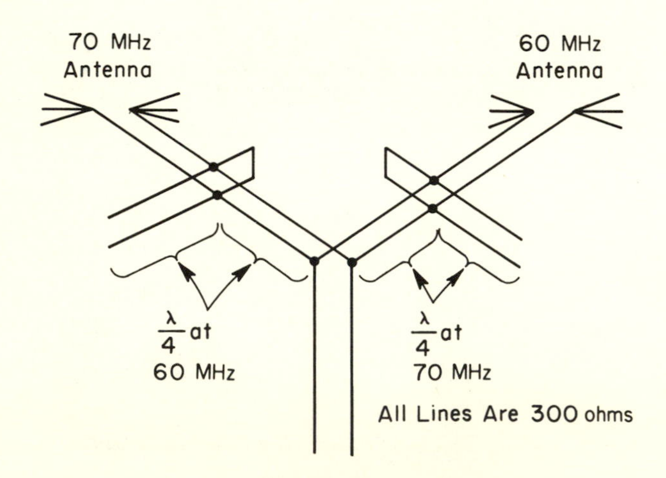

Prob. 6–3.

6–4. Show that the input impedance of a quarter-wavelength shorted stub is infinite even though the impedance is measured across the line at any point between the open and shorted ends but exclusive of the shorted end. Repeat for a three-quarter wavelength shorted stub.

6–5. Suppose that in the construction of a parallel-tuned circuit from a quarter wavelength of transmission line the line is deliberately made only 0.120λ at the desired center frequency of 100 MHz. If the characteristic impedance of the line is 300 ohms, determine the type and size of reactance (in picofarads for a capacitor and in microhenries for an inductor) needed to tune the line to parallel resonance.

6–6. Find the type and size of the reactance necessary to tune the system of Problem 6–5 if it is connected at the shorted end of the line (in series with the short). If two solutions are possible, give both of them.

6–7. A coaxial transmission line used at 200 MHz might have an attenuation of 1 dB/100 ft. Assume the line to be insulated with polyethylene, with $\epsilon_r = 2.26$. Find the Q of 0.25 m of this line used as a quarter-wavelength resonant circuit.

6–8. Show that the Q of a three-quarter wavelength circuit resonant at 200 MHz, using the same line as in Problem 6–7, will also have exactly the same Q. Explain this result qualitatively.

6–9. Suppose that ten sections of the tee network shown in the accompanying illustration are cascaded and terminated in characteristic resistance. Determine the time delay of such a lumped-constant delay line and express the result in microseconds.

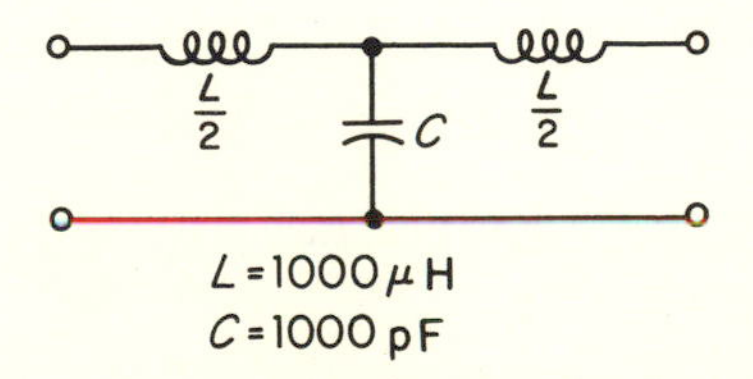

Prob. 6–9.

The frequency response of one section of such a network terminated in its characteristic resistance is known to be given approximately by the expression

$$G(j\omega) = \frac{1}{[1 - \omega^2(LC/2)] + j\omega\sqrt{LC}}$$

6–10. The angular cutoff frequency of the low-pass tee circuit used as a delay circuit in Problem 6–9 is known to be given by

$$\omega_0 = \frac{2}{\sqrt{LC}}$$

Show that the time delay per section of the network may then be expressed as $2/\omega_0$ or $1/\pi f_0$ seconds. The cutoff frequency of the tee network of Problem 6–9 is $1/\pi$ MHz, which equals 318,000 Hz, and hence the delay per section is 1 μs, as predicted before.

6–11. Check by conventional circuit theory to see that the pad given in Fig. 6–20 has a characteristic impedance of 50 ohms and produces a loss in a matched system of 20 dB.

6–12. Determine the values of a pi section for a 300-ohm system required to produce a loss of 10 dB in a matched circuit.

6–13. Given a 60-hertz power line with constants of $Z = 0.260 + j0.920$ ohm/mile and $Y = 0 + j5 \times 10^{-6}$ mho/mile. These values correspond to a 250,000-circular-mil power line with a spacing of 35 ft. Find Z_0 and γ. Determine the tee section for a 300-mile length. Find the velocity of propagation, as a percentage of that of light. Determine the load voltage at the end of the 300-mile section if the input voltage is 100 kV and the input power flow equals $50 + j50$-MVA. Also determine the load power and thus the line loss.

6–14. Explain how you would construct a circuit that would produce an output when the input pulses were 1 μs wide but would not produce an output for pulses 2 μs wide.

6–15. By general network theory the characteristic impedance of a network is related to the open- and short-circuit input impedances by the formula

$$Z_0^2 = Z_{i(\mathrm{OC})} Z_{i(\mathrm{SC})}$$

Check the tee and pi equivalent circuits to see that the characteristic impedance is R_0.

6–16. Find the tee and pi equivalent circuits for a half-wavelength of transmission line. Draw a transformer equivalent of the tee network.

6–17. Find the tee and pi equivalent circuits for a full wavelength of transmission line.

6–18. Determine the βd for which the approximation of Eq. (6–25) is in error by 10 per cent.

6–19. Prove either of Kuroda's identities shown in Fig. 6–28. *Hint:* Determine the *ABCD* matrix representations of the equivalences.

6–20. Redesign the filter of Fig. 6–33 so that the band edge occurs at $f = 0.25f_0$ instead of at $f = 0.5f_0$.

7

The Field Equations

7–1. INTRODUCTION

The problem of transmitting energy from one point to another by means of a pair of wires was discussed in some detail in the preceding chapters. This is an important problem in its own right but is by no means the complete story with regard to energy transmission. As mentioned in Chapter 1, there are many other ways of propagating energy. The purpose of the subsequent chapters is to discuss some of these other modes of energy transmission. A thorough understanding of transmission-line phenomena is essential before proceeding, however, because many of the basic concepts such as reflection coefficient, characteristic impedance, and propagation constant carry over directly to the more sophisticated forms of energy transmission.

As a starting point for generalization, let us look at the classical transmission-line problem from a field rather than a circuit viewpoint. From elementary field theory it is known that there will be electric and magnetic fields associated with the line voltage and current, as shown in Fig. 7–1. Just as the voltage and current appear to travel down the line in the direction of propagation, so do the associated fields, much as a shadow travels along with its object. Furthermore, since it is strictly a matter of viewpoint, we may think of the fields as inducing the currents in the wires, rather than vice versa. Thus, from this viewpoint, we see electric and magnetic fields which appear to travel together at the same

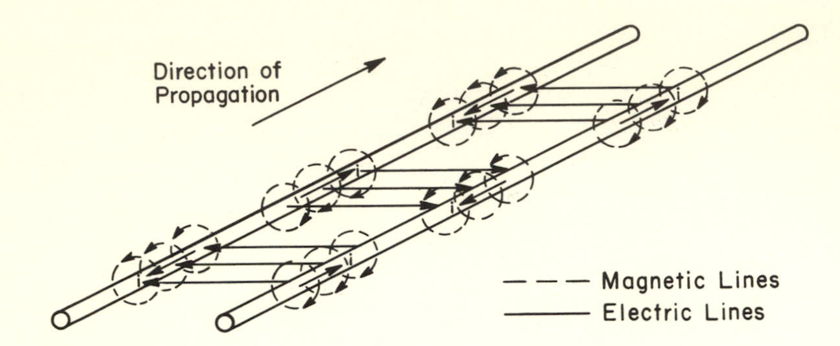

Fig. 7–1. Traveling waves on open-wire transmission line.

velocity in space, with the wires merely serving to "guide" the wave in a particular direction.

We may go one step further now and consider the energy as residing in the field and energy propagation as taking place in space surrounding the conductors rather than within the wires themselves. Again this is just a matter of viewpoint, but the field viewpoint is enlightening in that other possible modes of energy transmission are immediately suggested. For example, if an electromagnetic wave can be guided in a particular direction with a pair of wires, can this be done with a single wire? Or with parallel planes? Or with a hollow pipe? The answer to these questions is an emphatic "yes." Waves can be guided by most any sort of metallic structure and even by dielectric boundaries under certain conditions. The field configurations involved differ for different conductor geometries, but all are electromagnetic waves nevertheless.

This short introduction has been qualitative so far, and as such it cannot be satisfying to the engineer, who must have a quantitative picture as well as a qualitative one.

The quantitative answers to these questions must come from field theory, since they are conceptually outside the realm of circuit theory. The remainder of this chapter will therefore be devoted to the basic principles of electromagnetic field theory. Subsequent chapters will deal with some of the simpler and more important modes of wave propagation.

7–2. MAXWELL'S EQUATIONS

The fundamental differential equations relating time-varying electric and magnetic fields are known as Maxwell's equations. They are appropriately named for the British physicist James Clerk Maxwell (1831–79), who was to a large extent responsible for the mathematical formulation of electromagnetic field theory as it is known today. Strictly speaking, these equations are postulates and cannot be derived from any laws that are more basic. However, one can show that Maxwell's equations are simply extensions of Faraday's and Ampere's laws, and while this tie to more elementary concepts is not rigorous in the strict sense of the word, it is helpful in gaining an understanding of Maxwell's equations.

Maxwell's First Equation. Maxwell's first equation evolves from a generalization of Faraday's law—from the realm of circuit theory to field theory. Faraday's law states that the voltage induced in a circuit is equal to the time rate of change of the magnetic flux linkage in the circuit; or put mathematically,

$$e = -\frac{d\phi}{dt} \tag{7–1}$$

where ϕ is the magnetic flux linking the circuit. When this concept is extended to a three-dimensional space situation, the resultant equation is

$$\oint_C \boldsymbol{\mathcal{E}} \cdot \mathbf{dl} = -\frac{d}{dt}\iint_S \boldsymbol{\mathcal{B}} \cdot \mathbf{da} \tag{7–2}$$

where C is an arbitrary closed curve in space and S is any surface bounded by C. The term on the left is simply a mathematical way of expressing the "total voltage" around curve C, and the one on the right is the time derivative of the flux enclosed by C. It is at this point that the "derivation" of Maxwell's first equation from Faraday's law is not rigorous, since it must be assumed that this generalization is valid. Of course, this leads to results that may be verified experimentally, and since this is the ultimate goal of any theory, it may be assumed that the generalization is valid. It should be noted that Faraday's law, as applied to circuits, is a special case of Eq. (7–2), where the path of integration C is restricted along a closed conducting path of the circuit.

The connection between circuit and field theories is often confusing to the beginning student. An example of this point will help clarify the situation.

Example 7–1

Consider the circuit shown in the accompanying figure to be laid out physically in a plane with dimensions as shown. Then consider the circuit subjected to a uniform

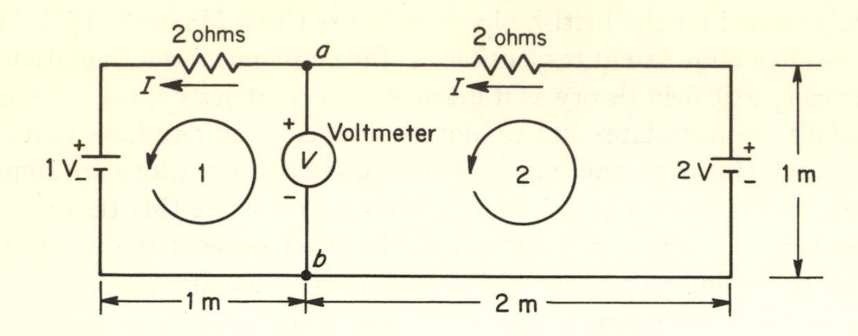

Example 7–1.

magnetic field directed out of the paper and varying linearly with time according to

$$\mathcal{B} = t \text{ weber/m}^2$$

The wires connecting the various elements will be assumed to be perfect conductors and the elements to be of negligible size physically. The voltmeter, labeled V in the diagram, is also of negligible size and has relatively high internal resistance. Three questions will be posed in connection with this circuit: (1) What are the currents? (2) What does the voltmeter read? (3) What is the voltage from a to b?

The answers to questions 1 and 2 may be found by applying Faraday's law (Eq. 7–2) to the closed loops labeled 1 and 2 and following rectangular paths within the conductors. Since the current in the voltmeter branch is negligible, the currents in the resistors will be the same. They will be denoted by I and will be positive in the direction of integration. Then, remembering that $\mathcal{E}$ is zero within a perfect conductor and that $\int \mathcal{E} \cdot \mathbf{dl}$ through an element is simply the voltage drop across the element, the following equations result:

$$2I + 1 - V = -\frac{d}{dt}(t)$$

$$2I + V - 2 = -\frac{d}{dt}(2t)$$

Solving these equations for V and I yields

$$V = 1 \text{ V}$$

$$I = -\frac{1}{2} \text{ A}$$

Now one might immediately jump to the conclusion that the answer to (3) is 1 V, since the voltmeter reads 1 V. However, in checking, it will be noticed that summing voltages around the 1-V battery branch gives zero for an answer; doing likewise for the 2-V battery branch yields 3 V. Which is correct? The answer is that all three are correct, in a sense, because the line integral of $\mathcal{E} \cdot \mathbf{dl}$ is dependent on the path of integration in time-varying magnetic field situations. Thus the term "voltage between two points" is ambiguous and has no meaning unless the path is specified, or at least implied. It may seem that this is in direct contradiction with ac circuit theory, where we speak of the voltage across an inductance with no qualms whatsoever. In this case, though, the path of integration is implied to be in the dielectric medium between the two terminals, rather than within the conductor, as this would lead to zero for the voltage between the two terminals. Basically it is just a matter of viewpoint as to whether the voltage-inducing effect of the magnetic field is taken into account on the left side of the equation as a voltage drop or on the right side as a driving function. Both viewpoints give the correct answer when properly applied.

Equation (7-2) is an integral equation relating $\mathcal{E}$ and $\mathcal{B}$, and while it is correct enough, it is often more convenient to have this relation in the form of a differential equation instead of an integral equation. It may be converted to differential form by moving the time-derivative sign inside the integral sign and then converting the integral on the left to a surface integral by means of Stokes' theorem. This leads to the equation

$$\iint_S (\nabla \times \mathcal{E}) \cdot \mathbf{da} = \iint_S \left[-\frac{\partial \mathcal{B}}{\partial t}\right] \cdot \mathbf{da} \tag{7-3}$$

It can be argued now that if this is to be true for all S, it must be true for an arbitrarily oriented, infinitesimal surface element. This can be true only if

$$\nabla \times \mathcal{E} = -\frac{\partial \mathcal{B}}{\partial t} \tag{7-4}$$

This is known as Maxwell's first equation. It relates the space rate of change of the electric field to the time rate of change of the magnetic flux density. It is a vector equation, and thus three scalar equations may be obtained from it by equating the respective components on both sides of the equation. Writing both sides in terms of rectangular components and equating the respective x, y, and z components lead to[1]

$$\begin{aligned}
\frac{\partial \mathcal{E}_z}{\partial y} - \frac{\partial \mathcal{E}_y}{\partial z} &= -\frac{\partial \mathcal{B}_x}{\partial t} \\
\frac{\partial \mathcal{E}_x}{\partial z} - \frac{\partial \mathcal{E}_z}{\partial x} &= -\frac{\partial \mathcal{B}_y}{\partial t} \\
\frac{\partial \mathcal{E}_y}{\partial x} - \frac{\partial \mathcal{E}_x}{\partial y} &= -\frac{\partial \mathcal{B}_z}{\partial t}
\end{aligned} \tag{7-5}$$

At this point one can see the beauty of vector notation. It enables one to write three scalar equations, Eqs. (7-5), in the form of one concise vector equation.

Maxwell's Second Equation. Maxwell's second equation evolves from Ampere's law, just as the first equation evolved from Faraday's law.

[1] Maxwell's first equation in the form of Eqs. (7-5) may also be obtained by direct application of Eq. (7-2) to differential rectangles in each of the xy, yz, and zx planes. See Problem 7-1 at the end of this chapter.

Ampere's law states that the scalar magnetic potential around a closed curve C is equal to the current enclosed; or, mathematically,

$$\oint_C \mathcal{H} \cdot \mathbf{dl} = \iint_S \mathcal{J} \cdot \mathbf{da} \tag{7-6}$$

where $\mathcal{J}$ indicates current density. Now, when this basic law of statics is extended to the time-varying situation, the concept of current must be modified to include displacement current as well as conduction current. Visualization of displacement current is somewhat elusive, as witnessed by the fact that this concept eluded investigators prior to Maxwell. A crude way of looking at it is this: A changing electric field produces a magnetic field in the same sort of way as does a flow of electric charges. Thus a time-varying electric flux constitutes displacement current. This is illustrated in Fig. 7–2, which shows the flow of both conduction and

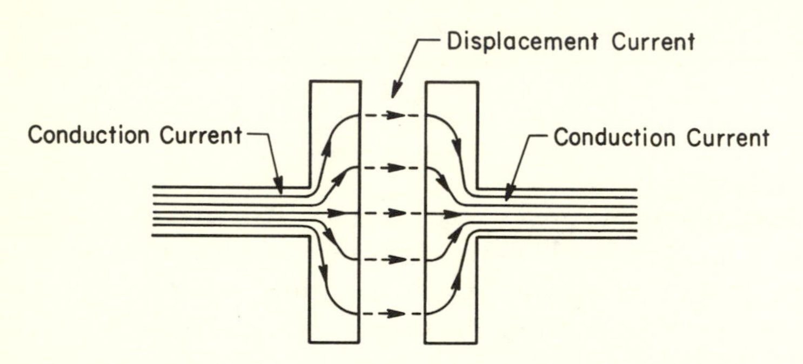

Fig. 7–2. Displacement current in parallel-plate capacitor.

displacement currents in a parallel-plate capacitor. The dimensions of the conducting regions are purposely exaggerated in order to show the conduction-current flow lines within the conductors. Solid lines indicate conduction current, and dotted lines indicate displacement current. There is a one-to-one correspondence between them because the charge must equal the electric flux in the rationalized MKS system of units.

The concept of displacement current is an important one, and the student should make every effort to understand its significance. Prior to Maxwell's work, electric and magnetic theory consisted of detached theories of electrostatics, magnetostatics, magnetic induction, and electrokinetics, all of which were only loosely connected at best. The missing link in all this theory was displacement current, and Maxwell provided this along with the unified electromagnetic theory we know today. The theory as presented here differs in notation from Maxwell's and has been stripped of some of its less essential features, as far as electromagnetic waves are concerned, but it is nevertheless Maxwell's theory.

Maxwell deduced that current must be continuous, in the same sense that magnetic flux is continuous, if electromagnetic theory is to be consistent. Thus he postulated in his theory that true current was the sum of conduction and displacement currents. This, then, gave meaning to what is now known as Ampere's law when applied in a time-varying situation. After all, in a circuit containing capacitance, what meaning could one give to "current enclosed" if current were not continuous? It would depend on the particular surface area chosen to be bounded by the curve along which $\mathcal{H} \cdot \mathbf{dl}$ is to be integrated. Where one area could be chosen to go between the capacitor plates, another could be so chosen as to be pierced by the current—thus an irreconcilable inconsistency.

Returning now to Eq. (7–6), $\mathcal{J}$ must include both conduction and displacement-current densities. Stated mathematically,

$$\mathcal{J} = \mathcal{J}^c + \mathcal{J}^d = \mathcal{J}^c + \frac{\partial \mathfrak{D}}{\partial t} \tag{7–7}$$

The first term of this equation represents conduction-current density. The second term represents displacement-current density. This can be seen to be consistent with the picture of displacement current given in Fig. 7–2, if it is remembered that electric flux density is $\epsilon\mathcal{E}$ in a simple dielectric.

Example 7–2

The correctness of the displacement-current-density term $\epsilon(\partial\mathcal{E}/\partial t)$ can be easily checked for the case of a parallel-plate capacitor. Consider such a capacitor with area A and a separation distance d. The capacitance is then

$$C = \frac{\epsilon A}{d}$$

If an ac voltage of the form $V_m \cos \omega t$ is applied to the capacitor, the input current will be

$$\mathcal{I}_{\text{input}} = -\omega C V_m \sin \omega t = -\frac{\omega V_m \epsilon A}{d} \sin \omega t$$

The total displacement current in the region between the two plates can be computed as

$$\mathcal{I}_{\text{displacement}} = \left(\epsilon \frac{\partial \mathcal{E}}{dt}\right) \times (\text{area})$$

$$= \epsilon A \frac{\partial}{\partial t}\left(\frac{V_m \cos \omega t}{d}\right) = -\frac{\omega V_m \epsilon A}{d} \sin \omega t$$

Thus it can be seen that when one considers both conduction and displacement current as making up the total current picture, the current lines are continuous in a simple capacitive circuit.

When Eq. (7–7) is substituted into Eq. (7–6), Ampere's law becomes

$$\oint_C \mathcal{H} \cdot d\mathbf{l} = \iint_S \left(\mathcal{J}^c + \frac{\partial \mathcal{D}}{\partial t} \right) \cdot d\mathbf{a} \tag{7–8}$$

Note the similarity between this and Faraday's law. Again we have a line integral on the left side and a surface integral on the right side; and as before this integral equation may be converted to differential form by applying Stokes' theorem. The result is

$$\nabla \times \mathcal{H} = \mathcal{J}^c + \frac{\partial \mathcal{D}}{\partial t} \tag{7–9}$$

which is known as Maxwell's second equation. Since it is a vector equation, three scalar equations relating the rectangular components may be derived from it. They are

$$\begin{aligned} \frac{\partial \mathcal{H}_z}{\partial y} - \frac{\partial \mathcal{H}_y}{\partial z} &= \mathcal{J}^c_x + \frac{\partial \mathcal{D}_x}{\partial t} \\ \frac{\partial \mathcal{H}_x}{\partial z} - \frac{\partial \mathcal{H}_z}{\partial x} &= \mathcal{J}^c_y + \frac{\partial \mathcal{D}_y}{\partial t} \\ \frac{\partial \mathcal{H}_y}{\partial x} - \frac{\partial \mathcal{H}_x}{\partial y} &= \mathcal{J}^c_z + \frac{\partial \mathcal{D}_z}{\partial t} \end{aligned} \tag{7–10}$$

Again note the beauty and conciseness of the vector notation.

Continuity Equations. In addition to Maxwell's first and second equations, which relate space and time rates of change of the electric and magnetic fields, two other equations relate to the continuity of the respective fields.[2] The concept of continuous magnetic lines of flux from magnetostatics carries over to the time-varying case; i.e.,

$$\oint\!\!\oint_S \mathcal{B} \cdot d\mathbf{a} = 0 \tag{7–11}$$

This is merely a mathematical way of stating that magnetic lines of flux are continuous and always close on themselves. Or, the net magnetic flux leaving a closed volume is zero. Electric lines of flux, however, may or may not close on themselves, depending on the situation. It will be remembered from static field theory that if charges are present, electric lines of flux originate with positive charges and terminate on negative ones. Thus the lines can be discontinuous if charges are present. In other

[2] Strictly speaking, the two continuity equations given in this article are redundant in that they can be derived from Maxwell's first and second equations and the principle of conservation of charge. See J. A. Stratton, *Electromagnetic Theory* (New York: McGraw-Hill Book Co., Inc., 1941), p. 6.

words, the net electric flux emanating from a closed volume is equal to the positive charge enclosed. The formal mathematical statement of this is

$$\oiint_S \mathfrak{D} \cdot \mathbf{da} = \iiint_V \rho \, dv \tag{7-12}$$

where $\mathfrak{D}$ is electric flux density and ρ is the volume charge density.

The differential forms of Eqs. (7-11) and (7-12) can be found by application of Gauss' theorem to the surface integrals of Eqs. (7-11) and (7-12). The results are

$$\nabla \cdot \mathfrak{B} = 0 \tag{7-13}$$

and

$$\nabla \cdot \mathfrak{D} = \rho \tag{7-14}$$

Thus, we arrive at a set of four equations which we can use as a basis for explaining electromagnetic phenomena. The dependent variables are the electric and magnetic field intensities and flux densities ($\mathcal{E}$, $\mathfrak{D}$, $\mathcal{H}$, $\mathfrak{B}$). The sources or independent variables are impressed current density $\mathcal{J}^i$ and impressed charge density ρ^i. Maxwell's equations are enumerated below in differential form:

$$\begin{aligned}
&\text{(a)} \qquad \nabla \times \mathcal{E} = -\frac{\partial \mathfrak{B}}{\partial t} \\
&\text{(b)} \qquad \nabla \times \mathcal{H} = \mathcal{J}^c + \frac{\partial \mathfrak{D}}{\partial t} + \mathcal{J}^i \\
&\text{(c)} \qquad \nabla \cdot \mathfrak{B} = 0 \\
&\text{(d)} \qquad \nabla \cdot \mathfrak{D} = \rho + \rho^i
\end{aligned} \tag{7-15}$$

Note that the impressed current density $\mathcal{J}^i$ and the impressed charge density ρ^i have been separated out from the $\mathcal{J}^c$ and ρ of Eqs. (7-9) and (7-14). These source terms have been included in order to present a complete picture of Maxwell's equations. We will not actually consider situations where impressed sources are present until Chapter 12. The number of dependent variables can be reduced by means of the constitutive parameters which relate the flux densities and the field intensities.

7-3. THE CONSTITUTIVE EQUATIONS

The constitutive equations characterize the media in which the electric and magnetic fields are established. For example, one can relate electric flux density $\mathfrak{D}$ to the electric field intensity $\mathcal{E}$ by a constitutive equation:

$$\mathfrak{D} = \epsilon \mathcal{E} \tag{7-16}$$

In a similar manner we can write

$$\mathcal{B} = \mu \mathbf{H} \tag{7-17}$$

and

$$\mathcal{J}^c = \sigma \mathcal{E} \tag{7-18}$$

The terms ϵ, μ, and σ are the constitutive parameters, called *permittivity*, *permeability*, and *conductivity*, respectively. The properties of the constitutive parameters are described by use of the following terms:

1. *Time dependent or time independent*
2. *Linear or nonlinear*
3. *Isotropic or anisotropic*
4. *Homogeneous or inhomogeneous*

All material media are characterized by some combination of descriptive terms from this list. For example, simple loss-free dielectrics are characterized by a permittivity (ϵ) which is time-independent, independent of electric field intensity, isotropic, and independent of space coordinates. The meanings of the descriptive terms are given below.

Time Dependence. We seldom encounter a situation where a material parameter is varying with time at a rate high enough to warrant considering the time dependence. However, one can visualize situations where such behavior could occur. In these cases, the product of two sinusoidal signals, for example, would produce new frequency components.

Linearity. A medium is termed linear or nonlinear depending upon the applicability of the principle of superposition to the fields within the medium. For example, in an iron-core transformer, operating below the saturation region of the *B-H* curve, increasing the magnetic field intensity does not produce a proportionate increase in the magnetic flux density. Consequently, in this region, the principle of superposition does not apply, the medium is nonlinear, and the permeability is a function of the magnetic field intensity.

Anisotropy. A medium is termed isotropic or anisotropic depending on whether or not the relationship between the flux density and the field intensity is dependent upon the orientation of the field in the medium. For example, the permeability of a ferrite medium subjected to a z-directed magnetic bias field[3] has the general matrix form

[3] See Appendix G for details on the permeability tensor of a ferrite medium.

$$\mu = \begin{bmatrix} \mu_{xx} & \mu_{xy} & 0 \\ \mu_{yx} & \mu_{yy} & 0 \\ 0 & 0 & \mu_0 \end{bmatrix}$$

where μ_{xx}, μ_{xy}, μ_{yx}, and μ_{yy} are scalars and μ_0 is the permeability of free space. If the applied magnetic field intensity is y-directed, that is, $\mathcal{H}_{app} = \mathbf{a}_y\mathcal{H}_y$, the magnetic flux density is related to the magnetic field intensity by the matrix equation

$$\begin{bmatrix} \mathcal{B}_x \\ \mathcal{B}_y \\ \mathcal{B}_z \end{bmatrix} = \begin{bmatrix} \mu_{xx} & \mu_{xy} & 0 \\ \mu_{yx} & \mu_{yy} & 0 \\ 0 & 0 & \mu_0 \end{bmatrix} \begin{bmatrix} 0 \\ \mathcal{H}_y \\ 0 \end{bmatrix}$$

The magnetic flux density vector is

$$\mathcal{B} = \mathbf{a}_x u_{xy}\mathcal{H}_y + \mathbf{a}_y u_{yy}\mathcal{H}_y$$

However, if a z-directed magnetic field intensity vector were applied, one could easily show that

$$\mathcal{B} = \mathbf{a}_z u_0 \mathcal{H}_z$$

Thus, we see that the magnetic flux density bears a different relationship to the magnetic field intensity for a z-directed magnetic field than for a y-directed magnetic field. The ferrite medium is said to be anisotropic.

Homogeneity. This term reflects the degree to which the spatial variation of the flux density or current density differs from the spatial variation of the associated field intensity. For example, the permittivity of the earth's atmosphere is a function of the pressure, temperature, and water vapor content of the atmosphere. One finds these quantities in varying amounts as one moves around in the atmosphere. As a result, the permittivity of the earth's atmosphere is a function of the spatial coordinates and the medium is said to be inhomogeneous.

Using the constitutive parameters, one can express Maxwell's equations in terms of the field intensities as shown below.

(a) $$\nabla \times \mathcal{E} = -\frac{\partial}{\partial t}(\mu\mathcal{H})$$

(b) $$\nabla \times \mathcal{H} = \sigma\mathcal{E} + \frac{\partial}{\partial t}(\epsilon\mathcal{E}) + \mathcal{J}^i \qquad (7\text{-}19)$$

(c) $$\nabla \cdot \mu\mathcal{H} = 0$$

(d) $$\nabla \cdot \epsilon\mathcal{E} = \rho + \rho^i$$

7-4. BOUNDARY CONDITIONS

Maxwell's two equations and the two continuity equations constitute the basic equations which describe all electromagnetic wave phenomena.

Since the equations are in the form of differential equations rather than algebraic equations, boundary conditions must be applied if a specific solution for a given problem is to be obtained. The boundary conditions will be discussed in general at this point and more specifically later on when particular geometric situations are considered.

Consider the electric field first. Two basic rules apply at the surface between two different materials:

1. The tangential component of the electric field intensity $\mathcal{E}$ must be continuous at the boundary.
2. The normal component of electric flux density $\mathfrak{D}$ will be discontinuous at the boundary by an amount equal to the surface-charge density on the boundary.

The first statement is justified by applying Faraday's law to a differential rectangular loop as shown in Fig. 7–3(a), and the second statement is substantiated by applying Gauss' law to the differential volume shown in Fig. 7–3(b).

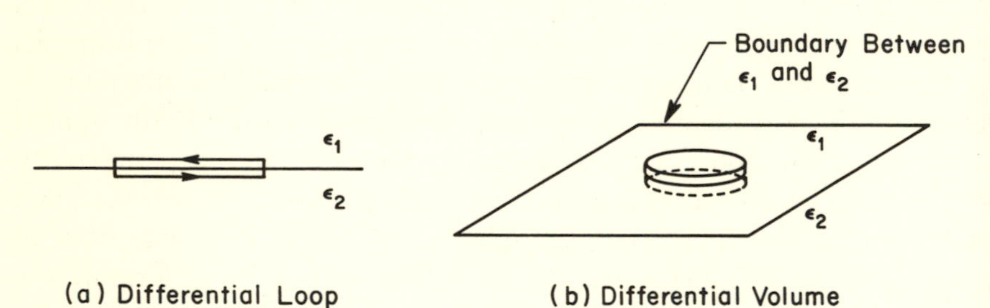

Fig. 7–3. Boundaries between two dielectric regions.

If the boundary is formed by the surface of a perfect conductor ($\sigma \rightarrow \infty$), the tangential electric field must be zero. This can be seen by considering the current density expression $\mathcal{J}^c = \sigma\mathcal{E}$. Since $\mathcal{J}^c$ must be finite even though $\sigma \rightarrow \infty$, $\mathcal{E}$ must approach zero. This problem will be considered in more detail in the next chapter, where we will find it useful to represent the current flow on the surface of a perfect conductor as a sheet of surface current which would represent an infinite current density. This is a useful model and will not mean that the tangential electric field is non-zero on a perfect conductor.

Similar rules apply to the magnetic field except that the magnetic flux lines are never discontinuous in that isolated magnetic poles do not exist. The two rules are:

1. The tangential component of magnetic field intensity $\mathcal{H}$ is continuous at the boundary (providing the current density is finite at the boundary).
2. The normal component of magnetic flux density $\mathcal{B}$ is continuous at the boundary.

These statements may be justified in a manner analogous to that used in the electric field case, using in this case Ampere's law and $\iint \mathcal{B} \cdot \mathbf{da} = 0$.

7-5. ENERGY CONSIDERATIONS

The same concept of energy storage used in static field theory carries over to the dynamic case. That is, the energy densities associated with the electric and magnetic fields, in simple media, respectively, are

$$w_e = \tfrac{1}{2}\epsilon \mathcal{E} \cdot \mathcal{E} = \frac{1}{2}\epsilon \mathcal{E}^2 \tag{7-20}$$

$$w_h = \tfrac{1}{2}\mu \mathcal{H} \cdot \mathcal{H} = \frac{1}{2}\mu \mathcal{H}^2 \tag{7-21}$$

where $\mathcal{E}$ and $\mathcal{H}$ denote the magnitudes of the respective field intensities. Thus the total energy stored in a given volume V is given by

$$W = \iiint_V (\tfrac{1}{2}\epsilon \mathcal{E}^2 + \tfrac{1}{2}\mu \mathcal{H}^2)\, dv \tag{7-22}$$

Since this chapter will be dealing with dynamic situations, it is of interest to derive a formula for the time rate of change of energy within a volume V. Differentiating both sides of Eq. (7-22) with respect to time (and noting that $\mathcal{E}(\partial \mathcal{E}/\partial t)$ is equivalent to $\mathcal{E} \cdot (\partial \mathcal{E}/\partial t)$ in vector notation) leads to

$$\frac{dW}{dt} = \iiint_V \left[\mathcal{E} \cdot \left(\epsilon \frac{\partial \mathcal{E}}{\partial t}\right) + \mathcal{H} \cdot \left(\mu \frac{\partial \mathcal{H}}{\partial t}\right)\right] dv \tag{7-23}$$

The quantities within parentheses may be replaced with their equivalents $\nabla \times \mathcal{H} - \sigma\mathcal{E} - \mathcal{J}^i$ and $-\nabla \times \mathcal{E}$ from Maxwell's equations. The result is

$$\frac{dW}{dt} = \iiint_V [\mathcal{E} \cdot \nabla \times \mathcal{H} - \mathcal{H} \cdot \nabla \times \mathcal{E} - \sigma\mathcal{E} \cdot \mathcal{E} - \mathcal{J}^i \cdot \mathcal{E}]\, dv \tag{7-24}$$

Taking advantage of the vector identity (see Appendix D),

$$\nabla \cdot (\mathbf{A} \times \mathbf{B}) = \mathbf{B} \cdot \nabla \times \mathbf{A} - \mathbf{A} \cdot \nabla \times \mathbf{B}$$

and converting the volume integral to a surface integral by the divergence theorem leads to

$$+\frac{dW}{dt} = -\iiint_V \sigma\mathcal{E} \cdot \mathcal{E}\, dv - \iiint_V \mathcal{J}^i \cdot \mathcal{E}\, dv - \oiint_S \mathcal{E} \times \mathcal{H} \cdot \mathbf{da} \tag{7-25}$$

The first two terms on the right-hand side of the equality sign can be identified as

$$-\iiint_V \mathcal{J}^i \cdot \mathcal{E}\, dv = \text{instantaneous power supplied by the source } \mathcal{J}^i$$

$$\iiint_V \sigma\mathcal{E} \cdot \mathcal{E}\, dv = \text{instantaneous power dissipated within the volume due to conduction currents}$$

It can be seen that the quantity $\mathcal{E} \times \mathcal{H}$ might be interpreted as power density, since integration of this quantity over the surface S gives the total power (rate of energy flow) flowing out of the closed surface. This concept is quite useful, and the quantity $\mathcal{E} \times \mathcal{H}$ will be denoted as $\mathcal{P}$ and referred to as the Poynting vector; i.e.,

$$\mathcal{P} = \mathcal{E} \times \mathcal{H} \tag{7–26}$$

Note that $\mathcal{P}$ is a vector quantity and gives the direction of power density flow as well as its magnitude.

Equation (7–26) is then interpreted as an instantaneous expression of conservation of energy, as can be seen if it is rewritten as

$$-\iiint_V \mathcal{J}^i \cdot \mathcal{E}\, dv = \frac{dW}{dt} + \iiint_V \sigma\mathcal{E} \cdot \mathcal{E}\, dv + \oint\!\!\oint_S \mathcal{P} \cdot \mathbf{da} \tag{7–27}$$

time rate of energy supplied to the volume by the source = time rate of energy stored within the volume + time rate of energy dissipated within the volume + time rate of energy flow out of the volume

A word of caution is in order at this point. Care must be used when applying this concept of power density to a surface which is not closed, since Eq. (7–27) is valid only for a closed surface. As an example of one of the paradoxes resulting from improper application of Poynting's theorem, consider the situation shown in Fig. 7–4. A casual integration of

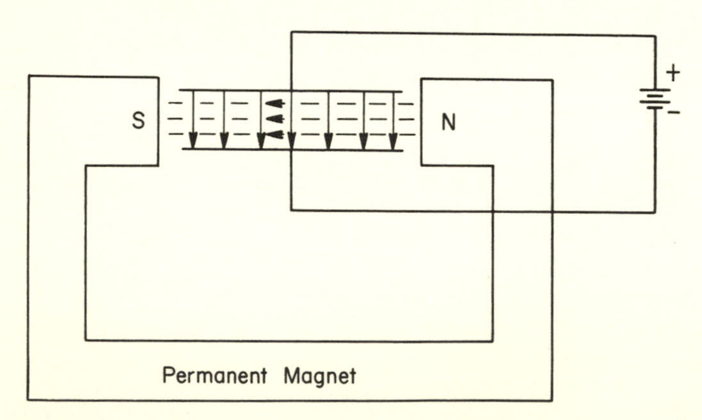

Fig. 7–4. Permanent magnet and charged parallel-plate capacitor.

$\mathcal{E} \times \mathcal{H}$ over a rectangular area in the plane of the paper and between the two capacitor plates leads to the conclusion that there is continuous flow of power into the paper. This is an obvious falsehood and results because the surface area considered is not closed.

7-6. MAXWELL'S EQUATIONS IN PHASOR FORM

Since the fields considered in this text are characterized by a sinusoidal time dependence, it is most convenient to express Maxwell's equations in phasor notation. This is easily accomplished by applying the phasor transformation discussed in Chapter 1 as it applies to time-dependent vectors. An example of this follows. Consider the electric field intensity vector

$$\mathcal{E} = \mathbf{a}_x|E_x| \cos(\omega t + \theta_x) + \mathbf{a}_y|E_y| \cos(\omega t + \theta_y) + \mathbf{a}_z|E_z| \cos(\omega t + \theta_z)$$

where $|E_x|$, $|E_y|$, and $|E_z|$ are scalars which are functions of the spatial coordinates only.

$$\begin{aligned}
\mathcal{E} &= \text{Re}\,\{\mathbf{a}_x|E_x|e^{j(\omega t+\theta_x)} + \mathbf{a}_y|E_y|e^{j(\omega t+\theta_y)} + \mathbf{a}_z|E_z|e^{j(\omega t+\theta_z)}\} \\
&= \text{Re}\,\{(\mathbf{a}_x|E_x|e^{j\theta_x} + \mathbf{a}_y|E_y|e^{j\theta_y} + \mathbf{a}_z|E_z|e^{j\theta_z})e^{j\omega t}\} \\
&= \text{Re}\,\{(\mathbf{a}_xE_x + \mathbf{a}_yE_y + \mathbf{a}_zE_z)e^{j\omega t}\} \\
&= \text{Re}\,\{\mathbf{E}e^{j\omega t}\}
\end{aligned}$$

Using this notation, the phasor $\mathbf{E}$ is a peak quantity. Applying this identity to Maxwell's first equation results in

$$\nabla \times \text{Re}\,\{\mathbf{E}e^{j\omega t}\} = -\mu \frac{\partial}{\partial t} \text{Re}\,\{\mathbf{H}e^{j\omega t}\}$$

or

$$\text{Re}\,\{(\nabla \times \mathbf{E})e^{j\omega t}\} = -\text{Re}\,\{\mu\mathbf{H}j\omega e^{j\omega t}\}$$

which reduces to

$$\nabla \times \mathbf{E} = -j\omega\mu\mathbf{H} \tag{7-28}$$

The rest of Maxwell's equations are written in complex phasor form as

$$\nabla \times \mathbf{H} = \sigma\mathbf{E} + j\omega\epsilon\mathbf{E} + \mathbf{J}^i \tag{7-29}$$

$$\nabla \cdot \mu\mathbf{H} = 0 \tag{7-30}$$

$$\nabla \cdot \epsilon\mathbf{E} = \rho + \rho^i \tag{7-31}$$

The constitutive parameters become

$$\mathbf{J}^c = \sigma\mathbf{E} \tag{7-32}$$

$$\mathbf{D} = \epsilon\mathbf{E} \tag{7-33}$$

$$\mathbf{B} = \mu\mathbf{H} \tag{7-34}$$

In the case of lossy materials, the constitutive parameter becomes complex. For example, a lossy dielectric would be characterized as: $\epsilon = \epsilon' - j\epsilon''$. The boundary condition statements are unaffected for the fields in phasor form.

Now consider the instantaneous Poynting vector $\mathcal{P} = \mathcal{E} \times \mathcal{H}$. It can be shown that the time-averaged power density is

$$\frac{1}{T}\int_0^T \mathcal{P}\, dt = \frac{1}{2}\,\mathrm{Re}\,\{\mathbf{E} \times \mathbf{H}^*\} = \frac{1}{2}\,\mathrm{Re}\,\{\mathbf{P}\} \tag{7-35}$$

where

$$\mathbf{P} = \mathbf{E} \times \mathbf{H}^* \tag{7-36}$$

is called the *complex Poynting vector.*

Equation (1–12) is used in the proof of Eq. (7–35). A conservation of complex power expression can now be developed by integrating the complex Poynting vector over a closed surface:

$$\oint\!\!\!\oint_S \mathbf{P} \cdot \mathbf{da} = \oint\!\!\!\oint_S \mathbf{E} \times \mathbf{H}^* \cdot \mathbf{da} = \iiint_V \nabla \cdot \mathbf{E} \times \mathbf{H}^*\, dv$$

The vector identity $\nabla \cdot (\mathbf{A} \times \mathbf{B}) = \mathbf{B} \cdot \nabla \times \mathbf{A} - \mathbf{A} \cdot \nabla \times \mathbf{B}$ can be used to write

$$\oint\!\!\!\oint_S \mathbf{P} \cdot \mathbf{da} = \iiint_V [\mathbf{H}^* \cdot \nabla \times \mathbf{E} - \mathbf{E} \cdot \nabla \times \mathbf{H}^*]\, dv$$

Substituting Maxwell's equations results in

$$-\iiint_V \mathbf{E} \cdot \mathbf{J}^{i*}\, dv = \iiint_V \sigma|\mathbf{E}|^2\, dv + j\omega \iiint_V (\mu|\mathbf{H}|^2 - \epsilon|\mathbf{E}|^2)\, dv + \oint\!\!\!\oint_S \mathbf{P} \cdot \mathbf{da}$$

complex power supplied by the source within the volume = power dissipated within the volume + net reactive power stored in the volume + complex power flow out of the volume

(7–37)

7–7. THE COAXIAL TRANSMISSION LINE

The coaxial transmission line can now be analyzed from the viewpoint of electromagnetic field theory as an example of the application of Maxwell's equations. The coordinate system is shown in Fig. 7–5. The guiding structure is characterized by perfectly conducting surfaces at $\rho = a$ and $\rho = b$. We will apply Maxwell's equations to the region between the conductors; that is, $a \leq \rho \leq b$. In this region Maxwell's equations become

$$\begin{aligned} -\nabla \times \mathbf{E} &= j\omega\mu\mathbf{H} \\ \nabla \times \mathbf{H} &= j\omega\epsilon\mathbf{E} \\ \nabla \cdot \mu\mathbf{H} &= 0 \\ \nabla \cdot \epsilon\mathbf{E} &= 0 \end{aligned} \tag{7-38}$$

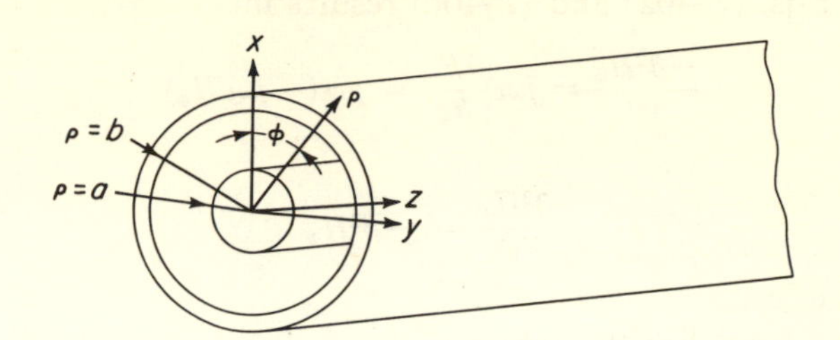

Fig. 7–5. Coaxial-transmission-line coordinate system.

Expanding Maxwell's first two equations in cylindrical coordinates (Appendix D) results in

$$\text{(a)} \qquad \begin{bmatrix} \dfrac{1}{\rho}\mathbf{a}_\rho & \mathbf{a}_\phi & \dfrac{1}{\rho}\mathbf{a}_z \\ \dfrac{\partial}{\partial \rho} & \dfrac{\partial}{\partial \phi} & \dfrac{\partial}{\partial z} \\ E_\rho & \rho E_\phi & E_z \end{bmatrix} = -j\omega\mu(\mathbf{a}_\rho H_\rho + \mathbf{a}_\phi H_\phi + \mathbf{a}_z H_z)$$

$$\text{(b)} \qquad \begin{bmatrix} \dfrac{1}{\rho}\mathbf{a}_\rho & \mathbf{a}_\phi & \dfrac{1}{\rho}\mathbf{a}_z \\ \dfrac{\partial}{\partial \rho} & \dfrac{\partial}{\partial \phi} & \dfrac{\partial}{\partial z} \\ H_\rho & \rho H_\phi & H_z \end{bmatrix} = j\omega\epsilon(\mathbf{a}_\rho E_\rho + \mathbf{a}_\phi E_\phi + \mathbf{a}_z E_z) \qquad (7\text{–}39)$$

The first problem is to look for a solution to these equations in the region $a \leq \rho \leq b$. The static field solution is $E_\phi = E_z = H_\rho = H_z = 0$. Substituting these values into Maxwell's equations results in four equations:

$$\text{(a)} \qquad \frac{\partial E_\rho}{\partial z} = -j\omega\mu H_\phi$$

$$\text{(b)} \qquad \frac{\partial E_\rho}{\partial \phi} = 0 \qquad (7\text{–}40)$$

$$\text{(c)} \qquad -\frac{\partial H_\phi}{\partial z} = j\omega\epsilon E_\rho$$

$$\text{(d)} \qquad \frac{1}{\rho}\frac{\partial}{\partial \rho}(\rho H_\phi) = 0$$

From Eq. (7–40d), it is apparent that

$$H_\phi = \frac{C}{\rho} f(\phi, z) \qquad (7\text{–}41)$$

Combining Eqs. (7–40a) and (7–40c) results in

$$\frac{-\partial^2 H_\phi}{\partial z^2} = j\omega\epsilon\frac{\partial E_\rho}{\partial z} = j\omega\epsilon(-j\omega\mu H_\phi)$$

or

$$\frac{\partial^2 H_\phi}{\partial z^2} = -\beta^2 H_\phi \tag{7–42}$$

where $\beta^2 = \omega^2\mu\epsilon$.

Equation (7–42) has the solution

$$H_\phi = Ae^{-j\beta z} + Be^{+j\beta z} \tag{7–43}$$

which represents waves traveling in the $+z$-direction and $-z$-direction. Substituting the ρ-dependence from Eq. (7–41) results in

$$H_\phi = \frac{C^+g(\phi)}{\rho}e^{-j\beta z} + \frac{C^-g(\phi)}{\rho}e^{+j\beta z} \tag{7–44}$$

Using Eq. (7–40c):

$$E_\rho = -\frac{1}{j\omega\epsilon}\left[-j\frac{\beta C^+g(\phi)}{\rho}e^{-j\beta z} + j\frac{\beta C^-g(\phi)}{\rho}e^{+j\beta z}\right]$$

Now, the ϕ-dependence can be evaluated by means of Eq. (7–40b):

$$\frac{\partial E_\rho}{\partial \phi} = -\frac{1}{j\omega\epsilon}\left[-j\frac{\beta C^+e^{-j\beta z}}{\rho} + j\frac{\beta C^-e^{+j\beta z}}{\rho}\right]\frac{\partial g(\phi)}{\partial \phi} = 0$$

The only general solution is for $\dfrac{\partial g(\phi)}{\partial \phi} = 0$; therefore, $g(\phi) =$ const. and

$$H_\phi = \frac{C^+e^{-j\beta z}}{\rho} + \frac{C^-e^{+j\beta z}}{\rho} \tag{7–45}$$

Now consider just the wave traveling in the $+z$-direction (assume that we are considering a transmission line of infinite length). At $\rho = a$, the integral form of Maxwell's second equation states that

$$\oint \mathbf{H} \cdot \mathbf{dl} = \text{current flowing on the center conductor}$$

Say that the current is I at $z = 0$; then

$$2\pi a\frac{C^+}{a} = I$$

or

$$C^+ = \frac{I}{2\pi}$$

and

$$H_\phi = \frac{I}{2\pi\rho}e^{-j\beta z} \tag{7–46}$$

The associated electric field is

$$E_\rho = +\frac{\omega\sqrt{\mu\epsilon}}{\omega\epsilon}\frac{I}{2\pi\rho}e^{-j\beta z} = \sqrt{\frac{\mu}{\epsilon}}\frac{I}{2\pi\rho}e^{-j\beta z} \tag{7-47}$$

At this point one should substitute E_ρ and H_ϕ back into the equations derived from Maxwell's equations, Eqs. (7-40), to be certain that they are indeed solutions of Maxwell's equations. One should also be sure that the boundary conditions are met. In this case, the boundary conditions require that there be no tangential electric field at $\rho = a$ and $\rho = b$. It is easy to show that the solution developed here satisfies both Maxwell's equations and the boundary conditions. This is all one needs to be assured that this is a realizable set of fields for this structure.

The voltage between the inner and outer conductor is

$$\begin{aligned} V = \int \mathbf{E}\cdot d\mathbf{l} = \sqrt{\frac{\mu}{\epsilon}}\frac{I}{2\pi}e^{-j\beta z}\int_a^b \frac{d\rho}{\rho} &= \sqrt{\frac{\mu}{\epsilon}}\frac{I}{2\pi}e^{-j\beta z}\ln b/a \\ &= \rho E_\rho \ln b/a \end{aligned} \tag{7-48}$$

Now, consider Eq. (7-40a) again. Using the results of Eqs. (7-48) and (7-46), we can write (letting V and I now include $e^{-j\beta z}$)

$$\frac{1}{\rho \ln b/a}\frac{\partial}{\partial z}V = -j\omega\mu\frac{I}{2\pi\rho}$$

which, since $L = \dfrac{\mu}{2\pi}\ln b/a$, reduces to

$$\frac{-\partial V}{\partial z} = j\omega LI \tag{7-49}$$

In a similar fashion, Eq. (7-40c) can be rewritten as

$$\frac{-\partial}{\partial z}\frac{I}{2\pi\rho} = j\omega\epsilon\frac{V}{\rho\ln b/a}$$

which, since $C = \dfrac{2\pi\epsilon}{\ln b/a}$, reduces to

$$\frac{-\partial I}{\partial z} = j\omega CV \tag{7-50}$$

These are the familiar differential equations of transmission-line theory. It is interesting to note also that

$$\beta = \omega\sqrt{\mu\epsilon} = \omega\sqrt{LC} \tag{7-51}$$

which says that $LC = \mu\epsilon$; this turns out to be true regardless of the nature of the transmission line, providing that there are no longitudinal components of E or H.

Now consider the time-averaged power flow on the line:

$$P = \int_0^{2\pi} \int_a^b \frac{1}{2} \text{Re} (\mathbf{E} \times \mathbf{H}^*) \cdot \mathbf{da}$$

$$P = \frac{1}{2} \text{Re} \int_0^{2\pi} \int_a^b E_\rho H_\phi^* \rho \, d\rho \, d\phi$$

$$P = \frac{1}{2} \text{Re} \int_0^{2\pi} \int_a^b \sqrt{\frac{\mu}{\epsilon}} \frac{I}{2\pi\rho} e^{-j\beta z} \frac{I^*}{2\pi\rho} e^{+j\beta z} \rho \, d\rho \, d\phi \tag{7-52}$$

$$P = \sqrt{\frac{\mu}{\epsilon}} \frac{|I|^2}{4\pi} \ln b/a$$

Since $L = \dfrac{\mu}{2\pi} \ln b/a$ and $C = \dfrac{2\pi\epsilon}{\ln b/a}$, one can recall that

$$Z_0 = \sqrt{\frac{L}{C}} = \sqrt{\frac{\mu}{\epsilon}} \frac{\ln b/a}{2\pi} \tag{7-53}$$

Thus, the integral of the real part of the complex Poynting vector over the cross-section of the coaxial transmission line results in

$$P = \tfrac{1}{2}|I|^2 Z_0$$

which is the same expression that one would get by application of the transmission-line theory developed in the preceding chapters. This example merely emphasizes the statements made in the introduction to this chapter, where it was asserted that the only distinction between the circuit theory approach to this problem and the field theory approach is one of viewpoint. We will soon encounter situations where circuit theory simply will not be applicable because one cannot define a voltage in an unambiguous manner. In such cases, field theory is the only method by which the problem can be solved.

PROBLEMS

7–1. Show that the first of Eqs. (7–5) may be derived by applying Faraday's law directly to a differential rectangle in the yz-plane. It should be clear that the other two equations may be obtained by cyclic permutation of the subscripts.

7–2. A square loop of side d is oriented in the xy-plane with its center at (x_0, y_0). The magnetic field in this region is given by

$$\mathcal{H}_x = \mathcal{H}_y = 0$$
$$\mathcal{H}_z = A \sin \beta y \cos \omega t$$

and the region has zero conductivity. (a) Find the electric field in the region. (b) Find the induced voltage in the loop, i.e., $\oint \boldsymbol{\mathcal{E}} \cdot \mathbf{dl}$. (c) Show that β must equal

$\omega\sqrt{\mu\epsilon}$ if Maxwell's equations are to be satisfied. (d) Write the expression for the instantaneous Poynting vector at the point (x_0, y_0).

7–3. Consider a voltage $\mathcal{V} = |V| \cos \omega t$ applied to a simple parallel-plate capacitor with plates that are circular in shape. Let the radius of the plates be R and the separation d; it may be assumed that $R \gg d$. Also assume the capacitor is fed symmetrically at the center points. (a) Find the total displacement current within the region between the two capacitor plates. (b) Write the expression for the magnetic field in the region between the plates as a function of the cylindrical coordinate ρ. It may be assumed that the wavelength is long with respect to the physical dimensions of the problem, and thus the charge distribution on the capacitor plates will be approximately uniform.

7–4. Consider a thin sheet of surface current of infinitesimal thickness, having a density of J amperes per unit width. Show that the normal component of **B** is continuous at such a surface and that the tangential component of **H** is discontinuous by an amount J. (*Hint:* Apply Ampere's law to a differential rectangular path which encloses and is normal to a portion of the surface current.)

7–5. Half of the rectangular box shown by dotted lines in the accompanying figure lies between the capacitor plates; the other half lies outside the plates. A casual application of Poynting's theorem would seem to indicate that there is continuous power flowing in one end of the box but none coming out the opposite end. Thus it would appear that there is a net flow of power into the box—and obvious falsehood. Give a rational explanation of why this is not so.

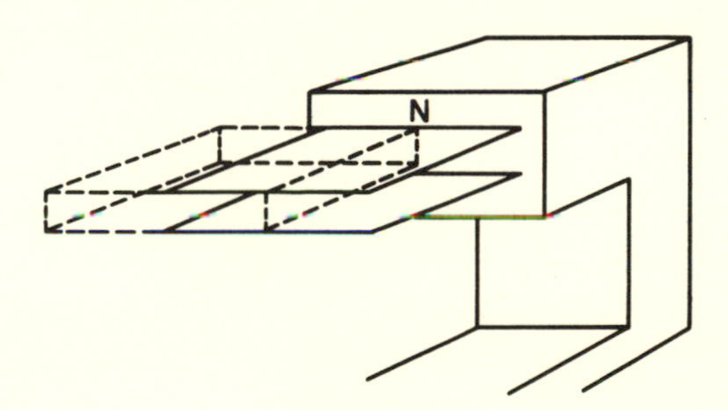

Prob. 7–5.

8

Propagation of Plane Waves

8–1. DESCRIPTION OF A UNIFORM PLANE WAVE

The fundamental differential equations describing electromagnetic wave phenomena were established in the preceding chapter. These equations are partial differential equations rather than ordinary ones, and thus one cannot write down a simple explicit solution which will apply to all possible situations. Therefore we must be content to examine a few special cases, and in particular we wish to look at those of immediate practical importance. The first of these special cases to be investigated in detail will be the uniform plane wave.

The plane wave is probably the most common of all the different forms of wave propagation. Light waves and ordinary radio waves are examples of this phenomenon. As a matter of fact, electromagnetic waves from any source become essentially plane waves as the distance from the source becomes large. The simplest type of plane wave is called a *uniform plane wave.* It is characterized by uniformity in a plane normal to the direction of propagation and by electric and magnetic fields which are mutually perpendicular to each other and the direction of propagation. A non-uniform plane wave is similar except that the amplitude (and not phase) may vary within a plane normal to the direction of propagation. For the time being we shall focus our attention on the uniform plane wave and defer discussion of the nonuniform one to later. In particular we shall be concerned with the propagation of a uniform plane wave in a lossless dielectric region.

The simplest approach to a mathematical description of such a wave is to postulate its existence and then proceed to show that the resulting wave will satisfy Maxwell's equations. With this in mind we arbitrarily let

$$E_y = E_z = 0 \qquad H_x = H_z = 0$$
$$\frac{\partial E_x}{\partial x} = \frac{\partial E_x}{\partial y} = 0 \qquad \frac{\partial H_y}{\partial x} = \frac{\partial H_y}{\partial y} = 0 \tag{8–1}$$

This assures uniformity (i.e., no variation) in the xy-plane and dictates an electric field in the x-direction only and a magnetic field in the y-direction only, as shown in Fig. 8–1. If the conditions of Eqs. (8–1) are

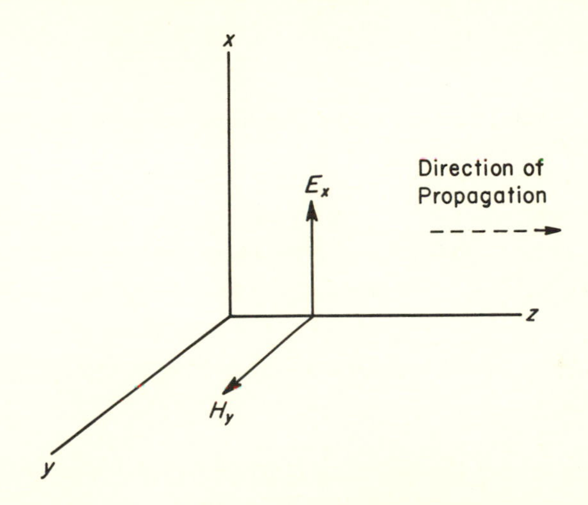

Fig. 8–1. Coordinate system for plane wave.

substituted into Maxwell's equations for nonconducting space ($\sigma = 0$), the following equations result:

$$\frac{dE_x}{dz} = -j\omega\mu H_y \qquad \frac{dH_y}{dz} = -j\omega\epsilon E_x \tag{8–2}$$

Note the similarity between these and the lossless transmission-line equations:

$$\frac{dV}{dx} = -j\omega LI \qquad \frac{dI}{dx} = -j\omega CV$$

In fact, they are identical in form. Thus all the theory built up around the lossless transmission line will carry over to the plane-wave problem by direct analogy.

Before discussing this similarity in detail, the matter of boundary conditions must be investigated. Consider the analogous discontinuities shown in Fig. 8–2. In the transmission-line case, V and I must be con-

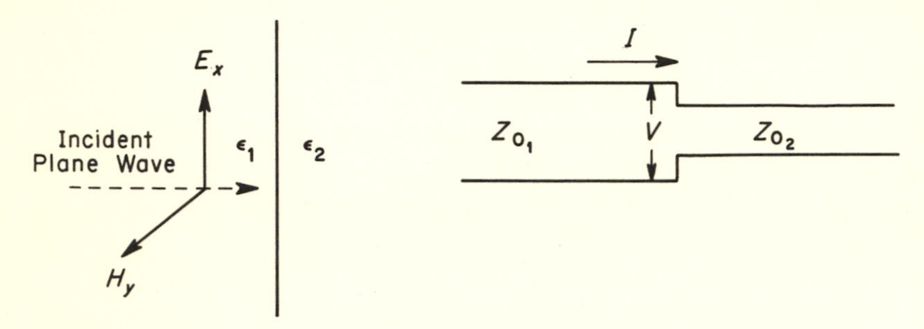

Fig. 8–2. Analogous plane-wave and transmission-line discontinuities.

tinuous at the discontinuity in characteristic impedance. In the plane-wave case, E_x and H_y must be continuous at the boundary because they are tangent to the boundary. Thus both differential equations and boundary conditions are analogous, and the analogy is now complete.

8–2. SUMMARY OF PLANE-WAVE–TRANSMISSION-LINE ANALOGY

A summary of the plane-wave–transmission-line analogy is given in Table 8–1. A few comments about this analogy are in order before proceeding with an example. First note that the plane-wave coordinate system was chosen such that the direction of propagation is in the positive z-direction. This is, of course, arbitrary but is the custom in most present-day literature. The use of z as a distance coordinate was avoided in transmission lines because of possible confusion with impedance. This problem does not exist in wave theory, since "impedance" is denoted by the Greek letter η. Note that the quantity in wave theory which is analogous to characteristic impedance is referred to as intrinsic impedance, and it is defined basically as the ratio of the incident E and H waves. It is called impedance simply because it has the dimensions of impedance; it is not the ratio of voltage to current.

TABLE 8-1

Transmission-Line-Plane-Wave Analogy

Transmission-Line Quantity	*Symbol or Equation*	*Plane-Wave Quantity*	*Symbol or Equation*
Voltage	V	Electric field intensity	E_x
Current	I	Magnetic field intensity	H_y
Inductance per unit length	L	Permeability	μ
Capacitance per unit length	C	Permittivity	ϵ
Characteristic impedance	$Z_0 = \sqrt{\dfrac{L}{C}}$	Intrinsic impedance	$\eta = \sqrt{\dfrac{\mu}{\epsilon}}$
Phase-shift constant	$\beta = \omega\sqrt{LC}$	Phase-shift constant	$\beta = \omega\sqrt{\mu\epsilon}$
Velocity of propagation	$v = \dfrac{1}{\sqrt{LC}}$	Velocity of propagation	$v = \dfrac{1}{\sqrt{\mu\epsilon}}$
Reflection coefficient (at load Z_L)	$K_L = \left(\dfrac{Z_L - Z_0}{Z_L + Z_0}\right)$	Reflection coefficient (at boundary between ϵ_1 and ϵ_2)	$K = \left(\dfrac{\eta_2 - \eta_1}{\eta_2 + \eta_1}\right)$
Incident wave power	$P^+ = \dfrac{\lvert V^+\rvert^2}{2Z_0}$	Incident wave power density	$P^+ = \dfrac{\lvert E_x^+\rvert^2}{2\eta}$

Example 8-1

Consider a plane wave in air impinging normally on a relatively large, thick slab of polyethylene, as shown in Fig. 8-2. Let the peak value of the incident electric field intensity be 10 V/m. It is desired to find the transmitted and reflected power densities. First the intrinsic impedances for air and polyethylene are found to be

$$\eta_{\text{air}} = \sqrt{\frac{\mu_0}{\epsilon_0}} = \sqrt{\frac{4\pi \times 10^{-7}}{(1/36\pi) \times 10^{-9}}} = 377 \text{ ohms}$$

$$\eta_{\text{polyethylene}} = \sqrt{\frac{\mu_0}{\epsilon_r \epsilon_0}} = \frac{1}{\sqrt{\epsilon_r}} \sqrt{\frac{\mu_0}{\epsilon_0}} = \frac{377}{1.5} \text{ ohms}$$

These intrinsic impedances correspond to the characteristic impedances Z_{01} and Z_{02} in the transmission-line analogy shown in Fig. 8-2. The reflection coefficient at the boundary is then

$$K = \frac{\eta_{\text{polyethylene}} - \eta_{\text{air}}}{\eta_{\text{polyethylene}} + \eta_{\text{air}}} = \frac{377/1.5 - 377}{377/1.5 + 377} = -\frac{1}{5}$$

The minus sign indicates 180-degree phase difference between incident and reflected waves at the boundary. The reflected wave then has a peak amplitude of

$$|E^-| = |K||E^+| = (0.2)(10) = 2\text{ V/m}$$

$$P_{\text{incident}} = \frac{(10)^2}{(2)(377)} = 0.133\text{ W/m}^2$$

$$P_{\text{reflected}} = \frac{(2)^2}{(2)(377)} = 0.0053\text{ W/m}^2$$

$$P_{\text{transmitted}} = P_{\text{incident}} - P_{\text{reflected}} = 0.128\text{ W/m}^2$$

8–3. PLANE WAVE PROPAGATION IN A LOSSY DIELECTRIC

We shall now consider a plane wave propagating in a lossy dielectric region. Maxwell's equations for a lossy region are

$$\begin{aligned} \nabla \times \mathbf{E} &= -j\omega\mu\mathbf{H} \\ \nabla \times \mathbf{H} &= j\omega\epsilon\mathbf{E} + \sigma\mathbf{E} \end{aligned} \tag{8–3}$$

It will be noticed that the second equation may be written in the form

$$\nabla \times \mathbf{H} = j\omega\left(\epsilon + \frac{\sigma}{j\omega}\right)\mathbf{E} \tag{8–4}$$

which tells us immediately what we must do to our lossless solution in order to generalize it to the lossy case. It is simply necessary to replace ϵ by $\epsilon + (\sigma/j\omega)$ in the lossless formulas, and the result will be the appropriate formulas for the lossy case. The quantity $\epsilon + (\sigma/j\omega)$ is referred to as the *complex permittivity*. Similarly, when $\epsilon + (\sigma/j\omega)$ is written in the form

$$\epsilon + \frac{\sigma}{j\omega} = \epsilon_0\left(\epsilon_r - \frac{j\sigma}{\omega\epsilon_0}\right) \tag{8–5}$$

the quantity in parentheses is called the *complex dielectric constant*. Also, the ratio

$$\frac{\sigma/\omega\epsilon_0}{\epsilon_r} = \frac{\sigma}{\omega\epsilon} = \theta \tag{8–6}$$

is the *loss tangent*, which will be denoted by θ. It can be seen that this is the ratio of conduction to displacement current and thus is a measure of the quality of the dielectric.

The preceding remarks are general and apply to any lossy region, providing μ is real, which is normally the case. We now wish to look at the slightly lossy case, and the loss tangent is the mathematical criterion for determining whether a material is "slightly" lossy or not. If this factor is much less than unity, then the displacement current is large with respect to the conduction current, and by definition we say the material is slightly lossy. Just how small this quantity must be is, of course, a question of desired accuracy. Usually, 0.01 is small enough to give good results, the error being of the order of a fraction of a per cent. The expressions for intrinsic impedance and propagation constant may now be written as

$$\eta = \sqrt{\frac{\mu}{\epsilon(1 - j\theta)}} \approx \sqrt{\frac{\mu}{\epsilon}}\left[\left(1 - \frac{3}{8}\theta^2\right) + j\frac{\theta}{2}\right] \tag{8–7}$$

$$\gamma = \alpha + j\beta = j\omega\sqrt{\mu\epsilon(1 - j\theta)} \tag{8–8}$$

or

$$\alpha \approx (\omega\sqrt{\mu\epsilon})\left(\frac{1}{2}\theta\right) = \frac{\sigma}{2}\sqrt{\frac{\mu}{\epsilon}} \tag{8–9}$$

$$\beta \approx (\omega\sqrt{\mu\epsilon})\left(1 + \frac{1}{8}\theta^2\right) \tag{8–10}$$

A number of things are evident from these equations. To begin with, the addition of small losses affects the intrinsic impedance and phase-shift constant only in a second-order way. The magnitude of η is not changed appreciably, but its phase angle goes slightly positive. The phase-shift constant is increased slightly, and as a result, the velocity of propagation is reduced by a corresponding amount. For most calculations these small changes in η and β may be neglected, and only the change in α need be taken into account. As could be expected, this is consistent with transmission-line theory.

A word of caution is in order at this point. From Eq. (8–9) it may appear at first glance that the attenuation factor α for a given dielectric is independent of frequency. However, this is not the case, since most dielectrics become increasingly lossy with an increase in frequency. The precise cause of this is a complex matter, but experimentally it is a known fact which can be readily verified in the laboratory. From a macroscopic viewpoint it is usually accounted for by saying the conductivity is a function of frequency. Therefore one should be careful not to use a conductivity figure which was obtained by dc measurements for anything other than very low frequency situations. For high-frequency applications the loss tangent rather than the conductivity is usually given in tables of physical constants, and it is normally specified for a number of frequencies within the useful range of the dielectric material.[1] For example, the loss tangent of polystyrene is approximately 0.7×10^{-4} at 1 MHz, 1×10^{-4} at 100 MHz, and 4.3×10^{-4} at 10 GHz. Its relative dielectric constant

[1] Loss tangents for a few common dielectric materials are given in Appendix F. For a more elaborate table of loss tangents see A. R. von Hippel (ed.), *Dielectric Materials and Applications* (New York: John Wiley & Sons, Inc., 1954), pp. 291–425.

is approximately 2.5 throughout this frequency range. It can be verified directly from Eq. (8–9) that the corresponding attenuation factors for these three frequencies are 0.46×10^{-6}, 0.66×10^{-4}, and 2.85×10^{-2} Np/m, respectively. Thus, even for polystyrene (which is considered to be a high-quality dielectric), the attenuation constant is very much dependent on frequency.

8–4. PLANE WAVE PROPAGATION IN A GOOD CONDUCTOR

A slightly lossy or "good" dielectric has just been defined as one in which the displacement current is large with respect to the conduction current. A good conductor is defined to be just the reverse of this; i.e., the conduction current term is large with respect to the displacement one, or in mathematical language, $\sigma/\omega\epsilon \gg 1$. This is an important special case, since most of the commonly used conductors such as copper, brass, and aluminum fall into this category even at very high frequencies. For example, consider copper at 10 MHz:

$$\sigma = 58 \times 10^{6} \text{ mhos/m}$$
$$\omega\epsilon = (2\pi)(10^{7})(8.85 \times 10^{-12}) = 55.5 \times 10^{-5} \text{ mhos/m}$$

Thus, at 10 MHz it is a very good approximation to say that $\sigma \gg \omega\epsilon$ for copper. It will be seen presently that this approximation is useful in that it leads to considerable simplification of the pertinent equations.

As mentioned in Art. 8–3, all we have to do to obtain the appropriate lossy formulas from the lossless ones is to replace ϵ with $\epsilon + (\sigma/j\omega)$, and in the good-conductor case we shall neglect ϵ with respect to $\sigma/j\omega$. When this is done, the equations for γ and η become

$$\gamma = j\omega\sqrt{\mu\left(\frac{\sigma}{j\omega}\right)} = \sqrt{\frac{\omega\mu\sigma}{2}} + j\sqrt{\frac{\omega\mu\sigma}{2}} = \alpha + j\beta \tag{8–11}$$

or

$$\alpha = \beta = \sqrt{\frac{\omega\mu\sigma}{2}} \text{ (Np/m, rad/m)} \tag{8–12}$$

and

$$\eta = \sqrt{\frac{\mu}{\sigma/j\omega}} = \sqrt{\frac{\omega\mu}{2\sigma}} + j\sqrt{\frac{\omega\mu}{2\sigma}} = \sqrt{\frac{\omega\mu}{\sigma}}\,\underline{/45^\circ} \text{ ohms} \tag{8–13}$$

Notice that the propagation constant has both a real and imaginary part. Thus there is attenuation as well as phase shift as the wave proceeds through the conducting region. In order to get some feel for the magnitude of those quantities, consider again the example of copper at 10 MHz:

$$\eta = \sqrt{\frac{(j2\pi \times 10^7)(4\pi \times 10^{-7})}{58 \times 10^6}} = 0.00117 \underline{/45^\circ} \text{ ohms}$$

$$\alpha = \sqrt{\frac{(2\pi \times 10^7)(4\pi \times 10^{-7})(58 \times 10^6)}{2}} = 47{,}800 \text{ Np/m}$$

$$\beta = 47{,}800 \text{ rad/m}$$

It can be seen that metals are characterized by very low intrinsic impedance and very high attenuation.

The following example will illustrate the usefulness of the transmission-line analogy in handling plane wave problems involving losses.

Example 8–2

Consider a plane wave in air impinging normally on a large, thick sheet of copper. The frequency is 10 MHz and the rms value of the electric field intensity is 1 V/m. Find (a) the rms value of the electric field intensity at the surface of the copper, (b) the distance from the surface (within the copper) at which the field is only 1 per cent of its value at the surface, and (c) the fraction of the incident power which is absorbed by the copper.

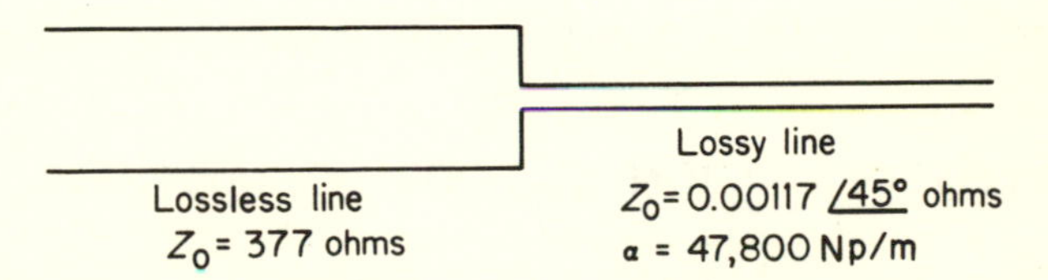

Fig. 8–3. Transmission-line analogy for wave impinging on copper at 10 MHz.

The transmission-line analogy for this problem is shown in Fig. 8–3. The reflection coefficient at the boundary may be computed from transmission-line theory as

$$K = \frac{\eta_{\text{Cu}} - \eta_{\text{air}}}{\eta_{\text{Cu}} + \eta_{\text{air}}}$$

Since η_{Cu} is small, this is approximately -1. However, we dare not say it is exactly -1 or we shall lose the very thing we are looking for. This is the problem. The expression for the total electric field intensity at the discontinuity is

$$E = E^+(1 + K) = E^+\left(1 + \frac{\eta_{\text{Cu}} - \eta_{\text{air}}}{\eta_{\text{Cu}} + \eta_{\text{air}}}\right) = E^+ \frac{2\eta_{\text{Cu}}}{\eta_{\text{Cu}} + \eta_{\text{air}}}$$

or, to a high degree of approximation,

$$|E| \approx \left|E^+ \frac{2\eta_{\text{Cu}}}{\eta_{\text{air}}}\right| = (1)\left(\frac{2 \times 0.00117}{377}\right) = 6.2 \times 10^{-6} \text{ V/m}$$

In order to find the distance within the copper required for the wave to be attenuated by a factor of 100, we need only look at the attenuation factor in the lossy line part of the analogy. Since α is 47,800 Np/m, we must find a distance d such that

$$e^{-47,800d} = \frac{1}{100}$$

or

$$d = 0.098 \text{ mm}$$

It can be seen that the penetration of the wave into the copper is quite small even at this modest frequency. It is essentially dissipated in just a fraction of a millimeter.

The power absorbed by the copper can also be found readily from the analogy. If we think of the total voltage at the boundary as the driving voltage on the lossy portion of the line, the input power would be (assuming the line is relatively "long" so there will be no reflected wave)

$$P_{\text{Cu}} = \frac{|E|^2}{|\eta_{\text{Cu}}|} \cos 45° = \frac{(6.2 \times 10^{-6})^2}{0.00117}(0.707) = 23.2 \times 10^{-9} \text{ W/m}^2$$

The incident power density is

$$P_{\text{incident}} = \frac{(1)^2}{377} = 2.65 \times 10^{-3} \text{ W/m}^2$$

and therefore

$$\frac{P_{\text{Cu}}}{P_{\text{incident}}} = 8.75 \times 10^{-6}$$

or only about 0.001 per cent of the incident power is absorbed by the copper.

8–5. WAVE PROPAGATION IN POOR CONDUCTORS

There are some conducting materials having low conductivity which normally cannot be considered as either good conductors or good dielectrics. Seawater is an example. It has a conductivity of 4 mhos/m and a relative dielectric constant of 81. At relatively low frequencies the conduction current will be greater than the displacement current, whereas at reasonably high frequencies just the reverse will be true. About all that can be said for such problems is that the transmission-line analogy is valid in any event, but one may not be able to make simplifying assumptions. In general, one can always write the general equations for η and γ as

$$\eta = \sqrt{\frac{\mu}{\epsilon + (\sigma/j\omega)}} = \sqrt{\frac{j\omega\mu}{\sigma + j\omega\epsilon}} \tag{8–14}$$

$$\gamma = j\omega\sqrt{\mu\left(\epsilon + \frac{\sigma}{j\omega}\right)} = \sqrt{j\omega\mu(\sigma + j\omega\epsilon)} \tag{8–15}$$

and then resort to lossy transmission-line theory to work the problem.

8–6. SKIN EFFECT, DEPTH OF PENETRATION, SURFACE RESISTANCE

It was shown in Art. 8–4 that a wave is attenuated very rapidly within a good conductor. The distance required for the wave to reach a value of

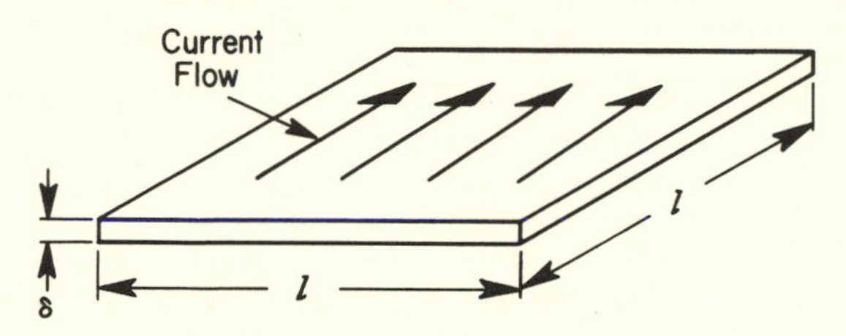

Fig. 8–4. Surface resistance.

$1/e$ of that at the surface is called the *depth of penetration* and will be denoted by δ. It can be easily verified that

$$\delta = \frac{1}{\alpha} = \sqrt{\frac{2}{\omega\mu\sigma}} \tag{8–16}$$

Also, as the current density within the conductor is equal to σE, it must taper off exponentially at the same rate as E. The concept of depth of penetration is very useful, as one can think of the current as being uniformly distributed within the conductor down to a depth δ and zero from there on, and this will lead to the correct result for total power loss within the conductor. Proof of this is a straightforward problem of integral calculus and is left as an exercise (Problem 8–5). One word of caution, though—one must remember in summing differential currents within the conductor that they will differ in phase because there is phase shift as well as attenuation within the conductor.

The resistance of the layer of conducting material through which the equivalent uniform current is flowing is known as the surface resistance and is useful in determining the power dissipated in the conductor. The surface resistance R_s is defined as the resistance of a square of the conducting material of thickness δ, as shown in Fig. 8–4. It is, of course, dependent on frequency as δ varies with frequency. It is not dependent on the dimensions of the square, though, as can be seen when the expression for R_s is written out explicitly as

$$R_s = \frac{l}{\sigma(l\delta)} = \frac{1}{\sigma\delta} = \sqrt{\frac{\omega\mu}{2\sigma}} \tag{8–17}$$

The dimension for R_s is ohms, but it is common to say ohms per square rather than just ohms, thus indicating that this is the resistance of a square of material of thickness δ. The power dissipated per unit area is now given by the equation

$$P \text{ (per unit area)} = \tfrac{1}{2}|\mathbf{J}_s|^2 R_s \tag{8-18}$$

where $\mathbf{J}_s$ indicates the peak surface current per unit width and R_s is the surface resistance.[2]

The problem just considered, that of a plane wave impinging normally on a flat metallic surface, is not quite the same as that encountered in guided wave problems because the conducting surfaces may be curved. However, the two are similar and the depth-of-penetration formula, Eq. (8–16), may be used with reasonable accuracy as long as δ is small with respect to the radius of curvature of the surface being considered. Also, it should be pointed out that whether one considers this "crowding of the current toward the surface" as depth of penetration or skin effect is just a question of viewpoint. That is, from one viewpoint we think of the current inducing the field, and from the other, the field inducing the current. Actually we have no right to associate cause and effect with either field or current; we can only say that the two must be associated in such a way as to be compatible with Maxwell's equations.

Example 8–3

As an example of just how small the depth of penetration actually is for a good conductor, consider its value for copper at 1, 100, and 10,000 MHz. From Eq. (8–16)

$$\delta \text{ (at 1 MHz)} = \sqrt{\frac{2}{(2\pi \times 10^6)(4\pi \times 10^{-7})(58 \times 10^6)}} = 0.066 \text{ mm}$$

$$\delta \text{ (at 100 MHz)} = 0.0066 \text{ mm}$$
$$\delta \text{ (at 10 GHz)} = 0.00066 \text{ mm}$$

It can be seen from these calculations that the depth of penetration is indeed a very small distance. In microwave work it is a common practice to silver-plate the inner surfaces of waveguides in order to reduce the losses. It can be seen that the thickness of the plating need be only a few ten-thousandths of an inch to achieve effective reduction of the losses.

[2] Actually the value of $\mathbf{J}_s$ is the total current per unit width to an infinite depth in the conductor, but it is commonly called surface current because most of it flows within a small region near the surface. One should always keep in the back of his mind, though, that the current is actually distributed exponentially within the conductor.

8–7. NORMAL INCIDENCE ON A PERFECT CONDUCTOR

When a plane wave impinges normally on a perfectly conducting surface, complete reflection takes place. This can be justified easily in terms of the transmission-line analogy, since a perfectly conducting surface is analogous to a perfect short on a transmission line. In the respective cases the voltage or voltage gradient is constrained to be zero at the boundary. Thus 100 per cent reflection with 180-degree phase reversal must take place at the boundary.

We shall digress for a moment and look at an example of wave reflection which illustrates the usefulness of attacking a problem from two different viewpoints. When making measurements in an open-wire transmission system, there is often occasion to place a perfect (or as near perfect as possible) short on the end of the line connecting to the load. If one conductor is merely connected to the other by means of a single wire, as shown in Fig. 8–5(a), the "short" may be far from perfect, since this

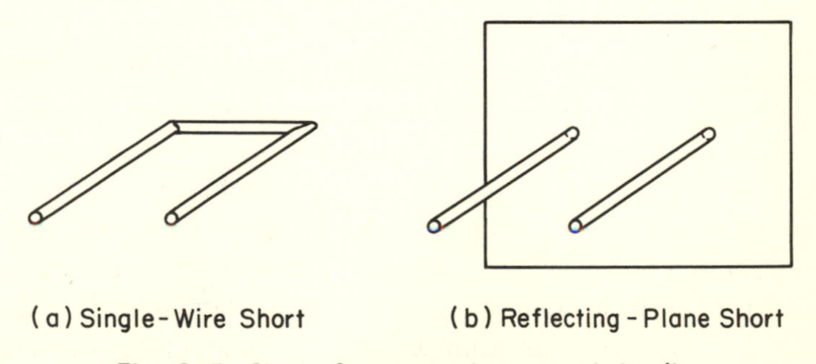

Fig. 8–5. Shorts for open-wire transmission line.

segment of wire looks like an inductive reactance. This is particularly true at relatively high frequencies where the length of the shorting segment is appreciable with respect to the wavelength. The solution to this problem is obvious when viewed in terms of field theory. The wave propagation that takes place in an open-wire transmission system is actually a plane-wave phenomenon,[3] and thus a noninductive short may be obtained by terminating the line with a conducting plane surface, as shown in Fig. 8–5(b).

[3] A rigorous justification of this is given in Chapter 9.

An interesting way to look at the effect of a normally incident uniform plane wave incident on a perfectly conducting plane surface is to visualize the surface as being sliced into an array of thin strips electrically insulated from one another. If the incident wave is such that the electric field is parallel to the long axis of the strips, currents will flow in the strips just as though the surface were an infinite perfectly conducting plane. If the incident electric field is at right angles to the long axis of the strips, current cannot flow because of the insulation between the strips. This surface is an excellent example of an anisotropic material; the current density depends on the orientation of the electric field.

The question of the effect of such a surface on the incident plane wave leads to some insights which will prove useful in later chapters. We began by stating that one could cut a perfectly conducting surface into an array of strips without affecting the fields as long as the cuts were parallel to the electric field, and the resultant current flow. We could separate the fields into the incident field and a field produced by the interaction of the incident field with the strips. This field is said to be scattered by the strips and is termed the *scattered field*. Thus, we can consider the resultant fields to be the superposition of an incident uniform plane wave and a scattered field due to the currents in the strips. The boundary conditions are such that the net tangential electric field must be zero on the strips. We know how the fields should behave when the incident electric field is parallel to the long axis of the strip; all of the energy should be reflected with a reflection coefficient of -1. From a scattered-wave viewpoint we would say that scattered waves would propagate outward on both sides of the array of strips. On the incident side the scattered wave would be identical to the reflected wave from an infinite perfectly conducting plane. On the other side we would say that the incident wave is transmitted through the gaps, where it combines with the scattered field to produce zero net field; which is consistent with the absence of fields inside a perfectly conducting plane surface.

For the case of an incident plane wave with its electric field vector at right angles to the long axis of the strips, currents will not be able to flow, a scattered field will not be generated, and the wave will be transmitted through the array of strips with little attenuation.

This view of reflection from a perfectly conducting surface seems somewhat artificial at this point but will become a valuable concept in the understanding of radiation from wire antennas.

8-8. OBLIQUE INCIDENCE OF PLANE WAVES

In all that has been said to this point regarding reflection of plane waves, normal incidence on the boundary was assumed. It was shown that this gives rise to a straightforward transmission-line analogy, and problems of this nature (including those involving multiple layers of dielectrics) can be solved readily by using this technique. We now wish to consider the case of oblique incidence at the boundary between two different dielectric materials. First of all, it can be seen from Fig. 8-6 that

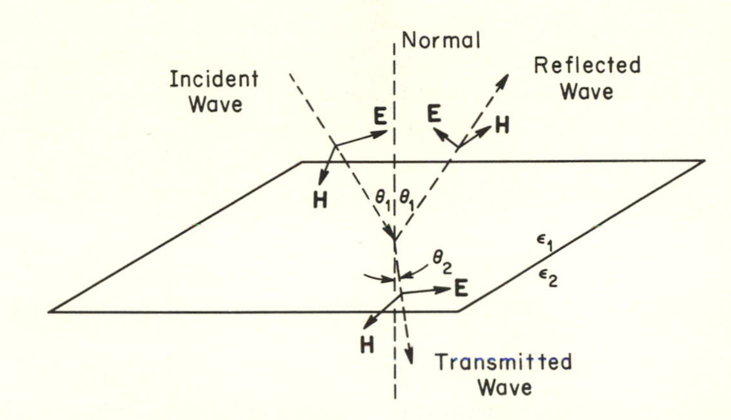

Fig. 8-6. Plane wave impinging obliquely on dielectric boundary.

the **E** and **H** fields of the impinging wave will not necessarily be tangent to the boundary. This immediately destroys the previous analogy, since the boundary conditions are no longer similar. In the oblique incidence case only the tangential components of **E** and **H** have to be continuous at the boundary, whereas in the normal incidence situation the tangential components and total values are one and the same thing. Also, in the normal incidence case the orientation of the **E** vector within a plane normal to the direction of propagation is not critical because it is tangent to boundary regardless of its direction within this plane. Thus there was no loss in generality by assuming **E** always in the x-direction for the normal incidence case. However, when a wave impinges obliquely on a plane boundary, the orientation of the **E** field is important and very much affects the solution of the problem. As a matter of terminology the direction of the **E** vector of a plane wave is known as the *direction of polarization*. Also, the simple type of plane wave considered thus far is said to be linearly polarized because the **E** vector is always directed along the same line. More complicated forms of polarization will be considered in Art. 8–11.

The general procedure for solving the problem of oblique incidence will be to resolve the incident wave into two component waves, one of which is polarized in the plane of incidence (which is defined by the direction of propagation and a line normal to the boundary) and the other normal to the plane of incidence. It should be evident from Fig. 8–6 that any linearly polarized wave can be considered as a superposition of these two types of waves, and thus the general problem is solved if we can solve these two special cases of polarization. The next step in the general solution is to exploit a wave analogy involving only the tangential components of **E** and **H**, since these are the components that must be continuous at the boundary. And finally, after the tangential components of the incident, reflected, and transmitted waves have been determined by the analogy, the total waves can be reconstructed from the geometry of the problem. We shall now consider the two special cases of polarization.

8–9. OBLIQUE INCIDENCE; E IN THE PLANE OF INCIDENCE

Consider a linearly polarized plane wave impinging obliquely on a boundary between two dielectric materials, as shown in Fig. 8–7. The **E** field lies in the plane of incidence and the associated **H** field is normal to plane and directed outward for the direction of propagation shown. First

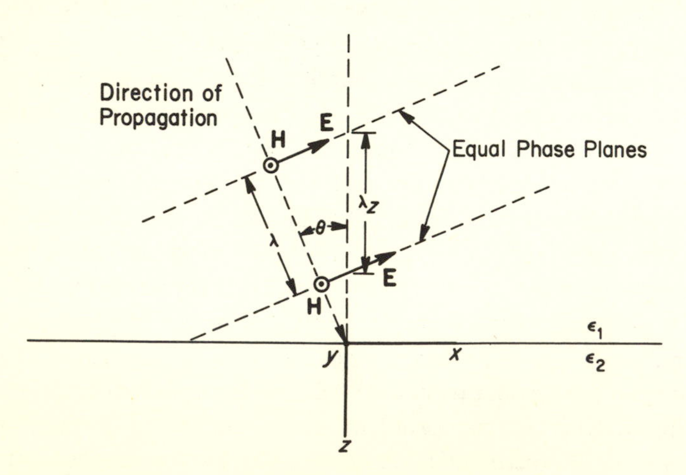

Fig. 8–7. Plane wave with electric field in plane of incidence.

we look for a valid analogy involving only the tangential components of **E** and **H**. From Fig. 8–7 it can be seen that these quantities vary sinusoidally in the z-direction and that the wavelength, when referred to the z-axis, is greater than that seen in the direction of propagation by a factor of $\sec\theta$. This means that the velocity of wave propagation seen along the z-axis is also greater, by the same factor, than that along the axis of propagation.

The fact that the apparent velocity along the z-axis is greater than the "free velocity" of the wave often troubles the beginning student, who asks, "How can anything move faster than the speed of light?" The answer is that only the apparent motion of an arbitrarily defined mathematical quantity is being considered, and not the motion of particles. When approached from a particle viewpoint, the photons are moving with the speed of light in the direction of propagation of the incident wave. An analogy that is often cited is that of an ocean wave impinging obliquely on the beach. The point of the wave contact with the beach moves along the beach faster than the original wave motion in the water.

The fact that the tangential components exhibit apparent wave properties in the z-direction immediately suggests a transmission-line analogy. That such an analogy exists can be justified formally by noting that, for

$$H_x = H_z = E_y = 0$$

and no field variation in the y-direction, Maxwell's equations (Eqs. 7–28 and 7–29) reduce to

$$\frac{\partial H_y}{\partial z} = -j\omega\epsilon E_x \tag{8–19}$$

$$\frac{\partial H_y}{\partial x} = j\omega\epsilon E_z \tag{8–20}$$

$$\frac{\partial E_x}{\partial z} - \frac{\partial E_z}{\partial x} = -j\omega\mu H_y \tag{8–21}$$

Next, from the geometry of Fig. 8–6 and remembering that there is no field variation in a plane normal to the direction of propagation,

$$E_z = -E_x \tan\theta$$
$$\frac{\partial}{\partial x} = \tan\theta \frac{\partial}{\partial z} \tag{8–22}$$

where θ could be either θ_1 or θ_2, depending on the region being considered. If Eqs. 8–22 are now substituted into Eqs. (8–19) and (8–21), these equations reduce to

$$\frac{\partial E_x}{\partial z} = -(j\omega\mu \cos^2 \theta)H_y$$
$$\frac{\partial H_y}{\partial z} = -(j\omega\epsilon)E_x \tag{8-23}$$

These equations are identical in form to those for the lossless transmission line, with E_x being analogous to voltage and H_y analogous to current. The analogy is now complete, since both the differential equations and boundary conditions are similar for the two situations. The expressions for characteristic impedance, propagation constant, and velocity of propagation may now be written as

$$Z_{0z} = \sqrt{\frac{j\omega\mu \cos^2 \theta}{j\omega\epsilon}} = \eta \cos \theta \tag{8-24}$$

$$\gamma = \sqrt{(j\omega\mu \cos^2 \theta)(j\omega\epsilon)} = j\omega\sqrt{\mu\epsilon} \cos \theta \tag{8-25}$$

$$v = \frac{\omega}{\beta} = \frac{1}{\sqrt{\mu\epsilon}} \sec \theta \tag{8-26}$$

Note that these parameters are similar in form to those obtained in the normal incidence case except for a factor of $\cos \theta$.

One more item must be discussed before looking at an example. It is necessary to know the direction of propagation of the reflected and transmitted waves if we are to be able to reconstruct the complete wave picture. The appropriate relationships can be found by observing the wave phenomenon which takes place in the x-direction at the boundary. If we are to match boundary conditions for all values of x, the phase-shift constants with respect to the x-direction of all three component waves (incident, reflected, and transmitted) must be the same. Otherwise a "shearing" effect would take place, and the continuity requirement could not possibly be met at all points along the boundary. If the phase-shift constants are the same for all three component waves, then so are the apparent velocities in the x-direction along the boundary. Since the transmission medium is the same for both incident and reflected waves, it is immediately obvious that the incident and reflected angles are equal. Also, equating appropriate apparent velocities in the two media gives

$$\frac{1}{\sqrt{\mu_1\epsilon_1}} \frac{1}{\sin \theta_1} = \frac{1}{\sqrt{\mu_2\epsilon_2}} \frac{1}{\sin \theta_2}$$

or

$$\frac{\sin \theta_1}{\sin \theta_2} = \sqrt{\frac{\mu_2\epsilon_2}{\mu_1\epsilon_1}} \tag{8-27}$$

This is Snell's law of optics, which relates the incident and refracted angles. The fact that this law carries over to electromagnetic theory should not be surprising, since light waves are simply electromagnetic radiation with an extremely short wavelength. Also, it can be seen from Eq. (8–27) that the index of refraction of a dielectric material should be equal to the square root of its relative dielectric constant. A casual check in a table of physical constants will not always verify this for materials which are transparent to both radio frequency and visible light waves. However, one should remember that the relative dielectric constant usually cited in tables is measured over a range of frequencies which are many orders of magnitude less than those involved in the visible light range. This accounts for the discrepancy, since the relative dielectric constant is not independent of frequency when extremely wide frequency ranges are considered.

Example 8–4

Consider a plane wave in air impinging obliquely on a large, thick slab of polyethylene. The polarization of the wave is in the plane of incidence, as shown in Fig. 8–7. It is desired to determine the reflected and transmitted waves as a function of the angle of incidence, θ_1. First construct the transmission-line analogy shown in the (a) portion of the accompanying figure. It can be seen now that the reflection coefficient will vary as shown in the (b) portion as the incident angle is varied from 0 to 90 degrees. This figure will be recognized as a Smith chart with the impedance contours

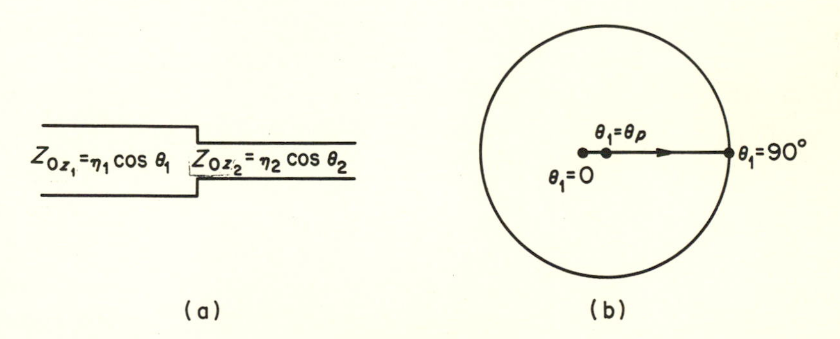

Example 8–4.

suppressed. When the incident angle is zero, we have just the normal incidence case previously analyzed and $K = -1/5$. As θ_1 is increased, K moves to the right along the real axis and ends up at unity for $\theta_1 = 90$ degrees. Note that there will be a value of θ_1 for which the reflection coefficient is zero. No reflection will occur for this value of θ_1, which is known as the "polarizing angle." The reason for this name will become apparent in Example 8–5 in the following article. The polarizing angle is found by equating Z_{0z1} to Z_{0z2}. For the case of **E** in the plane of incidence and $\mu_1 = \mu_2$,

$$\theta_p = \tan^{-1}\sqrt{\frac{\epsilon_2}{\epsilon_1}} \tag{8–28}$$

Now, for any given incident angle θ_1, the refracted angle θ_2 may be computed from Snell's law. Then the resultant component waves may be evaluated. For example, if the incident wave has a peak value of 10 V/m and the incident angle is 60 degrees,

$$\theta_2 = \sin^{-1}\left(\sqrt{\frac{1}{2.25}}\sin 60°\right) = 35.3°$$

and

$$Z_{0z1} = (377)(\cos 60°) = 188.5 \text{ ohms}$$

$$Z_{0z2} = \left(\frac{377}{\sqrt{2.25}}\right)(\cos 35.3°) = 205 \text{ ohms}$$

Therefore

$$K = \frac{205 - 188.5}{205 + 188.5} = 0.042$$

and

$$|E_{x1}^+| = (10)(\cos 60°) = 5 \text{ V/m}$$
$$|E_{x1}^-| = |K|\,|E_{x1}^+| = 0.21 \text{ V/m}$$
$$|E_{x2}^+| = |1 + K|\,|E_{x1}^+| = 5.21 \text{ V/m}$$

The complete wave picture may now be reconstructed from the tangential components and the known directions of propagation. That is,

$$|\mathbf{E}_1^+| = (5)(\sec 60°) = 10 \text{ V/m}$$
$$|\mathbf{E}_1^-| = (0.21)(\sec 60°) = 0.42 \text{ V/m}$$
$$|\mathbf{E}_2^+| = (5.21)(\sec 35.3°) = 6.38 \text{ V/m}$$

The phase relationships may also be computed from the transmission-line analogy. In this case it is fairly obvious that the incident, reflected, and transmitted waves are all in phase at any given point on the boundary. Also, the magnetic field intensities associated with the above electric field intensities can be obtained from the intrinsic impedance relationship. In the example just considered, the angle for which total reflection occurred was the limiting case of tangential incidence. An interesting situation leading to total reflection is found by consideration of Eq. (8–27). Suppose that $\mu_2\epsilon_2$ is smaller than $\mu_1\epsilon_1$, resulting in $\sqrt{\mu_2\epsilon_2/\mu_1\epsilon_1} < 1$. This means that for all angles of incidence greater than

$$\theta_1 = \theta_c = \sin^{-1}\sqrt{\frac{\mu_2\epsilon_2}{\mu_1\epsilon_1}} \tag{8–29}$$

$\sin\theta_2$ will be forced to take on values greater than unity. Let us say that $\sin\theta_2 = a$, where $a \geq 1$. Then, using the relationship $\sin^2\theta + \cos^2\theta = 1$, we see that

$$\cos^2 \theta_2 = 1 - a^2$$

which is a negative number. Let us call the negative number $-b^2$ (b being a positive real number), so that $\cos \theta_2 = \pm jb$. Now let us see what this does to the transmission-line analogy Eqs. (8–24), (8–25), and (8–26). These become

$$Z_{0_{z_2}} = \eta_2 \cos \theta_2 = \eta_2(\pm jb) \tag{8–30}$$

$$\gamma_2 = j\omega\sqrt{\mu_2\epsilon_2}(\pm jb) \tag{8–31}$$

$$v_2 = \frac{1}{\sqrt{\mu_2\epsilon_2}(\pm jb)} \tag{8–32}$$

Note that the z-directed wave impedance has become imaginary, the propagation constant is now completely real, and the z-directed phase velocity is imaginary. If we look at the propagation constant and choose the wave that corresponds to a decaying exponential in region 2 (we do this because there is no source in region 2 and the fields should be zero at infinitely large z), it is apparent that the proper sign in Eqs. (8–30) through (8–32) is the negative one. This leads to the equations

$$Z_{0_{z_2}} = -j\eta b \tag{8–33}$$

$$\gamma_2 = \omega b\sqrt{\mu_2\epsilon_2} \tag{8–34}$$

Since the wave impedance in region 2 is imaginary, it is evident that the reflection coefficient is unity because this impedance serves as the load on the transmission-line analog of region 1.

It is interesting to examine the complex Poynting vector at the interface. The power density flowing into region 2 is

$$P_{2_z} = \tfrac{1}{2}(E_x H_y^*)$$

Say that $E_x = E_0 e^{-\gamma_2 z}$; then substituting into Eq. (8–23b) results in

$$H_y = \frac{+j\omega\epsilon_2}{\gamma_2} E_0 e^{-\gamma_2 z}$$

Using $\gamma_2 = \omega b\sqrt{\mu_2\epsilon_2}$ and evaluating P_{2_z} at $z = 0$,

$$P_{2_z} = \frac{|E_0|^2}{2}\left(\frac{-j}{b\eta_2}\right) \text{ watts/meter}^2$$

The power flow is completely imaginary at the interface, and if the z-dependence had been retained, would be decreasing exponentially as the wave penetrated into region 2. Actually, we should have anticipated the imaginary Poynting vector when we saw that the z-directed wave impedance was imaginary. A wave that has the characteristics described in the preceding paragraphs, that is, imaginary wave impedance and a real propagation constant, is called an *evanescent wave*. The angle of incidence at which this behavior becomes evident is called the *critical angle*, θ_c. It is very important to remember the condition $\mu_2\epsilon_2 < \mu_1\epsilon_1$, which must be satisfied in order to realize total reflection for angles of incidence less than 90 degrees.

8–10. OBLIQUE INCIDENCE; E NORMAL TO THE PLANE OF INCIDENCE

A plane wave with its polarization normal to the plane of incidence is shown in Fig. 8–8. The analysis of this case is similar to the preceding

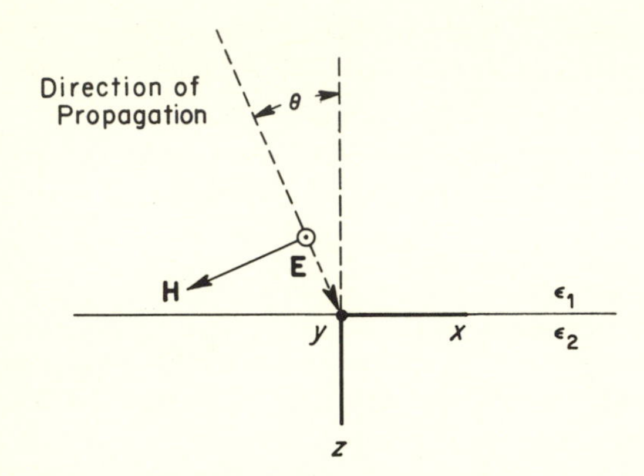

Fig. 8–8. Plane wave with electric field normal to plane of incidence.

case, and the expressions for velocity of propagation and propagation constant are the same as before, i.e., Eqs. (8–25) and (8–26). However, the formula for characteristic impedance differs from that of the previous case and is

$$Z_{0z} = \eta \sec \theta \tag{8–35}$$

This can be seen intuitively in Fig. 8–8: the tangential component of **H** is obtained by multiplying the magnitude of **H** by $\cos \theta$. Since the characteristic impedance is basically the ratio of E_x^+ to H_y^+, it can be seen that the $\cos \theta$ factor shows up in the denominator rather than in the numerator as before. This can also be shown formally, but because the derivation is so similar to that of the other case, it will not be repeated. An example will illustrate how this one dissimilarity between the two transmission-line analogies makes a vast difference in the solutions for the two respective cases of polarization.

Example 8–5

Again consider Example 8–4, but with the polarization normal to the plane of incidence rather in the plane of incidence. The resulting transmission-line analogy and reflection coefficient locus are shown in the (a) and (b) portions, respectively, of the accompanying illustration. The reflection coefficient K is $-\frac{1}{5}$ as before for $\theta_1 = 0$.

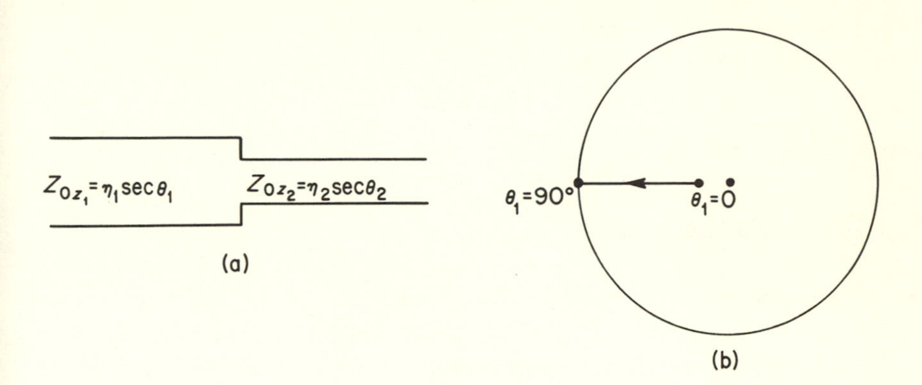

Example 8–5.

This time, though, K moves to the left on the real axis as θ_1 is increased and ends up at -1 for $\theta = 90$ degrees. Note that it never goes through the $K = 0$ point, and thus there is no incident angle for which total absorption takes place. The reason in Example 8–4 for distinguishing the incident angle which gives rise to total absorption by calling it the polarizing angle should now be clear. If the incoming wave consists of a superposition of both types of polarization, only one type will be reflected for this angle of incidence, and the resulting reflected wave will be polarized with a known direction of polarization.

The analysis for a specific angle of incidence is similar to that of Art. 8–9. For the same numerical values used before (i.e., $|E_1^+| = 10$ V/m and $\theta_1 = 60$ degrees), the numerical solution is as follows:

$$\theta_2 = \sin^{-1}\left(\sqrt{\frac{1}{2.25}}\sin 60°\right) = 35.3°$$

$$Z_{0z_1} = (377)(\sec 60°) = 754 \text{ ohms}$$

$$Z_{0z_2} = \left(\frac{377}{\sqrt{2.25}}\right)(\sec 35.3°) = 307 \text{ ohms}$$

Therefore

$$K = \frac{307 - 754}{307 + 754} = -0.42$$

Since the E fields of all three (incident, reflected, and transmitted) component waves are tangent to the boundary, they may be computed directly as

$$|E_1^+| = 10 \text{ V/m}$$
$$|E_1^-| = (10)(0.42) = 4.2 \text{ V/m}$$
$$|E_2^+| = (10)|1 + K| = 5.8 \text{ V/m}$$

Note that these are different from the values obtained for the other type of polarization. As before, the associated magnetic field intensities may be obtained from the intrinsic impedance relationship.

The critical angle discussion included in the last section is also applicable to the situation described in this section. The only difference is in the characteristic impedance. In this case

$$Z_{0z_2} = j\frac{\eta}{b} \qquad (8\text{–}36)$$

The propagation constant is given by Eq. (8–34) and the critical angle by Eq. (8–29).

8–11. POLARIZATION

The simple uniform plane waves considered so far in this chapter have been linearly polarized. This means that an observer at some point in space would see the electric field vector oriented parallel to some line in space. We can assure ourselves that this is indeed true by considering the electric field $\mathbf{E} = \mathbf{a}_x E_0 e^{-j\beta z}$ in real time. The time-dependent electric field is $\mathcal{E} = \mathbf{a}_x E_0 \cos(\omega t - \beta z)$. At $z = 0$, the electric field is

$$\mathcal{E} = \mathbf{a}_x E_0 \cos \omega t$$

which is parallel to the x-axis.

A more complex wave is formed by the superposition of two uniform linearly polarized plane waves with electric field vectors at right angles to each other in space and differing in time phase by θ degrees. The electric field expression in complex phasor form is

$$\mathbf{E} = \mathbf{a}_x E_1 e^{-j\beta z} + \mathbf{a}_y E_2 e^{j\theta} e^{-j\beta z}$$

E_1 and E_2 are considered to be real positive numbers. The time-dependent electric field is

$$\mathcal{E} = \mathbf{a}_x E_1 \cos(\omega t - \beta z) + \mathbf{a}_y E_2 \cos(\omega t + \theta - \beta z)$$

At $z = 0$, the electric field expression becomes

$$\mathcal{E} = \mathbf{a}_x E_1 \cos \omega t + \mathbf{a}_y E_2 \cos(\omega t + \theta).$$

The tip of the electric field vector traces an ellipse in the x-y-plane. Such a wave is said to be *elliptically polarized.* As an example of the foregoing, consider the case where $\theta = \pi/4$ and $E_1 = 2E_2$. Looking in the direction of the $+z$-axis, one sees that the electric field is as shown in Fig. 8–9 for the values of ωt indicated. As time increases, the tip of the electric field vector rotates in a counterclockwise direction when looking in the direc-

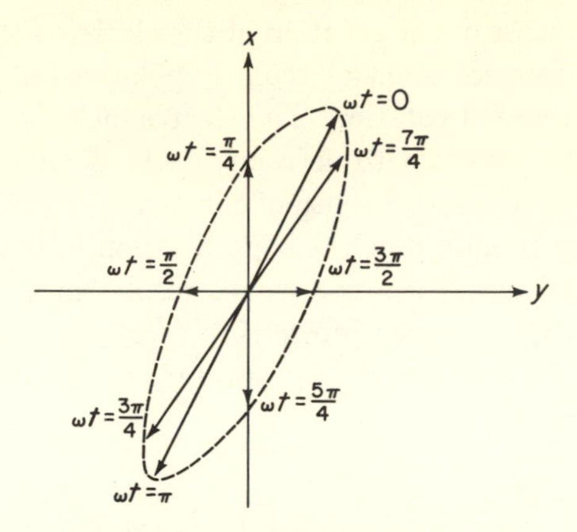

Fig. 8–9. Elliptical polarization.

tion of propagation. For this reason the wave is referred to as an elliptically polarized uniform plane wave with a counterclockwise sense of rotation. A special case of elliptical polarization occurs when $E_1 = E_2$ and $\theta = \pm\pi/2$. For this situation, the electric field at $z = 0$ is

$$\mathcal{E} = \mathbf{a}_x E_1 \cos \omega t + \mathbf{a}_y E_1 \cos \left(\omega t \pm \frac{\pi}{2}\right)$$

or

$$\mathcal{E} = \mathbf{a}_x E_1 \cos \omega t \mp \mathbf{a}_y E_1 \sin \omega t.$$

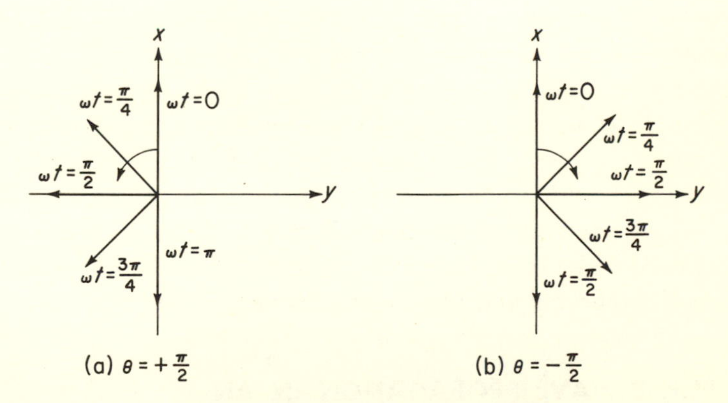

Fig. 8–10. Circular polarization.

The electric field vector at $z = 0$ is plotted for values of ωt in Figs. 8–10(a) and 8–10(b). Consideration of the plots shows that, in either case, the tip of the electric field vector traces a circle in the x-y-plane. In one case

$(\theta = -\pi/2)$, the sense of the rotation of the electric field vector is clockwise. This field is referred to as a circularly polarized uniform plane wave with a clockwise sense of rotation. The electric field vector of the second case $(\theta = +\pi/2)$ is referred to as a circularly polarized uniform plane wave with a counterclockwise sense of rotation.

It is interesting to note that the superposition of two counterrotating circularly polarized uniform plane waves results in a linearly polarized uniform plane wave. Another interesting facet of circular polarization becomes evident if we consider the instantaneous Poynting vector $\mathcal{P}$ for a circularly polarized wave:

$$
\begin{aligned}
\mathcal{P} &= \mathcal{E} \times \mathcal{H} \\
&= [\mathbf{a}_x E_0 \cos(\omega t - \beta z) \pm \mathbf{a}_y E_0 \sin(\omega t - \beta z)] \\
&\quad \times \left[\mp \mathbf{a}_x \frac{E_0}{\eta} \sin(\omega t - \beta z) + \mathbf{a}_y \frac{E_0}{\eta} \cos(\omega t - \beta z)\right] \\
&= \mathbf{a}_z \left[\frac{E_0^2}{\eta} \cos^2(\omega t - \beta z) + \frac{E_0^2}{\eta} \sin^2(\omega t - \beta z)\right] \\
&= \mathbf{a}_z \frac{E_0^2}{\eta}
\end{aligned}
$$

The instantaneous power density flow is constant at any point in space. This is in marked contrast to a linearly polarized wave, which exhibits a $\cos^2(\omega t - \beta z)$ or $\sin^2(\omega t - \beta z)$ dependence in its instantaneous Poynting vector.

Since the elliptically polarized waves can be considered to be the superposition of two linearly polarized waves of the appropriate amplitude and phase which are oriented at right angles in space, the transmitted and reflected waves arising from an elliptically polarized wave incident upon a material interface can be determined by considering the reflected and transmitted waves due to the constituent linearly polarized waves. The reflected and transmitted waves due to these components are then combined with appropriate amplitude and phase relationships to get the composite reflected and transmitted waves.

8–12. PLANE WAVE PROPAGATION IN AN ANISOTROPIC MEDIUM

Knowledge of the propagation characteristics of a uniform plane wave in an anisotropic medium is essential in understanding many microwave and radio-wave propagation phenomena. As pointed out in Chapter 7, anisotropic media are characterized by tensor constitutive parameters. Plasmas and ferrites are two frequently encountered examples of anisotropic media. The tensor constitutive parameters for plasmas and ferrites are developed in Appendix G. We will study the propagation characteristics of an anisotropic media by considering the specific case of a linearly polarized uniform plane wave in a ferrite medium. A ferrite medium is characterized by a scalar permittivity ϵ, zero conductivity, and a tensor permeability $\bar{\bar{\mu}}$. The tensor permeability is defined by the matrix

$$\bar{\bar{\mu}} = \mu_0 \begin{bmatrix} \mu & jK & 0 \\ -jK & \mu & 0 \\ 0 & 0 & 1 \end{bmatrix} \tag{8–37}$$

for a z-directed magnetic bias field. The elements of the permeability matrix, μ and K, are, at most, complex scalars which involve the physical properties of the ferrite. Maxwell's equations are immediately written as

$$\begin{aligned} -\nabla \times \mathbf{E} &= j\omega\bar{\bar{\mu}}\mathbf{H} \\ \nabla \times \mathbf{H} &= j\omega\epsilon\mathbf{E} \\ \nabla \cdot \bar{\bar{\mu}}\mathbf{H} &= 0 \\ \nabla \cdot \epsilon\mathbf{E} &= 0 \end{aligned}$$

We will look for a solution representing waves propagating along the z-axis and having $E_z = H_z = 0$ and $\dfrac{\partial}{\partial x} = \dfrac{\partial}{\partial y} = 0$.

Expanding the first two of Maxwell's equations and applying the constraints results in

$$\begin{aligned} &\text{(a)} & \frac{\partial E_y}{\partial z} &= j\omega\mu\mu_0 H_x - \omega\mu_0 K H_y \\ &\text{(b)} & -\frac{\partial E_x}{\partial z} &= \omega\mu_0 K H_x + j\omega\mu\mu_0 H_y \\ &\text{(c)} & -\frac{\partial H_y}{\partial z} &= j\omega\epsilon E_x \\ &\text{(d)} & \frac{\partial H_x}{\partial z} &= j\omega\epsilon E_y \end{aligned} \tag{8–38}$$

There are four equations and four unknowns. If we assume that the waves propagate as $e^{-\gamma z}$, we can further reduce the set of equations to

$$\begin{aligned} -\gamma E_y &= j\omega\mu\mu_0 H_x - \omega\mu_0 K H_y \\ \gamma E_x &= \omega\mu_0 K H_x + j\omega\mu\mu_0 H_y \\ \gamma H_y &= j\omega\epsilon E_x \\ -\gamma H_x &= j\omega\epsilon E_y \end{aligned}$$

Substituting and reducing result in two equations:

$$\begin{aligned} &\text{(a)} \quad (\gamma^2 + \omega^2\mu\mu_0\epsilon)H_x + j\omega^2\mu_0 K\epsilon H_y = 0 \\ &\text{(b)} \quad (-j\omega^2\mu_0 K\epsilon)H_x + (\gamma^2 + \omega^2\mu\mu_0\epsilon)H_y = 0 \end{aligned} \tag{8–39}$$

There will be a solution (nontrivial H_x and H_y) if the following characteristic equation is satisfied:

$$\gamma^4 + 2\gamma^2\omega^2\mu\mu_0\epsilon + \omega^4\epsilon^2\mu^2_0(\mu^2 - K^2) = 0 \tag{8–40}$$

The roots are

$$\gamma^2 = -\omega^2\mu\mu_0\epsilon \pm \omega^2\epsilon\mu_0 K \tag{8–41}$$

or

$$\gamma = \pm j\omega\sqrt{\mu_0\epsilon}\sqrt{\mu \pm K} \tag{8–42}$$

Thus, there are four roots to the characteristic equation:

$$\begin{aligned} &\text{(a)} \quad \gamma_1 = j\omega\sqrt{\mu_0\epsilon}\sqrt{\mu + K} \\ &\text{(b)} \quad \gamma_2 = j\omega\sqrt{\mu_0\epsilon}\sqrt{\mu - K} \\ &\text{(c)} \quad \gamma_3 = -j\omega\sqrt{\mu_0\epsilon}\sqrt{\mu + K} \\ &\text{(d)} \quad \gamma_4 = -j\omega\sqrt{\mu_0\epsilon}\sqrt{\mu - K} \end{aligned} \tag{8–43}$$

We see that γ_1 and γ_2 indicate waves traveling in the $+z$-direction while γ_3 and γ_4 indicate two waves traveling in the $-z$-direction. Now consider just the waves traveling in the $+z$-direction and examine the form of the waves. Using γ_1, substitute back into Maxwell's equations to get

$$\begin{aligned} &\text{(a)} \quad -j\omega\sqrt{\mu_0\epsilon}\sqrt{\mu + K}E_y = j\omega\mu\mu_0 H_x - \omega\mu_0 K H_y \\ &\text{(b)} \quad j\omega\sqrt{\mu_0\epsilon}\sqrt{\mu + K}E_x = \omega\mu_0 K H_x + j\omega\mu\mu_0 H_y \\ &\text{(c)} \quad j\omega\sqrt{\mu_0\epsilon}\sqrt{\mu + K}H_y = j\omega\epsilon E_x \\ &\text{(d)} \quad -j\omega\sqrt{\mu_0\epsilon}\sqrt{\mu + K}H_x = j\omega\epsilon E_y \end{aligned} \tag{8–44}$$

These reduce to

$$\begin{aligned} &\text{(a)} \quad j(\mu + K)H_x = j\mu H_x - K H_y \\ &\text{(b)} \quad j(\mu + K)H_y = K H_x + j\mu H_y \end{aligned} \tag{8–45}$$

These equations are satisfied if

$$H_x = jH_y \tag{8–46}$$

Substituting back into Maxwell's equations results in

$$E_y = -jE_x \tag{8–47}$$

Looking back to Art. 8–11, we see that this is a circularly polarized wave with a clockwise sense of rotation. It appears to be propagating in a medium characterized by a relative permeability of $\mu + K$.

Now consider the other root, $\gamma_2 = j\omega\sqrt{\mu_0\epsilon(\mu - K)}$. In this case, substituting into Maxwell's equations results in

$$\begin{aligned}
&\text{(a)} & -j\omega\sqrt{\mu_0\epsilon}\sqrt{\mu - K}E_y &= j\omega\mu\mu_0 H_x - \omega\mu_0 K H_y \\
&\text{(b)} & j\omega\sqrt{\mu_0\epsilon}\sqrt{\mu - K}E_x &= \omega\mu_0 K H_x + j\omega\mu\mu_0 H_y \\
&\text{(c)} & j\omega\sqrt{\mu_0\epsilon}\sqrt{\mu - K}H_y &= j\omega\epsilon E_x \\
&\text{(d)} & -j\omega\sqrt{\mu_0\epsilon}\sqrt{\mu - K}H_x &= j\omega\epsilon E_y
\end{aligned} \tag{8-48}$$

which reduce to

$$\begin{aligned}
&\text{(a)} & j(\mu - K)H_x &= j\mu H_x - K H_y \\
&\text{(b)} & j(\mu - K)H_y &= K H_x + j\mu H_y
\end{aligned} \tag{8-49}$$

In this case,

$$H_x = -jH_y \tag{8-50}$$

and

$$E_y = jE_x \tag{8-51}$$

This wave is seen to be a circularly polarized wave with a counterclockwise sense of rotation which appears to be propagating in a medium characterized by a relative permeability of $\mu - K$.

A physical picture of what we have just developed mathematically can be attained by considering the expression for the torque on a magnetic dipole $\mathbf{m}$ subjected to a dc magnetic field $\mathcal{B}_0$. This is Eq. (G–3) of Appendix G:

$$\mathcal{T} = \mathbf{m} \times \mathcal{B}_0$$

The origin of the magnetic dipole is a spinning bound electron which, possessing mass, has an angular momentum vector $\mathcal{J}_m$. As stated in Appendix G, the angular momentum vector is antiparallel to the magnetic dipole vector. The torque causes the dipole to precess about the magnetic field with a precessional frequency $\boldsymbol{\omega}_0$, defined by

$$\mathcal{T} = \boldsymbol{\omega}_0 \times \mathcal{J}_m$$

Consideration of the foregoing statements shows that the vector $\boldsymbol{\omega}_0$ is parallel to $\mathcal{B}_0$. Thus the tip of the magnetic dipole moves in a clockwise sense about $\mathcal{B}_0$ when looking in the direction of $\mathcal{B}_0$, and traces a circle in a plane normal to $\mathcal{B}_0$.

When a circularly polarized RF magnetic field interacts with an assemblage of dipoles, the field will be most strongly coupled to the dipoles when the sense of rotation of the RF field is the same as the sense of the precession of dipoles. For this case the effective relative permeability $\mu_{r_{eff}} = \mu^+ = \mu + K$. When the sense of rotation of the RF field is opposite to the sense of precession of the dipoles, the coupling is weaker and $\mu_{r_{eff}} = \mu^- = \mu - K$.

Now we can consider the propagation of a linearly polarized uniform plane wave in this medium. We will consider an electric field at $z = 0$ to be

$$\mathbf{E} = \mathbf{a}_x E_0$$

Further, we will consider this linearly polarized wave to be made up of two circularly polarized waves designated as $\mathbf{E}_1 = \mathbf{a}_x \frac{E_0}{2} - j\mathbf{a}_y \frac{E_0}{2}$ and $\mathbf{E}_2 = \mathbf{a}_x \frac{E_0}{2} + j\mathbf{a}_y \frac{E_0}{2}$ at $z = 0$. These are circularly polarized waves with clockwise and counterclockwise sense, respectively, so they propagate in the $+z$-direction with γ_1 describing the z-dependence of $\mathbf{E}_1$ and γ_2 describing the z-dependence of $\mathbf{E}_2$. Thus, the field in the ferrite medium is

$$\mathbf{E} = \mathbf{a}_x \frac{E_0}{2} (e^{-\gamma_1 z} + e^{-\gamma_2 z}) - j\mathbf{a}_y \frac{E_0}{2} (e^{-\gamma_1 z} - e^{-\gamma_2 z}) \tag{8-52}$$

Since γ_1 and γ_2 are not equal, it is evident that a y-directed component of the electric field will result. Let us consider this field at some point $z = l$. It will simplify matters if we make the following definitions:

$$\begin{aligned} &\text{(a)} \quad j\theta_1 = \gamma_1 l = j\omega\sqrt{\mu_0\epsilon}\sqrt{\mu + K}\, l = j(\theta_0 + \Delta) \\ &\text{(b)} \quad j\theta_2 = \gamma_2 l = j\omega\sqrt{\mu_0\epsilon}\sqrt{\mu - K}\, l = j(\theta_0 - \Delta) \end{aligned} \tag{8-53}$$

The electric field expression becomes

$$\mathbf{E} = \mathbf{a}_x E_0 e^{-j\theta_0} \cos \Delta - \mathbf{a}_y E_0 e^{-j\theta_0} \sin \Delta \tag{8-54}$$

It is apparent that the electric field vector is still linearly polarized but it is physically reoriented in space, making an angle of $-\Delta$ with respect to the x-axis. This phenomenon is known as *Faraday rotation.* The rotation angle is Δ, where

$$\Delta = j \frac{\omega\sqrt{\mu_0\epsilon}(\sqrt{\mu + K} - \sqrt{\mu - K})l}{2} = -\frac{(\theta_1 - \theta_2)}{2} \tag{8-55}$$

This is a sensible answer, for we recognize that, in the case of circular polarization, the time required for the electric field vector to rotate through 360 degrees is exactly equal to the time required for the wave to travel one wavelength. That is, if you, as an observer at some point in space, could flag a particular wave front as the reference, then in the time that it takes for the field vector to rotate 360 degrees, the reference phase front would have traveled one wavelength. Since the propagation constants are different for the two senses of rotation, the wavelengths will be different. Thus, the distance l is a different fraction of a wavelength for the two senses of rotation and the net $\mathbf{E}$ will rotate in space as it propagates.

It is interesting to consider the waves traveling in the $-z$-direction also, because one sees that the Faraday rotation phenomenon is nonreciprocal. That is, if the electric field vector rotates counterclockwise with respect to the direction of propagation when traveling in the $+z$-direction, then it will rotate clockwise with respect to the direction of propagation when traveling in the $-z$-direction. (The proof of this is left as an exercise.) These results are summarized in Fig. 8–11.

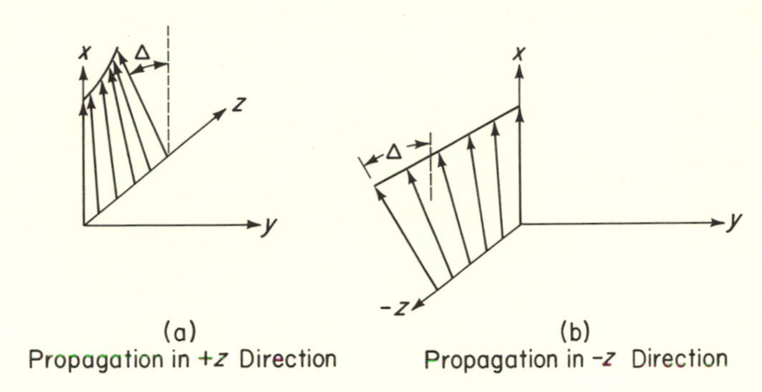

Fig. 8–11. Faraday rotation.

PROBLEMS

8–1. It was tacitly assumed in Art. 8–1 that a uniform plane wave of the form

$$E_x = E_0 e^{-j\beta z}$$
$$H_y = (E_0/\eta) e^{-j\beta z}$$

represents a solution of Maxwell's equation. Show by actual substitution in Maxwell's equations that this is so if $\beta = \omega\sqrt{\mu\epsilon}$.

8–2. A uniform plane wave is traveling in the z-direction in free space. The electric field of this wave is polarized in the x-direction and has an rms value of 10 V/m. The frequency is 300 MHz. Find (a) the magnitude and direction of the magnetic field intensity of this wave; (b) the average power density associated with this wave; and (c) the wavelength and velocity of propagation.

8–3. A square loop 10 cm on a side is oriented so that the plane wave of Problem 8–2 impinges on it with the magnetic field normal to the plane of the loop. Find the induced voltage in the loop.

8–4. Let us arbitrarily say that a ratio of 100:1 between σ and $\omega\epsilon$ is the criterion for determining whether a dielectric or a conductor is "good." Find the frequency range for which seawater may be classed as a good conductor and the range for which it may be considered to be a good dielectric. Assume $\sigma = 4$ mhos/m and $\epsilon_r = 81$.

8–5. Show that the power dissipated per unit area as given by $\frac{1}{2}|\mathbf{J}_s|^2R_s$ is the same as that obtained by using the actual current distribution and summing the effect of this throughout the volume involved.

8–6. Find the surface resistance and depth of penetration for copper at 60 Hz, 60 MHz, and 6000 MHz. On the basis of these calculations would you expect the current distribution to be uniform in large 60-Hz power cables which may be of the order of 1 in. in diameter?

8–7. In Example 8–2 the power density transmitted into the copper was computed by using the transmission-line analogy. If our theory is correct, we should obtain the same result from the $\frac{1}{2}|\mathbf{J}_s|^2R_s$ formula. Show that this is the case.

8–8. Consider the case of a uniform plane wave obliquely incident upon a perfectly conducting surface (with H in the plane of incidence) at angle θ. Show that if an electric field null occurs at a distance $d = \lambda/2 \cos\theta$ back from the surface this means that one could place a second conducting surface at $z = -d$ without violating any boundary conditions. This forms a structure which will guide energy in the x-direction. Consider θ to be variable and determine the lowest frequency for which this parallel plate structure will guide energy in the x-direction. What is the phase velocity for the component of the waves traveling in the x-direction?

8–9. A uniform plane wave impinges normally on a plane polyethylene-copper boundary from the polyethylene side. The incident wave has a peak electric field of 1 V/m and the frequency is 300 MHz. Assuming the copper to be a perfect conductor, (a) find the electric and magnetic field intensities at the surface of the copper, and (b) describe the current flow in the copper, giving both density and direction of flow.

8–10. Consider a uniform plane wave in air impinging normally on a large slab of polyethylene which is 10 cm thick. The frequency of the incident wave is 300 MHz and the peak value of the electric field is 10 V/m. (a) Find the electric field intensity at the incident surface. It may be assumed that the air space is infinite in extent on either side of the polyethylene slab. (b) Find the fraction of incident power density which is transmitted through the slab. (c) Show that energy is conserved (i.e., incident power − reflected power = transmitted power).

8–11. A dielectric coating is to be applied at an air–pyrex glass ($\epsilon_r = 4.5$) boundary in order to eliminate reflections at the boundary. The frequency is 3000 MHz. By using the quarter-wave transformer analogy, determine the thickness and dielectric constant of the matching coating.

8–12. Consider a uniform plane wave impinging obliquely on an air-to-glass ($\epsilon_r = 4.0$) boundary from the air side of the boundary. The boundary is flat and the dielectrics may be considered to be infinite in extent on both sides of the boundary. The incident wave is polarized in the plane of incidence. Find the incident angle for which total absorption takes place. Would this also apply for a wave which is polarized normal to the plane of incidence?

8–13. Consider the incident wave in Problem 8–12 impinging on the boundary from the glass side rather than vice versa. (a) Sketch the reflection coefficient locus as the incident angle (in glass) varies from 0 to 90 degrees and (b) find the range of incident angles for which total reflection takes place.

8–14. Consider the air-glass interface of Problem 8–12 with a circularly polarized (clockwise sense of rotation) normally incident upon it. Write explicit expressions for the reflected and transmitted waves. Assume that the amplitude of the incident wave is E_0 volts/meter.

8–15. Show that a linearly polarized uniform plane wave propagating in the negative z-direction in the ferrite medium of Art. 8–12 exhibits Faraday rotation in the same sense as the Faraday rotation exhibited by a similar wave propagating in the $+z$-direction.

9

Guided Waves

9–1. GENERAL REMARKS

We shall now look at a general class of waves loosely referred to as guided waves. They derive their name from the metallic structures which serve to "guide" the waves in a particular direction. A few examples of such structures are shown in Fig. 9–1. Note that conventional transmission lines fall into this general category.

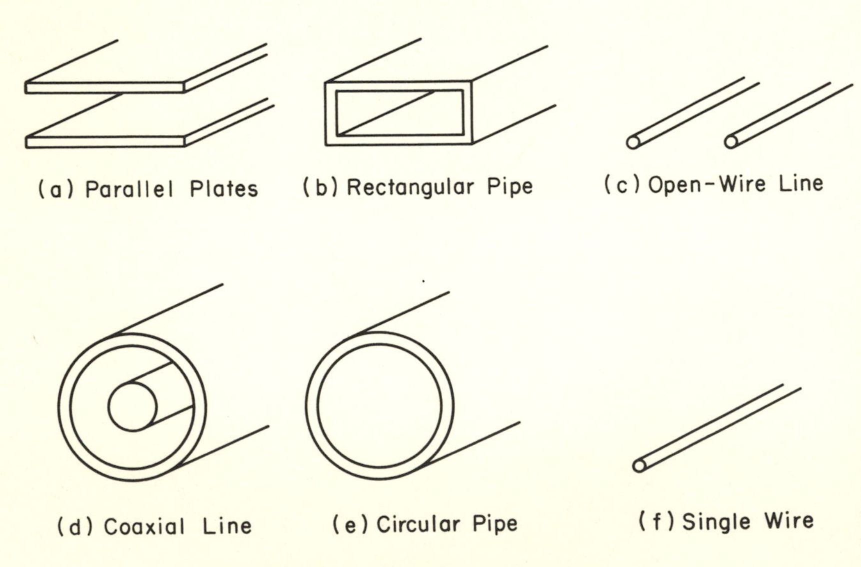

Fig. 9–1. Guided wave systems.

9-2. BASIC EQUATIONS

All guided wave systems have one thing in common: they exhibit propagation properties in one direction. As a matter of convention this will be denoted as the z-direction. Thus, before discussing specific examples, some general equations may be derived which will apply to all waveguide systems irrespective of the cross-sectional geometry. As was the case in transmission-line theory, we shall begin with the study of lossless systems. All conductors and dielectrics will be considered perfect, and the effect of losses will be considered later. The starting point for this discussion will be Maxwell's equations, since they are the equations which govern the behavior of the electric and magnetic fields within the region of wave propagation. When written in rectangular form for a perfect dielectric, they are

$$\begin{aligned}
&(a) \quad \frac{\partial \mathcal{E}_z}{\partial y} - \frac{\partial \mathcal{E}_y}{\partial z} = -\mu \frac{\partial \mathcal{H}_x}{\partial t} \qquad &&(d) \quad \frac{\partial \mathcal{H}_z}{\partial y} - \frac{\partial \mathcal{H}_y}{\partial z} = \epsilon \frac{\partial \mathcal{E}_x}{\partial t} \\
&(b) \quad \frac{\partial \mathcal{E}_x}{\partial z} - \frac{\partial \mathcal{E}_z}{\partial x} = -\mu \frac{\partial \mathcal{H}_y}{\partial t} \qquad &&(e) \quad \frac{\partial \mathcal{H}_x}{\partial z} - \frac{\partial \mathcal{H}_z}{\partial x} = \epsilon \frac{\partial \mathcal{E}_y}{\partial t} \qquad (9\text{-}1) \\
&(c) \quad \frac{\partial \mathcal{E}_y}{\partial x} - \frac{\partial \mathcal{E}_x}{\partial y} = -\mu \frac{\partial \mathcal{H}_z}{\partial t} \qquad &&(f) \quad \frac{\partial \mathcal{H}_y}{\partial x} - \frac{\partial \mathcal{H}_x}{\partial y} = \epsilon \frac{\partial \mathcal{E}_z}{\partial t}
\end{aligned}$$

At this point $\boldsymbol{\mathcal{E}}$ and $\boldsymbol{\mathcal{H}}$ are functions of x, y, z, and t, and following the previous notation, are denoted by boldface script letters.

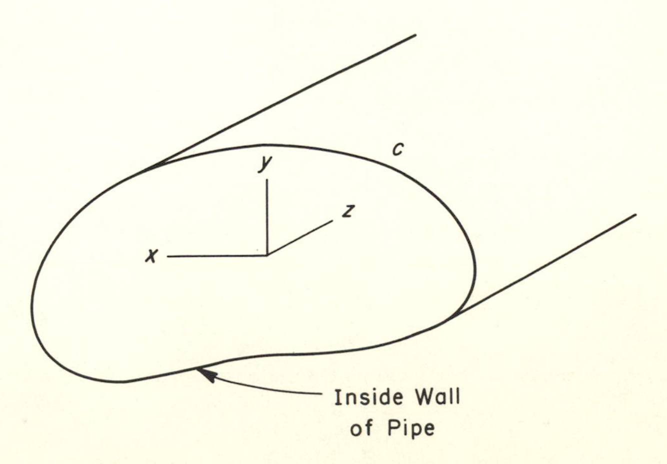

Fig. 9-2. Waveguide coordinate system.

As mentioned before, one cannot write a simple explicit solution for these equations because there are an infinite number of them. However, in the problem at hand we are looking for a particular class of solutions, namely, those which have a sinusoidal variation with time and propagation properties in the z-direction. This enables us to narrow the possible solutions to the point where we can enumerate them and attach physical significance to them. The first step then will be to assume sinusoidal form timewise and switch to the complex phasor notation of ac circuit theory. Equations (9–1) may be written in terms of phasors by merely replacing the $\partial/\partial t$ operator with $j\omega$. The equations then become

$$\begin{aligned}
&\text{(a)}\quad \frac{\partial E_z}{\partial y} - \frac{\partial E_y}{\partial z} = -j\omega\mu H_x &&\text{(d)}\quad \frac{\partial H_z}{\partial y} - \frac{\partial H_y}{\partial z} = j\omega\epsilon E_x \\
&\text{(b)}\quad \frac{\partial E_x}{\partial z} - \frac{\partial E_z}{\partial x} = -j\omega\mu H_y &&\text{(e)}\quad \frac{\partial H_x}{\partial z} - \frac{\partial H_z}{\partial x} = j\omega\epsilon E_y \qquad (9\text{–}2) \\
&\text{(c)}\quad \frac{\partial E_y}{\partial x} - \frac{\partial E_x}{\partial y} = -j\omega\mu H_z &&\text{(f)}\quad \frac{\partial H_y}{\partial x} - \frac{\partial H_x}{\partial y} = j\omega\epsilon E_z
\end{aligned}$$

The capital letters E and H are used here to denote complex phasor quantities. They are functions of x, y, and z (and *not* time).[1]

Now we shall restrict the field components further by assuming propagation in the longitudinal direction of the pipe. This is the $+z$-direction, as shown in Fig. 9–2. If there is to be propagation in the z-direction, the field quantities must all be of the form

$$\begin{aligned}
E_x(x, y, z) &= E_x'(x, y)e^{-\gamma z} \\
E_y(x, y, z) &= E_y'(x, y)e^{-\gamma z} \qquad (9\text{–}3) \\
&\cdots\cdots\cdots \\
H_z(x, y, z) &= H_z'(x, y)e^{-\gamma z}
\end{aligned}$$

where γ is the propagation constant. When Eqs. (9–3) are substituted into Eqs. (9–2), they reduce to

$$\begin{aligned}
&\text{(a)}\quad \frac{\partial E_z'}{\partial y} + \gamma E_y' = -j\omega\mu H_x' &&\text{(d)}\quad \frac{\partial H_z'}{\partial y} + \gamma H_y' = j\omega\epsilon E_x' \\
&\text{(b)}\quad -\gamma E_x' - \frac{\partial E_z'}{\partial x} = -j\omega\mu H_y' &&\text{(e)}\quad -\gamma H_x' - \frac{\partial H_z'}{\partial x} = j\omega\epsilon E_y' \qquad (9\text{–}4) \\
&\text{(c)}\quad \frac{\partial E_y'}{\partial x} - \frac{\partial E_x'}{\partial y} = -j\omega\mu H_z' &&\text{(f)}\quad \frac{\partial H_y'}{\partial x} - \frac{\partial H_x'}{\partial y} = j\omega\epsilon E_z'
\end{aligned}$$

[1] It will be recalled that the relationship between the phasor and time-domain quantities is

$$\begin{aligned}
\mathcal{E}_x &= \text{Re}\,(E_x e^{j\omega t}) \\
\mathcal{E}_y &= \text{Re}\,(E_y e^{j\omega t}) \\
&\cdots\cdots \\
\mathcal{H}_z &= \text{Re}\,(H_z e^{j\omega t})
\end{aligned}$$

At this point it is possible to solve these six equations for the transverse components (x- and y-directions) in terms of the longitudinal ones (z-direction). The result is

$$
\begin{aligned}
&\text{(a)} \qquad H'_x = \frac{1}{\gamma^2 + \omega^2\mu\epsilon}\left(j\omega\epsilon\,\frac{\partial E'_z}{\partial y} - \gamma\,\frac{\partial H'_z}{\partial x}\right) \\
&\text{(b)} \qquad H'_y = -\frac{1}{\gamma^2 + \omega^2\mu\epsilon}\left(j\omega\epsilon\,\frac{\partial E'_z}{\partial x} + \gamma\,\frac{\partial H'_z}{\partial y}\right) \\
&\text{(c)} \qquad E'_x = -\frac{1}{\gamma^2 + \omega^2\mu\epsilon}\left(\gamma\,\frac{\partial E'_z}{\partial x} + j\omega\mu\,\frac{\partial H'_z}{\partial y}\right) \\
&\text{(d)} \qquad E'_y = \frac{1}{\gamma^2 + \omega^2\mu\epsilon}\left(-\gamma\,\frac{\partial E'_z}{\partial y} + j\omega\mu\,\frac{\partial H'_z}{\partial x}\right)
\end{aligned}
\qquad (9\text{–}5)
$$

Proof of these equations is a routine matter of algebraic substitution. For example, (a) of Eq. (9–5) is obtained by combining (a) and (e) of Eq. (9–4). These equations are an important link in the solution of the waveguide problem, since they determine uniquely the four transverse components, once the longitudinal ones have been obtained. Thus the problem has been reduced to solving for just two of the six components, namely, E'_z and H'_z.

The general procedure now is to consider three separate cases as follows:

1. TM Mode (Tranverse Magnetic): As the name implies, the magnetic field for this mode is transverse to the direction of propagation, i.e., $H'_z = 0$. This mode can exist in either a hollow pipe or open-wire system.
2. TE Mode (Tranverse Electric): This mode has its electric field transverse to the direction of propagation, i.e., $E'_z = 0$. It also can exist in either a hollow pipe or open-wire system.
3. TEM Mode (Transverse Electric and Magnetic): This mode is characterized by no electric or magnetic field in the direction of propagation and is the mode generally encountered in transmission-line work. This mode cannot exist within a hollow pipe with no center conductor, since there would be no current (either conduction or displacement) for transverse magnetic lines to link, as required by Ampere's law. Thus this mode is possible only in an open-wire system where there are two or more separate conductors or in a hollow pipe with a center conductor or conductors.

Now any general wave propagating in a single direction (the z-direction for our coordinate system) may be considered as a superposition of the above three cases.

9–3. DIFFERENTIAL EQUATIONS FOR E'_z AND H'_z

The problem has been reduced to that of finding E'_z and H'_z. Once these have been found, the four transverse components may be obtained by direct substitution into Eqs. (9–5).[2] The differential equations for E'_z and H'_z are obtained directly from Maxwell's equations as follows:

$$\begin{aligned} \nabla \times \mathbf{E} &= -j\omega\mu\mathbf{H} \\ \nabla \times \mathbf{H} &= j\omega\epsilon\mathbf{E} \end{aligned} \tag{9–6}$$

Now taking the curl of both sides and substituting one into the other leads to

$$\begin{aligned} \nabla \times \nabla \times \mathbf{E} &= \omega^2\mu\epsilon\mathbf{E} \\ \nabla \times \nabla \times \mathbf{H} &= \omega^2\mu\epsilon\mathbf{H} \end{aligned} \tag{9–7}$$

The vector identity

$$\nabla \times \nabla \times \mathbf{A} = \nabla(\nabla \cdot \mathbf{A}) - \nabla^2\mathbf{A} \tag{9–8}$$

along with the fact that the divergence of both **E** and **H** is zero for a charge-free region, makes it possible to write Eqs. (9–7) in the form

$$\begin{aligned} \nabla^2\mathbf{E} + \omega^2\mu\epsilon\mathbf{E} &= 0 \\ \nabla^2\mathbf{H} + \omega^2\mu\epsilon\mathbf{H} &= 0 \end{aligned} \tag{9–9}$$

Since the preceding equations are vector equations, similar equations apply to each of the separate rectangular components. In particular, the equations for E_z and H_z are

$$\begin{aligned} \nabla^2 E_z + \omega^2\mu\epsilon E_z &= 0 \\ \nabla^2 H_z + \omega^2\mu\epsilon H_z &= 0 \end{aligned} \tag{9–10}$$

Finally, as $E_z = E'_z e^{-\gamma z}$ and $H_z = H'_z e^{-\gamma z}$, Eqs. (9–10) reduce to

$$\begin{aligned} \nabla_t^2 E'_z + k_c^2 E'_z &= 0 \\ \nabla_t^2 H'_z + k_c^2 H'_z &= 0 \end{aligned} \tag{9–11}$$

where

$$k_c^2 = \gamma^2 + \omega^2\mu\epsilon \tag{9–12}$$

and ∇_t^2 indicates the two-dimensional Laplacian operator; i.e.,

$$\nabla_t^2 = \frac{\partial^2}{\partial x^2} + \frac{\partial^2}{\partial y^2}$$

Equations (9–11) are the basic differential equations which E'_z and H'_z must satisfy. It will be noted that the differential equations are the same

[2] This is, of course, not true in case of the TEM mode, where both E'_z and H'_z are zero. In this case the transverse fields are obtained by the usual static methods. More will be said of this in Art. 9–14.

for both E'_z and H'_z. However, the resulting solutions are not identical, since different boundary conditions must be satisfied in each case. If perfectly conducting walls are assumed, it should be apparent that E'_z must be zero along the boundary. Thus

$$\begin{aligned} \nabla_t^2 E'_z + k_c^2 E'_z &= 0 \\ E'_z &= 0 \quad \text{on } C \end{aligned} \tag{9–13}$$

comprise the differential equation and boundary condition which must be satisfied in the TM case.

The boundary condition for H'_z is not quite so obvious as that for E'_z, as H'_z automatically satisfies the requirement that it be parallel to the boundary. Figure 9–3 is helpful at this point. The differential rectangular

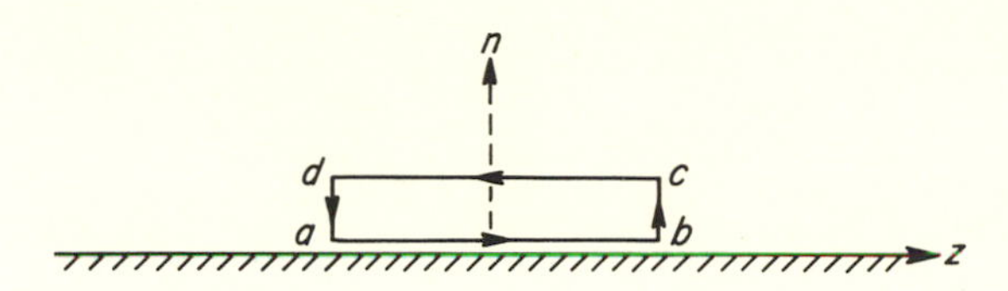

Fig. 9–3. Side view of waveguide wall.

path shown is in a plane normal to the wall of the guide and directed along the z-axis. The electric field at the conducting surface will be normal, and thus no electric flux (and thus no displacement current) will link the rectangle. Therefore, according to Ampere's law, the H along cd must be the same as that along ab. This implies that the rate of change of H_z in the normal direction must be zero at the surface of the wall. Thus the differential equation and boundary condition for the TE mode become

$$\begin{aligned} \nabla_t^2 H'_z + k_c^2 H'_z &= 0 \\ \frac{\partial H'_z}{\partial n} &= 0 \quad \text{on } C \end{aligned} \tag{9–14}$$

9–4. RECTANGULAR WAVEGUIDE: TM MODE

The rectangular waveguide is of considerable practical importance, being perhaps the most common of all of the different types of waveguides. Its solution is relatively simple, and thus it will be considered first. Figure 9–4 shows the coordinate system which will be used for this problem. The TM mode will be dealt with first, and thus a solution of

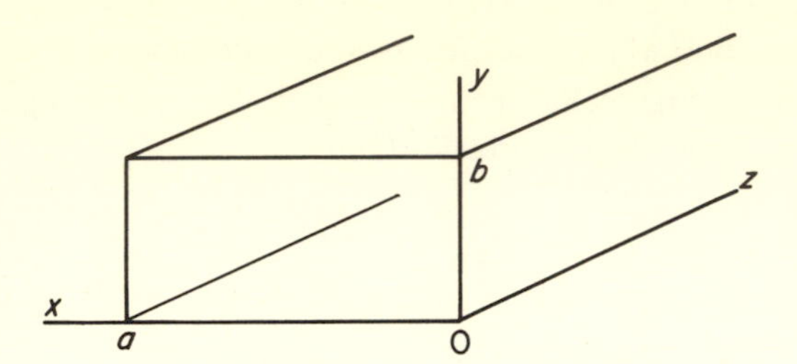

Fig. 9–4. Coordinate system for rectangular waveguide.

Eq. (9–13) is desired. When the ∇_t^2 operator is written out explicitly, this equation becomes

$$\frac{\partial^2 E_z'}{\partial x^2} + \frac{\partial^2 E_z'}{\partial y^2} + k_c^2 E_z' = 0 \tag{9–15}$$

where

$$k_c^2 = \gamma^2 + \omega^2 \mu \epsilon \tag{9–16}$$

The usual procedure for solving this type of differential equation is to assume a product-type solution in x and y and then make this solution fit the differential equation and the boundary condition. Proceeding along this line, let

$$E_z' = X(x)Y(y) \tag{9–17}$$

Then, substituting into Eq. (9–15) yields

$$Y\frac{d^2X}{dx^2} + X\frac{d^2Y}{dy^2} + k_c^2 XY = 0$$

or

$$\frac{1}{X}\frac{d^2X}{dx^2} + \frac{1}{Y}\frac{d^2Y}{dy^2} = -k_c^2 \tag{9–18}$$

Note that the first term of Eq. (9–18) is a function of x alone and the second one of y alone. Thus, in order for the two to sum to a constant, each must be a constant. Letting the first term equal $-k_x^2$ and the second equal $-k_y^2$, we have

$$\frac{d^2X}{dx^2} + k_x^2 X = 0 \tag{9–19}$$

$$\frac{d^2Y}{dy^2} + k_y^2 Y = 0 \tag{9–20}$$

where

$$k_x^2 + k_y^2 = k_c^2 \tag{9–21}$$

We now have reduced the problem to two simple, ordinary differential equations, and their solutions may be written by inspection as

$$X = A \sin k_x x + B \cos k_x x \tag{9–22}$$

$$Y = C \sin k_y y + D \cos k_y y \tag{9–23}$$

where A, B, C, and D are constants. The boundary conditions may now be used to evaluate the constants. As E_z' must be zero along the boundary, the following four constraints must be placed on the solution:

$$\begin{aligned} &\text{(a)} \quad E_z' = 0 \quad \text{along } x = 0 \\ &\text{(b)} \quad E_z' = 0 \quad \text{along } y = 0 \\ &\text{(c)} \quad E_z' = 0 \quad \text{along } x = a \\ &\text{(d)} \quad E_z' = 0 \quad \text{along } y = b \end{aligned} \tag{9–24}$$

The first of these constraints forces B to be zero and the second requires D to equal zero. Thus the solution reduces to

$$E_z' = XY = E_0 \sin k_x x \sin k_y y \tag{9–25}$$

where E_0 is an arbitrary amplitude constant. Now, along $x = a$, E_z' must be zero for all values of y within the interval of interest. The only way this can happen without having a trivial solution is for

$$\sin k_x a = 0$$

or

$$k_x a = m\pi \tag{9–26}$$

where $m = 1, 2, 3, \cdots$.

A similar argument applies along the $y = b$ boundary, leading to the conclusion

$$k_y b = n\pi \tag{9–27}$$

where $n = 1, 2, 3, \cdots$.

The final solution for E_z' may now be written as

$$E_z' = E_0 \sin \frac{m\pi x}{a} \sin \frac{n\pi y}{b} \tag{9–28}$$

where m and n are arbitrary integers (excluding zero) and may be chosen independently. Notice that the solution to the problem is not one simple unique solution. Rather there are an infinite number of solutions which will satisfy the differential equation and boundary conditions. That is, for each different set of integers chosen for m and n, there will be a completely different solution. This is characteristic of waveguide solutions in general, and each different solution is referred to as a mode. The ones described by Eq. (9–28) will be referred to as TM_{mn} modes.

Now the complete solution for all six components may be obtained by substituting $H'_z = 0$ and Eq. (9–28) into Eqs. (9–5). The results are summarized as follows:

Rectangular Waveguide: TM_{mn} Mode

$$\begin{aligned}
E'_z &= E_0 \sin\frac{m\pi x}{a}\sin\frac{n\pi y}{b} \\
H'_z &= 0 \\
H'_x &= \frac{j\omega\epsilon n\pi E_0}{bk_c^2}\sin\frac{m\pi x}{a}\cos\frac{n\pi y}{b} \\
H'_y &= -\frac{j\omega\epsilon m\pi E_0}{ak_c^2}\cos\frac{m\pi x}{a}\sin\frac{n\pi y}{b} \\
E'_x &= -\frac{\gamma m\pi E_0}{ak_c^2}\cos\frac{m\pi x}{a}\sin\frac{n\pi y}{b} \\
E'_y &= -\frac{\gamma n\pi E_0}{bk_c^2}\sin\frac{m\pi x}{a}\cos\frac{n\pi y}{b}
\end{aligned} \tag{9–29}$$

where

$$k_c^2 = \gamma^2 + \omega^2\mu\epsilon = \left(\frac{m\pi}{a}\right)^2 + \left(\frac{n\pi}{b}\right)^2 \tag{9–30}$$

9–5. WAVELENGTH, VELOCITY OF PROPAGATION, WAVE IMPEDANCE, CUTOFF PROPERTIES

From a strictly mathematical viewpoint the problem at hand has been completely solved. However, there is still much to be said about the TM_{mn} mode from a physical viewpoint. First of all, the solution exhibits propagation properties in the z-direction, and thus γ deserves closer scrutiny. Referring to Eq. (9–30),

$$\gamma = \sqrt{\left(\frac{m\pi}{a}\right)^2 + \left(\frac{n\pi}{b}\right)^2 - \omega^2\mu\epsilon} \tag{9–31}$$

Notice that for a given m and n there will exist a frequency for which γ is zero. For frequencies above this critical value, γ will be imaginary, and wave propagation will take place without attenuation. If the frequency is less than the critical value, γ is real and attenuation takes place. This critical frequency is referred to as the *cutoff frequency* and is given by

$$\omega_c = \frac{1}{\sqrt{\mu\epsilon}}\sqrt{\left(\frac{m\pi}{a}\right)^2 + \left(\frac{n\pi}{b}\right)^2} \tag{9–32}$$

It can be seen that a close analogy exists between a waveguide and an ideal high-pass filter.

The "pass band" ($\omega > \omega_c$) is of primary concern in waveguide work, since the system frequency must be above cutoff if propagation is to take place in the ordinary sense of the word. If $\omega > \omega_c$, then

$$\gamma = j\beta = j\sqrt{\omega^2\mu\epsilon - \omega_c^2\mu\epsilon} \tag{9-33}$$

or

$$\beta = \omega\sqrt{\mu\epsilon}\sqrt{1 - \left(\frac{\omega_c}{\omega}\right)^2}$$

and the velocity of propagation is

$$v_p = \frac{\omega}{\beta} = \frac{v}{\sqrt{1 - (\omega_c/\omega)^2}} \tag{9-34}$$

where v denotes the free-wave velocity within the guide $\left(\text{i.e., } v = \dfrac{1}{\sqrt{\mu\epsilon}}\right)$. More specifically, v_p is the *phase velocity*, since it represents the velocity at which the whole field picture appears to move in the z-direction.[3] The wavelength within the guide can now be written as

$$\lambda_g = \frac{v_p}{f} = \lambda\frac{1}{\sqrt{1 - (\omega_c/\omega)^2}} \tag{9-35}$$

where λ_g denotes the wavelength within the guide, and λ represents the "free" wavelength, or that which would exist for plane wave propagation at this frequency. Note that they are not equal.

The wave impedance concept is useful in waveguide work, just as it is in the study of plane waves. It can be verified from Eqs. (9-29) that the transverse electric and magnetic field components are always at right angles and that their ratio is constant throughout the cross-section of the guide. Thus a characteristic wave impedance may be defined as

$$\begin{aligned} Z_{0(\mathrm{TM})} &= \frac{\text{transverse electric field}}{\text{transverse magnetic field}} \\ &= \frac{\gamma}{j\omega\epsilon} = \eta\sqrt{1 - \left(\frac{\omega_c}{\omega}\right)^2} \end{aligned} \tag{9-36}$$

This impedance has the same significance in waveguide theory as the characteristic impedance Z_0 has in transmission-line theory, and many of the concepts from transmission lines carry over directly to waveguides.

The physical distribution of the field for the TM_{mn} mode is, of course, different for each of the different modes within the group. Figure 9-5

[3] See Appendix E for a discussion of the distinction between phase and group velocity. Also included in Appendix E is a brief discussion of an alternate approach to the rectangular waveguide problem, in which the concept of group velocity plays an important part.

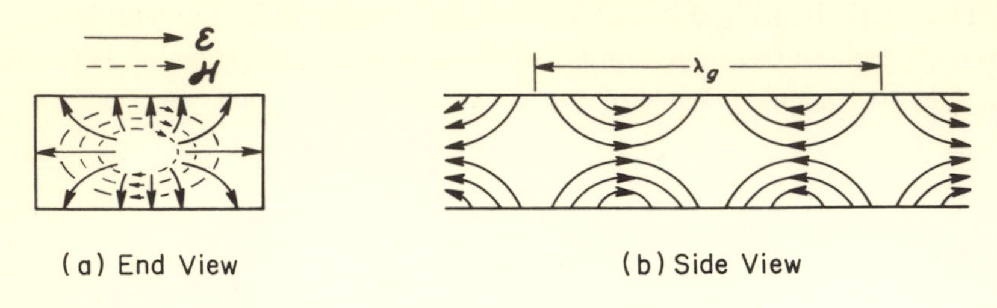

Fig. 9–5. Field distribution for TM_{11} mode.

shows the field distribution for the simplest possible TM mode, the TM_{11} mode.

Example 9–1

Consider a rectangular waveguide which is 2 cm by 1 cm (a and b dimensions, respectively) and is air-filled. Find (a) the mode with the lowest cutoff frequency; (b) the phase velocity, phase-shift constant, wavelength within the guide, and characteristic wave impedance; (c) the explicit expressions in phasor form for the six field components.

The cutoff frequency for any particular mode is given by Eq. (9–32). Note that we cannot let either m or n be zero without having a trivial situation for which all components of E and H are zero. Therefore, the lowest possible value for ω_c is obtained when both m and n are unity. This corresponds to the TM_{11} mode, and its cutoff frequency is

$$\omega_c \text{ (for TM}_{11}\text{ mode)} = \frac{1}{\sqrt{\mu\epsilon}}\sqrt{\left(\frac{\pi}{a}\right)^2 + \left(\frac{\pi}{b}\right)^2} = 105.5 \times 10^9 \text{ rad/s}$$

or

$$f_c = \frac{\omega_c}{2\pi} = 16.8 \text{ GHz}$$

For a system operating in the TM_{11} mode at a frequency of $1.5\,\omega_c$, the phase velocity, phase-shift constant, wavelength within the guide, and characteristic wave impedance can be found from Eqs. (9–33 through 9–36). They are

$$v_p = (3 \times 10^8)\frac{1}{\sqrt{1 - (1/1.5)^2}} = 4.03 \times 10^8 \text{ m/s}$$

$$\beta = \frac{(1.5)(105.5 \times 10^9)}{4.03 \times 10^8} = 392 \text{ rad/m}$$

$$\lambda_g = \frac{4.03 \times 10^8}{(1.5)(16.8 \times 10^9)} = 0.016 \text{ m}$$

$$Z_{0(\text{TM}_{11})} = 377\sqrt{1 - (1/1.5)^2} = 280 \text{ ohms}$$

Finally, the explicit expressions (in phasor form) for all six field components are obtained by directly substituting into Eqs. (9–29) and multiplying by $e^{-j\beta z}$ in order to reinsert the z-dependence. Assuming the peak value of E_z to be E_0, the result is

$$
\begin{aligned}
E_z &= E_0 \sin(50\pi x) \sin(100\pi y) e^{-j\beta z} \\
H_z &= 0 \\
H_x &= j(3.57 \times 10^{-3}) E_0 \sin(50\pi x) \cos(100\pi y) e^{-j\beta z} \\
H_y &= -j(1.79 \times 10^{-3}) E_0 \cos(50\pi x) \sin(100\pi y) e^{-j\beta z} \\
E_x &= -j0.50 E_0 \cos(50\pi x) \sin(100\pi y) e^{-j\beta z} \\
E_y &= -j1.0 E_0 \sin(50\pi x) \cos(100\pi y) e^{-j\beta z}
\end{aligned}
$$

9–6. RECTANGULAR WAVEGUIDE: TE MODE

The solution of the TE mode problem differs from the TM one only in the boundary conditions, since the differential equation which must be satisfied is the same for both cases. Thus the general solution for H_z' may be written immediately as

$$H_z' = (A' \sin k_x x + B' \cos k_x x)(C' \sin k_y y + D' \cos k_y y) \tag{9–37}$$

Now the constants A', B', C', D', k_x, and k_y must be chosen such that the normal derivative of H_z' will be zero along the boundary. Then, referring to Fig. 9–4,

$$
\begin{aligned}
&\text{(a)} \qquad \frac{\partial H_z'}{\partial x} = 0 \qquad \text{along } x = 0 \\
&\text{(b)} \qquad \frac{\partial H_z'}{\partial y} = 0 \qquad \text{along } y = 0 \\
&\text{(c)} \qquad \frac{\partial H_z'}{\partial x} = 0 \qquad \text{along } x = a \\
&\text{(d)} \qquad \frac{\partial H_z'}{\partial y} = 0 \qquad \text{along } y = b
\end{aligned}
\tag{9–38}
$$

Applying the above constraints and using the same line of reasoning used in the TM case lead to

$$A' = C' = 0$$

$$B'D' = \text{arbitrary amplitude constant} = H_0$$

$$\left.\begin{aligned} k_x &= \frac{m\pi}{a} \\ k_y &= \frac{n\pi}{b} \end{aligned}\right\} \quad \text{where } m \text{ and } n \text{ are arbitrary integers, except both are not equal to zero.}$$

Thus the final solution for H_z' is

$$H_z' = H_0 \cos \frac{m\pi x}{a} \cos \frac{n\pi y}{b} \tag{9–39}$$

This, along with $E_z' = 0$, may now be substituted into Eq. (9–6) in order to obtain the four transverse components. The resultant six field components are summarized as follows:

Rectangular Waveguide: TE_{mn} Mode

$$
\begin{aligned}
H_z' &= H_0 \cos \frac{m\pi x}{a} \cos \frac{n\pi y}{b} \\
E_z' &= 0 \\
H_x' &= \frac{\gamma m\pi H_0}{a k_c^2} \sin \frac{m\pi x}{a} \cos \frac{n\pi y}{b} \\
H_y' &= \frac{\gamma n\pi H_0}{b k_c^2} \cos \frac{m\pi x}{a} \sin \frac{n\pi y}{b} \\
E_x' &= \frac{j\omega\mu n\pi H_0}{b k_c^2} \cos \frac{m\pi x}{a} \sin \frac{n\pi y}{b} \\
E_y' &= -\frac{j\omega\mu m\pi H_0}{a k_c^2} \sin \frac{m\pi x}{a} \cos \frac{n\pi y}{b}
\end{aligned}
\tag{9–40}
$$

where

$$k_c^2 = \gamma^2 + \omega^2\mu\epsilon = \left(\frac{m\pi}{a}\right)^2 + \left(\frac{n\pi}{b}\right)^2 \tag{9–41}$$

It can be seen at this point that the expressions for cutoff frequency, phase-shift constant, phase velocity, and wavelength are identical to those derived for the TM case. For convenience these are repeated here:

$$\omega_c = \frac{1}{\sqrt{\mu\epsilon}} \sqrt{\left(\frac{m\pi}{a}\right)^2 + \left(\frac{n\pi}{b}\right)^2} \tag{9–32}$$

$$\beta = \omega\sqrt{\mu\epsilon} \sqrt{1 - \left(\frac{\omega_c}{\omega}\right)^2} \tag{9–33}$$

$$v_p = v \frac{1}{\sqrt{1 - (\omega_c/\omega)^2}} \tag{9–34}$$

$$\lambda_g = \lambda \frac{1}{\sqrt{1 - (\omega_c/\omega)^2}} \tag{9–35}$$

The expression for characteristic impedance, however, is different from that of the TM case. Referring to Eqs. (9–40) and remembering that Z_0 is the ratio of the transverse electric and magnetic fields, the $Z_{0(\mathrm{TE})}$ is derived as

$$Z_{0(\mathrm{TE})} = \frac{j\omega\mu}{\gamma} = \eta \frac{1}{\sqrt{1 - (\omega_c/\omega)^2}} \tag{9–42}$$

A sketch of $Z_{0(\mathrm{TE})}$ along with $Z_{0(\mathrm{TM})}$ versus frequency is shown in Fig. 9–6. Note that one is always above the intrinsic impedance of the dielectric η, that the other is below η, and that they both approach η as $\omega \to \infty$.

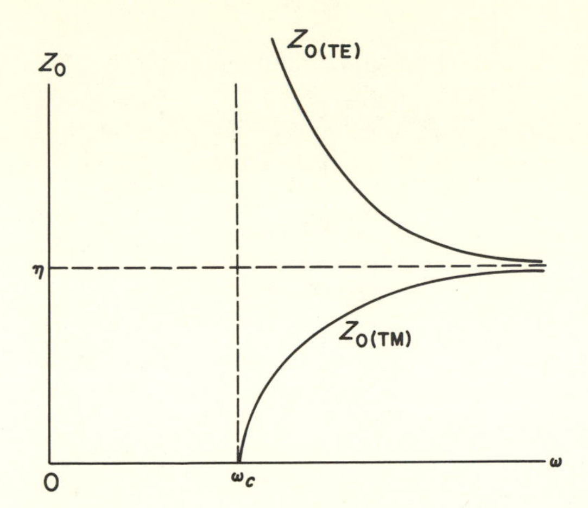

Fig. 9–6. Characteristic wave impedance of TE and TM modes versus frequency.

9–7. NOTE

It is easy for the beginning student to become completely lost in the maze of equations which seem to be unavoidable in waveguide theory. The key to keeping things straight in this analysis is to "stand back" and look at groups of equations rather than individual ones, to see the overall picture rather than the details. As a help in doing this, a summary of the solution of the rectangular waveguide problem is shown in Fig. 9–7 along with references to appropriate equations.

9–8. RECTANGULAR WAVEGUIDE: TE_{10} MODE

The simplest and most frequently used mode in rectangular waveguide work is the TE_{10} mode. Thus it warrants special attention. The field expressions for this mode are obtained by letting $m = 1$ and $n = 0$ in Eqs. (9–40). The result is

$$\begin{aligned} E_y' &= E_0 \sin \frac{\pi x}{a} \\ H_x' &= -\frac{E_0}{Z_{0(TE)}} \sin \frac{\pi x}{a} \\ H_z' &= j\frac{E_0}{\eta}\left(\frac{\lambda}{2a}\right) \cos \frac{\pi x}{a} \\ E_x' &= E_z' = H_y' = 0 \end{aligned} \tag{9–43}$$

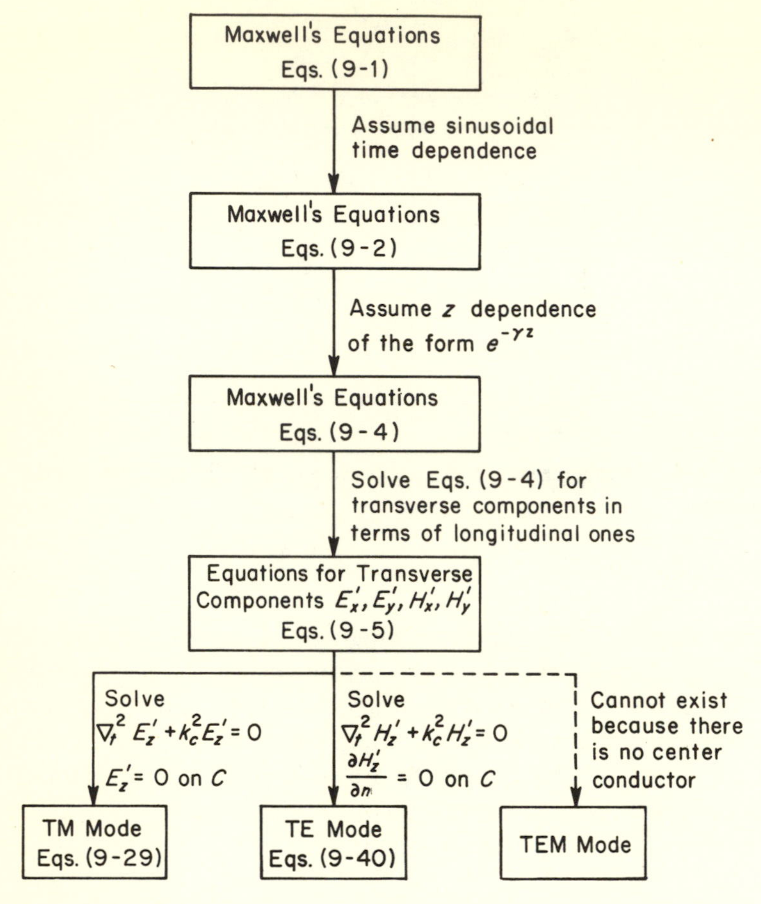

Fig. 9–7. Summary of the solution of the rectangular waveguide problem.

where λ indicates the "free" wavelength and E_0 the arbitrary amplitude coefficient associated with the E_y' term. The field distribution for this mode is shown in Fig. 9–8.

Now if we assume $a > b$ (and we can always orient the coordinate system so that this is the case),[4] it can be seen that this mode has the lowest cutoff frequency of any of the possible modes of transmission, including both TM and TE types. Such a mode is called the *dominant* mode. It is of special interest because there will exist a frequency range

[4] If $a = b$, the TE_{10} and TE_{01} modes have the same cutoff frequency. This situation is referred to as a *degenerate* case, and this must be excluded in the present discussion because there is no separation of the two lowest-order modes.

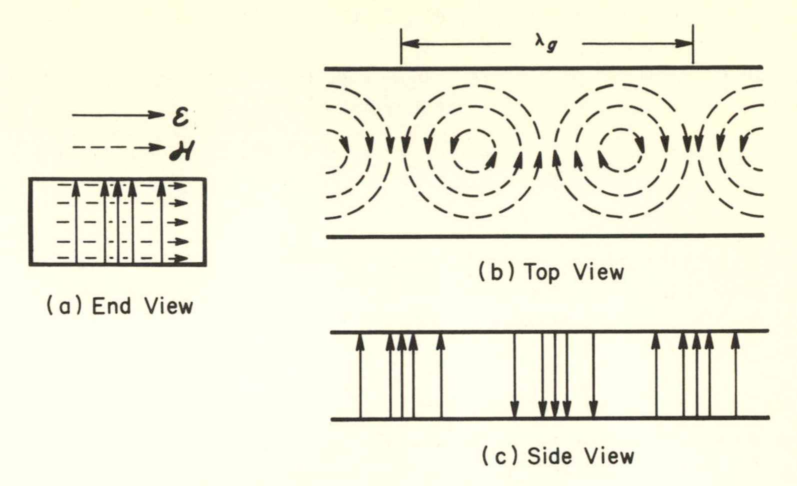

Fig. 9–8. Field distribution for TE_{10} mode.

between its cutoff frequency and that of the next higher-order mode in which this is the only possible mode of transmission. Thus, if a waveguide is excited within this frequency range, energy propagation must take place in the dominant mode, regardless of the way in which the guide is excited. Control of the mode of operation is important in any practical transmission system, and thus the TE_{10} mode has a distinct advantage over the other possible modes in a rectangular waveguide. A specific example illustrating this special frequency range of interest will be helpful at this point.

Example 9–2

Consider an air-filled rectangular waveguide whose a and b dimensions are 0.9 and 0.4 in., respectively. Find (a) the cutoff frequency of the dominant mode and also that of the next higher-order mode, and (b) the attenuation factor associated with the next higher-order mode at a frequency midway between the two cutoff frequencies.

The cutoff frequency for the dominant mode can be computed from Eq. (9–32). Letting $m = 1$ and $n = 0$ gives for a result

$$f_c = \frac{1}{2\pi\sqrt{\mu\epsilon}}\sqrt{\left(\frac{\pi}{a}\right)^2 + (0)^2} = \frac{3 \times 10^8}{(2)(0.9)(2.54 \times 10^{-2})} = 6.56 \text{ GHz}$$

After trying a few values of m and n, it can be seen that the next higher-order mode is the TE_{20} mode. Its cutoff frequency is

$$f_c = \frac{1}{2\pi\sqrt{\mu\epsilon}}\sqrt{\left(\frac{2\pi}{a}\right)^2 + (0)^2} = 13.1 \text{ GHz}$$

Thus, within the frequency range from 6.56 to 13.1 GHz, only the TE_{10} mode can propagate in the ordinary sense of the word.

Looking next at the propagation constant for the TE_{20} mode, we find from Eq. (9–33) that it is (for a mid-range frequency of 9.84 GHz)

$$\gamma = j\beta = j\omega\sqrt{\mu\epsilon}\sqrt{1 - \left(\frac{\omega_c}{\omega}\right)^2} = j\,\frac{(2\pi)(9.84 \times 10^9)}{3 \times 10^8}\sqrt{1 - \left(\frac{13.1}{9.84}\right)^2}$$

$$= 181 \text{ Np/m}$$

This will be recognized as a sizeable attenuation factor—over 1500 dB/m. Furthermore, if we look at the higher-order modes beyond the TE_{20} one, we find that they are attenuated at even a higher rate. This can be seen by rearranging Eq. (9–33) as

$$\gamma = j\omega\sqrt{\mu\epsilon}\sqrt{1 - \left(\frac{\omega_c}{\omega}\right)^2} = \omega_c\sqrt{\mu\epsilon}\sqrt{1 - \left(\frac{\omega}{\omega_c}\right)^2}$$

and noting that each successive higher-order mode has a higher cutoff frequency.

Strictly speaking, it should be noted that we are not correct when we say that higher-order modes "cannot" exist. They can, but the point is that they attenuate very rapidly in the z-direction. Normally a waveguide is excited by means of coupling holes, loops, or probes which excite higher-order modes as well as the dominant mode. It can be seen from this example, though, that for all practical purposes only the TE_{10} mode is left after a few centimeters of travel from the source.

It was observed in the foregoing example that a rectangular waveguide whose width is roughly twice the height will propagate only the TE_{10} mode over a 2:1 frequency range. In waveguide transmission work it is usually desired to have only one mode present, so this frequency is the one of interest insofar as energy propagation is concerned. Actually, for reasons which will be seen later in the chapter when losses are considered, it is better not to operate near either extreme of the range. The usual frequency range recommended for a particular size of guide is about 1.5:1 rather than the theoretical 2:1. For instance, in the example just cited, the dimensions given were the actual ones for RG–52/U, which is one of the standard sizes used in the X-band frequency range. The recommended range for this waveguide size is 8.2 to 12.4 GHz, which is, of course, within the theoretical limits established in the example. As a result of this limited range of usefulness, standard sizes of waveguides have been established, each having a specified frequency range. These can be found in most any modern handbook of electronic engineering.

There is much more to be said about engineering applications of rectangular waveguides, and this discussion will be continued from a circuit viewpoint in Chapter 11. However, while looking at the theoretical aspects of waveguides, we shall consider the cylindrical waveguide before pursuing the rectangular one further.

9–9. CYLINDRICAL WAVEGUIDE: BASIC EQUATIONS

We shall now consider wave propagation inside a hollow pipe of circular cross-section. The general theory discussed previously is applicable regardless of the shape of the boundary, since it is a direct consequence of assumed propagation in the z-direction. Thus these concepts carry over to cylindrical guides directly. More specifically, the basic differential equation to be solved is the same for all waveguides; the only difference lies in the shape of the boundary along which the boundary conditions must be applied. However, this seemingly innocent difference affects the form of the solutions considerably, as will be seen presently.

In solving the cylindrical problem, one could start with the solutions as found in rectangular coordinates and try to make a superposition of these satisfy the new boundary conditions. However, this is quite a difficult task. A more natural approach is to choose a coordinate system which will lend itself well to the symmetry of the problem at hand. The obvious choice in this case is a cylindrical coordinate system.

As mentioned previously, the basic differential equations to be solved are the same as in the rectangular case:

$$\left.\begin{aligned} &\nabla_t^2 E_z' + k_z^2 E_z' = 0 \\ &E_z' = 0 \quad \text{on } C \end{aligned}\right\} \quad \text{TM mode} \qquad (9\text{–}13)$$

$$\left.\begin{aligned} &\nabla_t^2 H_z' + k_c^2 H_z' = 0 \\ &\frac{\partial H_z'}{\partial n} = 0 \quad \text{on } C \end{aligned}\right\} \quad \text{TE mode} \qquad (9\text{–}14)$$

However, we wish to effect the solution in cylindrical coordinates as shown in Fig. 9–9, and thus ∇_t^2 must be written in cylindrical coordinates.

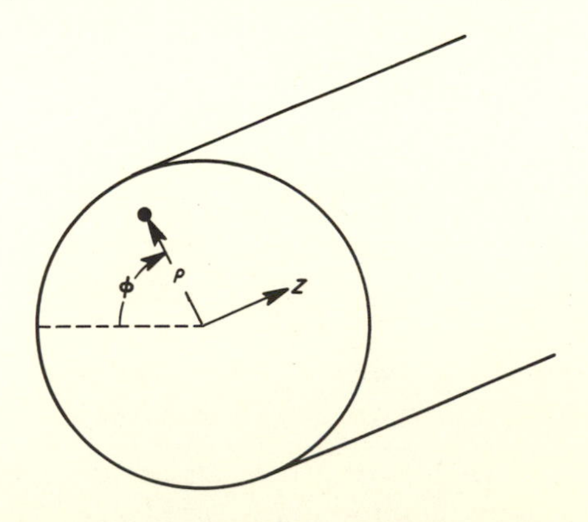

Fig. 9–9. Cylindrical coordinate system.

It may be verified that the two-dimensional Laplacian operator ∇_t^2 in cylindrical coordinates is given by

$$\nabla_t^2 = \frac{\partial^2}{\partial \rho^2} + \frac{1}{\rho}\frac{\partial}{\partial \rho} + \frac{1}{\rho^2}\frac{\partial^2}{\partial \phi^2} \tag{9–44}$$

As before, the transverse components of the field may be written in terms of the longitudinal ones. This is done by merely converting Eqs. (9–5) to polar form. The result is

$$\begin{aligned}
&\text{(a)} \quad H'_\rho = \frac{1}{\gamma^2 + \omega^2\mu\epsilon}\left(\frac{j\omega\epsilon}{\rho}\frac{\partial E'_z}{\partial \phi} - \gamma\frac{\partial H'_z}{\partial \rho}\right)\\
&\text{(b)} \quad H'_\phi = -\frac{1}{\gamma^2 + \omega^2\mu\epsilon}\left(j\omega\epsilon\frac{\partial E'_z}{\partial \rho} + \frac{\gamma}{\rho}\frac{\partial H'_z}{\partial \phi}\right)\\
&\text{(c)} \quad E'_\rho = -\frac{1}{\gamma^2 + \omega^2\mu\epsilon}\left(\gamma\frac{\partial E'_z}{\partial \rho} + \frac{j\omega\mu}{\rho}\frac{\partial H'_z}{\partial \phi}\right)\\
&\text{(d)} \quad E'_\phi = \frac{1}{\gamma^2 + \omega^2\mu\epsilon}\left(-\frac{\gamma}{\rho}\frac{\partial E'_z}{\partial \phi} + j\omega\mu\frac{\partial H'_z}{\partial \rho}\right)
\end{aligned} \tag{9–45}$$

This can be verified readily if one imagines the rectangular and polar coordinate systems oriented in such a way that the unit vectors $\mathbf{a}_x$ and $\mathbf{a}_y$ are in the same directions as the $\mathbf{a}_\rho$ and $\mathbf{a}_\phi$ vectors, respectively.

The general procedure to be followed now is the same as in the rectangular case. We look at the TM and TE cases separately and find the solution for the appropriate longitudinal component. Then the transverse components follow directly from Eqs. (9–45).

9–10. CYLINDRICAL WAVEGUIDE: TM MODE

The differential equation to be solved for this case is

$$\frac{\partial^2 E'_z}{\partial \rho^2} + \frac{1}{\rho}\frac{\partial E'_z}{\partial \rho} + \frac{1}{\rho^2}\frac{\partial^2 E'_z}{\partial \phi^2} + k_c^2 E'_z = 0 \tag{9–46}$$

Again we look for a product-type solution of the form

$$E'_z = R(\rho)\Phi(\phi) \tag{9–47}$$

Substituting this into Eq. (9–46) yields

$$\left(\frac{\rho^2}{R}\frac{d^2R}{d\rho^2} + \frac{\rho}{R}\frac{dR}{d\rho} + \rho^2 k_c^2\right) + \left(\frac{1}{\Phi}\frac{d^2\Phi}{d\phi^2}\right) = 0 \tag{9–48}$$

The variables in Eq. (9–48) have been separated, and thus each term within parentheses may be set equal to a constant. Also, since the two terms must sum to zero, the two constants must be equal in magnitude and opposite in sign. Thus, denoting the constant as $-n^2$,

$$\frac{1}{\Phi}\frac{d^2\Phi}{d\phi^2} = -n^2 \tag{9–49}$$

$$\frac{\rho^2}{R}\frac{d^2R}{d\rho^2} + \frac{\rho}{R}\frac{dR}{d\rho} + \rho^2 k_c^2 = n^2 \tag{9–50}$$

or

$$\frac{d^2\Phi}{d\phi^2} + n^2\Phi = 0 \tag{9–51}$$

$$\frac{d^2R}{d\rho^2} + \frac{1}{\rho}\frac{dR}{d\rho} + \left(k_c^2 - \frac{n^2}{\rho^2}\right)R = 0 \tag{9–52}$$

Equation (9–51) will be recognized as the simple harmonic equation. Its general solution is

$$\Phi = A \sin n\phi + B \cos n\phi \tag{9–53}$$

Equation (9–52) is almost Bessel's equation and would be exactly if k_c^2 were replaced by unity. A simple change of variable will bring the equation into appropriate form. Let

$$R(\rho) = R_1(k_c\rho) \tag{9–54}$$

Substituting this into Eqs. (9–52) leads to

$$\frac{d^2R_1}{d(k_c\rho)^2} + \frac{1}{(k_c\rho)}\frac{dR_1}{d(k_c\rho)} + \left(1 - \frac{n^2}{(k_c\rho)^2}\right)R_1 = 0 \tag{9–55}$$

This is precisely Bessel's equation of order n with $k_c\rho$ as the argument of the equation. Its general solution may be written as

$$R_1(k_c\rho) = R(\rho) = CJ_n(k_c\rho) + DN_n(k_c\rho) \tag{9–56}$$

where J_n and N_n are Bessel functions of the first and second kind, respectively. The Bessel functions for $n = 0$, 1, and 2 are shown in Fig. 9–10.

The solution for E_z' may now be written as

$$E_z' = (A \sin n\phi + B \cos n\phi)[CJ_n(k_c\rho) + DN_n(k_c\rho)] \tag{9–57}$$

It remains now to evaluate the constants. First of all, A may be set equal to zero arbitrarily, since this merely amounts to rotation of the coordinate system to a point where the harmonic variation in ϕ appears as a simple cosine function rather than a combination of sine and cosine terms. Next, the solution must be periodic with respect to ϕ with a period of 2π if E_z' is to be a single-valued function of ρ and ϕ. Thus n must be restricted to integral values (zero included). Finally, Bessel functions of the second kind are all infinite at the origin, and thus D must be zero if our solution

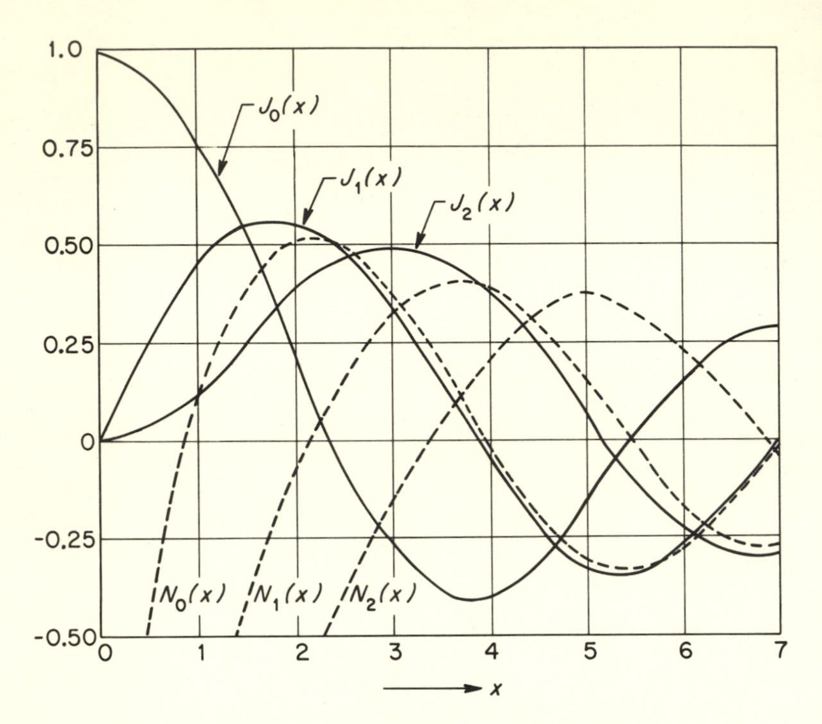

Fig. 9–10. Zero-, first-, and second-order Bessel functions.

is to remain finite at $\rho = 0$. It can be seen then that the form of E'_z must reduce to

$$E'_z = E_0 J_n(k_c\rho) \cos n\phi \tag{9–58}$$

where E_0 is an arbitrary amplitude constant and n is an integer.

One boundary condition remains to be satisfied

$$E'_z = 0 \qquad \text{for } \rho = a$$

This must be true for all values of ϕ and thus

$$J_n(k_c a) = 0$$

This restricts $k_c a$ to discrete values, namely, the roots of $J_n = 0$. Table 9–1 gives the lower-order roots, which are the ones of primary interest. Index n is associated with the order of Bessel function, and l with the particular root of J_n, beginning with the first nonzero root. A subscript notation will be used to denote the roots: τ_{nl} will denote the lth root of the nth-order Bessel function. The permissible values of k_c are now given by

$$k_c = \frac{\tau_{nl}}{a} \tag{9–59}$$

TABLE 9-1

Roots of $J_n = 0$ (TM Modes)

	n			
l	0	1	2	3
1	2.40	3.83	5.14	6.38
2	5.52	7.02	8.42	9.76
3	8.65	10.17	11.62	13.02

It can be seen that there exists a "double infinity" of values for k_c here, just as in the rectangular case.

The final expression for E'_z may now be written as

$$E'_z = E_0 J_n\left(\frac{\tau_{nl}\rho}{a}\right)\cos n\phi \tag{9-60}$$

The transverse components may now be determined by substituting Eq. (9-60) and $H'_z = 0$ into Eqs. (9-45). The results are summarized as follows:

$$\begin{aligned} H'_\rho &= -\frac{j\omega\epsilon n E_0}{k_c^2\rho} J_n\left(\frac{\tau_{nl}\rho}{a}\right)\sin n\phi \\ H'_\phi &= -\frac{j\omega\epsilon\tau_{nl}E_0}{k_c^2 a} J'_n\left(\frac{\tau_{nl}\rho}{a}\right)\cos n\phi \\ E'_\rho &= -\frac{\gamma\tau_{nl}E_0}{k_c^2 a} J'_n\left(\frac{\tau_{nl}\rho}{a}\right)\cos n\phi \\ E'_\phi &= \frac{\gamma n E_0}{k_c^2\rho} J_n\left(\frac{\tau_{nl}\rho}{a}\right)\sin n\phi \end{aligned} \tag{9-61}$$

In the above equations J'_n denotes the derivative of J_n and should not be confused with the primes on the field components, which indicate that these are functions of ρ and ϕ only.

9-11. WAVELENGTH, VELOCITY OF PROPAGATION, CHARACTERISTIC IMPEDANCE, CUTOFF PROPERTIES

The constant k_c is the key to determining the cutoff frequency in any waveguide problem. That is,

$$k_c^2 = \gamma^2 + \omega^2\mu\epsilon$$

and ω_c is the particular ω which makes γ zero in the above equation. Thus ω_c is always given by

$$\omega_c = \frac{1}{\sqrt{\mu\epsilon}} k_c \tag{9-62}$$

For the TM_{nl} mode in a cylindrical waveguide, the cutoff frequency is then

$$\omega_c = \frac{\tau_{nl}}{a\sqrt{\mu\epsilon}} \tag{9-63}$$

This is an important parameter, since once it is found, the characteristic impedance, velocity of propagation, and wavelength are obtained by

$$Z_{0(\text{TM})} = \frac{\gamma}{j\omega\epsilon} = \eta\sqrt{1 - \left(\frac{\omega_c}{\omega}\right)^2} \tag{9-64}$$

$$v_p = \frac{v}{\sqrt{1 - (\omega_c/\omega)^2}} \tag{9-65}$$

$$\lambda_g = \lambda \frac{1}{\sqrt{1 - (\omega_c/\omega)^2}} \tag{9-66}$$

These equations may be verified in the usual way. Note that they are the same equations encountered in the rectangular problem. As a matter of fact, these equations for $Z_{0(\text{TM})}$, v_p, and λ_g are valid for any TM mode, regardless of the waveguide shape.

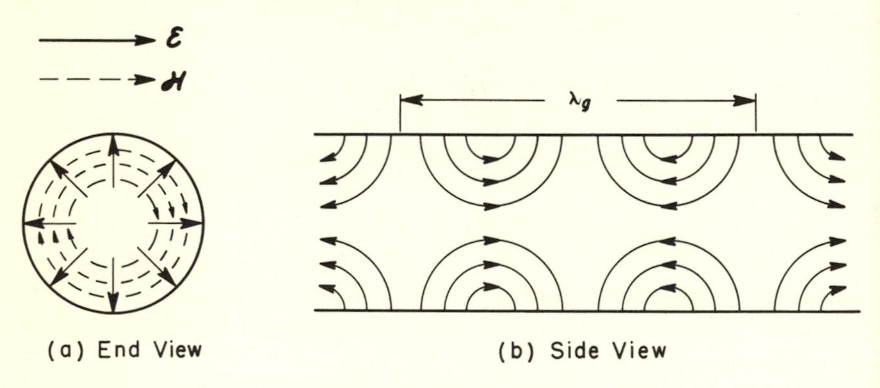

Fig. 9–11. Field distribution for TM_{01} mode.

The TM_{01} mode is the simplest possible TM mode in a cylindrical pipe and is of considerable practical importance. The field distribution for this mode is shown in Fig. 9–11. Note the similarity between this mode and the TM_{11} mode in a rectangular guide.

9–12. CYLINDRICAL WAVEGUIDE: TE MODE

The solution for the TE mode is essentially the same as that for the TM mode up to the point where the boundary condition is applied at

$\rho = a$. Therefore we shall pick up the derivation at the point where we write H'_z in the form

$$H'_z = H_0 J_n(k_c\rho) \cos n\phi \tag{9-67}$$

Now the normal derivative of H'_z must be zero along the boundary; i.e.,

$$\frac{\partial H'_z}{\partial \rho} = 0 \qquad \text{along } \rho = a \tag{9-68}$$

This must be true for all values of ϕ. Thus

$$J'_n(k_c a) = 0 \tag{9-69}$$

where J'_n indicates the derivative of J_n. Thus the roots of $J'_n = 0$ become the permissible values of $k_c a$. These will be denoted as τ'_{nl} and are tabulated in Table 9-2.

TABLE 9-2

Roots of $J'_n = 0$ (TE Modes)

	n			
l	0	1	2	3
1	3.83	1.84	3.05	4.20
2	7.02	5.33	6.71	8.02
3	10.17	8.54	9.97	11.35

The allowable values of k_c may now be written explicitly as

$$k_c = \frac{\tau'_{nl}}{a} \tag{9-70}$$

and thus H'_z may be written as

$$H'_z = H_0 J_n\left(\frac{\tau'_{nl}\rho}{a}\right) \cos n\phi \tag{9-71}$$

Also, the transverse components may be found from Eqs. (9-45) and are

$$\begin{aligned}
H'_\rho &= -\frac{\gamma\tau'_{nl}H_0}{k_c^2 a} J'_n\left(\frac{\tau'_{nl}\rho}{a}\right) \cos n\phi \\
H'_\phi &= \frac{\gamma n H_0}{k_c^2 \rho} J_n\left(\frac{\tau'_{nl}\rho}{a}\right) \sin n\phi \\
E'_\rho &= \frac{j\omega\mu n H_0}{k_c^2 \rho} J_n\left(\frac{\tau'_n \rho}{a}\right) \sin n\phi \\
E'_\phi &= \frac{j\omega\mu\tau'_{nl}H_0}{k_c^2 a} J'_n\left(\frac{\tau'_{nl}\rho}{a}\right) \cos n\phi
\end{aligned} \tag{9-72}$$

The expressions for cutoff frequency, velocity of propagation, wavelength, and characteristic impedance are summarized below. Note that they are essentially the same as those for the TE mode in the rectangular case.

$$\omega_c = \frac{k_c}{\sqrt{\mu\epsilon}} = \frac{\tau'_{nl}}{a\sqrt{\mu\epsilon}} \tag{9-73}$$

$$v_p = \frac{v}{\sqrt{1 - (\omega_c/\omega)^2}} \tag{9-74}$$

$$\lambda_g = \frac{\lambda}{\sqrt{1 - (\omega_c/\omega)^2}} \tag{9-75}$$

$$Z_{0(\mathrm{TE})} = \frac{j\omega\mu}{\gamma} = \eta\frac{1}{\sqrt{1 - (\omega_c/\omega)^2}} \tag{9-76}$$

9–13. CYLINDRICAL WAVEGUIDE: TE_{11} MODE

It can be seen by inspection of Tables 9–1 and 9–2 that the dominant mode for cylindrical waveguides is the TE_{11} mode. The dominant mode is always of special interest and thus it will be described further. The field distribution is shown in Fig. 9–12. Note the similarity between this mode

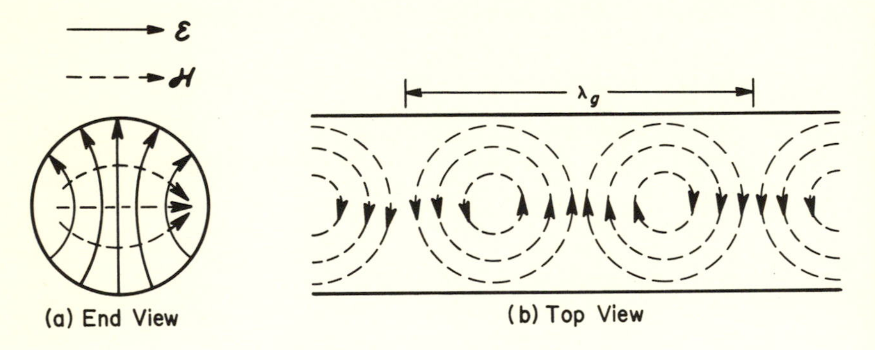

Fig. 9–12. Field distribution for TE_{11} mode.

and the TE_{10} mode of the rectangular case. They are so similar that one could imagine the cylindrical mode as evolving from the corresponding rectangular one by gradually distorting the sides of a square guide until it is circular in cross-section. Looking at it from this physical viewpoint, one might suspect that the two modes have approximately the same cutoff frequency; and, as a matter of fact, this is true. For example, consider an air-dielectric square guide, d by d, and a cylindrical one with diameter d. The respective cutoff frequencies are

Rectangular TE_{10} mode

$$f_c = \frac{c}{2d}$$

Cylindrical TE_{11} mode

$$f_c = \frac{c \times 1.84}{2\pi \times (d/2)} = \frac{c}{1.7d}$$

The cutoff frequency for the cylinder is slightly the higher of the two. This is to be expected because it has less cross-sectional area than the circumscribed square.

There is a subtle inference which may be drawn from the above example. A rectangular waveguide may be distorted considerably from its original shape without a radical effect on either the cutoff frequency or general field configuration. This means that manufacturing tolerances need not be extreme and that such things as bends and twists would be permissible as long as they take place gradually.

9–14. THE TEM MODE

We shall now justify some of the previous remarks made about the TEM or transmission-line mode. This mode is characterized by no component of either **E** or **H** in the direction of propagation. That is, the field is completely transverse to the direction of propagation. When the condition ($E_z' = H_z' = 0$) is substituted into the equations relating the transverse components to the longitudinal ones, Eqs. (9–5), it can be seen that $\gamma^2 + \omega^2\mu\epsilon$ must be zero if the transverse components are not to be zero. Thus

$$\beta = \omega\sqrt{\mu\epsilon} \tag{9–77}$$

and

$$v_p = \frac{\omega}{\beta} = \frac{1}{\sqrt{\mu\epsilon}} \tag{9–78}$$

These are precisely the expressions for phase-shift constant and velocity of propagation encountered in the plane wave case. Also, from (a) and (b) of Eqs. (9–4),

$$\frac{E_y'}{-H_x'} = \frac{E_x'}{H_y'} = \frac{j\omega\mu}{\gamma} = \frac{j\omega\mu}{j\omega\sqrt{\mu\epsilon}} = \sqrt{\frac{\mu}{\epsilon}} = \eta \tag{9–79}$$

and therefore the characteristic wave impedance for the TEM mode is

$$Z_{0(\text{TEM})} = \frac{\text{transverse } E}{\text{transverse } H} = \frac{\sqrt{E_x'^2 + E_y'^2}}{\sqrt{H_x'^2 + H_y'^2}} = \eta \tag{9–80}$$

This is just the intrinsic impedance encountered in plane wave theory. It can be seen by forming the dot and cross products of the transverse **E** and **H** that they are orthogonal and oriented so as to produce positive power in the direction of propagation. That is,

$$\mathbf{E}_t' \cdot \mathbf{H}_t' = (E_x'\mathbf{a}_x + E_y'\mathbf{a}_y) \cdot \left(-\frac{E_y'}{\eta}\mathbf{a}_x + \frac{E_x'}{\eta}\mathbf{a}_y\right) = 0$$

$$\mathbf{E}_t' \times \mathbf{H}_t' = \begin{vmatrix} \mathbf{a}_x & \mathbf{a}_y & \mathbf{a}_z \\ E_x' & E_y' & 0 \\ -\dfrac{E_y'}{\eta} & \dfrac{E_x'}{\eta} & 0 \end{vmatrix} = \mathbf{a}_z\left(\frac{E_x'^2}{\eta} + \frac{E_y'^2}{\eta}\right) = \mathbf{a}_z\frac{|E_t'|^2}{\eta}$$

Thus the TEM mode has all of the earmarks of a plane wave of the nonuniform variety. It is nonuniform because **E** and **H** may vary both in magnitude and direction within a plane normal to the direction of propagation.

It remains to be shown that this mode is exactly the one previously discussed in transmission-line theory. This can be deduced by noting that γ is imaginary for *all* frequencies and thus must apply at zero frequency. Since all higher-order modes (TE and TM) exhibit cutoff properties, the TEM mode is the only one which can exist at zero frequency. Therefore it and the transmission-line solution must be one and the same if our solution at low frequency is to be unique. Also the facts that η is a constant and β varies linearly with frequency tell us that the transmission-line solution (in which Z_0 is also a constant and β varies linearly with frequency) is valid for all frequencies with no upper limit.

It is important to recognize that the conventional transmission-line solution as developed from circuit theory and the TEM mode of waveguide theory are just two different ways of looking at the same phenomenon. In one case we see current and voltage waves as the essential propagating quantities, and in the other case electric and magnetic fields are the salient features of the propagation phenomenon. Actually, both traveling-wave pictures occur simultaneously and are inherently associated. That is, where we have voltage and current waves, we should expect to find associated electric and magnetic field waves and vice versa. The mathematical formalism used to describe the phenomenon can be either in terms of circuit theory or field theory, and either of these formalisms should lead to the same result when it is interpreted physically.

It was mentioned previously that it is often helpful to be able to look at a problem from two different viewpoints. An example of this is the balun problem which was discussed in the chapter on transmission-line matching. From strictly a circuit viewpoint there is no obvious need (for matching reasons, at least) for a transition section between a coaxial line and an open-wire line if they have identical characteristic impedances. By simple circuit theory there would be no reflection at the junction and nothing to worry about. However, when this same problem is viewed in terms of a traveling electromagnetic wave, we can immediately see that something drastic must happen at the junction. The field configurations called for by the two types of lines are radically different, and we cannot expect the field to change form abruptly at the discontinuity. The net result is a complicated combination of reflection and radiation at the junction, both of which are undesirable. The fallacy in the circuit approach lies in the L and C parameters which were computed on the basis of an infinite line of uniform cross-section, and thus we should not expect such an analysis to fit the problem at hand very well in the neighborhood of the discontinuity.

A possible solution to the problem is suggested by wave theory. If we are to avoid serious reflections, we could try making the change from coaxial to open-wire geometry gradually in order that the electric and magnetic fields may change their form gradually rather than abruptly. A simple scheme for doing this is shown in Fig. 9-13. Here the outer

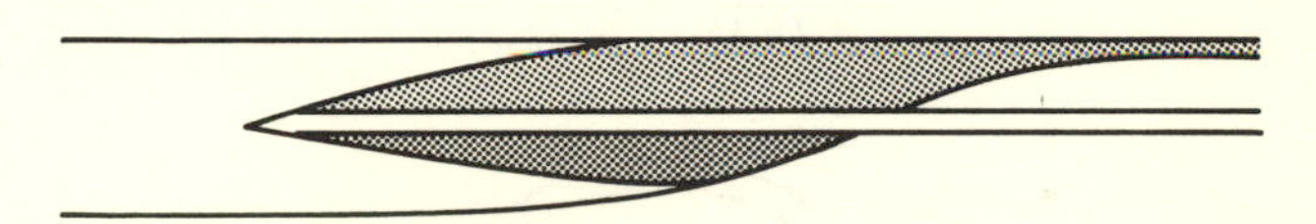

Fig. 9-13. A simple flared-out coaxial to open-wire transition.

conductor of the coaxial line is gradually flared out with a V-shaped opening until there is nothing left of it but a single, round conductor with appropriate diameter and spacing to give the open-wire line the same characteristic impedance as the coaxial one. The form of balun is not perfect, but it is much better than making the transition abruptly. It also has the advantage of being insensitive to frequency changes over a wide range, which is not true of the other baluns mentioned in Art. 4-4.

9-15. HIGHER-ORDER MODES IN COAXIAL AND OPEN-WIRE SYSTEMS

The TEM mode is, of course, the dominant mode in coaxial or open-wire systems because its cutoff frequency is zero. Higher-order or wave-guide types of modes are possible, though, and the analysis of these proceeds along the same general line as the previous waveguide problems. The coaxial line is the simpler of the two cases and it will be used to illustrate the procedure. The mathematical analysis of this case is identical with that of the cylindrical waveguide up to the point where the boundary conditions are applied.

In this case, however, the region of interest is that between the inner and outer conductors and thus does not include the origin (Fig. 9–14). This means that the Bessel function of the second kind must be retained in the solution because its singularity at the origin is of no consequence. Also, the boundary condition on E_z' (or H_z' as the case may be) must be applied along both $\rho = a$ and $\rho = b$. In the TM case this gives rise to a pair of equations which must be satisfied, of the form

$$\begin{aligned} CJ_n(k_c a) + DN_n(k_c a) &= 0 \\ CJ_n(k_c b) + DN_n(k_c b) &= 0 \end{aligned} \qquad (9\text{–}81)$$

or, dividing to eliminate C and D,

$$\frac{J_n(k_c a)}{J_n(k_c b)} = \frac{N_n(k_c a)}{N_n(k_c b)} \qquad (9\text{–}82)$$

Similarly, for the TE mode, the equation to be satisfied is

$$\frac{J_n'(k_c a)}{J_n'(k_c b)} = \frac{N_n'(k_c a)}{N_n'(k_c b)} \qquad (9\text{–}83)$$

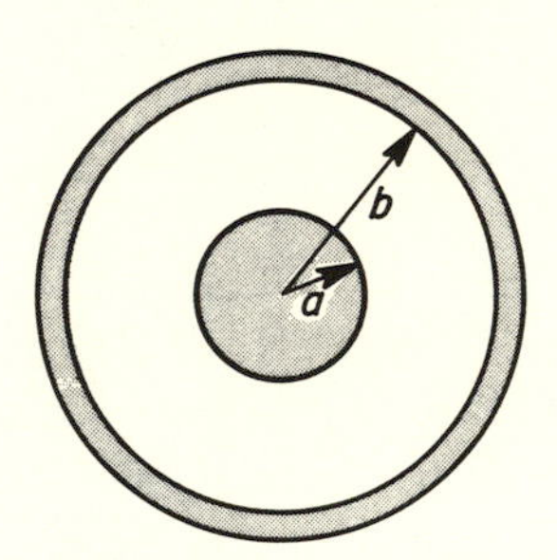

Fig. 9–14. Coaxial system.

These equations have been solved by numerical means, and solutions have been tabulated for various ratios of $a\!:\!b$.[5] The tables are somewhat involved and will not be repeated here. The first of the higher-order modes is of particular interest, though, since it determines the upper limit of the frequency range for which coaxial cable may be used without fear of mode

[5] See, for example, N. Marcuvitz (ed.), *Waveguide Handbook*, Vol. 10, MIT Radiation Laboratory Series (New York: McGraw-Hill Book Company, Inc., 1951), p. 74.

switching. This is a TE mode and is similar to the cylindrical TE_{11} mode except that the electric lines of flux are intercepted by the center conductor in the coaxial case. The cutoff frequency of this mode can be approximated by thinking of the field in the upper-half annular region as being a distortion of the TE_{10} rectangular mode, as shown in Fig. 9–15. The mean width of the "equivalent" rectangular guide would then be

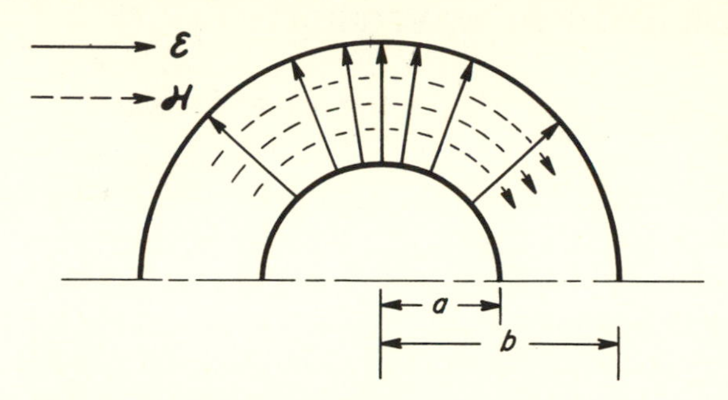

Fig. 9–15. Higher-order mode in coaxial system.

$$\text{Mean width} \approx \pi\left(\frac{a+b}{2}\right)$$

and the approximate cutoff frequency is (from Example 9–2)

$$f_c \approx \frac{\text{free velocity}}{2 \times \text{width}} = \frac{v}{\pi(a+b)} \tag{9–84}$$

An example will illustrate the use of this approximate formula.

Example 9–3

Find the cutoff frequency of the first higher-order mode for an air-dielectric 50-ohm coaxial line whose inner conductor has a radius of 2 mm.

Indirectly the characteristic impedance of the line specifies the outer radius b through the equation

$$Z_0 = 60 \ln \frac{b}{a}$$

Therefore

$$b = ae^{Z_0/60} = (2 \times 10^{-3})e^{5/6} = 4.6 \text{ mm}$$

The approximate cutoff frequency of the first of the higher-order modes may now be computed from Eq. (9–84) as

$$f_c \approx \frac{3 \times 10^8}{\pi(2 + 4.6)10^{-3}} = 14.5 \text{ GHz}$$

It can be seen from this example that the higher-order modes in coaxial cable are usually not bothersome, since their cutoff frequencies are much higher than those in the frequency range where coaxial cable is normally used.

9–16. ATTENUATION IN WAVEGUIDES

We shall begin by restricting ourselves to the slightly lossy case and assuming that the introduction of slight losses will not appreciably distort the field picture obtained in the ideal case. This assumption is almost an absolute necessity, the analysis being extremely complex otherwise. No formal justification of this will be given other than to say that the assumption seems reasonable in the light of our previous experience with plane wave and transmission-line theory. The general procedure will be to evaluate the expression $\alpha = P_L/2P_T$ and use the fields as obtained in the lossless case for computing P_L and P_T. The transmitted power P_T can be obtained by integrating $\mathcal{E} \times \mathcal{H}$ over the cross-section of the guide, and the power lost per unit distance in the z-direction can be found by summing the total losses in a differential section dz and dividing this by dz.

The TE_{10} mode is of considerable practical importance, and thus it will be used to illustrate this method of computing attenuation. We shall also assume that the only losses are wall losses as this fits the physical situation for air-filled waveguides quite well. For convenience the coordinate system for a rectangular guide is repeated in Fig. 9–16, along

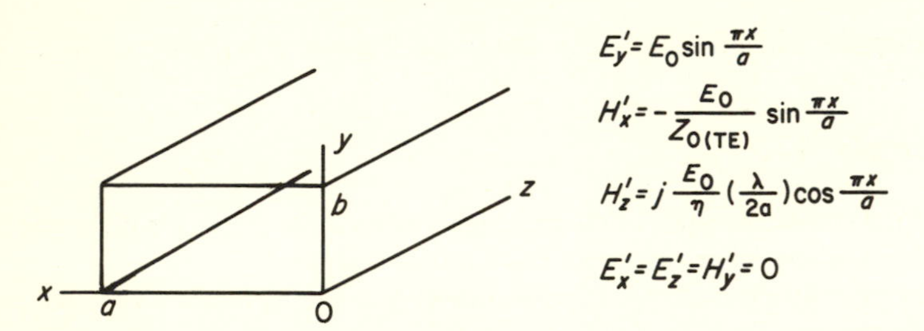

Fig. 9–16. Rectangular waveguide, TE_{10} mode.

with the appropriate equations. It will be recalled from Chapter 7 that the expression for average (timewise) power density, when written in terms of phasors, is

$$\mathbf{P} = \tfrac{1}{2}\,\text{Re}\,(\mathbf{E} \times \mathbf{H}^*) \tag{9–85}$$

where $\mathbf{H}^*$ denotes the complex conjugate of $\mathbf{H}$. The factor of 1/2 is present in the equation because $\mathbf{E}$ and $\mathbf{H}^*$ refer to peak rather than rms values.

Equation (9–85) will now be used to find the time-averaged power flowing in the z-direction. It can be seen from the equations of Fig. 9–16 that the z component of **P** is given by

$$P(z \text{ direction}) = \tfrac{1}{2} \text{Re} \left[(E_y' e^{-j\beta z})(-H_x'^* e^{j\beta z})\right]$$

$$= \frac{1}{2} \frac{E_0^2}{Z_{0(\text{TE})}} \sin^2 \frac{\pi x}{a} \tag{9–86}$$

Integrating this expression over the cross-section of the guide yields, for the total time-averaged transmitted power,

$$P_T = \frac{1}{2} \frac{|E_0|^2}{Z_{0(\text{TE})}} \int_0^a \int_0^b \sin^2 \left(\frac{\pi x}{a}\right) dy\, dx$$

$$= \frac{|E_0|^2 ab}{4Z_{0(\text{TE})}} \tag{9–87}$$

It will next be necessary to find the expression for wall losses per unit length. This is done by integrating the loss per unit area over a section of guide of differential length dz. It will be recalled that the loss per unit area is given by

$$P \text{ (per unit area)} = \tfrac{1}{2} |\mathbf{J}_s|^2 R_s \tag{9–88}$$

The magnitude of the current density $\mathbf{J}_s'$ is equal to the magnitude of **H** at the wall surface and is thus obtained directly from the equations given in Fig. 9–16. By referring to Fig. 9–17, the losses are summarized as follows:

1. Area A:

$$(dP)_A = \frac{1}{2} R_s \frac{|E_0|^2}{\eta^2} \left(\frac{\lambda}{2a}\right)^2 b\, dz$$

2. Area B: Same as A, because of symmetry.
3. Area C:

$$(dP)_C = \frac{1}{2} R_s \int_0^a (|H_x'|^2 + |H_z'|^2)\, dx\, dz$$

$$= \frac{1}{2} R_s\, dz \left[\frac{|E_0|^2}{Z_{0(\text{TE})}^2} \int_0^a \sin^2 \frac{\pi x}{a}\, dx + \frac{|E_0|^2}{\eta^2} \left(\frac{\lambda}{2a}\right)^2 \int_0^a \cos^2 \frac{\pi x}{a}\, dx\right]$$

$$= \frac{1}{4} R_s\, dz |E_0|^2 a \left[\frac{1}{Z_{0(\text{TE})}^2} + \frac{1}{\eta^2} \left(\frac{\lambda}{2a}\right)^2\right]$$

4. Area D: Same as C, because of symmetry.

Summing the total losses and noting that P_L is

$$\frac{(dP)_A + (dP)_B + (dP)_C + (dP)_D}{dz}$$

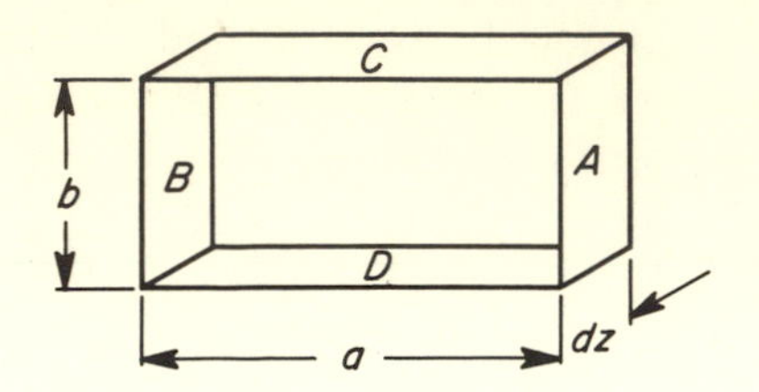

Fig. 9–17. Differential section of guide.

leads to

$$P_L = R_s|E_0|^2 \left[\frac{b}{\eta^2} \left(\frac{\lambda}{2a} \right)^2 + \frac{1}{2} \frac{a}{Z_{0(\mathrm{TE})}^2} + \frac{a}{2\eta^2} \left(\frac{\lambda}{2a} \right)^2 \right] \tag{9–89}$$

The attenuation is now given by

$$\alpha = \frac{P_L}{2P_T} = \frac{\frac{1}{2} R_s|E_0|^2 \left[\frac{2b}{\eta^2} \left(\frac{\lambda}{2a} \right)^2 + \frac{a}{Z_{0(\mathrm{TE})}^2} + \frac{a}{\eta^2} \left(\frac{\lambda}{2a} \right)^2 \right]}{\frac{|E_0|^2 ab}{2Z_{0(\mathrm{TE})}}}$$

$$= \frac{R_s Z_{0(\mathrm{TE})}}{ab} \left[\frac{2b}{\eta^2} \left(\frac{\lambda}{2a} \right)^2 + \frac{a}{\eta^2} \left(\frac{\lambda}{2a} \right)^2 + \frac{a}{Z_{0(\mathrm{TE})}^2} \right] \tag{9–90}$$

As could be expected, this equation may be written in a number of different forms. Another form which is somewhat more convenient from a computational viewpoint is

$$\alpha = \frac{R_s}{b\eta\sqrt{1 - (f_c/f)^2}} \left[1 + \frac{2b}{a} \left(\frac{f_c}{f} \right)^2 \right] \tag{9–91}$$

A sketch showing the general shape of the attenuation versus frequency curve is shown in Fig. 9–18. Note that α approaches infinity as $f \to f_c$ and $f \to \infty$. Thus there is an optimum range of operation from an

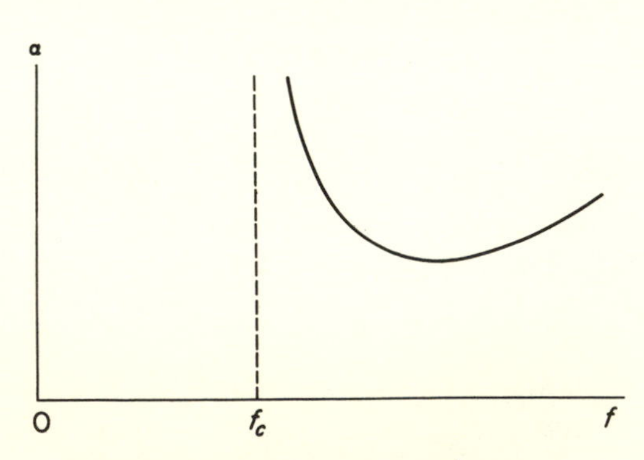

Fig. 9–18. Attenuation for TE_{10} mode.

attenuation viewpoint. Also, it can be seen from Eq. (9–91) that the attenuation decreases with an increase in b. The two basic factors which govern the choice of b/a for a rectangular waveguide have now been presented. From the viewpoint of band width (range in which only the TE_{10} mode is possible), b/a should be 1/2 or less. This gives a factor of 2 separating the TE_{10} and the next higher-order mode, the TE_{20} mode, and this is the maximum separation possible in a rectangular waveguide. On the other hand, b/a should be made as large as possible from the attenuation and maximum power capability viewpoint. In setting up standard sizes for rectangular waveguides, the band-width criterion has prevailed, and b/a is approximately 1/2 for all standard sizes. For example, RG–52/U (X-band size) has a b/a ratio of 4:9.

This same technique for finding attenuation may also be applied to other rectangular and circular waveguide modes. The procedure is straightforward, but the algebra is quite involved and will not be repeated here. A summary of the resulting formulas is given at the end of this article. With one exception, the attenuation curves have the same general shape as the one shown for the TE_{10} mode in Fig. 9–18. The exception is the TE_{01} circular waveguide mode and all similar higher-order modes possessing the circular symmetry property—in general, the TE_{0l} modes. These modes have attenuation versus frequency curves which decrease indefinitely with frequency, and this suggests obvious possibilities in long transmission systems where low attenuation is of prime importance. Particular attention has been given to the TE_{01} mode, it being the lowest order of these modes, and a considerable amount of experimental work has been done in an attempt to put this mode to practical use.[6] The basic problem, though, lies in mode switching, since the TE_{01} mode is not the dominant one in a circular waveguide. That is, any slight perturbations from perfect cylindrical geometry, such as dents or bends, cause some of the energy to switch from the TE_{01} mode to other modes, and this is highly undesirable.

For reference purposes, formulas for computing attenuation due to wall losses are on page 264. They may be verified by the same general technique which was illustrated for the TE_{10} rectangular mode. Dielectric losses are not included in the formulas, but they can be handled in a

[6] For a good discussion of the problems involved with this mode and some of the supporting experimental work, see S. E. Miller, "Waveguide as a Communication Medium," *Bell System Technical Journal*, vol. 33 (November 1954), pp. 1209–1265.

A more recent article describing this mode and the associated waveguide is the one by J. W. Carlin and P. D'Agostino, "Low-Loss Modes in Dielectric Lined Waveguide," *Bell System Technical Journal*, vol. 50, No. 5 (May–June 1971), pp. 1631–1638.

similar manner. When such losses are present, one simply adds the dielectric loss per unit length to the wall loss per unit length to obtain the correct P_L in the $\alpha = P_L/2P_T$ formula.

SUMMARY OF ATTENUATION FORMULAS

1. TE_{mn} Mode, Rectangular Waveguide (for m and $n \neq 0$):[7]

$$\alpha = \frac{2R_s}{b\eta\sqrt{1-\left(\frac{f_c}{f}\right)^2}}\left\{\left(1+\frac{b}{a}\right)\left(\frac{f_c}{f}\right)^2 + \left[1-\left(\frac{f_c}{f}\right)^2\right]\left[\frac{\frac{b}{a}\left(\frac{b}{a}m^2+n^2\right)}{\frac{b^2m^2}{a^2}+n^2}\right]\right\} \tag{9–92}$$

2. TM_{mn} Mode, Rectangular Waveguide:

$$\alpha = \frac{2R_s}{b\eta\sqrt{1-\left(\frac{f_c}{f}\right)^2}}\left[\frac{m^2\left(\frac{b}{a}\right)^3+n^2}{m^2\left(\frac{b}{a}\right)^2+n^2}\right] \tag{9–93}$$

3. TE_{nl} Mode, Circular Waveguide:

$$\alpha = \frac{R_s}{a\eta\sqrt{1-(f_c/f)^2}}\left[\left(\frac{f_c}{f}\right)^2+\frac{n^2}{\tau_{nl}'^2-n^2}\right] \tag{9–94}$$

4. TM_{nl} Mode, Circular Waveguide:

$$\alpha = \frac{R_s}{a\eta\sqrt{1-(f_c/f)^2}} \tag{9–95}$$

PROBLEMS

9–1. Consider two infinitely large, conducting, parallel planes separated by a distance a. All three types of modes (TM, TE, and TEM) can propagate in a direction parallel to the planes under the right conditions. (a) Which of the modes is the dominant mode and what is its cutoff frequency? (b) What is the cutoff frequency of the next higher-order mode in terms of the separation distance a and the free velocity in the dielectric medium? (c) Sketch the field distribution for the mode of (b).

9–2. An air-dielectric, rectangular waveguide has inner dimensions of 2 by 1 cm (a and b, respectively). Find the cutoff frequency for the TM_{21} mode and sketch the field distribution for this mode.

[7] This formula will not reduce to the correct formula for either m or n equal to zero. It is off by a factor of 2 in these cases because the field is uniform in either one or the other of the two directions. Note that Eq. (9–91) is valid, though, and may be used for the TE_{01} mode by simply interchanging the a's and b's.

9–3. X-band waveguide has a and b inner dimensions of 0.9 and 0.4 in., respectively. Assuming that the dielectric is air, list all the possible modes of propagation which can exist at a frequency of 20 GHz.

9–4. Sketch the current distribution in the walls for the TE_{10} rectangular mode and deduce from this why a longitudinal slot may be cut in the top of the guide without appreciably disturbing the field distribution. (*Hint:* Ampere's law applied to a small loop enclosing the boundary will give a relationship between the magnetic field at the surface and the current density in the wall.)

9–5. Derive an expression for the time-averaged power transmitted in a rectangular wave-guide operating in the TE_{10} mode with only an incident wave present.

9–6. Find the upper limit of power which can be propagated in X-band waveguide (0.9 by 0.4 in.) without causing voltage breakdown of the air dielectric. Assume that the frequency is 1.5 times the cutoff value and that air has a dielectric strength of 3×10^6 V/m.

9–7. A section of X-band waveguide is to be used as a below-cutoff-type waveguide attenuator at a frequency much below cutoff for the guide. What must be the length of the section to give 60-dB attenuation (i.e., the amplitude of the wave is to be reduced by a factor of 1000 within the section)?

9–8. A cylindrical waveguide is to be used as a TE_{11} mode below-cutoff-type attenuator at frequencies much below cutoff. (a) Sketch the field distribution within the attenuator section. (b) What is the upper limit of the frequency range of this attenuator expressed in terms of the ratio $f{:}f_c$ if the attenuation constant is not to deviate more than 5 per cent from the low-frequency value?

9–9. A cylindrical waveguide is air-filled and 10 cm in diameter. (a) Find the cutoff frequency for the TM_{11} mode. (b) Sketch the field distribution for this mode.

9–10. For the cylindrical waveguide of Problem 9–9 (diameter = 10 cm): (a) state the dominant mode and find its cutoff frequency; (b) tabulate the possible modes of propagation if the operating frequency is 3 GHz; (c) determine the cutoff frequency of the dominant mode if the guide were filled with a material with a dielectric constant of 2.25.

9–11. It is desired to propagate the TE_{11} mode in a cylindrical waveguide at a frequency of 10 GHz. What should be the radius of the guide if this frequency is to be 0.8 of the cutoff frequency of the next higher-order mode of operation?

9–12. Write the explicit expressions for the electric and magnetic field intensities for an air-dielectric coaxial line and show that the ratio of the two is η.

9–13. Show that the transverse electric field for the TEM mode may be derived from a scalar potential function ϕ which satisfies Laplace's equation; i.e., $\nabla^2\phi = 0$. Note that this is another way of showing that the TEM mode and the conventional transmission-line solution are one and the same because the field distribution assumed in computing L and C is the static one.

9–14. In the derivation of the wall losses for a rectangular waveguide, it was assumed that the timewise average of the square of the total magnetic field was equal to the sum of the average squares of the x and y components, even though these are 90 degrees out of phase timewise. Show that this is valid.

9–15. Show that the TE_{10} waveguide attenuation formula, Eq. (9–91), follows from Eq. (9–90).

9–16. Sketch a plot of theoretical attenuation versus frequency for a brass X-band waveguide (0.9 by 0.4 in.) within the X-band range (8.2 to 12.4 GHz). Plot α in decibels per 100 ft. Assume that the dielectric medium (air) is lossless and that brass has a conductivity of 17.3×10^6 mhos/m.

9–17. Compare the results of Problem 9–16 with those obtained for coaxial-type (TEM mode) transmission within this range. For purposes of comparison, assume an air-dielectric, brass, 50-ohm line with an outer conductor diameter of 1.5 cm. The two transmission systems will then be roughly the same size.

9–18. Compare the maximum power capabilities of the coaxial and rectangular waveguide systems of Problems 9–16 and 9–17. Assume the dielectric strength of air is 3×10^6 V/m.

9–19. It was pointed out in Art. 9–16 that the TE_{01} circular waveguide mode has possibilities as a practical mode of transmission even though it is not the dominant mode. It is desired to achieve a loss of only 2 dB/mi using copper pipe with an inner diameter of 2 in. and operating in the TE_{01} mode. (a) What must be the frequency of operation? (*Hint:* The frequency is much greater than cutoff and therefore $\sqrt{1 - (f_c/f)^2}$ is approximately unity.) (b) What would be the corresponding attenuation for the dominant mode at this frequency?

10

Resonant Cavities

10–1. GENERAL REMARKS ON RESONANT CAVITIES

It will be recalled from transmission-line theory that a transmission line shorted on both ends exhibits resonance properties at frequencies where the length is $\lambda/2$ or some multiple thereof. By direct analogy one would expect a similar phenomenon to take place when a section of waveguide is shorted on both ends, and this is the case. When shorting plates are placed over the ends of a waveguide section, there will exist a dielectric region completely surrounded by a conducting surface. This is known as a *resonant cavity*. It will have resonance properties very much the same as the shorted line. It can be seen that there are a great many different possible modes of resonance—in fact, an infinite number of them. For each different waveguide mode there will be an infinite number of multiples of a half-wavelength which may be fitted in the longitudinal direction between the two end faces. Thus a triple infinity of modes is possible. Usually only the lower-order modes are of interest, and the one with the lowest resonant frequency is referred to as the *dominant* mode.

Before proceeding to the mathematical solution of some simple geometric cases, it should be mentioned that this approach to resonant cavities via waveguides is somewhat restrictive. It will lead to a solution of only those geometric configurations having general cylindrical properties, i.e., having cross-sectional shape which is arbitrary but always the same when viewed at any point along the longitudinal axis. Actually, a dielectric region of any shape which is completely surrounded by a conducting surface will exhibit resonance properties, and this is the most general form of the resonant cavity. The solution of such general problems is quite difficult; so we shall confine our attention to the few simple cases which we can solve readily by using waveguide and transmission-line theory.

Oddly enough, the study of resonant cavities as an engineering science is relatively new, even though the fundamental theory goes back to Maxwell's time. The lack of engineering interest in resonant cavities was probably due to the extremely high frequencies required for resonating structures of reasonable size. Such high-frequency sources and associated test equipment were not available until recent times, and thus experimental work in this area was virtually impossible. The modern-day interest in resonant cavities began in the late 1930's with the publication of two papers on this subject by W. W. Hansen.[1] Hansen's papers have since become classics on the subject, and he deserves much of the credit for the state of the resonant cavity art as we know it today.

10–2. RECTANGULAR AND CYLINDRICAL RESONATORS

Rectangular and cylindrical resonators are two special cases which are readily solved in terms of waveguide theory. We shall use the same coordinate systems as in waveguide theory and shall denote length of the

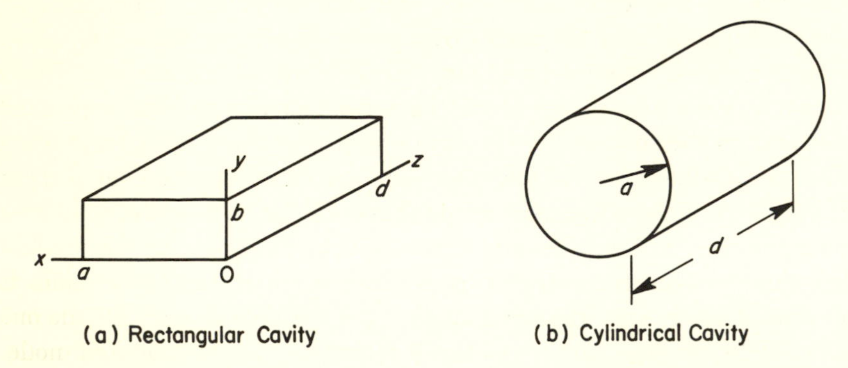

Fig. 10–1. Coordinate systems for rectangular and cylindrical cavities.

resonators by d, as shown in Fig. 10–1. Now, when shorting plates are placed over both ends of a section of waveguide, the reflection coefficient

[1] W. W. Hansen, "A Type of Electrical Resonator," *Journal of Applied Physics*, vol. 9 (1938), pp. 654–63; "On the Resonant Frequency of Closed Concentric Lines," *Journal of Applied Physics*, vol. 10 (1939), pp. 38–45.

is -1 at these points. Thus two waves will exist inside the resonator, an incident one and a reflected one, which are equal in magnitude and phased such that nodes in the transverse electric field will exist at both ends.[2] Therefore, it can be seen that βd must be some multiple of π; i.e.,

$$\beta d = p\pi \tag{10–1}$$

where p is an integer. Now, from waveguide theory, k_c^2 is related to the frequency by

$$k_c^2 = \gamma^2 + \omega^2\mu\epsilon = -\beta^2 + \omega^2\mu\epsilon \tag{10–2}$$

Thus combining Eqs. (10–1) and (10–2) leads to the equation for resonant frequency:

$$\omega_r = \frac{1}{\sqrt{\mu\epsilon}}\sqrt{k_c^2 + \left(\frac{p\pi}{d}\right)^2} \tag{10–3}$$

This formula applies to any waveguide type of cavity. It will be recalled from waveguide theory that k_c is a constant which depends only on the mode and the cross-sectional geometry involved. For the rectangular case,

$$k_c^2 = \left(\frac{m\pi}{a}\right)^2 + \left(\frac{n\pi}{b}\right)^2 \qquad \text{(for both TM and TE modes)} \tag{10–4}$$

For the cylindrical case,

$$k_c^2 = \begin{cases} \left(\dfrac{\tau_{nl}}{a}\right)^2 & \text{(for TM mode)} \\ \left(\dfrac{\tau'_{nl}}{a}\right)^2 & \text{(for TE mode)} \end{cases} \tag{10–5}$$

Note that there are three degrees of arbitrariness in the choice of modes, two coming from the waveguide mode and the third from the number of half-wavelengths in the axial direction. A subscript notation is useful for identifying the different possible modes in a resonant cavity, just as in the case of waveguides. The usual procedure is to carry over the two subscripts from the waveguide mode from which the resonant cavity mode is derived and then add a third subscript to denote the number of half-wavelengths involved in the axial direction. For example, the TE_{101} mode in a rectangular resonator is derived from the TE_{10} waveguide mode

[2] This discussion of resonant cavities assumes that the cavities are completely closed. Actually there must be some small opening (or openings) to get energy into (or out of) the cavity. This is usually accomplished by means of small probes or loops, and will be discussed in Art. 10–6. At this point the effect of the coupling system on the fields will be neglected. It will suffice to say that this is justified if the coupling loops or probes are small in comparison with the other dimensions involved.

and has one half-cycle variation in the z-direction. The resonant frequency for such a mode would be given by

$$\omega_r = \frac{1}{\sqrt{\mu\epsilon}}\sqrt{\left(\frac{1\cdot\pi}{a}\right)^2 + \left(\frac{0\cdot\pi}{b}\right)^2 + \left(\frac{1\cdot\pi}{d}\right)^2} \quad \text{(for TE}_{101}\text{ mode)} \quad (10\text{–}6)$$

The resultant field configuration is shown in Fig. 10–2.

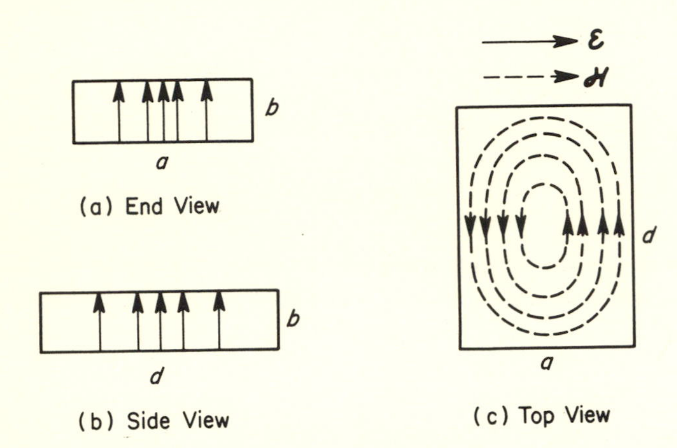

Fig. 10–2. Field configuration for TE_{101} mode.

Example 10–1

At this point it is instructive to develop explicit expressions for the fields of the TE_{101} mode in a rectangular resonator formed by placing short-circuit plates over the ends of a section of rectangular waveguide which is d meters long. The waveguide dimensions are a meters by b meters. The starting point is the field expressions for a TE_{10} mode in a rectangular waveguide. These can be abstracted from Eqs. (9–43):

$$E'_y = E_0 \sin\frac{\pi x}{a}$$

$$H'_x = -\frac{E_0}{Z_{0(\text{TE})}} \sin\frac{\pi x}{a}$$

$$H'_z = j\frac{E_0}{\eta}\frac{\lambda}{2a}\cos\frac{\pi x}{a}$$

$$E'_x = E'_z = H'_y = 0$$

Now we can insert the z-dependence $e^{-j\beta z}$ for the waves traveling in the $+z$-direction and $e^{+j\beta z}$ for the waves traveling in the $-z$-direction. The resultant waves are

$$E_y^+ = E_0^+ \sin \frac{\pi x}{a} e^{-j\beta z}$$

$$H_x^+ = -\frac{E_0^+}{Z_{0(\text{TE})}} \sin \frac{\pi x}{a} e^{-j\beta z} \qquad \left\{\begin{array}{l}\text{non-zero components for} \\ \text{the wave traveling in the} \\ +z\text{-direction}\end{array}\right\}$$

$$H_z^+ = j\frac{E_0^+}{\eta}\frac{\lambda}{2a} \cos \frac{\pi x}{a} e^{-j\beta z}$$

$$E_y^- = E_0^- \sin \frac{\pi x}{a} e^{+j\beta z}$$

$$H_x^- = \frac{E_0^-}{Z_{0(\text{TE})}} \sin \frac{\pi x}{a} e^{+j\beta z} \qquad \left\{\begin{array}{l}\text{non-zero components for} \\ \text{the wave traveling in the} \\ -z\text{-direction}\end{array}\right\}$$

$$H_z^- = j\frac{E_0^-}{\eta}\frac{\lambda}{2a} \cos \frac{\pi x}{a} e^{+j\beta z}$$

For the wave traveling in the $-z$-direction, the direction of the electric field is assumed to be the same as for the wave traveling in the $+z$-direction. This forces the transverse component of the magnetic field (H_x) to be directed opposite to that of the transverse component of the magnetic field of the wave traveling in the $+z$-direction.

These waves exist simultaneously in the rectangular structure, so we write the composite waves as

$$E_y = (E_0^+ e^{-j\beta z} + E_0^- e^{+j\beta z}) \sin \frac{\pi x}{a}$$

$$H_x = (-E_0^+ e^{-j\beta z} + E_0^- e^{+j\beta z}) \frac{1}{Z_{0(\text{TE})}} \sin \frac{\pi x}{a}$$

$$H_z = \frac{j}{\eta}\frac{\lambda}{2a} (E_0^+ e^{-j\beta z} + E_0^- e^{+j\beta z}) \cos \frac{\pi x}{a}$$

The boundary conditions require the tangential electric field to be zero at $z = 0$ and $z = d$. Thus, $E_y = 0$ at $z = 0$, or,

$$(E_0^+ + E_0^-) \sin \frac{\pi x}{a} = 0$$

This leads one to choose $E^- = E_0$ and $E_0^+ = -E_0$. Substituting into the expression for E_y results in

$$E_y = E_0(e^{j\beta z} - e^{-j\beta z}) \sin \frac{\pi x}{a} = 2jE_0 \sin \beta z \sin \frac{\pi x}{a}$$

In order to satisfy the boundary condition that $E_y = 0$ at $z = d$, $\beta = \pi/d$. The other field components can now be determined to be

$$H_x = \frac{2E_0}{Z_{0(\text{TE})}} \cos \frac{\pi}{d} z \sin \frac{\pi x}{a}$$

$$H_z = -\frac{\lambda 2E_0}{\eta 2a} \sin \frac{\pi}{d} z \cos \frac{\pi x}{a}$$

Using the relationships $Z_{0(TE)} = \dfrac{\eta}{\sqrt{1 - (\omega_c/\omega_r)^2}}$ and $\dfrac{\lambda}{2a} = \dfrac{\lambda}{\lambda_c} = \dfrac{\omega_c}{\omega_r}$ (in this case $\omega = \omega_r$), we can write the field expressions as

$$E_y = 2jE_0 \sin \frac{\pi z}{d} \sin \frac{\pi x}{a}$$

$$H_x = \frac{2E_0\sqrt{1 - (\omega_c/\omega_r)^2}}{\eta} \cos \frac{\pi z}{d} \sin \frac{\pi x}{a}$$

$$H_z = -\frac{2E_0\omega_c}{\omega_r\eta} \sin \frac{\pi z}{d} \cos \frac{\pi x}{a}$$

In these equations, $\omega_c = \dfrac{1}{\sqrt{\mu_0\epsilon_0}}\left(\dfrac{\pi}{a}\right)$, $\omega_r = \dfrac{1}{\sqrt{\mu_0\epsilon_0}}\sqrt{\left(\dfrac{\pi}{a}\right)^2 + \left(\dfrac{\pi}{d}\right)^2}$, and $\eta = \sqrt{\mu_0/\epsilon_0}$.

A word or two is in order at this point about finding the dominant mode for a given physical situation. The dominant mode is the one with the lowest resonant frequency, and thus it is desired to choose the three subscripts such that ω_r as given by Eq. (10–3) will be as small as possible. An obvious possibility is to let the integer p be zero. However, this is only possible for a TM mode and not for a TE one. In the TE case the entire electric field is transverse and must be zero at the ends of the resonator. If no half-wavelengths are allowed, then **E** must be zero everywhere, and a trivial situation exists. In the TM case a longitudinal component of **E** may exist even though the transverse component is zero everywhere, and thus this does not lead to a trivial or no-field situation. For the rectangular box it amounts to this: Any one, but no more than one, of the subscripts may be set equal to zero. In the TE case either of the first two, but not the last, subscripts may be zero, and in the TM case neither of the first two may be zero but the last one may be zero.

You may be wondering by now about the generality of our solution. With reference in particular to the rectangular resonator, it may appear at first glance that by reorienting the box along the coordinate axes, one could obtain entirely new modes of operation. However, this is not so. All modes obtained in this way will be simply repetitions of modes obtained by another orientation. The only distinction will be that they are numbered differently. This is apparent when one remembers that the subscripts simply denote the number of half-cycle variations along the respective coordinate axes. All possible situations may then be enumerated for any one orientation of the box. This is an important point and some subtle inferences may be drawn from it. In waveguide theory we saw that all the modes in a cylindrical pipe could be obtained from corresponding ones in a rectangular pipe by a gradual distortion of the walls until they become cylindrical. Thus we conclude that the waveguide approach leads us to a complete set of solutions for the cylindrical resonator also. Finally, the distortion argument applies equally well, regardless of the cross-sectional geometry, and thus it may be inferred that the waveguide approach leads to the most general solution possible for any resonator possessing axial properties.

Example 10–2

As a specific example of resonant frequency calculations we shall look at two cavities, one of which is rectangular and the other cylindrical in shape, and we shall choose the sizes to be approximately the same. It is desired to find the dominant mode and its field configuration for each case. The rectangular cavity to be considered has dimensions of 10 by 4 by 10 cm (a, b, and d, respectively), and the cylindrical one is 10 cm in diameter and 4 cm in length. The two cavities are roughly the same size, then, and just from physical reasoning, we would expect them to have similar dominant modes.

Consider the rectangular cavity first. By combining Eqs. (10–3) and (10–4), we get the expression for resonant frequency:

$$f_r = \frac{1}{2\pi\sqrt{\mu\epsilon}}\sqrt{\left(\frac{m\pi}{0.1}\right)^2 + \left(\frac{n\pi}{0.04}\right)^2 + \left(\frac{p\pi}{0.1}\right)^2}$$

We wish to find the lowest possible value for f_r, and thus we shall choose m, n, and p as small as possible, consistent with the restriction that only one of these may be zero. It is apparent that letting n equal zero and m and p equal one gives the minimum frequency situation. The resulting mode therefore will be the TE_{101} mode, and the corresponding resonant frequency is

$$f_r = \frac{3 \times 10^8}{2\pi}\sqrt{(10\pi)^2 + (10\pi)^2} = 2.12 \text{ GHz}$$

Note that the mode must be a TE type because we cannot have a zero appearing in the first two subscripts of a TM mode.

Next consider the cylindrical cavity. Its resonant frequency is obtained from Eqs. (10–3) and (10–5) and is

$$f_r = \frac{1}{2\pi\sqrt{\mu\epsilon}}\sqrt{\left(\frac{\tau_{nl}}{0.05} \quad \text{or} \quad \frac{\tau'_{nl}}{0.05}\right)^2 + \left(\frac{p\pi}{0.04}\right)^2}$$

By referring to Tables 9–1 and 9–2, it can be seen that the lowest possible values for τ and τ' are 2.40 and 1.84, respectively. Therefore the minimum frequency is obtained by letting p be zero and choosing the lowest-order TM mode, which is the one corresponding to $\tau_{01} = 2.40$. The resulting mode is the TM_{010} mode and its resonant frequency is

$$f_r = \frac{3 \times 10^8}{2\pi}\sqrt{\left(\frac{2.40}{0.05}\right)^2} = 2.29 \text{ GHz}$$

Note that we have two zeros in the subscripts for this mode. This is permissible in the cylindrical case simply because the ordering of Bessel functions begins with zero rather than one.

The two resonant frequencies work out to be approximately equal, but it may appear at first glance that the modes are radically different. This apparent discrepancy is due to the choice of coordinate system for the rectangular cavity; it was oriented such that the short dimension was vertical rather than longitudinal, as in the cylindrical case. When the two cavities are oriented similarly, their field configurations are similar, as shown in Fig. 10–3.

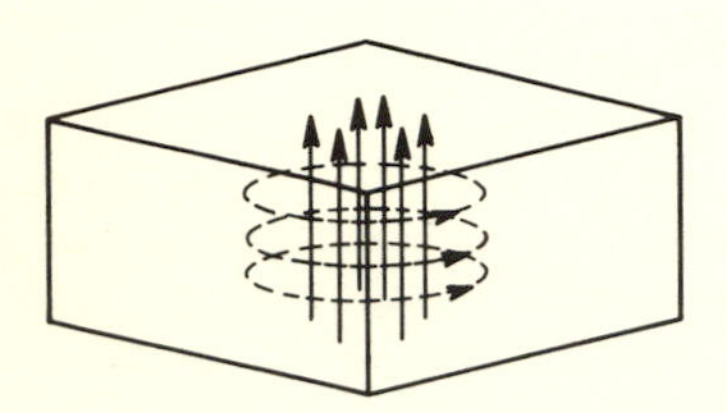
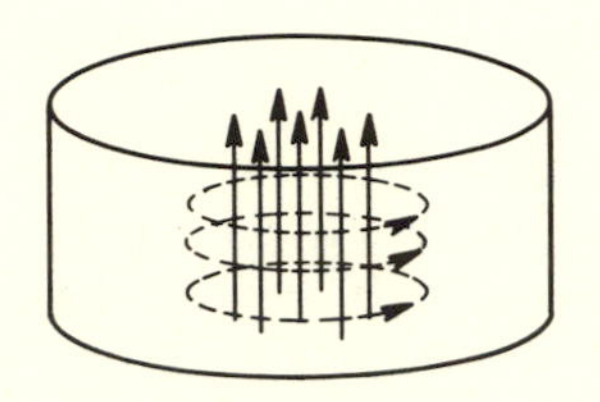

Fig. 10–3. Similar modes in rectangular and cylindrical cavities.

10–3. ENERGY CONSIDERATIONS IN RESONANT CAVITIES

The solution of Maxwell's equations for a cavity resonator is often referred to as a *source-free* solution since the fields are supported by the structure even though the sources which established the field have gone to zero. The conservation-of-energy expression developed in Chapter 7 (Eq. (7–37)) applied to a source-free region enclosed by a perfectly conducting surface reduces to

$$0 = j\omega_r \iiint_V (\mu|\mathbf{H}|^2 - \epsilon|\mathbf{E}|^2)\, dv \tag{10–7}$$

for a lossless structure. Since the resonant frequency is not zero, this equation states that, at resonance,

$$\iiint_V \mu|\mathbf{H}|^2\, dv = \iiint \epsilon|\mathbf{E}|^2\, dv \tag{10–8}$$

Or, the time-averaged energy stored in the electric field is equal to the time-averaged energy stored in the magnetic field. This is a characteristic of all cavity resonators; and, if we examine a specific case in detail, we would find that the energy oscillates back and forth between electric field and the magnetic field. The same behavior is noted in a resonant circuit made up of lumped-constant elements.

Example 10-3

Consider the rectangular resonator of Fig. 10-2. The field components for this structure are developed in Example 10-1 and are repeated here for reference:

$$E_x = E_z = H_y = 0$$

$$E_y = 2jE_0 \sin\frac{\pi z}{d} \sin\frac{\pi x}{a}$$

$$H_x = \frac{2E_0\sqrt{1 - (\omega_c/\omega_r)^2}}{\eta} \cos\frac{\pi z}{d} \sin\frac{\pi x}{a}$$

$$H_z = \frac{-2E_0\omega_c}{\eta\omega_r} \sin\frac{\pi z}{d} \cos\frac{\pi x}{a}$$

Now we shall determine the total energy stored in the electric field and the total energy stored in the magnetic field to assure ourselves that these quantities are indeed equal. The total time-averaged energy stored in the electric field is

$$\frac{1}{2}\iiint \frac{1}{2}\epsilon|\mathbf{E}|^2\,dv = \int_0^d \int_0^b \int_0^a \frac{\epsilon_0}{4} 4|E_0|^2 \sin^2\frac{\pi z}{d} \sin^2\frac{\pi x}{a}\,dx\,dy\,dz$$

$$= \epsilon_0 \frac{|E_0|^2}{4} abd$$

The total time-averaged energy stored in the magnetic field is

$$\frac{1}{2}\iiint \frac{1}{2}\mu|\mathbf{H}|^2\,dv = \int_0^d \int_0^b \int_0^a \frac{\mu_0}{4} \frac{4|E_0|^2}{\eta^2} \left\{\left[1 - \left(\frac{\omega_c}{\omega_r}\right)^2\right] \cos^2\frac{\pi z}{d} \sin^2\frac{\pi x}{a} + \left(\frac{\omega_c}{\omega_r}\right)^2 \sin^2\frac{\pi z}{d} \cos^2\frac{\pi x}{a}\right\} dx\,dy\,dz$$

$$= \frac{4\mu_0}{4}\frac{|E_0|^2}{\eta^2}\left[1 - \left(\frac{\omega_c}{\omega_r}\right)^2 + \left(\frac{\omega_c}{\omega_r}\right)^2\right]\frac{abd}{4}$$

$$= \frac{\mu_0|E_0|^2}{4\mu_0/\epsilon_0} abd$$

$$= \frac{\epsilon_0|E_0|^2}{4} abd$$

thus proving the conclusion arrived at in Eq. (10-8) that the total time-averaged energy stored in the electric field is equal to the total time-averaged energy stored in the magnetic field of the resonant cavity.

10-4. RESONANT CAVITY Q

As mentioned in Art. 6-4, the Q of any resonant circuit may be defined as

$$Q = 2\pi \frac{\text{maximum energy stored per cycle}}{\text{energy dissipated per cycle}}$$

and this equation affords the most direct approach for computing the Q of a resonant cavity. As an example of this procedure consider the relatively simple problem of computing the Q of a cubical, air-filled rectangular cavity operating in the TE_{101} mode, as shown in Fig. 10–4. The peak

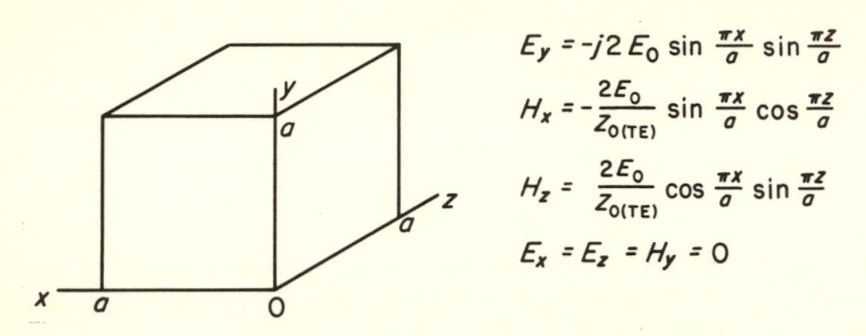

Fig. 10–4. Rectangular cavity, TE_{101} mode.

energy may be obtained by integrating $\frac{1}{2}\epsilon_0|\mathbf{E}|^2$ over the volume of the cavity. Using the equations given with Fig. 10–4 and integrating lead to

$$W\text{ (stored)} = \frac{1}{2}\epsilon_0\frac{4|E_0|^2a^3}{4} = \frac{\epsilon_0|E_0|^2a^3}{2} \tag{10–9}$$

The power dissipated in one of the sides is given by

$$P\text{ (side)} = \int_0^a \frac{1}{2}|\mathbf{J}_s|^2R_sa\,dz = \frac{1}{2}R_sa\int_0^a \frac{4|E_0|^2}{Z^2_{0(\text{TE})}}\sin^2\frac{\pi z}{a}\,dz$$

$$= \frac{R_sa^2|E_0|^2}{Z^2_{0(\text{TE})}} \tag{10–10}$$

Due to symmetry, the power dissipated in the four sides will be four times this. Next, the power dissipated in the top will be

$$P\text{ (top)} = \int_0^a\int_0^a \tfrac{1}{2}|\mathbf{J}_s|^2R_s\,dx\,dz = \tfrac{1}{2}R_s\int_0^a\int_0^a (|H_x|^2 + |H_y|^2)\,dx\,dz$$

$$= \frac{1}{2}R_s\frac{4|E_0|^2}{Z^2_{0(\text{TE})}}\int_0^a\int_0^a\left(\sin^2\frac{\pi x}{a}\cos^2\frac{\pi z}{a} + \cos^2\frac{\pi x}{a}\sin^2\frac{\pi z}{a}\right)dx\,dz$$

$$= \frac{R_sa^2|E_0|^2}{Z^2_{0(\text{TE})}} \tag{10–11}$$

Therefore the total power lost is

$$P\text{ (total)} = \frac{6R_sa^2|E_0|^2}{Z^2_{0(\text{TE})}} \tag{10–12}$$

and the Q is

$$Q = 2\pi f_r \frac{\dfrac{\epsilon_0|E_0|^2a^3}{2}}{\dfrac{6R_s a^2|E_0|^2}{Z^2_{0(\text{TE})}}} = \omega_r \frac{\epsilon_0 Z^2_{0(\text{TE})}a}{12R_s}$$

Now, noting that for a cubical cavity

$$\omega_r = \frac{1}{\sqrt{\mu_0\epsilon_0}}\sqrt{2}\frac{\pi}{a}$$

$$\omega_r = \sqrt{2}\omega_c$$

$$Z_{0(\text{TE})} = \sqrt{2}\eta$$

the final expression for Q is

$$Q = \frac{\sqrt{2}\pi\eta}{6R_s} \tag{10–13}$$

In the microwave range, R_s is of the order of a fraction of an ohm, and hence Q's of the order of many thousand are not unusual. Thus resonant cavities are quite useful in applications calling for a frequency-sensitive element such as oscillators and frequency meters.

The method of analysis just described also applies to other shapes and modes of operation. As in the case of waveguide attenuation, the method is straightforward, but the algebra is usually quite involved.

10–5. THE FORESHORTENED COAXIAL-LINE RESONATOR

The foreshortened coaxial-line resonator is one of the most useful of all cavities. It is particularly useful in klystron work and in other applications where particle acceleration is involved. The term *foreshortened* comes about by replacing a portion of the open end of a quarter-wave resonant line with a lumped capacitor, as shown in Fig. 10–5.

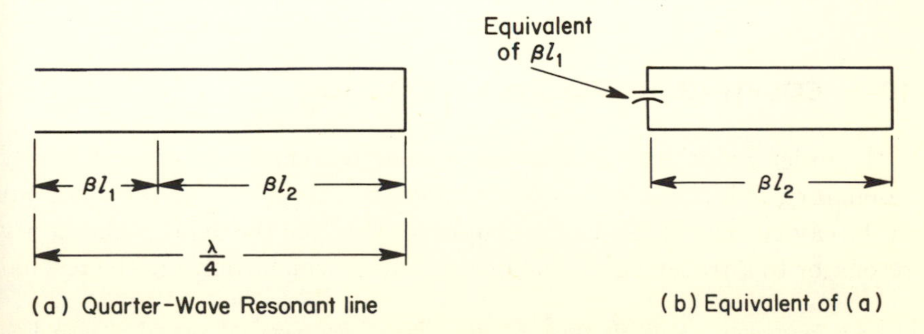

Fig. 10–5. Capacitively loaded line.

In the case of a coaxial line the capacitance loading is obtained by folding over the "open" end as shown in Fig. 10–6. The equivalent

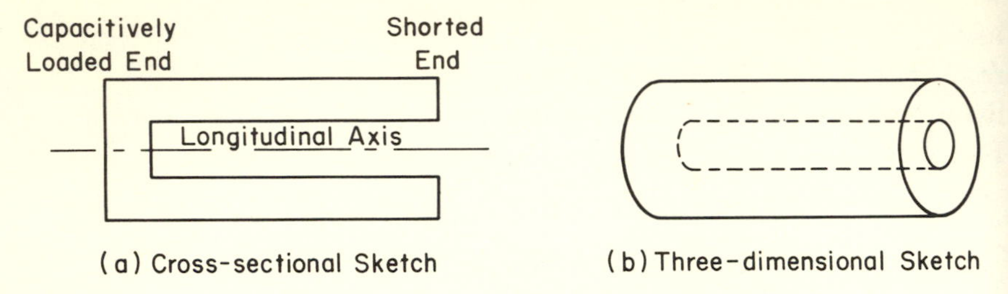

Fig. 10–6. Foreshortened coaxial line.

capacitance loading the line may be obtained by estimating the capacitance between the end faces of the inner and outer conductors. Once this capacitance is obtained, the resonant frequency may be obtained from transmission-line theory:

$$\frac{1}{\omega C} = Z_0 \tan \beta l \tag{10–14}$$

where l is the length of line, Z_0 the characteristic impedance, C the equivalent loading capacitance, and β the phase-shift constant given by ω/v. Note that when β is replaced by ω/v, Eq. (10–14) becomes transcendental in ω and must be solved by approximate techniques.

The approximate solution just described applies very well to long cavities of this type, such as shown in Fig. 10–6. However, as the line becomes shorter and shorter (with respect to radii involved), the solution becomes poorer and poorer and, of course, of less value. This is because the field for a short, stubby line is no longer distributed as assumed in transmission-line theory. However, solutions for the short-line case have been worked out and are presented in graphical form in a number of sources.[3]

10–6. COUPLING TO RESONANT CAVITIES

In order to utilize the frequency selective characteristics of cavity resonators, it is necessary to devise ways to couple energy into and out of the cavity. The problem is to couple energy from the fields of the cavity resonator to a transmission line or waveguide which connects the resona-

[3] See, for example, A. E. Harrison, *Klystron Tubes* (New York: McGraw-Hill Book Co., Inc., 1947), pp. 254–62.

tor to the rest of the system. Let us begin by considering a coaxial transmission line coupled to a cavity resonator. There are basically two ways in which the transmission line can couple to the fields of the resonator. First, the coaxial line can couple to the magnetic field of the resonator by inserting a wire loop into the resonator in such a way that it links the magnetic field in the resonator. This situation is shown in Fig. 10–7 as it might apply to a TE_{101} mode in a rectangular resonator.

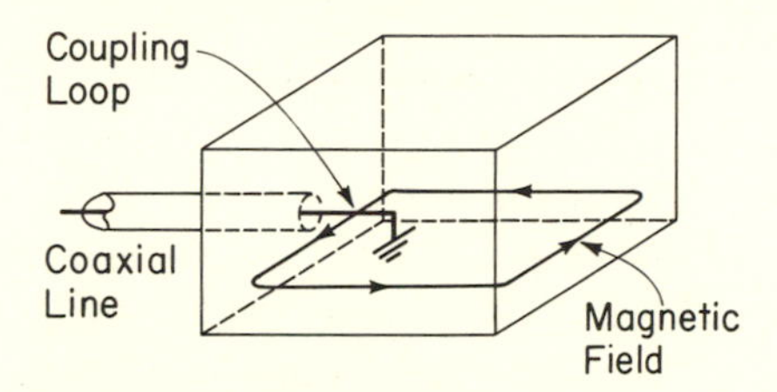

Fig. 10–7. Coupling to the magnetic field in a cavity resonator.

This form of coupling acts as a voltage source for which the open-circuit voltage can be determined by the application of Maxwell's equation:

$$\oint \mathbf{E} \cdot \mathbf{dl} = -j\omega\mu \iint \mathbf{H} \cdot \mathbf{da}$$

In the case where the magnetic field linking the loop is parallel to the surface vector, this reduces to

$$V_{oc} = -j\omega\mu \iint H\, da \tag{10–15}$$

In an application, a current will flow in the loop and the voltage at the input to the transmission line will be less than the open-circuit voltage.

The second method of coupling a transmission line to a cavity resonator is by coupling to the electric field. This method of coupling is accomplished by inserting a probe into the cavity resonator. A sketch of the physical configuration is shown in Fig. 10–8 as it might be applied to

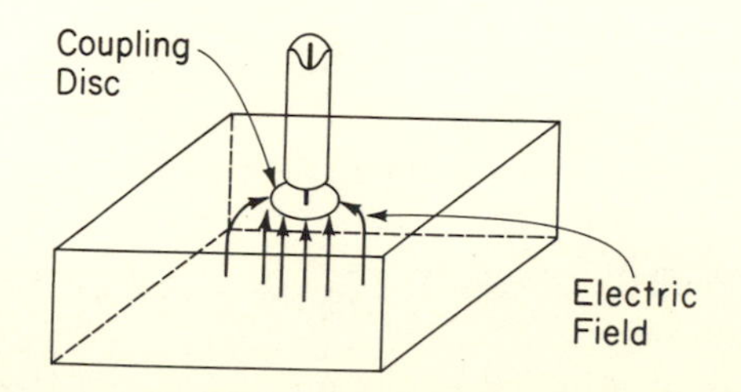

Fig. 10–8. Coupling to the electric field in a cavity resonator.

a rectangular resonator operating in the TE_{101} mode. This method of coupling acts as a current source with a short-circuit current determined by the principle of conservation of current. The current flow in the transmission line is equal to the displacement current flowing into the coupling disc due to the fields of the cavity. Neglecting the fringing fields, the expression for the short-circuit current is

$$I_{sc} = j\omega\epsilon \iint_{\text{disc}} \mathbf{E} \cdot \mathbf{da} \tag{10–16}$$

The disc loading the end of the coupling probe is not necessary to couple to the fields of the resonator. The probe shape shown in Fig. 10–8 was chosen mainly to facilitate the application of the theory which led to Eq. (10–16). This probe shape is occasionally encountered, however.

The most common form of the electric field probe is that in which the center conductor of the coaxial transmission line is simply extended into the resonator. It is important to note that the most effective coupling occurs when the electric field in the resonator and the surface vector of the loading disc of the probe are parallel.

Waveguides are coupled to resonators by means of apertures which will allow the fields of one structure to establish a field in the second structure. Consider the field coupling examples shown in Fig. 10–9.

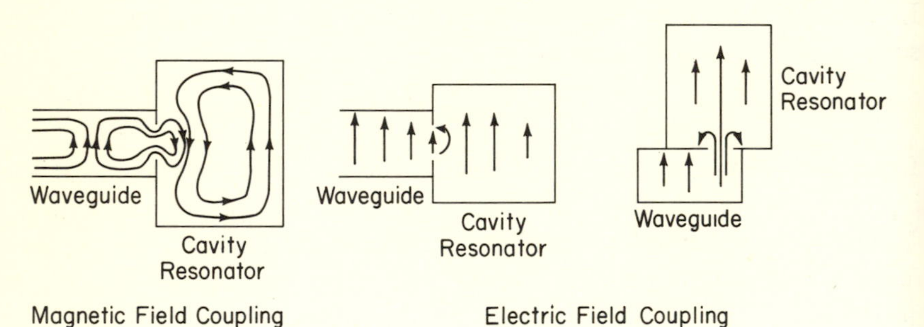

Fig. 10–9. Waveguide–cavity resonator coupling techniques.

Basically, all that is required is that the field in the waveguide be parallel to the desired field in the resonator. Care is usually taken to make the coupling primarily via either the electric field or the magnetic field.

10–7. THE EQUIVALENT CIRCUIT FOR A CAVITY RESONATOR

An equivalent circuit applicable to some point on a transmission line which is coupled into a cavity resonator can be developed by considering the complex power flow on the transmission line. An example of the structure is shown in Fig. 10-10. The complex power flowing into this transmission line can be determined by Eq. (7-37) for the case of no sources within the volume:

$$-\iint_S \mathbf{P} \cdot \mathbf{da} = \iiint_V \sigma|\mathbf{E}|^2\, dv + j\omega \iiint_V (\mu|\mathbf{H}|^2 - \epsilon|\mathbf{E}|^2)\, dv \qquad (10\text{-}17)$$

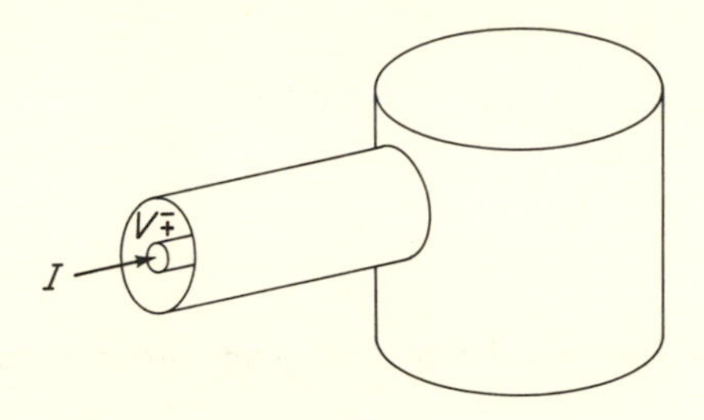

Fig. 10-10. Coaxial transmission line coupled to a cavity resonator.

The left-hand side of the equality of Eq. (10-17) is the complex power at some point on the transmission line. That is,

$$-\iint_S \mathbf{P} \cdot \mathbf{da} = VI^* = \iiint_V \sigma|\mathbf{E}|^2\, dV + j\omega \iiint_V (\mu|\mathbf{H}|^2 - \epsilon|\mathbf{E}|^2)\, dv \qquad (10\text{-}18)$$

The voltage and current of Eq. (10-18) are the values on the transmission line at the point where the surface of Eq. (10-17) cuts across the line.

Equation (7-37) can be used as the basis for a lumped-element circuit model of the conditions existing at the point on the transmission line where V and I are defined.

$$VI^* = |V|^2 Y^* = \iiint_V \sigma|\mathbf{E}|^2\, dv + j\omega \iiint_V (\mu|\mathbf{H}|^2 - \epsilon|\mathbf{E}|^2)\, dv \qquad (10\text{-}19)$$

or

$$Y^* = G - jB = \frac{\iiint_V \sigma|\mathbf{E}|^2\, dv + j\omega \iiint_V (\mu|\mathbf{H}|^2 - \epsilon|\mathbf{E}|^2)\, dv}{|V|^2} \qquad (10\text{-}20)$$

The equivalence is summarized in Fig. 10-11.

The lumped-element circuit model can be described in terms of the resonant frequency ω_0 and the quality factor Q_0, which are defined as

$$\omega_0 = \frac{1}{\sqrt{LC}}$$

$$Q_0 = \frac{\omega_0 C}{G}$$

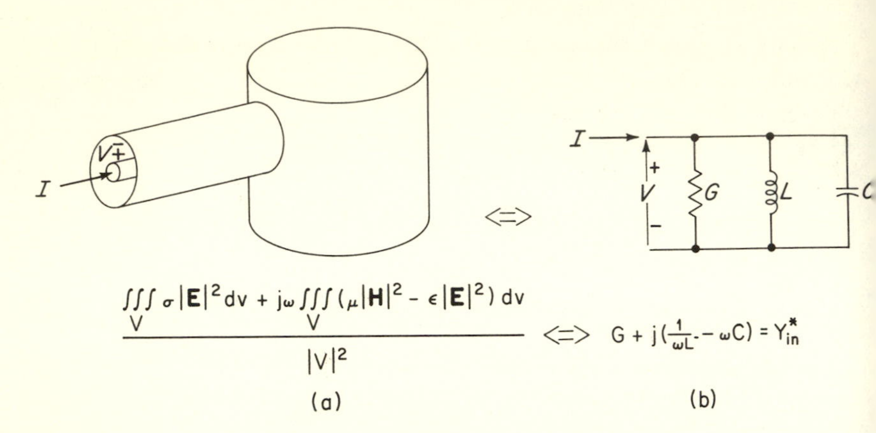

Fig. 10–11. The equivalent lumped-element circuit model of the cavity resonator.

The input admittance can be written as

$$Y_{in} = G\left(1 + j\frac{\omega C}{G}\left[1 - \left(\frac{\omega_0}{\omega}\right)^2\right]\right) = G\left[1 + jQ_0\left(\frac{\omega^2 - \omega_0^2}{\omega_0\omega}\right)\right]$$

Defining $\delta = \omega - \omega_0$, and assuming $\omega \cong \omega_0$, we can write

$$Y_{in} \cong G\left(1 + jQ_0\frac{2\delta}{\omega_0}\right).$$

The input admittance traces out a circle on the Smith chart as the frequency is varied from less than the resonant frequency to greater than the resonant frequency. A Smith chart plot of the input admittance versus frequency for a typical case is shown in Fig. 10–12. At the frequencies $\omega_0 \pm \delta_1$ where $Y_{in} = G(1 \pm j)$, we see that $Q_0 = \omega_0/2\delta_1$.

It is a fairly simple matter to determine frequency and impedance at microwave frequencies, so ω_0 and Q_0 are useful parameters in characterizing the resonant structure of Fig. 10–11(a).

The preceding discussion has been based on the assumption that the equivalent circuit is of a parallel-resonant nature. It should be clear that one could establish the reference plane one quarter-wavelength along the transmission line from the point that has been used in the foregoing discussion. This would result in an equivalent circuit which has the properties of a series-resonant circuit.

If the input plane were taken at the point where the transmission line couples to the cavity resonator, the condition of resonance occurs whenever the time-averaged energy stored in the magnetic field is equal to the time-averaged energy stored in the electric field. This condition results in

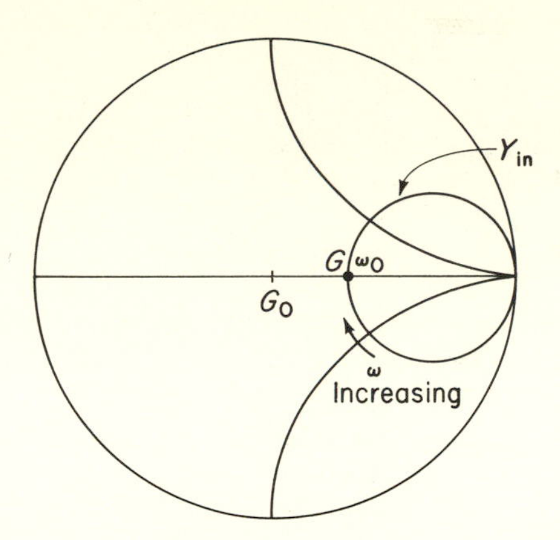

Fig. 10-12. Input admittance of a transmission line coupled to a cavity resonator.

a power flow into the resonant cavity that is real and the input impedance is purely resistive. The value of the input resistance changes as the coupling is changed. This input resistance serves as the basis in defining the coupling into the cavity resonator. If the transmission line coupled to the resonator has a characteristic impedance $R_0 = \frac{1}{G_0}$, and the input conductance looking into the cavity resonator at resonance is G, the coupling factor β is defined as

$$\beta = \frac{G_0}{G} \tag{10-21}$$

The system is said to be undercoupled when $\beta < 1$, critically coupled when $\beta = 1$, and overcoupled when $\beta > 1$.

In practice, coupling to a cavity resonator results in a load being coupled to the resonator and extracting energy from it. This load could be a passive termination connected to one end of the transmission line or it could be the impedance of a generator connected to the end of the transmission line. In either case, the equivalent circuit at the reference plane, for the case of a load of G_0, is shown in Fig. 10-13. The load G_0 changes the admittance at the reference plane from the value discussed in preceding paragraphs. The effect is to reduce the quality factor Q from Q_0 to a new value, Q_L, called the loaded Q. The loaded Q (Q_L) for this circuit model is easily calculated from Eq. (10-22), where Q_e is the external Q and Q_0 is the unloaded Q:

$$\frac{1}{Q_L} = \frac{G_0 + G}{\omega_0 C} = \frac{1}{Q_e} + \frac{1}{Q_0} \tag{10-22}$$

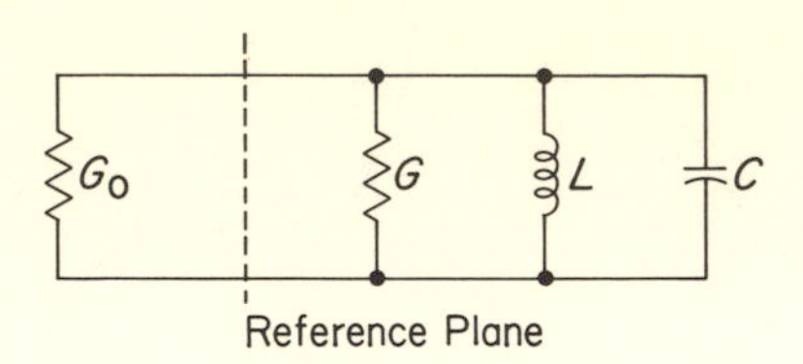

Fig. 10–13. Equivalent circuit for cavity resonator coupled to load G_0.

Equation (10–22) can be rewritten as

$$\frac{1}{Q_L} = \frac{\beta G + G}{\omega_0 C} = \frac{1}{Q_0}(1 + \beta) \tag{10–23}$$

The following relationship is apparent:

$$\frac{Q_0}{Q_L} = 1 + \beta \tag{10–24}$$

or

$$\beta = \frac{Q_0}{Q_e} = \frac{G_0}{G}$$

From the definition of β, one observes that $\beta = S$ (i.e., VSWR) if $\beta > 1$, that is, $G_0/G > 1$, or $\beta = \dfrac{1}{S}$ if $\beta < 1$. Equation 10–24 is useful in determining the unloaded Q of a resonator by measuring the loaded Q of the resonator and the VSWR on the transmission line when the cavity is resonant.

Again, it must be emphasized that a parallel-resonant circuit has been used as the equivalent circuit. This choice is made rather arbitrarily and one could redevelop the formulas using a series-resonant circuit as a lumped-element model.

PROBLEMS

10–1. Determine the resonant frequencies of the four lowest-order modes of a rectangular resonator which measures 10 cm × 20 cm × 30 cm.

10–2. Determine the resonant frequencies of the four lowest-order modes of a circular resonator which measures 10 cm in diameter by 20 cm in length.

10–3. Develop explicit expressions for the field components of the TM_{010} mode in a cylindrical resonator measuring a meters in radius by d meters in length.

10–4. Design a rectangular cavity resonator which will resonate in the TE_{101} mode at 10 GHz and resonate in the TM_{110} mode at 20 GHz.

10–5. A cylindrical resonator is designed to resonate in the TE_{011} mode over a frequency range of 8 GHz to 12 GHz. The tuning is accomplished by moving one end plate, which effectively changes the length of the resonator. If the radius of the cylindrical resonator is 3 cm, calculate the range displacement of the movable end piece which is required to meet the design specifications.

10–6. Show that the Q of the cubical resonator of Art. 10–4 can be expressed as

$$Q = \frac{1}{3}\sqrt{\frac{\pi\sigma}{2\epsilon f}}$$

10–7. Develop an expression for the Q of a rectangular resonator measuring a meters by b meters by d meters along the x, y, and z axes, respectively. The cavity is resonating in the dominant TE_{101} mode. The walls are characterized by a finite conductivity σ.

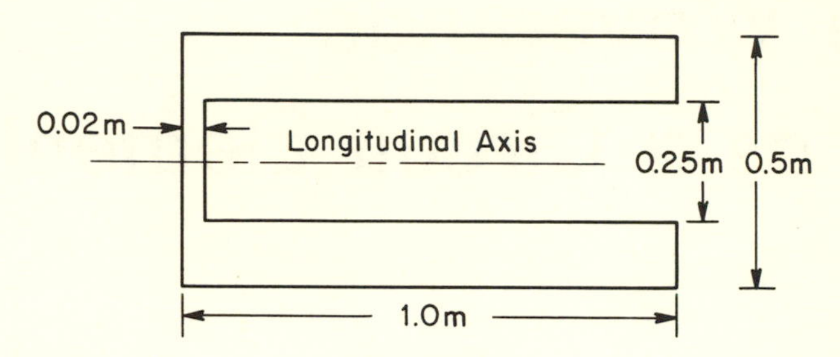

Prob. 10–8.

10–8. A foreshortened-line-type resonator has dimensions as shown in the accompanying figure. Compute the lowest resonant frequency for this cavity. The dielectric is air.

10–9. Design a coupling scheme for the resonator of Problem 10–4 which will assure that the probe or loop meant to couple to the TE_{101} mode will not couple to the TM_{110} mode, and vice versa.

11

The Waveguide as a Circuit Element

11–1. INTRODUCTION

The mathematical foundation for the analogy between waveguide-type transmission and that of conventional transmission lines has been presented in Chapter 9. This analogy is quite useful and will be elaborated on further at this point. Basically the waveguide is just another means of transmitting energy from one point to another. Thus many of the problems in transmission-line work, such as matching and termination, are also encountered in waveguide work. The transmission-line analogy often points the way to the solution of some of these problems.

The analogy is particularly close in the case of the TE_{10} mode in rectangular waveguides, and we shall confine our attention to this mode from this point on. The similarity between this mode and an open-wire transmission line can be seen from the field configurations shown in Fig. 11–1. Note that the transverse **E** and **H** fields are roughly similar and that the longitudinal wall currents in the top and bottom of the guide are similar to those in the two wires of the transmission line. As a matter of fact, one can go so far as to define a "voltage" and "current" in the waveguide in such a way that the total incident power in the guide is the product of the two, just as in the line. This leads to a characteristic impedance, though, which is different from the characteristic wave

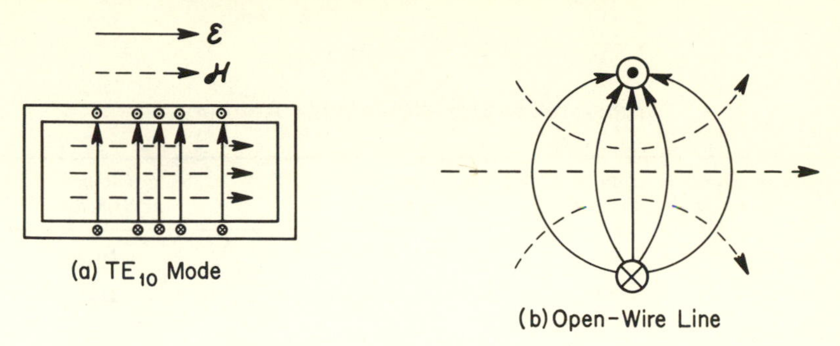

Fig. 11–1. Analogous waveguide and transmission-line modes.

impedance defined in Art. 9–5. Rather than have two different meanings for characteristic impedance we shall retain the wave impedance concept. The primary difference is that our formulas will lead directly to power density rather than total power.

A summary of the analogous formulas for the two cases is given in Table 11–1. An example will serve to illustrate the use of the analogy.

Example 11–1

It is desired to determine the wave impedance of a particular load in an X-band waveguide system (0.9 by 0.4 in.) by means of a slotted section of waveguide. With a shorting plane placed over the load end of the guide, adjacent nulls are found to be 2.44 cm apart. When the short is replaced with the load, the standing-wave ratio is found to be 2.0, and the nulls have shifted toward the load by 0.81 cm. We pose the following three questions: (1) What is the frequency of operation? (2) What is the reflection coefficient and wave impedance at the load? (3) If the incident power is 10 mW, what is the net power delivered to the load?

The solution to the first question closely parallels the corresponding transmission-line problem, but we must remember that the wavelength within the guide is not the same as the "free" wavelength. It is fundamental in wave theory that nulls appear at half-wavelength intervals, and therefore

$$\lambda_g = (2)(2.44) = 4.88 \text{ cm}$$

Now, noting that the free wavelength λ is just the free velocity v divided by the frequency f, the expression for λ_g given in Table 11–1 can be written as

$$\lambda_g = \frac{v}{f} \frac{1}{\sqrt{1 - (f_c/f)^2}}$$

This equation can be readily solved for frequency, and the result is

$$f = \sqrt{f_c^2 + \left(\frac{v}{\lambda_g}\right)^2}$$

TABLE 11–1

Transmission-Line–Waveguide Analogy

Transmission-Line Quantity	*Equation or Symbol*	*Waveguide Quantity*	*Equation or Symbol*
Voltage	V	Transverse electric field	$\mathbf{E}_t$
Current	I	Transverse magnetic field	$\mathbf{H}_t$
Characteristic impedance	$Z_0 = \sqrt{\frac{L}{C}}$	Characteristic wave impedance	$Z_{0(\mathrm{TE})} = \eta \frac{1}{\sqrt{1-\left(\frac{\omega_c}{\omega}\right)^2}}$ or $Z_{0(\mathrm{TM})} = \eta \sqrt{1-\left(\frac{\omega_c}{\omega}\right)^2}$
Phase-shift constant	$\beta = \omega\sqrt{LC}$	Phase-shift constant	$\beta = \omega\sqrt{\mu\epsilon}\sqrt{1-\left(\frac{\omega_c}{\omega}\right)^2}$
Phase velocity	$v = \frac{1}{\sqrt{LC}}$	Phase velocity	$v_p = \frac{v}{\sqrt{1-\left(\frac{\omega_c}{\omega}\right)^2}}$
Wavelength	$\lambda = \frac{v}{f}$	Wavelength (in waveguide)	$\lambda_g = \lambda \frac{1}{\sqrt{1-\left(\frac{\omega_c}{\omega}\right)^2}}$
Incident power*	$P^+ = \frac{\lvert V^+\rvert^2}{2Z_0}$	Incident power density*	$P^+ = \frac{\lvert \mathbf{E}_t^+\rvert^2}{2Z_{0(\mathrm{TE})}}$ (or $Z_{0(\mathrm{TM})}$)

* The presence of the factor of 2 in these expressions indicates that peak rather than rms value must be used for V^+ and $\mathbf{E}_t^+$.

The frequency may now be computed as (recalling that f_c for X-band waveguide is 6.56 GHz)

$$f = \sqrt{(6.56)^2 + \left(\frac{0.3}{0.0488}\right)^2} \times 10^9 = 9.0 \text{ GHz}$$

The reflection coefficient and per-unit wave impedance can be found with the aid of a Smith chart, just as in transmission theory. By referring to Fig. 11–2, we enter the chart at a point of minimum impedance and move back toward the source to a point known to be some multiple of a half-wavelength from the load. This takes place on a constant-radius circle corresponding to a standing-wave ratio of 2, and the distance moved is $(0.81/4.88)\lambda_g = 0.166\lambda_g$. The resulting reflection coefficient and wave impedance are

$$\text{Reflection coefficient} = 0.333 \underline{/\ 60^\circ}$$

$$\text{Per-unit wave impedance} = 1.12 + j0.72$$

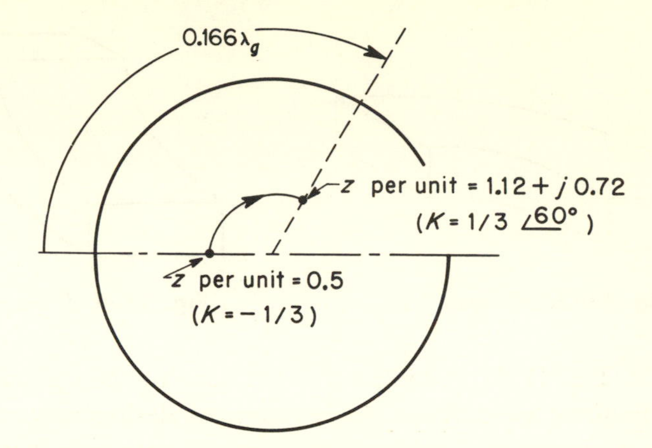

Fig. 11–2. Smith chart for Example 11–1.

The ohmic value of the wave impedance could be obtained, of course, by multiplying the per-unit value by the characteristic impedance. However, this is not normally of primary interest, so this will not be done.

Finally, if the incident power is 10 mW, then the net power delivered to the load is

$$
\begin{aligned}
P_{\text{load}} &= P_{\text{incident}} - P_{\text{reflected}} \\
&= 10(1 - |K|^2) = 8.9 \text{ mW}
\end{aligned}
$$

11–2. WAVEGUIDE COMPONENTS

Although waveguide hardware is, of course, quite different physically from that encountered in transmission-line work, many of the components, such as tuners, slotted sections, and attenuators, serve similar purposes. This is not true of all components because each method of transmission involves some problems which are peculiar to it and it alone. A brief description of some of the more common waveguide components follows. The analogies to transmission-line hardware should be obvious, where such analogies exist.

1. **H** *bend* (Fig. 11–3). This section is used to change the physical direction of propagation. It derives its name from the fact that the **H** lines are bent in the transition while the **E** lines remain vertical.

2. **E** *bend* (Fig. 11–4). This section is also used to change the physical direction of propagation. The **E** lines as well as the **H** lines are bent in this transition piece. The choice between an **H** bend and an **E** bend is usually governed by mechanical considerations, since neither will produce an appreciable discontinuity if the bends are reasonably gradual.

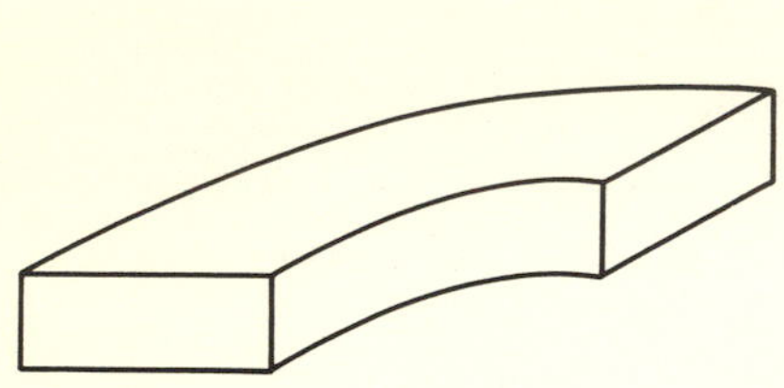

Fig. 11–3. H **bend.**

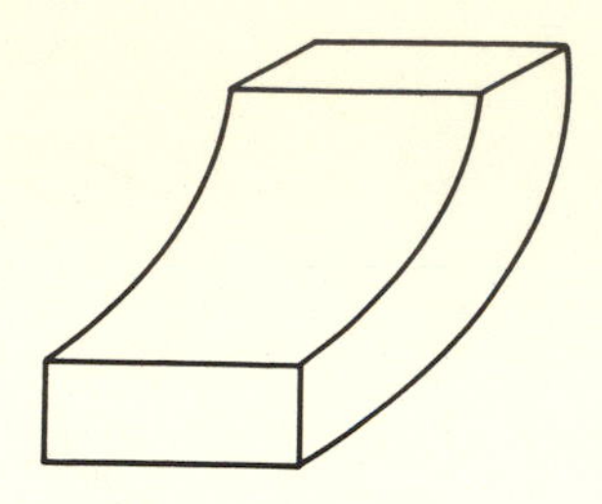

Fig. 11–4. E **bend.**

3. *Twist* (Fig. 11–5). A twist section is used to change the plane of polarization of the wave. It can be seen that any desired angular orientation of the wave may be obtained with an appropriate combination of the three types of sections (1), (2), and (3).

4. *Shunt Tee* (Fig. 11–6). A shunt tee is so named because the side arm is "shunting" the **E** field which is analogous to voltage in a transmission line. It can be seen that if two input waves at arms A and B are in phase, the portions transmitted into arm C will be in phase, and thus will be additive.

5. *Series Tee* (Fig. 11–7). It can be seen from the figure that the side arm is in "series" with the waveguide current in this case. Here, if two inputs at A and B are in phase, the portions transmitted into arm D will be 180 degrees out of phase and thus subtractive.

6. *Hybrid or Magic Tee* (Fig. 11–8). The hybrid tee is a combination of shunt and series tees and exhibits some of the properties of each. From previous consideration of the shunt and series tees, it can be seen that if two equal signals are fed into arms A and B in phase, there will be cancellation in arm D and reinforcement in arm C. Thus all the energy will be transmitted to C and none to D. Similarly, if energy is fed into C, it will divide evenly between A and B and none will be transmitted to D. The hybrid tee has many interesting applications and a few of these are mentioned in the problems at the end of the chapter.

7. *Slide-Screw Tuner* (Fig. 11–9). The slide-screw tuner consists of a screw or metallic object of some sort protruding vertically into the guide and adjustable both in depth and longitudinally. The effect of the protruding object is to produce shunting reactance across the guide. Thus it is analogous to a single-stub tuner in transmission-line theory.

8. *Double-Slug Tuner* (Fig. 11–10). This type of tuner involves placing two metallic objects, called *slugs*, in the waveguide. The necessary two degrees of freedom are obtained by making adjustable both the longi-

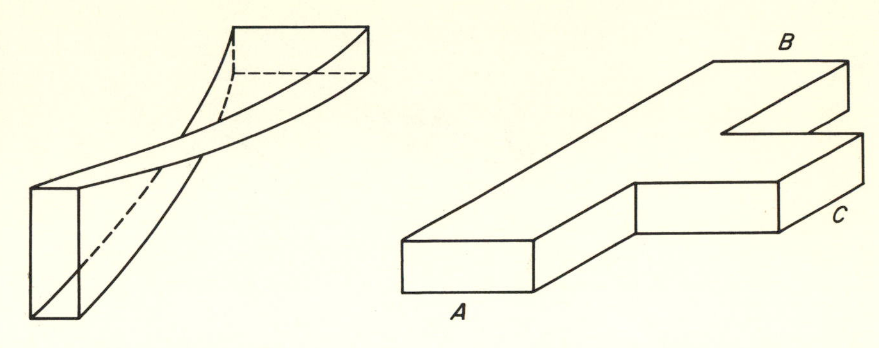

Fig. 11–5. Twist.

Fig. 11–6. Shunt tee.

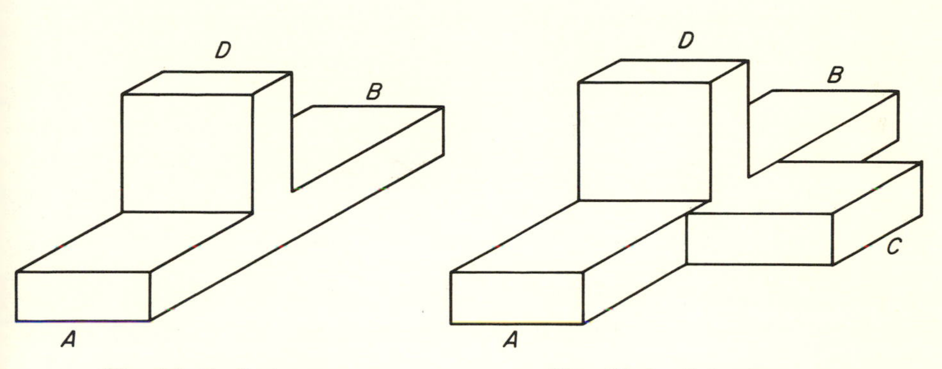

Fig. 11–7. Series tee.

Fig. 11–8. Hybrid tee.

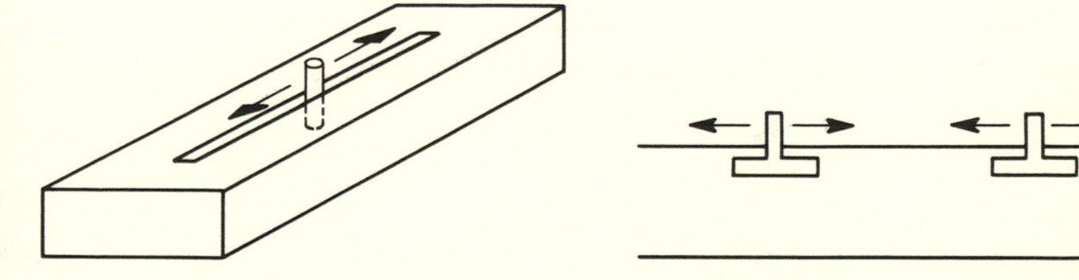

Fig. 11–9. Slide-screw tuner.

Fig. 11–10. Double-slug tuner.

tudinal position of the slugs and the spacing between them. The double-slug tuner is somewhat analogous to a double-stub tuner but not exactly, since the position of the slugs and not the effective shunting reactance is variable.

9. *Flap Attenuator* (Fig. 11–11). Attenuation is achieved by insertion of a thin card of resistive material through a slot in the top of the guide. The amount of insertion is variable, and the attenuation can be made

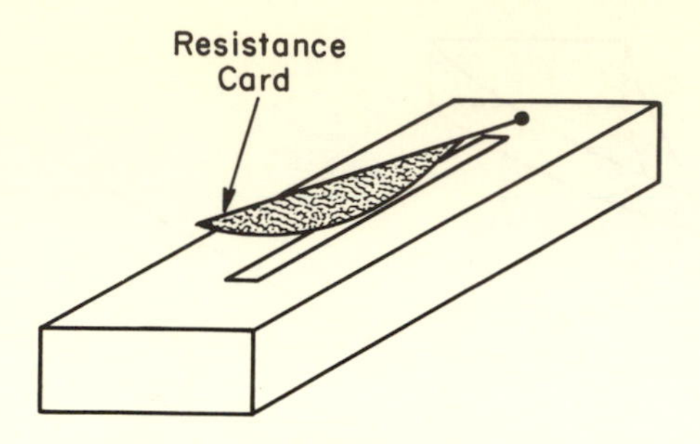

Fig. 11–11. Flap attenuator.

approximately linear with insertion by proper shaping of the resistance card.

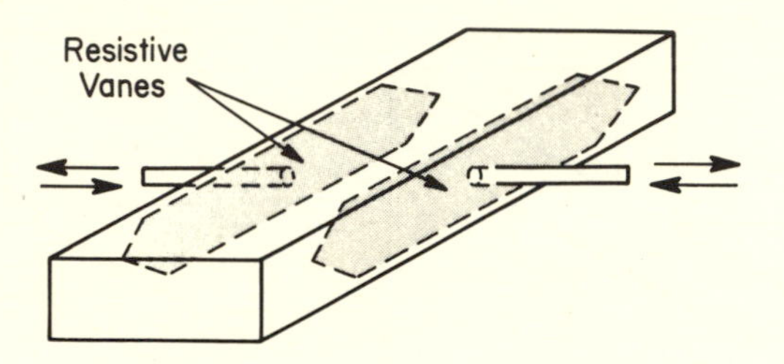

Fig. 11–12. Vane attenuator.

10. *Vane Attenuator* (Fig. 11–12). In this type of attenuator the resistance cards or vanes move in from the sides as shown in the figure. It can be seen that the losses (and thus attenuation) will be a minimum when the vanes are close to the side walls where **E** is small and maximum when the vanes are in the center.

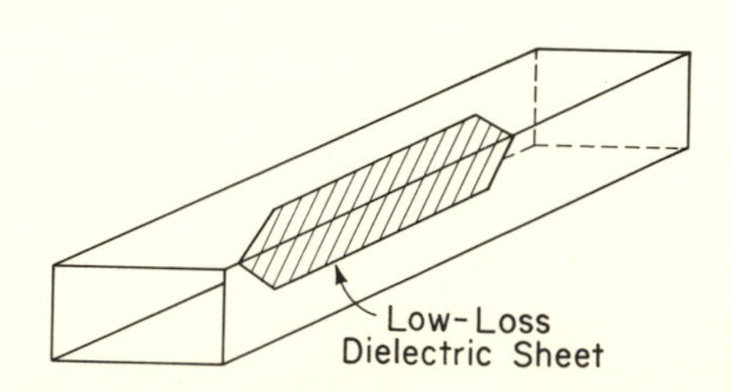

Fig. 11–13. Phase shifter.

11. *Phase Shifter* (Fig. 11–13). A phase shifter can be constructed in rectangular waveguide by inserting a sheet of low-loss dielectric material

in the waveguide so that the broad surface is parallel to the electric field.

The sheet of dielectric material, of length l, lowers the cutoff frequency in the section of waveguide which it occupies, thus changing the propagation constant (β) and introducing a phase shift of $\Delta\beta l$. The tapered ends of the dielectric sheet serve as impedance-matching sections to reduce reflections due to a change in wave impedance in the section of waveguide loaded with the dielectric sheet.

12. *Matched Vane-Type Termination* (Fig. 11–14). A single, tapered vane usually placed in the center of the guide is used in this type of matched termination. The tapering serves to minimize reflections.

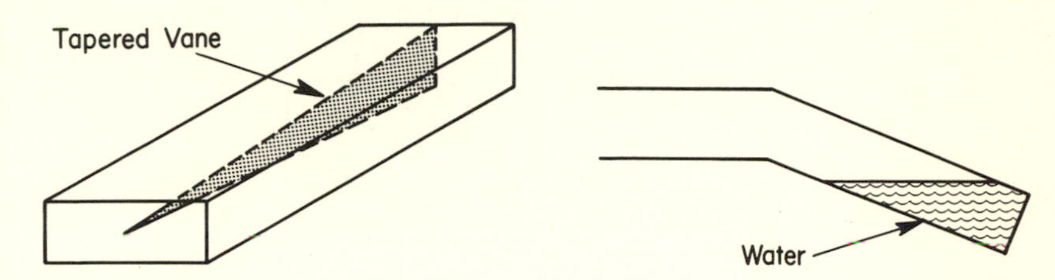

Fig. 11–14. Vane-type termination. **Fig. 11–15.** Side view of water load.

13. *Water Load* (Fig. 11–15). The water load termination is particularly useful in high-power applications. Again the taper principle is used to minimize reflections.

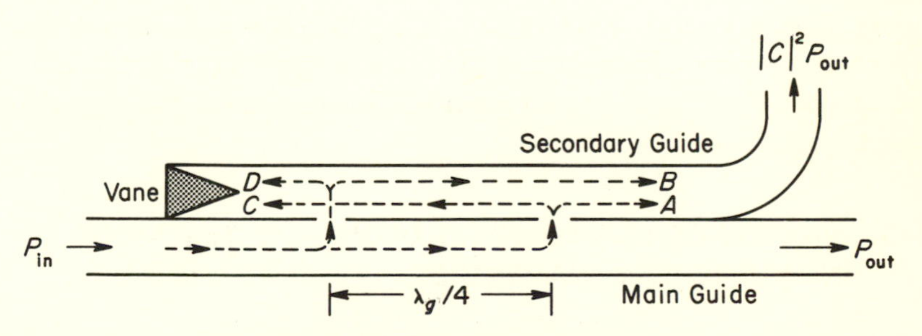

Fig. 11–16. Two-hole directional coupler.

14. *Two-Hole Directional Coupler* (Fig. 11–16). The two-hole directional coupler consists of two pieces of waveguide with one side common to both guides and two holes in the common side. These sections may be arranged physically either side by side or one over the other. The directional properties of such a device can be seen by looking at the wave paths labeled A, B, C, and D. Waves A and B follow equal-length paths and

thus combine in phase in the secondary guide. If the spacing between holes is $\lambda_g/4$, waves C and D are 180 degrees out of phase and thus cancel. Therefore, if the field within the main guide consists of a superposition of incident and reflected waves, a certain fraction of the wave moving from left to right will be coupled out through the secondary guide, and the same fraction of the right-to-left wave will be dissipated in the vane. This type of coupler is frequency-sensitive, since the spacing between holes must be $\lambda_g/4$ or an odd multiple thereof. The addition of more holes properly spaced can improve both the frequency response and directivity. This type of coupler is known as a multihole coupler, for obvious reasons.

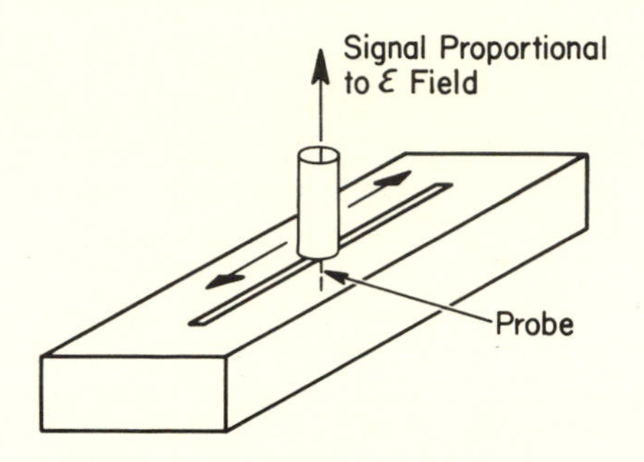

Fig. 11–17. Slotted section.

15. *Slotted Section* (Fig. 11–17). A slotted section serves the same purpose in a waveguide system as does a slotted coaxial line in a TEM system. It is, of course, different in mechanical construction, but electrically they are quite similar. The slot in the waveguide is cut longitudinally in the center of the top of the waveguide. This minimizes the disturbance caused by the slot.

11–3. WAVEGUIDE COMPONENTS USING THE TE_{11} MODE IN A CIRCULAR WAVEGUIDE

The TE_{11} mode in a circular waveguide is used in the realization of a number of interesting microwave devices. Among these are precision variable attenuators, precision variable phase shifters, and several devices using the Faraday rotation phenomena associated with a biased ferrite medium. The TE_{11} field expressions in a circular waveguide are shown in Fig. 11–18 for reference purposes.

In an application such as a variable attenuator, transitions between a rectangular waveguide supporting the TE_{10} mode and a section of circular

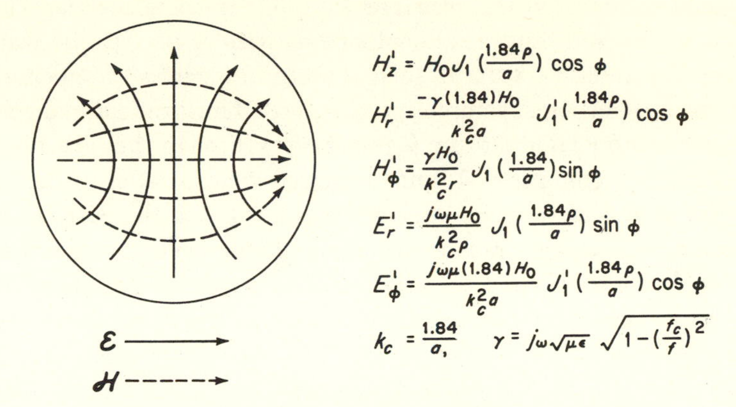

Fig. 11-18. TE_{11} mode in a circular waveguide.

waveguide supporting the TE_{11} mode are utilized. The sections of waveguide contain thin vanes of resistive material. A resistive vane attenuates components of electric field tangent to the surface while having little effect on components of electric field normal to the surface. A pictorial representation of a variable attenuator is given in Fig. 11-19.

The resistive vane in the input transition has little effect on the amplitude of the field because the electric field is normal to the surface of the vane.

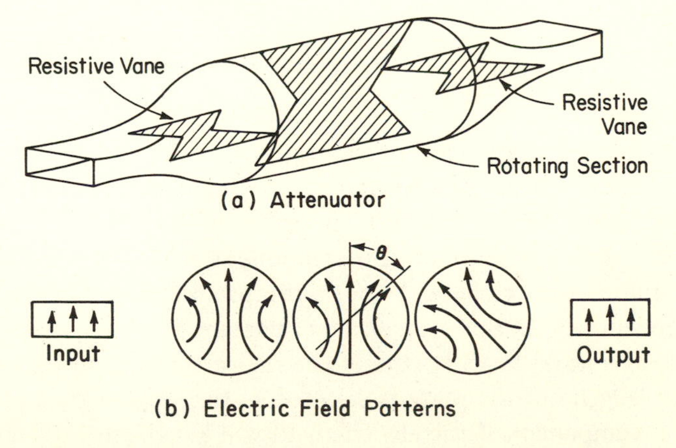

Fig. 11-19. The rotating-vane variable attenuator.

The field incident on the resistive vane in the center section can be split into normal and tangential components with respect to the plane of the vane. The resistive vane is such that the tangential component is, for all practical purposes, completely absorbed. Thus, the normal component, which is proportional to sin θ, propagates through the resistive vane with negligible attenuation. A third resistive vane parallel to the broad wall of the output rectangular waveguide is located in the output transition section. The signal transmitted through the center section has an electric field vector oriented at $180° - \theta$ with respect to the plane of the third resistive vane. The component normal to the vane (proportional to sin θ) appears in the output port. Thus, the electric field has been attenuated by a factor of $\sin^2 \theta$. If, as pointed out in Fig. 11–19, the center section can be rotated, a variable attenuator of high precision can be realized.

Another useful device using a section of circular cross-section waveguide is formed by orienting a sheet of dielectric material in the waveguide at an angle of 45 degrees with respect to the incident electric field. Such a structure is shown in Fig. 11–20.

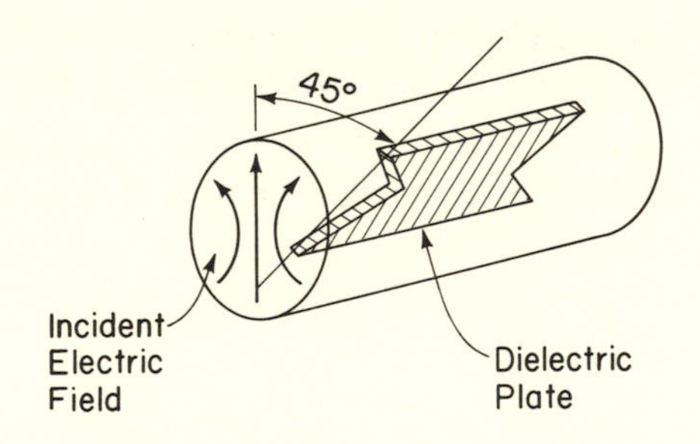

Fig. 11–20. The quarter-wave plate.

The incident electric field distribution can be split into a component normal to the plane of the dielectric plate plus a component tangent to the plane of the dielectric plate. The effect of the dielectric plate is to reduce the phase velocity of the component with electric field in the plane of the dielectric sheet. If β_T is the propagation constant for this component and β_N is the propagation constant for the component with electric field normal to the plane of the dielectric, β_T will be larger than β_N. If the length l of the plate is selected so that $(\beta_T - \beta_N)l = \pi/2$, the tangential component of the electric field will lag the normal component of the electric field by 90 degrees. This will result in a circularly polarized wave at the output side of the plate. That is:

$$\mathbf{E}_{\text{out}} = \mathbf{a}_N \frac{E_0}{\sqrt{2}} e^{-j\beta_N l} + \mathbf{a}_T \frac{E_0}{\sqrt{2}} e^{-j\beta_T l}$$

or

$$\mathbf{E}_{\text{out}} = \frac{E_0}{\sqrt{2}} e^{-j\beta_N l} (\mathbf{a}_N - j\mathbf{a}_T).$$

This sheet of dielectric material in a circular cross-section waveguide is termed a "quarter-wave plate."

The quarter-wave plate can be utilized in the construction of a variable phase-shifter. The phase-shifter consists of two quarter-wave plates, as previously described, with a half-wave plate located between the quarter-wave plates. The half-wave plate is free to rotate with respect to the quarter-wave plates, as shown in Fig. 11–21. When all of the plates are

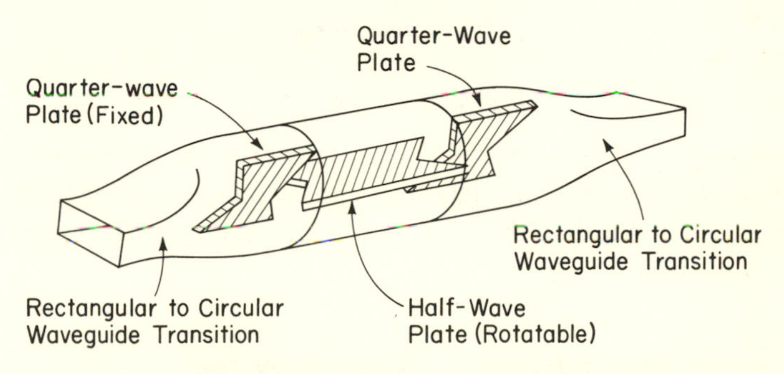

Fig. 11–21. **Rotating-plate variable phase-shifter.**

aligned, there is a total phase-shift difference of one wavelength and the output is linearly polarized. When the center section is rotated so that the plane of the plate makes an angle of θ with respect to the quarter-wave plates, the signal coming out of the input quarter-wave plate can be broken into normal and tangential components with respect to the half-wave plate. The result of this transformation is

$$\mathbf{E}_{\text{in}}\big|_{\lambda/2 \text{ section}} = \frac{E_0}{\sqrt{2}} e^{-j\beta_N l} [\mathbf{a}'_N (\cos\theta - j\sin\theta) - \mathbf{a}'_T (\sin\theta + j\cos\theta)]$$

$$\mathbf{E}_{\text{in}}\big|_{\lambda/2 \text{ section}} = \frac{E_0}{\sqrt{2}} e^{-j\beta_N l} e^{-j\theta} (\mathbf{a}'_N - j\mathbf{a}'_T)$$

where $\mathbf{a}'_N$ and $\mathbf{a}'_T$ are normal and tangential unit vectors for the half-wave plate. The output of the half-wave plate is easily formulated; the phase shift for the normal component is $2\beta_N l$ radians and the phase shift for the tangential component is $2\beta_N l + \pi$ radians. We can write directly

$$\mathbf{E}_{\text{out}}\Big|_{\lambda/2 \text{ section}} = \frac{E_0}{\sqrt{2}} e^{-j3\beta_N l} e^{-j\theta} (\mathbf{a}'_N + j\mathbf{a}'_T)$$

This wave is incident upon the output quarter-wave plate. Transforming the electric field to normal and tangential components with respect to the quarter-wave plate results in

$$E_{\text{in}}\Big|_{\lambda/4 \text{ section}} = \frac{E_0}{\sqrt{2}} e^{-j3\beta_N l} e^{-j\theta} [\mathbf{a}_N (\cos\theta - j\sin\theta) + j\mathbf{a}_T (\cos\theta - j\sin\theta)]$$

$$= \frac{E_0}{\sqrt{2}} e^{-j3\beta_N l} e^{-j2\theta} [\mathbf{a}_N + j\mathbf{a}_T]$$

The unit vectors $\mathbf{a}_N$ and $\mathbf{a}_T$ are as defined earlier.

After passing through the final quarter-wave plate the output is

$$\mathbf{E}_{\text{out}} = \frac{E_0}{\sqrt{2}} e^{-j4\beta_N l} e^{-j2\theta} (\mathbf{a}_N + \mathbf{a}_T)$$

$$\mathbf{E}_{\text{out}} = \mathbf{E}_{\text{in}} e^{-j(4\beta_N l + 2\theta)}$$

Thus, we see that the electric field at the output is exactly the same as the input: linearly polarized, but delayed in phase by an angle of $(4\beta_N l + 2\theta)$, where θ is the angle that the half-wave plate makes with respect to the quarter-wave plates.

11–4. MICROWAVE DEVICES USING FERRITES

A number of other useful microwave devices are constructed by inserting a biased ferrite rod in the waveguide and utilizing the Faraday rotation phenomenon described in Chapter 8. Perhaps the simplest device is the gyrator. The gyrator is a two-port device which introduces a phase shift of 180 degrees for a wave traveling through the structure in one direction when compared to a wave traveling through in the opposite direction. A typical gyrator requires a 90-degree spatial rotation via a twist in a waveguide and a biased ferrite rod capable of introducing an additional 90 degrees of spatial rotation. A typical gyrator and electric field orientation patterns are shown in Fig. 11–22.

For the wave traveling from port 1 to port 2 we see that the 90-degree twist rotates the field 90 degrees spatially and the biased ferrite rod rotates the field an additional 90 degrees to result in a net 180-degree spatial rotation at port 2. This 180-degree spatial rotation can be inter-

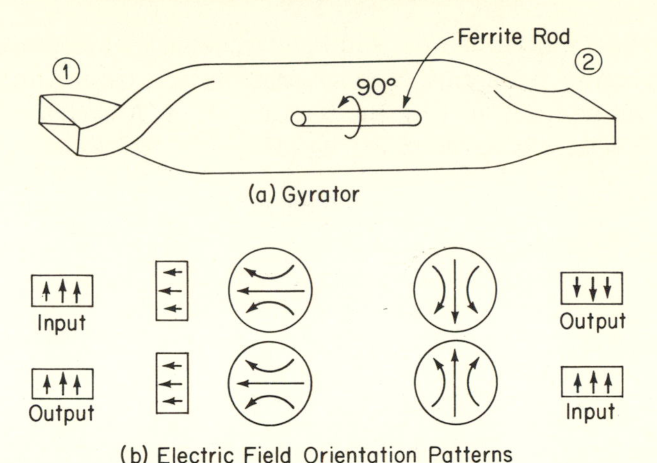

Fig. 11-22. The microwave gyrator. (After C. L. Hogan, "The Ferromagnetic Faraday Effect at Microwave Frequencies and Its Applications—The Microwave Gyrator," *Bell System Technical Journal,* vol. 31, no. 1, pp. 1-31, January 1952. Copyright 1952, American Telephone and Telegraph Co.; reprinted by permission.)

preted as a 180-degree phase change. The wave traveling from port 2 to port 1 experiences a 90-degree spatial rotation due to the ferrite rod. This rotation is in the same sense as that experienced by the wave passing from port 1 to port 2. The 90-degree twist now produces a spatial rotation of the wave in a sense opposite to Faraday rotation so the signal emerges from port 1 with the same spatial orientation that it had when it entered port 2. Thus a phase shift difference of 180 degrees has been produced for waves traveling in opposite directions through the device.

A device somewhat related to the gyrator is the Faraday rotation isolator. This device incorporates a 45-degree twist coupling the rectangular waveguide to the circular waveguide. A biased rod of ferrite is located in the circular waveguide section to rotate the electric field vector back to its original orientation so that it can exit through a circular-to-rectangular waveguide transition. For a wave propagating in the opposite direction, the Faraday rotation produced by the ferrite rod is such that the electric field leaves the ferrite oriented parallel to the broad wall of the rectangular waveguide through which it must exit. The waveguide is cut off to this orientation of electric field so the energy is reflected back toward the ferrite rod. To prevent a situation in which energy would bounce back and forth in the circular section, a resistive vane is placed in the transition with the 45-degree twist in such a way that the wave, having its electric field parallel to the broad wall and trying to propagate into the rectangular section, will be attenuated. A wave propagating in this section in the normal manner, electric field parallel to the narrow walls, will be unaffected by the resistive sheet. A sketch of a Faraday rotation isolator is shown in Fig. 11–23.

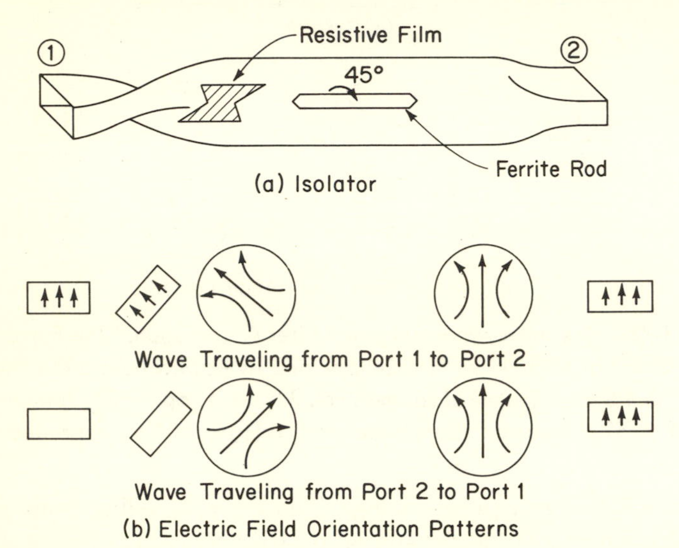

Fig. 11–23. The Faraday rotation isolator. (After C. L. Hogan, "The Ferromagnetic Faraday Effect at Microwave Frequencies and Its Applications—The Microwave Gyrator," *Bell System Technical Journal,* vol. 31, no. 1, pp. 1–31, January 1952. Copyright 1952, American Telephone and Telegraph Co.; reprinted by permission.)

Another common application of biased ferrites in waveguides is based on the fact that the magnetic field for the dominant mode in a rectangular waveguide is circularly polarized at certain positions across the waveguide. The behavior can be seen by considering the TE_{10} mode magnetic field pattern, as given in Eq. (9–43), assuming a z-dependence of $e^{-j\beta z}$:

$$H_x = -\frac{E_0}{Z_{0\mathrm{TE}}} \sin \frac{\pi x}{a} e^{-j\beta z}$$

$$H_z = j \frac{E_0}{\eta} \left(\frac{\lambda}{2a}\right) \cos \frac{\pi x}{a} e^{-j\beta z}$$

In real time, these expressions become

$$H_x = -\frac{E_0}{Z_{0\mathrm{TE}}} \sin \frac{\pi x}{a} \cos (\omega t - \beta z)$$

$$H_z = -\frac{E_0}{\eta} \left(\frac{\lambda}{2a}\right) \cos \frac{\pi x}{a} \sin (\omega t - \beta z)$$

These terms can be combined to determine the composite magnetic field vector, with the result

$$\mathbf{H} = -\frac{E_0}{\eta}\left[\mathbf{a}_x\sqrt{1-\left(\frac{f_c}{f}\right)^2}\sin\frac{\pi x}{a}\cos(\omega t-\beta z)+\mathbf{a}_z\left(\frac{f_c}{f}\right)\cos\frac{\pi x}{a}\sin(\omega t-\beta z)\right]$$

If we choose $x = x_1$ such that

$$\sqrt{1-\left(\frac{f_c}{f}\right)^2}\sin\frac{\pi x_1}{a} = \left(\frac{f_c}{f}\right)\cos\frac{\pi x_1}{a}$$

and let $z = 0$ to simplify things, we see that the magnetic field at this point would appear to be circularly polarized, with rotation counterclockwise when looking in the $+y$-direction. The value of x selected would be less than $a/2$ since both $\sin \pi x/a$ and $\cos \pi x/a$ are specified to be positive. There is another point along the x-axis where $x = x_2$, such that

$$\sqrt{1-\left(\frac{f_c}{f}\right)^2}\sin\frac{\pi x_2}{a} = -\left(\frac{f_c}{f}\right)\cos\frac{\pi x_2}{a}$$

At this point (again choosing $z = 0$ to simplify things) we see that the magnetic field is also circularly polarized, but now the rotation is clockwise when looking in the $+y$-direction. In this case x_2 lies between $a/2$ and a. Figure 11–24 shows the planes $x = x_1$ and $x = x_2$ and the rotation of the magnetic field vector for the case of a TE_{10} propagating in the $+z$-direction.

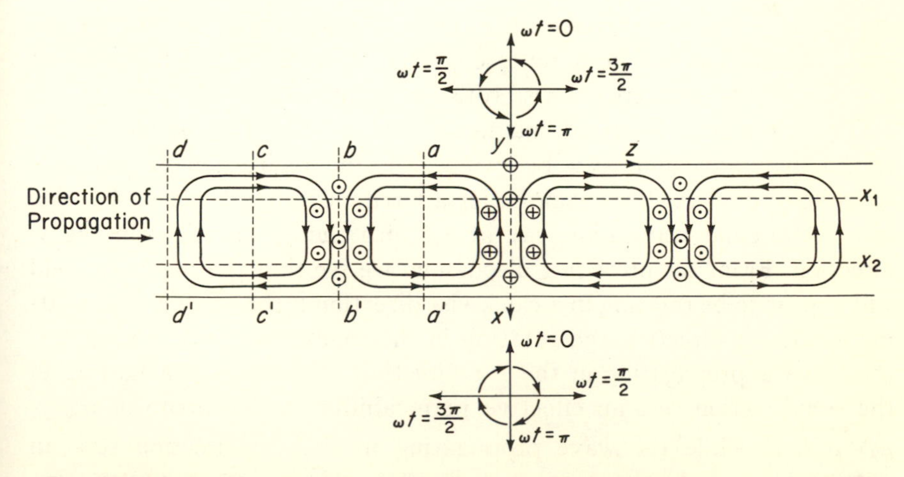

Fig. 11–24. Magnetic field of the TE_{10}-mode propagation in $+z$-direction.

It should be noted that we could draw the same conclusion regarding the sense of rotation by observing the change in the direction of the magnetic field as the field moves past the point of observation. For example, if the field pattern shown in Fig. 11–24 is that for $\omega t = 0$, then line $a - a$ passes the point of observation ($z = 0$) at $\omega t = \pi/2$, line $b - b'$ passes at $\omega t = \pi$, line $c - c'$ passes at $\omega t = 3\pi/2$, and line $d - d'$ passes at $\omega t = 2\pi$. We can use this concept to figure out how the magnetic field for a TE_{10} mode propagating in the $-z$-direction behaves. Consider Fig. 11–25.

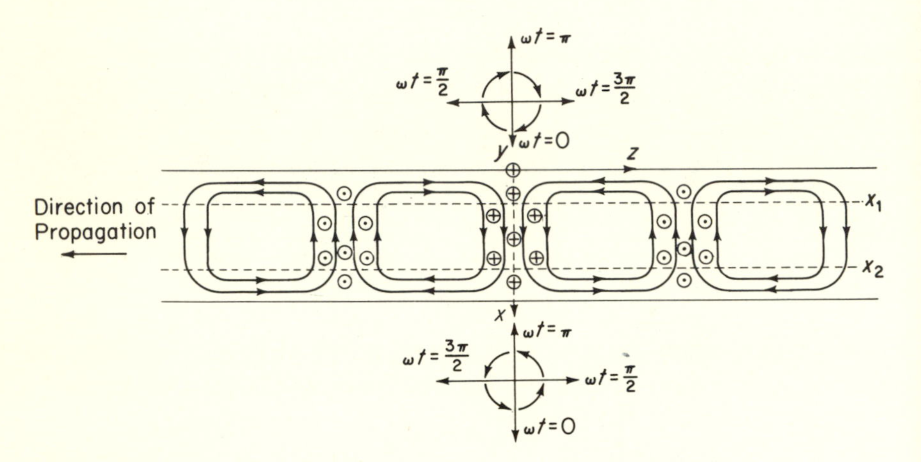

Fig. 11–25. Magnetic field for TE_{10}-mode propagation in $-z$-direction.

A study of the change in direction of magnetic field as the wave, moving in the $-z$-direction, passes the point of observation, shows that the sense of rotation has been reversed. The clockwise rotation occurs at $x = x_1$ and the counterclockwise rotation occurs at $x = x_2$.

The circularly polarized magnetic field of the TE_{10} mode in the rectangular waveguide can be utilized in the realization of practical microwave devices. For example, let a ferrite slab be placed at one side of a rectangular waveguide, say at x_2, with the dc bias field in the y-direction. From the preceding discussion, we see that the magnetic field will appear to be rotating in a clockwise direction if the wave is propagating in the $+z$-direction and rotating in the counterclockwise direction if the wave is propagating in the $-z$-direction. The wave propagating in the $+z$-direction sees an effective permeability in the ferrite of $\mu_{\text{eff}}^{+} = \mu_0\sqrt{\mu + K}$ while the wave propagating in the $-z$-direction sees an effective permeability in the ferrite of $\mu_{\text{eff}}^{-} = \mu_0\sqrt{(\mu - K)}$. This difference in permeability affects the propagation constant β so that β^+ (wave propagating in the $+z$-direction) is somewhat greater than β^- (wave propagating in the $-z$-direction). The difference in β's can be used to make a nonreciprocal phase shifter. That is, the electrical length (βl) of the waveguide section containing the ferrite is greater for the wave traveling in the $+z$-direction than it is for the wave traveling in the $-z$-direction.

The preceding discussion is based on the assumption that the amount of ferrite material introduced into the waveguide is relatively small, so that the field patterns are not significantly changed. If enough ferrite material is put into the waveguide, the field patterns will be drastically changed. Several useful microwave devices make use of this phenomenon. Probably the most common is the field displacement isolator. In this structure, the amount of ferrite and the bias field are chosen such that the electric field for the wave traveling in the $+z$-direction goes to zero at one edge of a ferrite slab while the electric field for the wave traveling in the $-z$-direction is maximum at that point. The field distributions and the position of the ferrite slab are shown in Fig. 11–26.

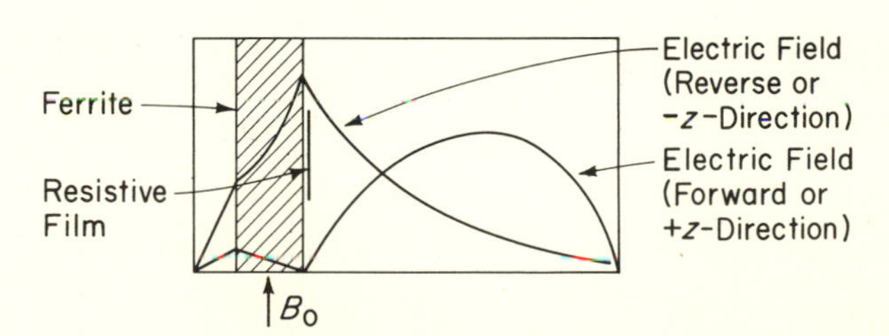

Fig. 11–26. Cross-sectional field patterns in the field displacement isolator. (After B. Lax and K. Button, *Microwave Ferrites and Ferrimagnetics* [McGraw-Hill Book Co., Inc., New York, 1962] page 631.)

The resistive film, placed as shown in Fig. 11–26, acts to attenuate the wave traveling in the $-z$-direction while energy traveling with $+z$-direction is unaffected. The result is an isolator. It should be pointed out that the cross-guide field patterns shown above are highly distorted TE_{10} mode patterns that are a result of the presence of the large amount of biased ferrite. The waves entering the ferrite region are TE_{10} as are the waves leaving the ferrite region. Some impedance matching is necessary in the transition region.

The counter-rotating nature of the magnetic field on opposite sides of a rectangular waveguide containing the TE_{10} mode is utilized in the

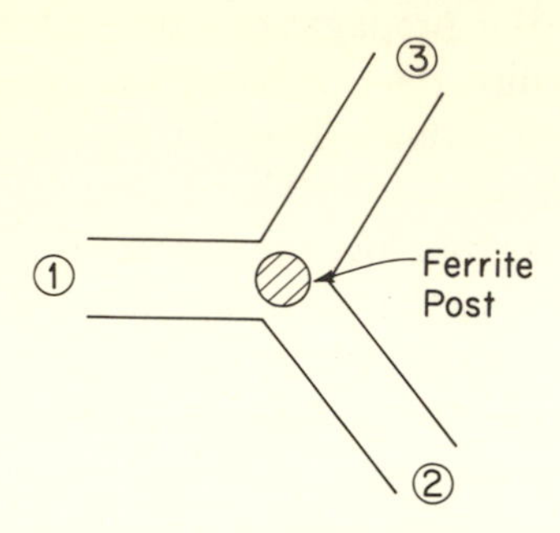

Fig. 11–27. The waveguide Y-junction circulator.

waveguide Y-junction circulator shown in Fig. 11–27. Energy entering port 1 is directed out of port 2; energy entering port 2 is directed out of port 3; and energy entering port 3 is directed out of port 1. This behavior can be deduced by visualizing a TE_{10} mode incident on the ferrite post from port 1. Assume that the bias field on the side toward port 2 sees an effective permeability of $\mu^+ = \mu_0\sqrt{\mu + K}$ and is slowed down. The magnetic field on the other side of the waveguide (the side toward port 3) sees an effective permeability of $\mu^- = \mu_0\sqrt{\mu - K}$ and is slowed a lesser amount. The net result is a bending of the Poynting vector toward port 2 and, if the parameters are chosen correctly, all of the energy entering port 1 will exit from port 2. Due to the symmetry of the structure, the results for energy entering ports 2 and 3 are similar to that for energy entering port 1.

A structure closely related to the waveguide Y-junction circulator is the stripline Y-junction circulator shown in Fig. 11–28. From our knowl-

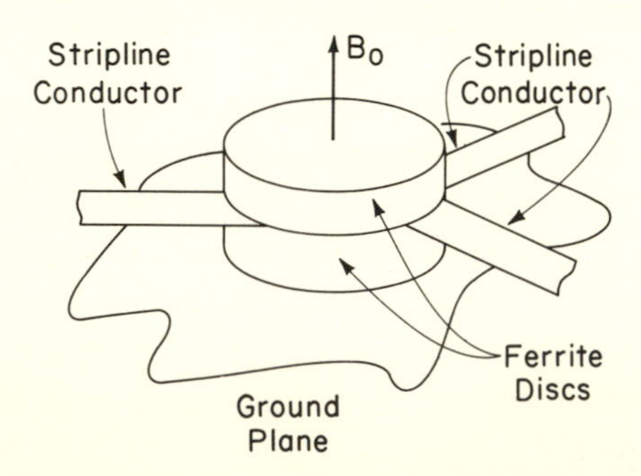

Fig. 11–28. The stripline Y-junction circulator.

edge of the fields of shielded stripline (TEM) it is not apparent that a rotating component of magnetic field is present, as was the case for the TE_{10} mode in the rectangular waveguide. This is correct; there is no rotating component of magnetic field on the transmission lines. However, Bosma[1] has shown that the ferrite discs can be treated as resonators with modes such that the magnetic field consists of two counterrotating components in the broad plane of the disc. When a dc bias field is applied, the effective permeability for one sense of rotation is different from the effective permeability for the other sense of rotation; the mode splits into two modes with slightly different resonant frequencies, symmetrical with respect to the operating frequency; and the field pattern is displaced so that the input energy is all directed to one of the other lines and the third line is isolated. The sense of circulation is dependent upon various factors, including the strength of the bias field, the frequency of operation, and the dimensions of the ferrite discs. Since it is usually desired to couple critically from the stripline to the ferrite disc, special attention is paid to to impedance matching.

PROBLEMS

11-1. The impedance and reflection coefficient of a certain waveguide load are to be measured by standing-wave techniques. The waveguide is X-band size (0.9 by 0.4 in.), and the mode may be assumed to be TE_{10}. When a shorting plate is placed over the end of the guide in place of the load, adjacent standing-wave minima are found to be 1.5 cm apart. When the short is replaced with the load, the standing-wave ratio is found to be 1.5 and the minima are found to have shifted 0.5 cm toward the load. (a) Find the frequency, wavelength within the guide, and velocity of propagation. (b) Find the reflection coefficient and wave impedance of the load.

11-2. It is desired to measure the amount of power delivered from a matched source to a matched load in a microwave system by means of a bolometer and a microwave power meter. How much percentage error is introduced in the power measurement if the bolometer is actually mismatched to the extent of giving rise to a VSWR of 1.5 rather than unity? The source may be assumed to be matched, and the desired power to be measured is that which would be delivered to the load (bolometer in this case) if it were perfectly matched.

11-3. The waveguide shown in the accompanying figure is operating at 10 GHz in the TE_{10} mode. The inside dimensions of the guide are 2 cm by 1 cm. What must

[1] H. Bosma, "On Stripline Circulation at UHF," *IEEE Transactions on Microwave Theory and Techniques*, January 1964, pp. 61–72.

be the length l and the dielectric constant ϵ_r' if the intermediate section is to act as a quarter-wave transformer? (*Note:* ϵ_r' is not simply $\sqrt{2.25}$.)

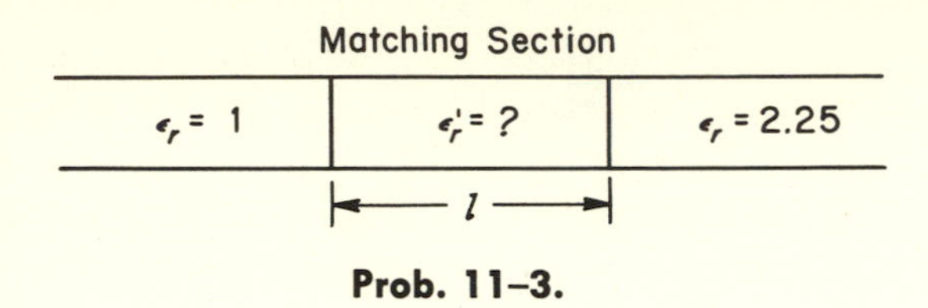

Prob. 11–3.

11–4. When a screw is inserted in the top of a waveguide system operating in the TE_{10} mode, it is noticed that this will introduce a VSWR varying from unity to 3.0 for the full range of depth of insertion. That is, with the screw out, the VSWR is one; when the screw protrudes the maximum amount, the VSWR is 3.0. (a) If one thinks of the screw as giving rise to the equivalent of shunt capacitive loading, what is the corresponding range of adjustment in the equivalent per-unit reactance X/Z_0? (b) Assuming the screw can be placed any arbitrary distance from the load, note on a Smith chart the range of load impedance which may be matched with this device.

11–5. Two examples of the use of the magic tee as a decoupling or isolating device are shown in the accompanying figure. Explain the operation of each. The points labeled A, B, C, and D refer to the respective ports of the magic tee as designated in Fig. 11–8.

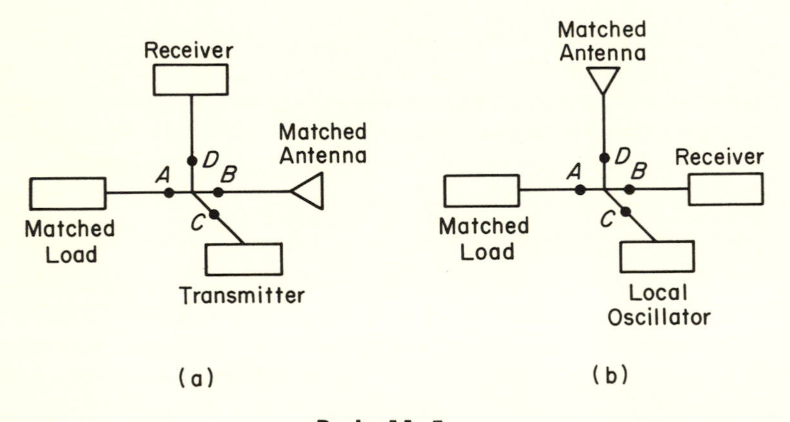

Prob. 11–5.

11–6. Explain how a magic tee along with matched and shorted terminations may be used to measure the reflection coefficient of an unknown termination.

12

Radiation of Electromagnetic Waves

12–1. INTRODUCTION

In the derivation of Maxwell's equations, the justification of the addition of a displacement current was primarily that its presence helped explain and predict certain hitherto unexplainable phenomena. Once this hurdle was taken, it was possible to derive the wave equations by completely rigorous and legitimate means. These wave equations, together with boundary conditions, then made possible the accurate prediction of wave behavior in all sorts of media, provided the waves already existed. This has been the substance of our work with waves so far.

Our main point of interest in the next two chapters now must shift somewhat. The new question is, "What gives rise to these waves?" In predicting the nature of the radiation, we shall use reasoning which is perhaps not rigorous. For example, we shall postulate solutions to vector partial differential equations which fortunately seem to work. Of course if we can somehow find a solution that fits both the differential equations and the boundary conditions we happen to establish, that solution is correct. That we may not have obtained it by a rigorous logical process is not really objectionable. It simply means that we have advanced a little beyond those very simplest cases where our familiarity is sufficient to mask the fact that they, too, were first solved by intelligent guessing. The so-called *logical development processes* came much later, and their logic hinged on the fact that the nature of the answer was already known.

Perhaps we must go even a step farther. In some cases we shall postulate solutions that "don't quite fit" the boundary conditions. This situation, where it occurs, will be discussed. It is sufficient to say now that such a procedure may be followed (when it is useful and experimentally justifiable) in lieu of an analytically unobtainable exact solution.

To this point we have been concerned with the transmission of electrical energy via one or more conductors, the two-wire transmission line and the hollow-pipe waveguide being excellent examples of such transmission structures. Generally speaking, the solutions obtained were "source-free" in the sense that we were concerned with some region of space of finite extent which did not contain sources. The source of electrical energy and the load were assumed to be outside this region since we were primarily concerned with the characteristics of the propagation of electrical energy through the region of interest. We then obtained solutions in terms of arbitrary excitation or amplitude factors which could be determined by matching boundary conditions at the source and load ends of the region.

Radiation can be thought of as the process of transmitting electrical energy in which there are no conductors or guiding structures linking the source and the load. The formulation of this problem must include the source. This source establishes electromagnetic waves in the medium in which it is imbedded. These electromagnetic waves propagate outward from the source. One might question whether or not the uniform plane wave, which was considered in Chapter 8, is an example of radiation. It is indeed an example of radiation, but an impractical example because a current sheet of infinite extent and capable of supplying an infinite amount of power would be required to support such a wave. This is not to say that the study of the uniform plane wave was a waste of time. We shall soon see that in a limited region of space, well removed from the source, the radiation fields can be quite accurately approximated by uniform plane waves.

The objective of the present chapter is to describe quantitatively the phenomenon of electromagnetic energy radiating from an isolated source in space. To facilitate this investigation we will consider the fields established by the simplest imaginable source since a more practical source distribution can be interpreted as a superposition of elemental sources.

12-2. FORMULATION OF THE PROBLEM

The problem is simply stated as the determination of the fields due to a source distribution in free space. The sources are characterized by some distribution of electric current and electric charge. Maxwell's equations for this situation become

$$\begin{aligned} &\text{(a)} \quad -\nabla \times \mathbf{E} = j\omega\mu_0\mathbf{H} \qquad &\text{(c)} \quad \nabla \cdot \mathbf{B} = 0 \\ &\text{(b)} \quad \nabla \times \mathbf{H} = j\omega\epsilon_0\mathbf{E} + \mathbf{J}^i \qquad &\text{(d)} \quad \nabla \cdot \mathbf{D} = \rho^i \end{aligned} \qquad (12\text{–}1)$$

The superscript i denotes impressed sources.

The constitutive equations are

$$\begin{aligned} &\text{(a)} \quad \mathbf{B} = \mu_0\mathbf{H} \\ &\text{(b)} \quad \mathbf{D} = \epsilon_0\mathbf{E} \end{aligned} \qquad (12\text{–}2)$$

The electric current and electric charge distributions are related in the time-varying situation by a continuity-of-current expression

$$\nabla \cdot \mathbf{J} = -j\omega\rho \qquad (12\text{–}3)$$

At first glance one might be inclined to manipulate these equations in much the same way that Maxwell's equations for the source-free case were treated to derive a vector wave equation. Indeed, performing these operations results in the inhomogeneous equations involving the electric and magnetic fields and source distributions

$$\nabla^2\mathbf{E} + \beta^2\mathbf{E} = \frac{1}{\epsilon_0}\nabla\rho^i + j\omega\mu_0\mathbf{J}^i, \qquad (12\text{–}4)$$

and

$$\nabla^2\mathbf{H} + \beta^2\mathbf{H} = -\nabla \times \mathbf{J}^i \qquad (12\text{–}5)$$

where $\beta^2 = \omega^2\mu_0\epsilon_0$. These equations are not particularly easy to work with since in some important cases the mathematical expression for the current can have undefined spatial derivatives. At any rate, these are not simple differential equations and one is led to look for a mathematical method which provides a more straightforward solution.

One such mathematical format is potential theory. The starting point for this method of solution is the fact the divergence of the magnetic field is zero. A solenoidal vector, that is, a vector with zero divergence, can be expressed as the curl of a second vector. Thus, we write

$$\mathbf{B} = \nabla \times \mathbf{A} \qquad (12\text{–}6)$$

where $\mathbf{A}$ is a vector called *vector magnetic potential*. Now, by substituting into Maxwell's equations Eqs. (12–1), one can write

$$\nabla \times (\mathbf{E} + j\omega\mathbf{A}) = 0 \tag{12–7}$$

An irrotational vector, that is, a vector with zero curl, can be expressed as the gradient of a scalar. Thus, we write

$$\mathbf{E} + j\omega\mathbf{A} = -\nabla\Phi \tag{12–8}$$

Φ is a scalar called *scalar electric potential.* It is seen that for $\omega = 0$, Eq. (12–8) reduces to the conventional definition of electric potential of static electric field theory.

At this point, Eqs. (12–2a), (12–6), and (12–8) are substituted into Eq. (12–1b) to get

$$\nabla \times \nabla \times \mathbf{A} = j\omega\mu_0\epsilon_0(-\nabla\Phi - j\omega\mathbf{A}) + \mu_0\mathbf{J}^i \tag{12–9}$$

Using the vector identity

$$\nabla \times \nabla \times \mathbf{A} = \nabla(\nabla \cdot \mathbf{A}) - \nabla^2\mathbf{A}$$

and collecting terms, Eq. (12–9) can be written as

$$\nabla^2\mathbf{A} + \beta^2\mathbf{A} = -\mu_0\mathbf{J}^i + \nabla(\nabla \cdot \mathbf{A} + j\omega\mu_0\epsilon_0\Phi) \tag{12–10}$$

This does not appear to be a particularly simple differential equation but at this point one of the real advantages of the potential-theory method of solution becomes apparent. Since there is no particular physical significance to vector magnetic potential, and since we have only specified its curl up to this point, we are free to make the divergence of **A** anything we want it to be. Specification of both the curl and divergence of a vector results in specification of the vector. In this case, set

$$\nabla \cdot \mathbf{A} = -j\omega\mu_0\epsilon_0\Phi \tag{12–11}$$

with the resulting equation for **A**

$$\nabla^2\mathbf{A} + \beta^2\mathbf{A} = -\mu_0\mathbf{J}^i \tag{12–12}$$

Substitution of Eq. (12–8) into another of Maxwell's equations, Eq. (12–1d), results in

$$\nabla^2\Phi + j\omega\nabla \cdot \mathbf{A} = -\rho^i/\epsilon_0 \tag{12–13}$$

By means of the gauge condition, Eq. (12–11), we can write

$$\nabla^2\Phi + \beta^2\Phi = -\rho^i/\epsilon_0 \tag{12–14}$$

Equations (12–12) and (12–14) are known as *inhomogeneous wave equations*, in vector and scalar form respectively. The solutions of these equations are well known, and we will consider the general solutions in later sections.

At this point it is instructive to consider the solution of the fields due to an element of electric current radiating in free space. As was pointed out in the introduction, the principle of linear superposition will allow us to treat the fields of an arbitrary source distribution as the vector sum of the fields due to an array of elemental currents.

12-3. POTENTIAL SOLUTIONS

In order to determine the fields due to a source distribution, it will be necessary to solve the differential equations

$$\nabla^2\mathbf{A} + \beta^2\mathbf{A} = -\mu_0\mathbf{J}^i \tag{12-12}$$

and

$$\nabla^2\Phi + \beta^2\Phi = -\rho^i/\epsilon_0 \tag{12-14}$$

These are basically the same differential equation expressed in vector and scalar form. We will solve this equation by considering the source to be an electric current distribution consisting of a current I flowing over a differential length l. Let us orient this current element in space so that it flows along the z-axis; thus, Eq. (12-14) can be written as

$$\nabla^2A_z + \beta^2A_z = -\mu_0J_z^i \tag{12-15}$$

To begin with, let us define a set of coordinate axes in which to describe the fields due to the source. A suitable set of coordinates is shown in Fig. 12-1. Notice that rectangular, cylindrical, and spherical polar coordinate systems are depicted in the diagram. The source is designated

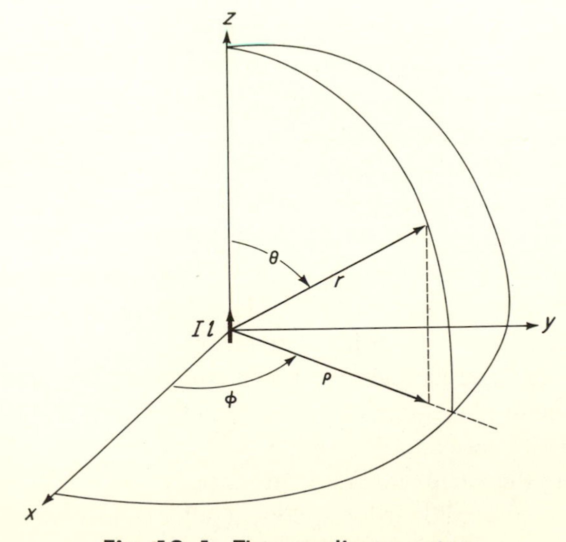

Fig. 12-1. The coordinate system.

as an element of current oriented along the z-axis as shown. A comment on the source must be made at this point. The current element corresponds to an infinite current density, but we define the source in terms of a volume integral encompassing the source. That is:

$$\iiint\limits_{\substack{\text{volume}\\\text{encompassing}\\\text{the source}}} J_z^i\,dv = Il \tag{12–16}$$

This particular source is called an *electric dipole.* The conservation-of-charge expression (Eq. (12–3)) tells us that the element of current must originate and terminate on charges. These charges give rise to an electric displacement current flowing in space, as shown in Fig. 12–2.

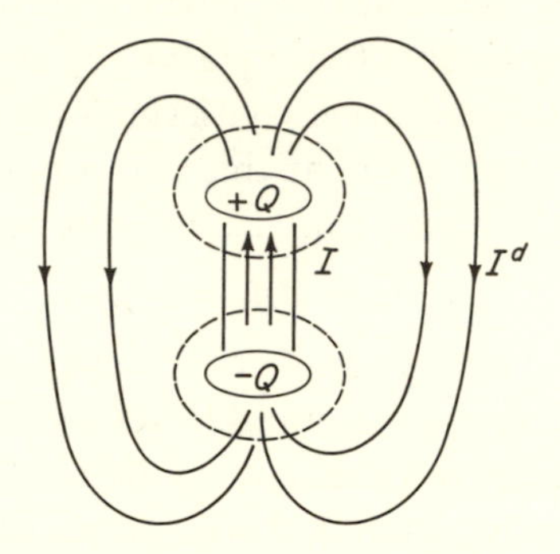

Fig. 12–2. The electric dipole.

The current I produces the charge distributions $+Q$ and $-Q$, which in turn give rise to the electric displacement current I^d. These currents are such that the displacement current flowing out of the volume surrounding the charge is equal to the current flowing into the volume.

The current distribution, which serves as a source for the magnetic potential vector, is a monopole distribution. The electric charge distribution, on the other hand, is a dipole distribution. The point monopole source distribution is much easier to manipulate mathematically than the point dipole source distribution, but in general we must be prepared to handle both in order to determine $\mathbf{A}$ and Φ. Since $\mathbf{J}^i$ and ρ^i are related by the conservation-of-charge equation, Eq. (12–3), $\mathbf{A}$ and Φ are not independent and the determination of either will suffice. We will choose the simpler case and determine $\mathbf{A}$.

Determining the vector magnetic potential $\mathbf{A}$ for this current element source will be facilitated by using spherical polar coordinates and splitting the region into two subregions. Subregion I (called volume I) will be a small sphere encompassing the source. Subregion II (called volume II) will be the rest of the region. This division makes volume II source-free, and the source is contained within volume I.

Thus we can write:

$$\begin{aligned} &\text{(a)} \quad \nabla^2 A_z^{\mathrm{I}} + \beta^2 A_z^{\mathrm{I}} = \mu_0 J_z^i \qquad \text{(Volume I)} \\ &\text{(b)} \quad \nabla^2 A_z^{\mathrm{II}} + \beta^2 A_z^{\mathrm{II}} = 0 \qquad \text{(Volume II)} \end{aligned} \tag{12–17}$$

The electric and magnetic fields must be continuous on the interface between the subregions. This forces the vector magnetic potential **A** to also be continuous on this surface.

Considering volume II first, we write Eq. (12–17b) in spherical-polar coordinates as

$$\frac{1}{r^2}\frac{\partial}{\partial r}\left(r^2 \frac{\partial A_z^{\mathrm{II}}}{\partial r}\right) + \beta^2 A_z^{\mathrm{II}} = 0 \tag{12–18}$$

We have recognized the fact that the vector magnetic potential is spherically symmetric, since the source is a point monopole; thus there are no variations with respect to θ and ϕ. The solution to this equation is

$$A_z^{\mathrm{II}} = C_1 \frac{e^{-j\beta r}}{r} + C_2 \frac{e^{+j\beta r}}{r} \tag{12–19}$$

The first term represents an outward-traveling wave, while the second term represents an inward-traveling wave. For physical reasons (namely, that the only source in all space is the current element at the origin and there are no reflecting surfaces present), we will discard the solution which represents inward-traveling waves. To prove that

$$A_z^{\mathrm{II}} = C_1 \frac{e^{-j\beta r}}{r} \tag{12–20}$$

is a solution of Eq. (12–17b), we substitute the expression into the differential equation. The result is

$$C_1\, e^{-j\beta r}\left(-\frac{\beta^2}{r} + \nabla^2 \frac{1}{r}\right) + \beta^2 C_1 \frac{e^{-j\beta r}}{r} = 0$$

Use of the vector identity, $\nabla^2 \dfrac{1}{r} = 0$, for $r \neq 0$, proves that we have a solution since $r = 0$ is not included in subregion II.

Now let us consider volume I. In this subregion, the differential equation which must be solved is

$$\frac{1}{r^2}\frac{\partial}{\partial r} r^2 \frac{\partial A_z^{\mathrm{I}}}{\partial r} + \beta^2 A_z^{\mathrm{I}} = -\mu_0 J_z^i \tag{12–21}$$

We will assume that the solution is represented by outward-traveling waves and has the same form as the solution for subregion II. That is,

$$A_z^{\mathrm{I}} = C_3 \frac{e^{-j\beta r}}{r} \tag{12–22}$$

Inserting this expression into the differential equation results in

$$C_3 e^{-j\beta r}\left(-\frac{\beta^2}{r} + \nabla^2 \frac{1}{r}\right) + \beta^2 C_3 \frac{e^{-j\beta r}}{r} = -\mu_0 J_z^i$$

This reduces to

$$C_3 e^{-j\beta r} \nabla^2 \frac{1}{r} = -\mu_0 J_z^i$$

We will now integrate this expression over the small sphere encompassing the source:

$$\underset{\text{volume I}}{\iiint} C_3 e^{-j\beta r} \nabla^2 \frac{1}{r}\, dv = -\underset{\text{volume I}}{\iiint} \mu_0 J_z^i\, dv \tag{12–23}$$

Since the volume of integration is a sphere of vanishingly small radius, we can use the approximation $e^{-j\beta r} \approx 1$, the vector identity

$$\nabla^2 \frac{1}{r} = \nabla \cdot \nabla \frac{1}{r}$$

and the divergence theorem to get

$$\underset{\substack{\text{surface}\\ \text{enclosing}\\ \text{volume I}}}{\oint\!\!\oint} C_3 \nabla \frac{1}{r} \cdot \mathbf{da} = -\iiint \mu_0 J_z^i\, dv = -\mu_0 I l$$

In evaluating the surface integral, the following statements can be made:

1. $C_3 = C_1$ (continuity of $\mathbf{A}$ at the surface)
2. $\mathbf{da} = \mathbf{a}_r r^2 \sin\theta\, d\theta\, d\phi$
3. $\nabla \dfrac{1}{r} = -\dfrac{\mathbf{a}_r}{r^2}$

Combining all of these statements into Eq. (12–22) results in

$$-\int_0^{2\pi}\int_0^{\pi} C_1 \frac{1}{r^2} r^2 \sin\theta\, d\theta\, d\phi = -\mu_0 I l \tag{12–24}$$

Evaluating the integral,

$$C_1 = \frac{\mu_0 I l}{4\pi}$$

This completes the solution for the vector magnetic potential of a z-directed current element:

$$\mathbf{A} = \mathbf{a}_z \frac{\mu_0 I l}{4\pi r} e^{-j\beta r} \tag{12–25}$$

Because of the basic similarity between the differential equations for **A** and Φ, we can write the solution for the scalar electric potential due to a point charge having a sinusoidal time dependence (monopole distribution):

$$\Phi = \frac{q}{4\pi\epsilon r} e^{-j\beta r} \tag{12–26}$$

Examination of Eq. (12–26) shows that it is just like the static potential expression for an isolated point charge except for the factor $e^{-j\beta r}$. The same situation holds true for Eq. (12–25). For this reason these potential expressions are called "retarded potentials" in recognition of the fact that it takes some time for a wave to propagate through space.

These potential expressions can be used to determine the fields due to sources whose dimensions are small with respect to wavelength. We shall consider such an example in the next article.

12–4. THE ELEMENTAL ANTENNA

Now let us consider a situation that is illustrated in Fig. 12–3. A high-frequency generator is connected to a two-wire transmission line. This line is of a conventional parallel-wire type. If the line is fairly remote from any metallic or other conducting object, essentially there will be equal and opposite currents in the two wires at any given position on the line. Therefore, at any significant distance from the line, the field effects of the two wires will substantially cancel each other. Now let us suppose that the end of the line remote from the generator is shorted with a straight segment of wire. The field effects of the current in this straight segment of wire may well be observable at distances remote from the line because there is no source of equal and opposite fields to cancel them.

If the antenna element has a current I flowing in it and we orient our coordinate system so that it flows for a distance l we will have a physical model to the current distribution discussed in Art. 12–3. The magnetic potential vector can be written directly as

$$\mathbf{A} = \mathbf{a}_z A_z = \mathbf{a}_z \frac{\mu_0 I l}{4\pi} e^{-j\beta r} \tag{12–27}$$

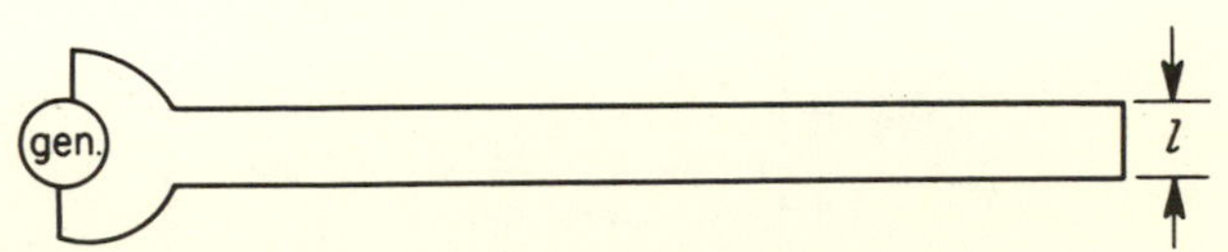

Fig. 12–3. An elemental antenna system.

At this point one might question whether or not the situation depicted in Fig. 12–3 is a legitimate physical realization of the mathematical model used in Art. 12–3. Obviously, there were no conductors present in that case. We postulated a current existing in free space and if you have retained any feel for Maxwell's equations you know that the presence of a perfect conductor means the absence of tangential electric fields on such surfaces. This is certainly not the case for a current distribution in free space.

One way to look at this problem is to consider that the source establishes fields on the transmission line which will propagate to the straight segment of wire. The segment of wire, being a perfect conductor, requires that the tangential electric field be zero on the surface. This means that there must be a scattered field established such that the scattered electric field added to the incident electric field results in the complete cancellation of the tangential component on the surface of the wire segment. The scattered field is actually due to conduction currents induced in the wire segment. It is this current and the resulting scattered field which we represent as a physical realization of the current element considered in the preceding articles.[1] It will soon be apparent that relatively little of the incident energy is radiated. Most of the energy is stored around the wire segment and in standing waves on the transmission line. The wire segment is much shorter than a wavelength so the induced current can be assumed to be uniform over its length.

It is now convenient to express the vector potential in spherical coordinates. The components are:

$$A_r = A_z \cos\theta = \frac{\mu_0 I l e^{-j\beta r} \cos\theta}{4\pi r} \tag{12–28}$$

$$A_\theta = -A_z \sin\theta = -\frac{\mu_0 I l e^{-j\beta r} \sin\theta}{4\pi r} \tag{12–29}$$

$$A_\phi = 0 \text{ (symmetry)} \tag{12–30}$$

From these vector potential components we can find the magnetic intensity $\mathbf{H}$, which is really one of the main points of interest. To do this, we must take the curl in spherical coordinates as in the following equation:

$$\nabla \times \mathbf{A} = \begin{bmatrix} \dfrac{\mathbf{a}_r}{r^2 \sin\theta} & \dfrac{\mathbf{a}_\theta}{r\sin\theta} & \dfrac{\mathbf{a}_\phi}{r} \\ \dfrac{\partial}{\partial r} & \dfrac{\partial}{\partial \theta} & \dfrac{\partial}{\partial \phi} \\ A_r & rA_\theta & r\sin\theta A_\phi \end{bmatrix} \tag{12–31}$$

[1] It might be helpful to reread Art. 8–7 at this point.

Recalling that **H** is by definition $(\nabla \times \mathbf{A})/\mu_0$, we can find the components of magnetic intensity due to our current element to be

$$H_r = 0 \qquad (12\text{-}32)$$

$$H_\theta = 0 \qquad (12\text{-}33)$$

$$H_\phi = \frac{Il e^{-j\beta r} \sin\theta}{4\pi r^2} + \frac{j\beta Il e^{-j\beta r} \sin\theta}{4\pi r} \qquad (12\text{-}34)$$

It is now expedient to examine the nature of the magnetic intensity field we have just derived in the light of the assumptions we made about retarded potentials. In particular, let us retrogress for a moment and see what would have happened if no retarded potential assumptions had been made. The only road open to us in finding **H** would then be simply to apply Ampere's law. From this we would then be forced to conclude that

$$H_\phi = \frac{Il \sin\theta}{4\pi r^2} \qquad (12\text{-}35)$$

Equation (12–35) is simply an expression of the well-known induction field of a current element. It is in phase with the exciting current and decreases in amplitude with distance according to the inverse square law. All these things are perfectly in accord with our magnetostatic intuitions.

Returning to the retarded potential case, examination of the first term of Eq. (12–34) shows this also to be the induction field. The only difference between this term and Eq. (12–35) is the retarded-potential implication. Even this is not difficult to accept if one believes that it always takes a finite amount of time for something to go somewhere. Anyway, the closer we get to the current element, the more we approach exact time-phase correlation between this component of H_ϕ and the exciting current I.

Now let us pass on to the second term of Eq. (12–34). This does not follow our intuitive picture in any respect. First, this term is a direct result of the retarded-potential assumption and does not occur at all in application of Ampere's law. Second, it falls off with the first power of distance rather than the second. Third, no matter how close we get to the current element, the term will always be in phase quadrature with the exciting current. In short, this term represents a new phenomenon with which we have heretofore been completely unfamiliar. We shall see later that this is the term that leads to the phenomenon of radiation. It is the term that keeps an antenna from just being another inductance.

It is now expedient to find the electric field associated with this magnetic field. There are several possible ways to attack the problem.

We could start with the solution for the scalar electric potential. The trouble with this is that we do not know for sure just what the charge function ρ is on the elemental wire. We could get this charge function from the current function with the continuity equation:

$$\nabla \cdot \mathbf{J} = -j\omega\rho \tag{12–36}$$

Having obtained the charge distribution function from Eq. (12–35) and thereafter the scalar electric potential function, we could proceed to find the electric field **E** from Eq. (12–8) since we already know **A**. However, there is an easier way. Let us examine the fundamental field equation, Eq. (12–1b), in a region of no current flow. It then becomes

$$\nabla \times \mathbf{H} = j\omega\epsilon_0\mathbf{E} \tag{12–37}$$

Let us solve Eq. (12–37) for the electric field. The result is

$$\mathbf{E} = \frac{1}{j\omega\epsilon_0}(\nabla \times \mathbf{H}) \tag{12–38}$$

Substituting Eqs. (12–32), (12–33), and (12–34) into Eq. (12–38) and solving for **E** yields

$$\begin{aligned}\mathbf{E} = \mathbf{a}_r \left(-j\frac{Ile^{-j\beta r}\cos\theta}{2\pi\omega\epsilon_0 r^3} + \frac{\beta Ile^{-j\beta r}\cos\theta}{2\pi\omega\epsilon_0 r^2}\right) \\ + \mathbf{a}_\theta \left(\frac{\beta Ile^{-j\beta r}\sin\theta}{4\pi\omega\epsilon_0 r^2} - j\frac{Ile^{-j\beta r}\sin\theta}{4\pi\omega\epsilon_0 r^3} + j\frac{\beta^2 Ile^{-j\beta r}\sin\theta}{4\pi\omega\epsilon_0 r}\right) + 0\mathbf{a}_\phi\end{aligned} \tag{12–39}$$

Examination of Eq. (12–39) reveals only one term which falls off as the first power of distance. For reasons which will be apparent shortly, this will be considered to be the radiation term of the **E** field. A plot of the electric field lines due to an element of current is shown in Fig. 12–4. Notice how the electric field lines close upon themselves as they move more than about one wavelength from the source. This is the essence of the phenomenon called "radiation."

Our major objective in obtaining both the **E** and **H** fields around the elemental antenna is to compute the Poynting vector. We then have the direction and amplitude of power flow density (watts per square meter) in the region around the antenna. In particular, since we have H_ϕ, E_r, and E_θ components only, we can write the time-averaged power flow density as[2]

$$\mathbf{P} = \frac{1}{2}\,\mathrm{Re}\,(\mathbf{E} \times \mathbf{H}^*) = \mathrm{Re}\left(-\mathbf{a}_\theta\frac{E_r H_\phi^*}{2} + \mathbf{a}_r\frac{E_\theta H_\phi^*}{2}\right) \tag{12–40}$$

[2] H_ϕ^* is the conjugate of H_ϕ.

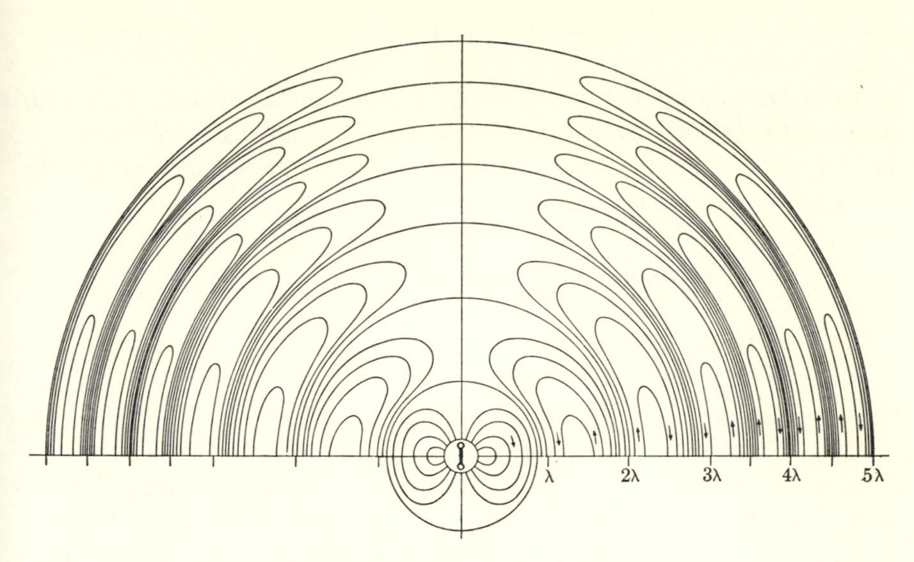

Fig. 12–4. Electric field lines produced by a current element. (After W. K. H. Panofsky and M. Phillips, *Classical Electricity and Magnetism*, second edition [Addison-Wesley Publishing Company, Inc., Reading, Mass., 1962], p. 259.)

The first term inside the parentheses of Eq. (12–40) can be shown to be identically zero. The entire outward power flow therefore comes from the second term. When evaluated, it is found that the total outward power flow density is[3]

$$\begin{pmatrix}\text{Time-averaged power radiated}\\ \text{per square meter}\end{pmatrix} = \frac{\omega^2|I|^2l^2\sin^2\theta}{32\pi^2\epsilon_0 v^3 r^2} \qquad (12\text{–}41)$$

Note that here is the first indication in all the multitude of mathematics that energy is really radiated into space when an alternating current is passed through a wire. Here is the first explanation we have had of the manner in which electromagnetic waves originate.

There is still another very important point. The only terms in the **E** and **H** fields which contributed to this spatial energy-release phenomenon were the so-called radiation terms mentioned above. Every other term, to be sure, contributed to the instantaneous Poynting vector, but all integrated out with time, leaving only that part of the Poynting vector made up of the radiation field terms. This means that considerable amounts of stored energy are in regions close to the antenna. These are

[3] Note that the symbol $|I|$ in this and subsequent equations for power refers to the peak value of the current.

sometimes called the *near fields*, whereas the radiated field is often called the *far field*.

If we were interested in the total average radiated energy from our elemental antenna, it would be necessary only to integrate the radiation field component of the Poynting vector over a sphere surrounding the antenna. This operation is indicated as follows:

$$P_{\text{avg radiated}} = \int_{\text{sphere}} \frac{\omega^2|I|^2l^2 \sin^2\theta}{32\pi^2\epsilon_0 v^3 r^2}\, ds$$

$$= \int_0^{\pi}\int_0^{2\pi} \frac{\omega^2|I|^2l^2 \sin^2\theta}{32\pi^2\epsilon_0 v^3 r^2}\, r \sin\theta\, d\phi\, r\, d\theta \tag{12–42}$$

$$P_{\text{avg radiated}} = \frac{\omega^2|I|^2l^2}{12\pi\epsilon_0 v^3} = \frac{\pi|I|^2l^2}{3\epsilon_0 v\lambda^2} \tag{12–43}$$

It is now seen that we can have a good idea of what is meant by the term "radiation resistance." When a current passes through a conventional resistor, there is an energy loss which is accounted for by heat dissipation. We now see that when a current flows in an antenna, there is also an energy loss. This in turn means that the transmission line feeding our elemental antenna must see a resistive component of load. If the load were all reactive, the average delivered power to the antenna would have to be zero, from fundamental ac circuit theory. It is unimportant that our energy is not dissipated in heat. There is still an energy loss to be accounted for, and the amount of energy lost is exactly that given in Eq. (12–43). Putting in numbers gives

$$P_{\text{avg radiated}} = \frac{789}{2}\left(\frac{l}{\lambda}\right)^2 |I|^2 \tag{12–44}$$

(Remember that I is a peak value, not an rms value.) The radiation resistance of our elemental antenna is obviously

$$R_{\text{radiation}} = 789\left(\frac{l}{\lambda}\right)^2 \tag{12–45}$$

The concept (or perhaps we should say fact) of radiation resistance is useful because it allows us to tie our circuit concepts and analysis methods in with antenna systems. The subject will be given further attention later.

12–5. POTENTIALS FOR EXTENDED SOURCE DISTRIBUTIONS

In Art. 12–3 we derived expressions for vector magnetic potential and scalar electric potential due to point sources located at the origin of the coordinate system. For example, for the current element located at the origin,

$$A_z = \frac{\mu_0 Il}{4\pi r} e^{-j\beta r}$$

where $r = \sqrt{x^2 + y^2 + z^2}$. Now imagine that the source is located at some other point, say (x', y', z') as shown in Fig. 12–5. The radial distance

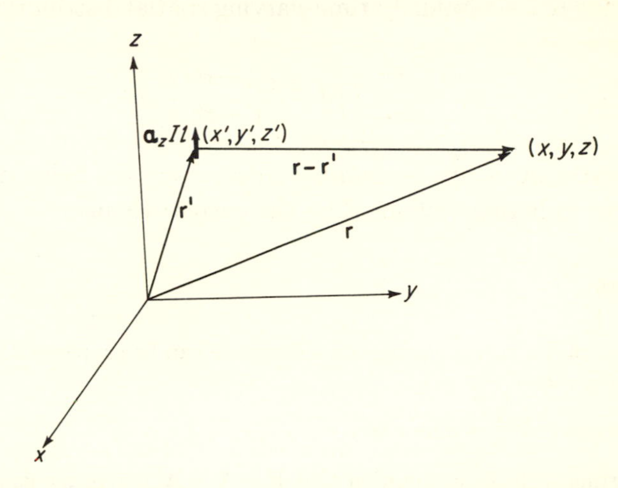

Fig. 12–5. Translation of the origin of the coordinate system.

from the source point (primed coordinates) to the field point (unprimed coordinates) is seen to be

$$|\mathbf{r} - \mathbf{r}'| = \sqrt{(x - x')^2 + (y - y')^2 + (z - z')^2} \tag{12–46}$$

This quantity corresponds to the radial distance r in the potential expressions developed for sources located at the origin of the coordinate system. Thus, for a z-directed current element located at (x', y', z') we write the vector magnetic potential as

$$\mathbf{A} = \mathbf{a}_z \frac{\mu_0 Il e^{-j\beta|\mathbf{r} - \mathbf{r}'|}}{4\pi|\mathbf{r} - \mathbf{r}'|} \tag{12–47}$$

Now let us consider the case where the current is not localized to a point in space, but rather, there is a finite current density over some region of space. Call it $\mathbf{J}(x, y, z)$ or $\mathbf{J}(\mathbf{r})$, where the tip of the vector $\mathbf{r}$ is the point (x, y, z). The vector magnetic potential for such a distribution can be considered to be the superposition of vector magnetic potentials caused by differential volumes of current. For example, when considering a differential volume centered at (x', y', z'), we can write

$$\mathbf{dA} = \frac{\mu_0 \mathbf{J}(\mathbf{r}')\, dv' e^{-j\beta|\mathbf{r} - \mathbf{r}'|}}{4\pi|\mathbf{r} - \mathbf{r}'|} \tag{12–48}$$

where we have replaced $\mathbf{a}_z Il$ with $\mathbf{J}(\mathbf{r}')\,dv'$. Integrating over the source distribution results in

$$\mathbf{A} = \frac{\mu_0}{4\pi} \iiint\limits_{\text{all space}} \frac{\mathbf{J}(\mathbf{r}')e^{-j\beta|\mathbf{r}-\mathbf{r}'|}}{|\mathbf{r}-\mathbf{r}'|}\,dv' \tag{12-49}$$

In a similar fashion one can develop the expression for the scalar electric potential due to a sinusoidally time-varying spatial distribution of charge in free space:

$$\Phi = \frac{1}{4\pi\epsilon_0} \iiint\limits_{\text{all space}} \frac{\rho(\mathbf{r}')e^{-j\beta|\mathbf{r}-\mathbf{r}'|}}{|\mathbf{r}-\mathbf{r}'|}\,dv' \tag{12-50}$$

These expressions are necessary to determine the fields due to more complicated radiating systems than the current element.

PROBLEMS

12–1. Show that a vector having zero divergence can be expressed as the curl of a second vector.

12–2. Prove that $\nabla^2 \dfrac{1}{r} = 0$ for $r \neq 0$.

12–3. Starting with the assumption that $\mathbf{H} = \nabla \times \mathbf{A}$, derive the equivalent equations of (12–11), (12–12), (12–13).

12–4. Obtain Eqs. (12–32), (12–33), and (12–34) from Eqs. (12–28), (12–29), (12–30), and (12–31).

12–5. Obtain Eq. (12–39) from Eq. (12–38).

12–6. Obtain Eq. (12–41) from Eq. (12–40). Show that the only contributing terms are the $1/r$ terms of the **E** and **H** fields.

12–7. Consider a piece of 300-ohm twin lead. The wires are spaced at 0.85 cm. The end is shorted with a perfect conductor. In what frequency range would the circuit concept of a short begin to be invalid, in your opinion?

12–8. Determine the total power radiated from an elemental antenna 1 m long if the uniform rms current is 1 A and the frequency is: (a) 60 Hz; (b) 1000 Hz; (c) 100 kHz; (d) 1 MHz; (e) 3 MHz.

12–9. What is the radiation resistance of the antenna of Problem 12–8 for each of the given frequencies?

12–10. How long would an elemental antenna have to be to have a radiation resistance of 1 ohm at a frequency of 60 Hz?

13

The Radiating Dipole Antenna

13-1. INTRODUCTION

In Chapter 12 the behavior of a sinusoidally varying current element was discussed in some detail. It was assumed that this antenna element was driven with a nonradiating transmission line of some kind and that each end of the current element was connected to one of the terminals of the transmission line. This established the concept of energy radiation, but for several reasons it is not practical to build an antenna in this manner. The resistive component of impedance in such a short antenna would be very low, and the total amount of energy radiated, though finite, would probably be insignificant. Obviously some other approach is necessary.

13-2. THE ELEMENTAL DIPOLE ANTENNA

As a start, let us take two pieces of straight wire, both of length $l/2$, and connect them to a two-wire balanced transmission line as shown in Fig. 13-1. Let us assume there is nothing in close proximity to the antenna except the transmission line, and let us hope that the transmission line does not affect the over-all situation much. It would appear that we can now drive the antenna by some means other than building the generator into it and still have a transmission line that behaves as an efficient line should. The two ends of the pieces of wire can be brought to within any reasonable distance of each other. If the impedance that the transmission-line termination sees is not that of the line itself, we could even accommodate it with a small matching transformer or tuning unit. It may seem at first that all we now have left to do is find the vector potential function, convert it to the **E** and **H** fields, and proceed from there. Technically, this is correct, but before we can write the vector potential function we must know something about the current in the two pieces of wire.

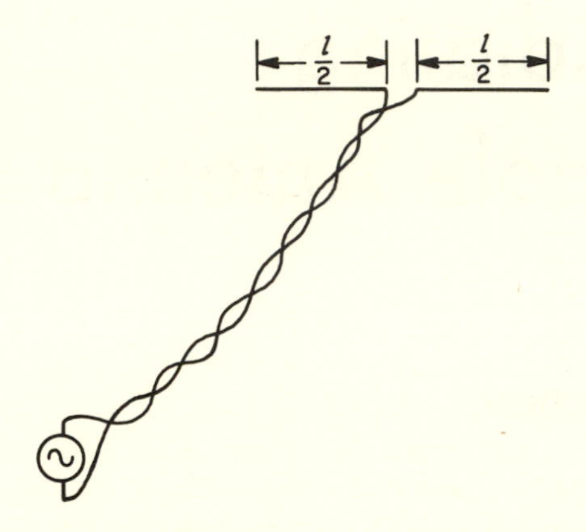

Fig. 13–1. The dipole antenna driven with a balanced transmission line.

Since the wires must have some individual inductance and also capacitance between them, we can be certain that some sort of current will flow at the transmission-line end and in the wire immediately adjacent to the end. However, it is fallacious to say that the current distribution is thereafter uniform over the entire length of the two wires, since the current at the far ends of the wires away from the transmission line must obviously be zero. The current, then, goes from finite value to zero as we traverse an individual wire element. To compute the vector potential function, we must know how the current amplitude varies along the wire. As a start, let us assume that the current decreases linearly to zero as a wire is traversed from the transmission line to the other end. In other words, we might plot the current distribution as in Fig. 13–2.

We do not know whether this distribution is correct or not, but let us assume that it is. Then, we can compute on that basis and compare the calculated radiation resistance with that measured on an r-f bridge. If the two values agree, it might be reasonable to assume that the assump-

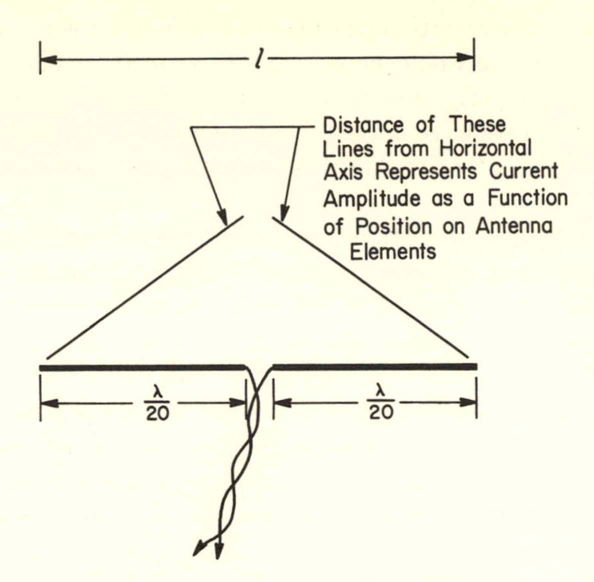

Fig. 13-2. Assumed current distribution.

tion of current distribution is good.[1] If not, we shall have to try something else. As our first try, it will be assumed that the length of each antenna element is equal to, or less than, about one-twentieth of a wavelength; thus

$$l \leqq \frac{\lambda}{10}$$

Since this is a small fraction of a wavelength, we could also say that the current phase is uniform along the wire. In Chapter 12 we found that the power radiated from an elemental antenna is

$$P = \frac{789}{2} \frac{(l)^2 |I|^2}{\lambda^2} \tag{13-1}$$

Since in our elements we have assumed that the current falls to zero linearly along the antenna element, the "average" current is half that at the driven end. The power radiated can therefore be approximated as

$$P = \frac{789}{2} \left(\frac{l}{\lambda}\right)^2 \left|\frac{I}{2}\right|^2 = \frac{197.5}{2} \left(\frac{l}{\lambda}\right)^2 |I|^2 \tag{13-2}$$

The radiation resistance of the antenna of Fig. 13-1 would then appear to be

$$R = 197.5 \left(\frac{l}{\lambda}\right)^2 \tag{13-3}$$

[1] Even so, this is risky. In a half-wave dipole, for example, a large number of widely different current distribution functions can give essentially similar radiation resistances.

If we measured the radiation resistance of such an antenna with a radio-frequency bridge, we would find it very close to the value given in Eq. (13–3). Our assumptions for current distribution are therefore probably correct, although we have not established this as an analytically proven fact. It would seem that we have the beginnings of a practical radiating antenna.

There would be several things we would not like about this two-element short antenna, however. First, we would find that while the radiation resistance predicted by Eq. (13–3) is essentially correct, it will turn out to be quite small compared with the driving point reactance at the antenna. Therefore we must either put a matching device between the transmission line and the antenna or be willing to put up with severe standing waves on the transmission line itself. In many practical cases, we shall be faced with these alternatives anyway, but working with longer antennas which have a greater ratio of radiation resistance to reactance will make the matching devices simpler and much more efficient.

13–3. THE HALF-WAVE DIPOLE

Now let us extend the length of the radiating elements still farther. In particular, let us make each element one quarter-wavelength, which makes the combination approximately one half-wavelength. This configuration is known as the half-wave dipole and is one of the most important types of antennas from the standpoint of both theoretical analysis and practical usage.

In the analysis of this antenna, we are again faced with the question of current distribution. Since the element length is no longer inappreciable, it may or may not be reasonable to assume that the current amplitude decreases linearly with distance along the element. If we think for a moment about an open-circuited transmission line, we recall that the current amplitude distribution is sinusoidal with distance. For one half-wavelength, the current is in time phase but varies sinusoidally in amplitude. For the next half-wavelength, the current phase reverses, but the amplitude continues to vary sinusoidally. We can extend our imagination to think of the half-wave dipole as a flared-out transmission line and assume a similar current distribution. The ultimate justification for this assumption will be that we can experimentally justify the calculated results. In truth, everything in science can be traced back to a similar situation. For analytical purposes, our half-wave dipole will be aligned

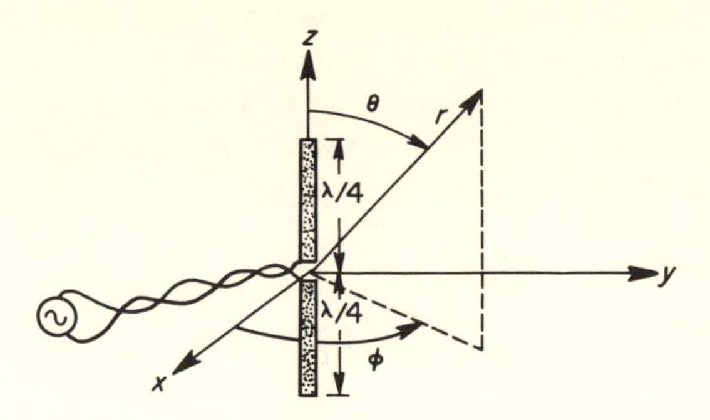

Fig. 13-3. Orientation of half-wave dipole in otherwise free space.

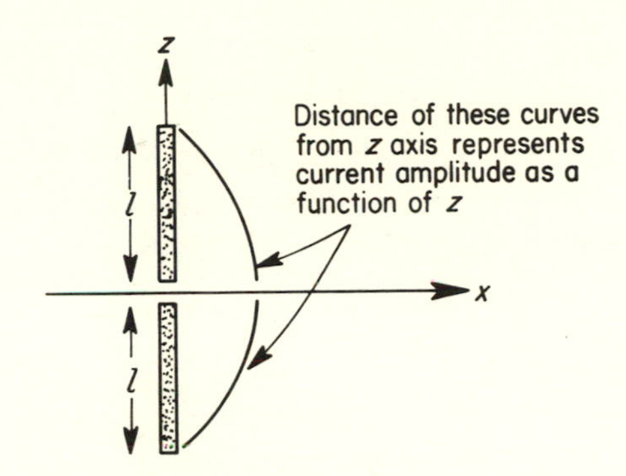

Fig. 13-4. Current distribution in half-wave dipole.

as shown in Fig. 13-3. The current distribution is illustrated in Fig. 13-4.[2]

The assumed current distributions can be expressed analytically as follows:

$$\mathcal{I} = |I| \cos \beta z \cos \omega t \qquad |z| \leq \frac{\lambda}{4} \tag{13-4}$$

$$\mathcal{I} = 0 \qquad |z| > \frac{\lambda}{4} \tag{13-5}$$

Note that the current is assumed to be "in phase" all along the antenna. This assumption is admittedly based on intuition, but it turns out to be

[2] We are begging the issue a little here. In theory it is possible to solve Maxwell's equations with the given boundary conditions and get the exact current distribution. The trouble is, nobody has been able to do this in general, although some simple cases have been solved. Fortunately this is one of those cases, and one finds that for infinitely thin straight antennas, the current distribution is essentially sinusoidal.

See Chapters 10 and 11 of E. C. Jordan and K. G. Balmain, *Electromagnetic Waves and Radiating Systems*, 2d ed. (Englewood Cliffs, N.J.: Prentice-Hall, Inc., 1968).

fairly good if the dipole legs do not get too long. In this case, β is the free-space propagation constant with no attenuation, in the sense that energy is lost through heat dissipation. From here on, as in Chapter 12, it will be convenient to express the fields and sources in complex phasor notation, and Eq. (13–4) will be rewritten as

$$\begin{aligned} I &= |I| \cos \beta z \qquad |z| \leq \frac{\lambda}{4} \\ I &= 0 \qquad\qquad\quad |z| > \frac{\lambda}{4} \end{aligned} \tag{13–6}$$

The magnetic potential vector is determined by application of Eq. 12–49, with $\mathbf{r}$ and $\mathbf{r}'$ defined as

$$\mathbf{r}' = \mathbf{a}_z z' \tag{13–7}$$

$$\mathbf{r} = \mathbf{a}_x x + \mathbf{a}_y y + \mathbf{a}_z z \tag{13–8}$$

The geometry of the situation is shown in Fig. 13–5. Visualize the source coordinates and the field coordinates as two identical coordinate systems superimposed on one another.

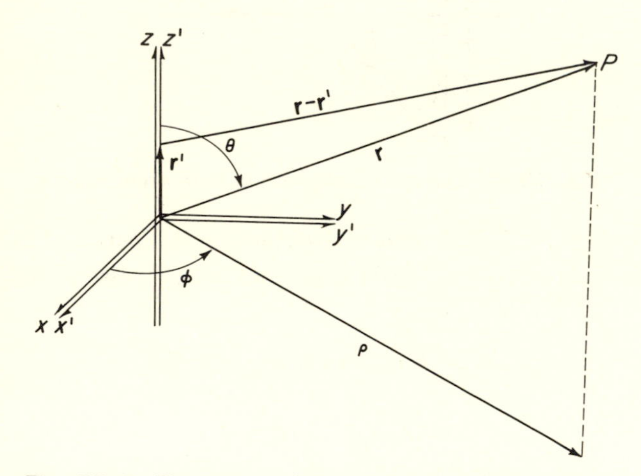

Fig. 13–5. Illustration of the source and field points.

Subtracting Eq. (13–7) from Eq. (13–8) results in

$$\mathbf{r} - \mathbf{r}' = \mathbf{a}_x x + \mathbf{a}_y y + \mathbf{a}_z (z - z') \tag{13–9}$$

The magnetic potential vector can now be written as

$$\mathbf{A} = \mathbf{a}_z \frac{\mu_0}{4\pi} \int_{-\lambda/4}^{\lambda/4} \frac{|I| \cos \beta z' e^{-j\beta\sqrt{x^2+y^2+(z-z')^2}}}{\sqrt{x^2 + y^2 + (z - z')^2}} \, dz' \tag{13–10}$$

Now we are faced with integrating Eq. (13–10). Let us remember, however, that our primary interest is only in the field far from the antenna. With this in mind we can approximate the term under the radical:

$$\sqrt{x^2 + y^2 + (z - z')^2} \approx r - z' \cos\theta \tag{13-11}$$

We will ignore the variation in the denominator as the integration is performed and include the phase effects caused by the exponential in the numerator. The resultant expression for the vector magnetic potential is

$$\mathbf{A} = \mathbf{a}_z \mu_0 \frac{|I|\, e^{-j\beta r}}{4\pi r} \int_{-\lambda/4}^{\lambda/4} (\cos\beta z') e^{j\beta z' \cos\theta}\, dz' \tag{13-12}$$

Let us again recall the geometry of Fig. 13-5 and remember that the point at which we are interested in A_z is very remote from the antenna. Under these conditions the angle θ is essentially constant, and the integration of Eq. (13-12) can be performed much more easily than would ordinarily be the case. This integration is left as an exercise. The final result is

$$A_z \approx \frac{\mu_0 |I| e^{-j\beta r} \cos[(\pi/2)\cos\theta]}{2\pi\beta r \sin^2\theta} \tag{13-13}$$

As you have probably noticed, with some dread, we have been working with a strange combination of spherical and cylindrical coordinates. The r and θ are from the spherical coordinates r, θ, and ϕ. The z on the other hand is from the cylindrical coordinates ρ, ϕ, and z. The r in spherical coordinates is, of course, defined somewhat differently from the ρ in cylindrical coordinates. This situation has not led to any serious difficulty so far because we did not let r and θ vary much during the integration along the z axis in cylindrical coordinates. Now, however, we must get the magnetic intensity, $\mathbf{H}$, at the point where we have A_z. This means we must take the curl because

$$\mathbf{H} = \frac{1}{\mu_0} \nabla \times \mathbf{A} \tag{13-14}$$

Therefore let us rewrite the vector potential function in spherical coordinates. It is[3]

$$\mathbf{A} = \frac{\mu_0 |I| e^{-j\beta r} \cos[(\pi/2)\cos\theta] \cos\theta}{2\pi\beta r \sin^2\theta} \mathbf{a}_r - \frac{\mu_0 |I| e^{-j\beta r} \cos[(\pi/2)\cos\theta]}{2\pi\beta r \sin\theta} \mathbf{a}_\theta \tag{13-15}$$

Taking the curl of Eq. (13-15) in spherical coordinates leads to two terms for H_ϕ only. The other two components of $\mathbf{H}$ are zero. One of the terms of

[3] The conversions here are

$$A_{r(\text{spherical})} = A_{z(\text{cylindrical})} \cos\theta_{(\text{spherical})}$$

and

$$A_{\theta(\text{spherical})} = -A_{z(\text{cylindrical})} \sin\theta_{(\text{spherical})}$$

H_ϕ falls off as the first power of the radius and represents the radiation field. The other term of H_ϕ falls off as the second power of the radius and represents the induction field. We are concerned only with the radiation field, and so we shall simply remove ourselves far enough from the antenna so that it is the only remaining term of significance.

Under these conditions, it is relatively easy to show that

$$H_\phi = j\frac{|I|e^{-j\beta r}\cos[(\pi/2)\cos\theta]}{2\pi r\sin\theta} \tag{13-16}$$

Equation (13–16) represents the radiation component of the magnetic intensity field surrounding a half-wave dipole. We may obtain the radiation component of the electric field in the same manner as was done for the current element in Chapter 12. Simply write the electric field as

$$\mathbf{E} = \frac{1}{j\omega\epsilon_0}\nabla \times \mathbf{H} \tag{13-17}$$

and then solve for the electric field, throwing out all terms which fall off as a second or higher power of the radius. The result is

$$E_\theta = j\frac{\beta|I|e^{-j\beta r}\cos[(\pi/2)\cos\theta]}{2\pi\omega\epsilon_0 r\sin\theta} \tag{13-18}$$

Note again that only an H_ϕ and an E_θ component survive as radiation terms. The outward power flow is again given by the Poynting vector $\frac{1}{2}\text{Re}[\mathbf{E}\times\mathbf{H}^*]$. In this case, however, we see again that E_θ and H_ϕ are in time and space phase. Further, a little thought reveals that the absolute value of the complex term $e^{-j\beta r}$ is always unity. The resulting time-averaged, net outward power flow in watts per square meter is given by

$$\begin{pmatrix}\text{Time-averaged power flow outward}\\ \text{in watts per square meter}\end{pmatrix} = \frac{E_\theta H_\phi^*}{2} = \frac{\beta I_{\text{rms}}^2\cos^2[(\pi/2)\cos\theta]}{4\pi^2\omega\epsilon_0 r^2\sin^2\theta} \tag{13-19}$$

Now let us take note of the fact that the intrinsic impedance of free space η is equal to $\sqrt{\mu_0/\epsilon_0}$, and the lossless propagation constant β is equal to $\omega\sqrt{\mu_0\epsilon_0}$. Equation (13–19) is then written as

$$\begin{pmatrix}\text{Power in watts}\\ \text{per square meter}\end{pmatrix} = \frac{\eta I_{\text{rms}}^2\cos^2[(\pi/2)\cos\theta]}{4\pi^2 r^2\sin^2\theta} \tag{13-20}$$

Recall that what we really want is the radiation resistance of the antenna so that we can determine the characteristics of the generator and transmission line required to drive the antenna. This means that we must find the *total power radiated* from the antenna in terms of the current that flows through it. As before, our best choice is to integrate the power flow in watts per square meter over a sphere completely surrounding the antenna. For simplicity, let r be constant. The resulting power integral is

$$\begin{pmatrix}\text{Total power}\\ \text{radiated}\end{pmatrix} = \int_{\substack{\text{sphere of large}\\ \text{constant radius}}} \mathbf{P} \cdot \mathbf{da}$$

$$= \int_0^{\pi}\int_0^{2\pi} P_r r^2 \sin\theta \, d\theta \, d\phi$$

$$= \frac{\eta I_{\text{rms}}^2}{2\pi}\int_0^{\pi} \frac{\cos^2[(\pi/2)\cos\theta]}{\sin\theta}\, d\theta \tag{13-21}$$

The integral of Eq. (13–21) presents a small problem. There is no easy inverse integration formula which can be applied. This is usually the case in real-life, nonclassroom problems anyway. The value of the integral can be determined fairly readily by either graphical or series approximation techniques. The actual integration is left as an exercise.[4] When it is evaluated, Eq. (13–21) becomes

$$\begin{pmatrix}\text{Total power}\\ \text{radiated}\end{pmatrix} = \frac{1.218\eta I_{\text{rms}}^2}{2\pi} = 73 I_{\text{rms}}^2 \tag{13-22}$$

Equation (13–22) indicates that if our assumption of a sinusoidal current distribution in a half-wave dipole is right, the radiation resistance (i.e., that component of the antenna driving-point impedance that absorbs power from the generator) is 73 ohms. At this point let us set up an experimental half-wave dipole at some reasonable frequency and measure the driving-point resistive component with a radio-frequency bridge. We find that this resistive component is indeed close to 73 ohms, and so it begins to look as though our assumption was a good one. Since we are still doubtful, however, we make a very small sampling loop and go along adjacent to the antenna wire to sample the adjacent magnetic field and hence the current.[5] We find the sinusoidal current distribution really does exist to a rather close degree. Philosophically, what does this mean? It means that we have correlated mathematical analysis with carefully chosen experimental measurements to obtain interesting and extremely useful engineering answers. Nevertheless, with mathematics alone we have not actually proved anything. This is really true of everything the engineer learns, but it is perhaps a little more apparent here than usual.

[4] See E. C. Jordan and K. G. Balmain, *Electromagnetic Waves and Radiating Systems*, 2d ed. (Englewood Cliffs, N.J.: Prentice-Hall, Inc., 1968), pp. 330–31.

[5] This experimental step is desirable because many other types of current distribution functions can be postulated which would lead to a 73-ohm radiation resistance.

Nothing has been said about the total driving-point impedance of the antenna. For example, what do we know about the reactive component? In the case of the current element in Chapter 12 we had out-of-phase components in both the electric and magnetic fields. These components represented the reactive fields. We could not have obtained the reactive power by a surface integration over any enclosing surface as we did the resistive power, however. The average value of the resistive power represented a continual energy dispersal by the antenna, and an integration of the outward-flowing Poynting vector over any surface enclosing the antenna was sufficient to evaluate it, irrespective of the size of the surface. Had we integrated the reactive power over that same surface, we would have obtained only the value of reactive energy that flows back and forth through the surface. Its average value would always be zero, and its peak value would not bear any particular relationship to the total reactive energy of the antenna that would be independent of the shape and size of the integrated surface. To obtain the reactive energy, we would have to use a surface extremely close to the antenna element such that essentially all the reactive energy would flow back and forth through it.

In the case of the half-wave dipole, all the preceding discussion about reactive fields also holds good. Since our main interest was in the determination of the radiation resistance, we did not even determine the reactive or induction fields. In Art. 13–4, on the induced EMF method, we shall discuss a versatile way to determine both the radiation resistance and the reactance of the antennas in question. Further, the method will be extendable (in Chapter 14) to allow the analysis of two or more antenna elements used in conjunction with each other.

In the case of the thin half-wave dipole just discussed, if we had evaluated the reactive component of the driving-point impedance, we would have found it to be $+j42.5$ ohms. In practice, this means that if the antenna were fed with a balanced transmission line of 73-ohm characteristic impedance, a matching network should be used at the antenna terminals if we are to be absolutely correct. In practice, two alternatives arise. The first is to shorten the dipole somewhat. If this is done properly, the reactance will go through zero as the antenna is shortened. At the resonant point of a dipole, then, the length is slightly less than half-wave and the radiation resistance is something less than 73 ohms. The second alternative is to simply drive the antenna with a 73-ohm balanced line and forget the $+j42.5$ ohms of reactance. For many years, standard transmission lines have been made in the 73- to 75-ohm characteristic impedance region as well as regions of several other values. Originally the choice of this numerical range came from the natural radiation resistance of the dipole. However, such transmission lines are now obviously used for many other things besides feeding this type of antenna.

There is one other very interesting fact that has come from this analysis. Whenever one cuts a thin half-wave dipole for any frequency at all, the radiation resistance is 73 ohms at that frequency. No other factors are involved. The reactance of the thin half-wave dipole is $+j42.5$ ohms at its specific frequency, irrespective of the value of the frequency itself. When the length of the dipole is other than half-wave, however, many other factors enter into the reactance determination.

13–4. THE INDUCED EMF METHOD

We are now going to give some attention to an alternative method of determining antenna impedance characteristics. It is called the "induced EMF method" and is credited primarily to P. S. Carter.[6] Let us reexamine the thin half-wave dipole and assign geometry as shown in Fig. 13–6. The antenna is oriented as before.

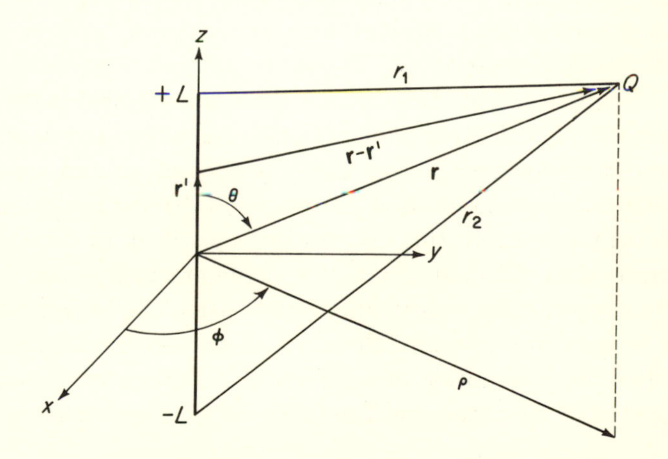

Fig. 13–6. Dipole antenna for induced EMF analysis.

This is exactly the same geometry that was used in the last section, with the distances r_1 and r_2 defined as shown to facilitate the analysis. We shall consider a current distribution extending along the z-axis from $-L$ to $+L$. The current distribution must go to zero at the ends, so the cur-

[6] "Circuit Relations in Radiating Systems and Applications to Antenna Problems," *Proc. Institute of Radio Engineers*, vol. 20 (June 1932), pp. 1004–41.

rent distribution will be assumed to be

$$I_z = I \sin \beta(L - |z|) \qquad |z| \leq L \tag{13-23}$$

$$I_z = 0 \qquad |z| > L \tag{13-24}$$

There is a value of electric field tangent to the antenna at its surface which we shall call E_z.

One may wonder how E_z at the antenna boundary can be anything but zero, since an antenna is usually assumed to be made of perfect conductors. Again it must be pointed out, as in Chapter 12, that we consider the fields around the wires to be the superposition of an incident field established by a current source connected to the antenna elements and a scattered field due to currents induced in the elements by the incident field. The boundary conditions imposed by the conductors require that the net tangential electric field be zero on the conductors. The current source establishes the current at the input terminals. In solving the radiation problem we consider only the scattered fields and the current distribution which establishes these fields. This problem is solved as though the current distribution existed in free space. This is acceptable since the perfectly conducting wires simply impose a boundary condition on the net fields. Besides, we are considering fields on or outside the conductor configuration.

It is difficult to accept the fact that the scattered field is the only one to be considered in calculating the radiation characteristics of an antenna but, when one considers the fact that the theoretical and experimental comparison of antenna radiation patterns and driving point impedances show excellent agreement, this assumption is justified. In fact it is easy to demonstrate that different conductor configurations when driven by identical sources display vastly different radiation and impedance characteristics. At any rate, virtually all calculations of antenna characteristics are based on the assumption of a current distribution in free space. The characteristics of the conductors are taken into account as a second-order effect.

For an antenna that is finite but extremely thin, the radiation fields encountered as one leaves the antenna are still essentially the same as though the antenna was of zero diameter. Moreover, the induction fields are also essentially as predicted for a finite wire size and an assumed sinusoidal current, even though the two cannot precisely coexist. That is, induction fields can be computed by taking finite wire size into account. As we shall see, however, the wire size does not enter into the calculation when the antennas are multiples of a quarter-wavelength. Since these are the overall effects we are after, we still use the induced EMF method even though it is not exactly correct. For further information on these questions, see the references listed below.[7] We shall return briefly to this subject after discussion of the induced EMF method.

Hallen, Synge, and King treat the antenna as a pure boundary value problem and as such are technically more correct. The final results do not differ significantly from those presented here.

If the current at z is $I_z(z)$ and the electric field at that point is $E_z(z)$, the complex power involved in allowing I_z to flow in the presence of E_z over the elemental length dz is

$$d(\text{power}) = -\frac{1}{2} E_z I_z^* \, dz \tag{13-25}$$

The total complex power involved over the entire length of the antenna is

$$\text{Complex power} = -\frac{1}{2} \int_{-L}^{L} E_z I_z^* \, dz \tag{13-26}$$

The minus sign on Eqs. (13-25) and (13-26) arises by convention because the antenna is an energy source rather than an energy sink. That is, if I_z is in the positive z-direction, then E_z must be in the negative direction if the power flow is outward. This is apparent from consideration of the Poynting vector at the antenna surface, or simply from sign considerations of voltage and current in a battery delivering energy to a circuit. Equation (13-23) can be substituted into Eq. (13-26). Furthermore, due to the physical symmetry of the problem, the integration interval can be halved and the remainder multiplied by two. The result is

$$\begin{aligned} \text{Complex power} &= -\frac{1}{2} \int_{-L}^{L} E_z I^* \cos \beta z \, dz \\ P &= -I^* \int_{0}^{L} E_z \cos \beta z \, dz \end{aligned} \tag{13-27}$$

where I^* is the conjugate of the driving-point current of the half-wave dipole and E_z is the value of the tangential field as a function of z. Since we have removed time dependence, the power P takes on a standard

[7] Erik Hallen, "Theoretical Investigations into the Transmitting and Receiving Qualities of Antennae," *Nova Acta Regiae Soc. Sci. Upsaliensis*, Ser. IV, 11, No. 4 (1938), pp. 1-44; J. L. Synge, "The General Problem of Antenna Radiation and the Fundamental Integral Equation with Application to an Antenna of Revolution," *Quarterly of Applied Mathematics*, vol. 6 (July 1948), pp. 133-56; L. W. King, "On the Radiation Field of a Perfectly Conducting, Base-Insulated Cylindrical Antenna over a Perfectly Conducting Plane Earth, and the Calculation of the Radiation Resistance and Reactance," *Philosophical Transactions of the Royal Society* (London), vol. 236 (1937), pp. 381-422.

interpretation we customarily associate with ac circuits. It will have a real and imaginary component. The real component will represent the time-averaged radiated power, and the imaginary component will represent the reactive power. Now these must be the real and reactive powers that are delivered to the driving-point terminals of the antenna. Let us designate the driving-point impedance of the antenna as $Z_{11} = R_{11} + jX_{11}$. We can then write

$$P = \frac{I^* I Z_{11}}{2} = \frac{|I|^2}{2}(R_{11} + jX_{11}) \tag{13–28}$$

From Eqs. (13–27) and (13–28) we can write an expression for the driving-point impedance of the half-wave dipole antenna:

$$Z_{11} = R_{11} + jX_{11} = -\frac{2}{I}\int_0^L E_z \cos \beta z \, dz \tag{13–29}$$

Equation (13–29) is a direct expression for the driving-point impedance, but it requires the determination of the tangential electric field at the surface of the antenna wire. In the Poynting vector analysis of the half-wave dipole, this field was not determined because it was not needed, so we shall now have to derive it. The steps of this derivation are outlined in the following paragraph. It is left for the student to fill in many of the mathematical details.

From the retarded vector potential solution obtained in Chapter 12, we can write the vector potential at a point Q (see Fig. 13–6):

$$A_z = \frac{\mu_0 I}{4\pi}\int_{-L}^{+L} \frac{\cos \beta z' e^{-j\beta s}}{s}\, dz' \tag{13–30}$$

where

$$s = |\mathbf{r} - \mathbf{r}'| = \sqrt{x^2 + y^2 + (z - z')^2} \tag{13–31}$$

The dipole has ϕ symmetry. Therefore in cylindrical coordinates we can write[8]

$$H_\phi = -\frac{1}{\mu_0}\frac{\partial A_z}{\partial \rho} \tag{13–32}$$

This derives from the fact that $\mathbf{H} = \dfrac{1}{\mu_0}\nabla \times \mathbf{A}$ and A_z is the only component of $\mathbf{A}$. The magnetic intensity can now be written as

[8] Remember, we are using cylindrical and spherical coordinates simultaneously. The cylindrical radius is written as ρ and the spherical radius is written as r. See Fig. 13–6 and note that s is a variable measured from a given point on the filamentary antenna to the point at which fields are being examined. The relationship between s and ρ is $s^2 = \rho^2 + (z - z')^2$. Then $\partial s/\partial \rho = \rho/s$, and $\partial s/\partial z' = (z' - z)/s$.

$$H_\phi = \frac{I}{8\pi}\left\{\int_{-L}^{L}\left[\frac{j\beta\rho e^{-j\beta(s-z')}}{s^2} + \frac{\rho e^{-j\beta(s-z')}}{s^3}\right]dz' + \int_{-L}^{L}\left[\frac{j\beta\rho e^{-j\beta(s+z')}}{s^2} + \frac{\rho e^{-j\beta(s+z')}}{s^3}\right]dz'\right\} \tag{13-33}$$

These integrands turn out to be perfect differentials. Therefore we can write

$$H_\phi = -\frac{I}{8\pi}\left[\frac{\rho e^{-j\beta(s-z')}}{s(-s+z'-z)}\bigg|_{-L}^{+L} + \frac{\rho e^{-j\beta(s+z')}}{s(s+z'-z)}\bigg|_{-L}^{+L}\right] \tag{13-34}$$

After Eq. (13-34) has been written out completely and the limits have been substituted in, it is expedient to take note of several useful facts, as follows:

(a) $$\frac{1}{-r_1 + L - z} = -\frac{r_1 + (L - z)}{\rho^2}$$

(b) $$\frac{1}{-r_2 - L - z} = -\frac{r_2 - (L + z)}{\rho^2}$$

(c) $$\frac{1}{r_1 + L - z} = \frac{r_1 - (L - z)}{\rho^2} \tag{13-35}$$

(d) $$\frac{1}{r_2 - L - z} = \frac{r_2 + L + z}{\rho^2}$$

(e) $$L = \frac{\lambda}{4} \qquad \cos\beta L = 0 \qquad \sin\beta L = 1$$

After making use of all these equations, the expression for H_ϕ finally reduces to

$$H_\phi = -\frac{I}{4\pi j}\left(\frac{e^{-j\beta r_1}}{\rho} + \frac{e^{-j\beta r_2}}{\rho}\right) \tag{13-36}$$

Now

$$\mathbf{E} = \frac{1}{j\omega\epsilon_0}\nabla \times \mathbf{H} \tag{13-37}$$

From Eqs. (13-36) and (13-37) we can write the electric field expressions

$$E_\rho = \frac{I\beta}{4j\pi\rho\omega\epsilon_0}\left(\frac{L - z}{r_1}e^{-j\beta r_1} - \frac{L + z}{r_2}e^{-j\beta r_2}\right) \tag{13-38}$$

$$E_z = \frac{I\beta}{4j\pi\omega\epsilon_0}\left(\frac{e^{-j\beta r_1}}{r_1} + \frac{e^{-j\beta r_2}}{r_2}\right) \tag{13-39}$$

Equations (13-36), (13-38), and (13-39) represent the complete electric and magnetic field equations for a filamentary half-wave dipole. The field for a general symmetrical dipole of any length can be developed in the same manner without too much added difficulty. It is interesting to note that, in the case of the E_z field at least, there appear to be two components. One originates from the upper end of the antenna and the other originates from the lower end. Each component expands spherically while falling off as the first power of the distance from the end points.

In Eq. (13–39), which is our main interest in the induced EMF method, the coefficient in front can be combined into a single numerical value (remember that β has an ω factor too). The result for the electric field in the z-direction is

$$E_z = -j30I\left(\frac{e^{-j\beta r_1}}{r_1} + \frac{e^{-j\beta r_2}}{r_2}\right) \tag{13–40}$$

Next, for the E_z field to be useful in the induced EMF method, it must be evaluated at the surface of the antenna itself. This needs to be done only in the interval from 0 to L on the antenna. See Eq. (13–29). The following relationships are useful in this case:

$$r_1 = L - z$$
$$r_2 = L + z$$

E_z at the antenna is then written as

$$E_z \text{ (at antenna)} = -j30I\left[\frac{e^{-j\beta(L-z)}}{L-z} + \frac{e^{-j\beta(L+z)}}{L+z}\right] \tag{13–41}$$

An $e^{-j\beta L}$ term can be factored. Also let us recall that βL is equivalent to 90 electrical degrees for a half-wave dipole, and consequently

$$e^{-j\beta L} = \cos\beta L - j\sin\beta L = -j \tag{13–42}$$

The tangential electrical field can then be written as

$$E_z \text{ (at antenna)} = -30I\left(\frac{e^{j\beta z}}{L-z} + \frac{e^{-j\beta z}}{L+z}\right) \tag{13–43}$$

Substituting Eq. (13–43) into Eq. (13–29) for driving-point impedance gives, after a little simplification,

$$Z_{11} = 30\left(\int_0^L \frac{1+e^{j2\beta z}}{L-z}\,dz + \int_0^L \frac{1+e^{-j2\beta z}}{L+z}\,dz\right) \tag{13–44}$$

In the first integral, let $v = 2\beta(L-z)$. Then $dv = -2\beta\,dz$, and this integral becomes

$$-\int_{2\beta L}^{0} \frac{1+e^{j(2\beta L - v)}}{v}\,dv = -\int_{2\beta L}^{0} \frac{1+e^{j2\beta L}e^{-jv}}{v}\,dv$$
$$= \int_0^{\pi} \frac{1-e^{-jv}}{v}\,dv \tag{13–45}$$

Note that $\beta L = \pi/2$ for the half-wave dipole. In the second integral of

Eq. (13–44), let $v = 2\beta(L + z)$. This integral then becomes

$$\int_{2\beta L}^{4\beta L} \frac{1 + e^{-j(v-2\beta L)}}{v} dv = \int_{\pi}^{2\pi} \frac{1 - e^{-jv}}{v} dv \tag{13–46}$$

Combining expressions (13–45) and (13–46) in Eq. (13–44) finally gives

$$Z_{11} = 30 \int_{0}^{2\pi} \frac{1 - e^{-jv}}{v} dv \tag{13–47}$$

Eq. (13–47) can be written as follows:

$$Z_{11} = 30 \int_{0}^{2\pi} \frac{1 - \cos v}{v} dv + j30 \int_{0}^{2\pi} \frac{\sin v}{v} dv \tag{13–48}$$

The integrals of Eq. (13–48) can be evaluated by series or graphical methods. However, they occur sufficiently often in nature to be given special symbols and they are tabulated in various function tables.[9] The first integral is symbolized as Cin (x), where x is the upper limit. The second integral is symbolized as Si (x), where x is again the upper limit. Thus one often sees Z_{11} written as

$$Z_{11} = 30\,\mathrm{Cin}(2\pi) + j30\,\mathrm{Si}(2\pi) \tag{13–49}$$

If we evaluate $\mathrm{Si}(2\pi)$ and $\mathrm{Cin}(2\pi)$, we find that

$$Z_{11} = 73 + j42.5 \text{ ohms} \tag{13–50}$$

Note that we have evaluated the reactance of the half-wave dipole and simultaneously corroborated our Poynting vector far-field analysis for the radiation resistance. As a matter of fact, the Poynting vector method can be derived from the induced EMF method. Suppose we enclose the antenna in a cylindrical sheath of very narrow radius. The complex Poynting vector on the surface of the cylinder is

$$\mathbf{P} = \frac{1}{2}\mathbf{E} \times \mathbf{H}^* \tag{13–51}$$

Integrating this Poynting vector over the entire sheath gives the total complex power:

$$\begin{pmatrix}\text{Total complex}\\ \text{power}\end{pmatrix} = \int_{\substack{\text{cylindrical}\\ \text{surface}}} \frac{1}{2}\mathbf{E} \times \mathbf{H}^* \cdot \mathbf{da}$$

$$= -\int_{-L}^{L} E_z H_\phi^*(2\pi\rho_\delta)\, dl \tag{13–52}$$

[9] See, for example, *Handbook of Mathematical Functions with Formulas, Graphs and Mathematical Tables*, National Bureau of Standards Applied Mathematics Series No. 55, U.S. Government Printing Office, Washington, D.C., 1964.

If the antenna element is an extremely thin filament and the radius of the enclosing cylinder (ρ_δ) is very small, the induction field term predominates over the radiation field term and

$$H_\phi \approx \frac{I_z}{2\pi\rho_\delta} \tag{13-53}$$

Substituting Eq. (13–53) into Eq. (13–52) yields Eq. (13–25), which is the starting point for the induced EMF method.

The induced EMF method can be used for many things besides simply cross-checking the Poynting vector method, however. It can be used to compute mutual impedances between antenna elements, and we shall use it for that a little later on. Also, it can be used to examine any linear antenna up to a wavelength or two long.[10] As a matter of fact, the figure in the next article, which gives radiation resistance for many lengths (Fig. 13–8), was actually computed by the induced EMF method. In short, the induced EMF method is the practical key to calculation of many types of antennas for which the Poynting vector method would be useless because of its extreme complication.

Let us momentarily return to the discussion of antenna reactance. In one sense, we were very fortunate in the preceding analysis. Because the antenna analyzed was exactly one half-wavelength long, certain terms canceled in the determination of the tangential field E_z. Had the antenna been some odd length, this would not have happened, and the subsequent integration process would have required a knowledge of the actual diameter of the antenna wire, since the inductance of a wire is a function of its size. The smaller the wire diameter, the larger is the inductance, and consequently the larger the reactance. The reactance referred to the driving point may be either inductive or capacitive according to antenna length in a manner somewhat (but not completely) analogous to the nature of improperly terminated transmission lines. All this means that if one wants a low Q, broad-band antenna, he had best make it of broad cross-section rather than narrow for any given length. Antenna reactance for a dipole of general length will be discussed in the next article.

13–5. RESISTANCE AND REACTANCE OF THE GENERALIZED CENTER-FED DIPOLE

It is interesting to discuss the behavior of the dipole antenna as its length varies. We can take some arbitrary length, assign a sinusoidal

[10] Above these lengths, the sinusoidal current assumption may begin to fail quite badly.

current distribution, and analyze the fields to obtain radiation resistance and perhaps driving-point reactance. Such an analysis would be very similar to that just completed except that it would be somewhat more involved mathematically. Rather than try to follow through such a procedure again, however, it is probably more worthwhile to spend the time and space available in attempting to develop an intuition about the practical nature of such antennas.

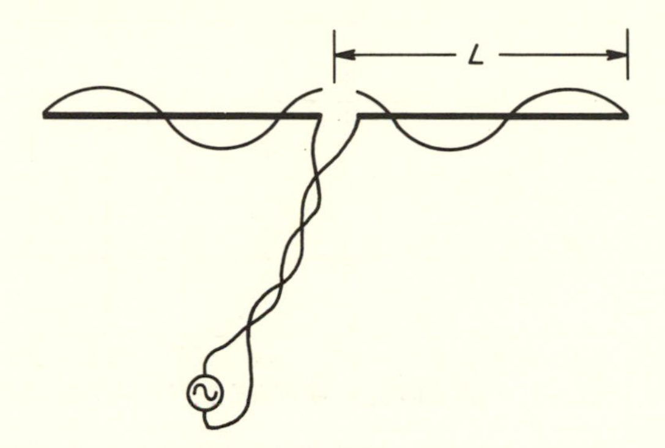

Fig. 13-7. Dipole antenna two and a fraction half-wavelengths on a leg.

Consider the dipole antenna illustrated in Fig. 13-7. This antenna is several half-wavelengths long on each leg plus perhaps another fraction of a half-wavelength. It is still geometrically symmetrical to the driving point. A sinusoidal current distribution is assumed as shown, and of course it must start with zero current at the ends. A little thought would lead to the intuitive (though essentially correct) conclusion that the driving-point impedance may be high or low, depending on whether the driving point is located at a current minimum or maximum, respectively. Indeed, the radiation resistance does follow such a general trend. Figure 13-8 is a plot of radiation resistance referred to a current maximum (this term will be discussed shortly) of a center-fed dipole antenna as the length is varied. Note that resistance maxima occur approximately (although not exactly) at numbers of quarter-wavelengths on a leg corresponding to current minima at the driving point. Furthermore, the resistance minima coincide roughly with the odd numbers of quarter-wavelengths on a leg corresponding to current maxima at the driving point. Note also that there is a gradual tendency for the over-all radiation resistance characteristics to increase as the length of a leg increases greatly.

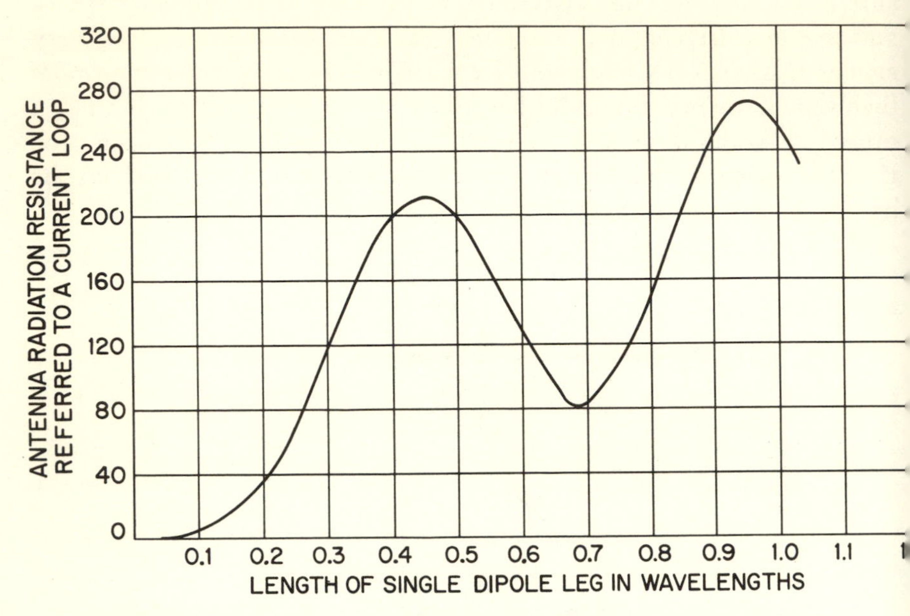

Fig. 13–8. Dipole radiation resistance referred to a current loop.

It is important to realize that the curve of Fig. 13–8 does not represent the radiation resistance at the driving point of the center-fed dipole. It will be recalled that in the half-wave dipole analysis, the final integration over the sphere surrounding the antenna had a factor of the current squared. This was the coefficient of the sinusoidal current distribution of Eq. (13–4) and represented the maximum value the current could have. It happened in the half-wave dipole that this current maximum occurred at the driving point. In the general dipole, this is not necessarily the case. Certainly the driving-point current in the antenna of Fig. 13–7 is not equivalent to the maximum current. Nevertheless, an analysis similar to that of the half-wave dipole would have led us to a power expression of the form

$$\text{Power radiated} = \frac{|I|^2 R(\lambda)}{2} \tag{13–54}$$

The $R(\lambda)$ obviously has dimensions of resistance, although it is a function of wavelength. I, on the other hand, is the current at a current maximum point just as it was for the half-wave dipole. $R(\lambda)$, then, is known as the *radiation resistance referred to a current loop* and is precisely the function given in Fig. 13–8. Unfortunately we may not want to drive the antenna at a current loop except in special cases. The determination of radiation resistance is easily resolved, however, by recalling the fundamental nature of the current distribution. Let us assign coordinates to the antenna as shown in Fig. 13-7. The total length of one leg of the dipole will be L. The origin will be taken at the center of the dipole, and it will be assumed that the dipole is situated along the x-axis. The current distribution in the antenna is then written as

$$I_x = I_{\text{loop}} \sin \beta(L - x) \tag{13-55}$$

for the right half-leg and as

$$I_x = I_{\text{loop}} \sin \beta(L + x) \tag{13-56}$$

for the left half-leg. In particular, at the origin, the current is

$$I_{\text{driving point}} = I_{\text{loop}} \sin \beta L \tag{13-57}$$

We now see that the radiation resistance referred to the driving point is related to the radiation resistance referred to a current loop as follows: The power radiated in terms of the driving-point current is

$$\text{Power radiated} = \frac{I^2_{\text{driving point (rms)}}}{\sin^2 \beta L} R(\lambda) \tag{13-58}$$

and consequently the radiation resistance referred to the driving point is

$$\begin{pmatrix} \text{Radiation resistance} \\ \text{referred to driving point} \end{pmatrix} = \frac{R(\lambda)}{\sin^2 \beta L} \tag{13-59}$$

For example, suppose we have a dipole of $3\lambda/8$ on each leg. From Fig. 13-8, we find $R(\lambda)$ to be 184 ohms. The value of the $\sin \beta L$ for L equal to $3\lambda/8$ is 0.707. The radiation resistance referred to the driving point is then found to be 368 ohms.

The next logical step is to plot a curve of dipole radiation resistance referred to the driving point versus length of a single dipole leg in wavelengths. This is done in Fig. 13-9. If we want to design an antenna system for which the elements are not a quarter-wavelength, however, we shall find that certain of the reactance terms will not cancel out. Except for elements which are multiples of quarter-wavelengths, the reactance is dependent upon antenna thickness and is usually computed with the induced EMF method on the assumption that all current flows at the antenna surface. Figure 13-10 gives cylindrical dipole-antenna reactances referred to the driving point for various lengths of radiating elements and various radii of the antenna cylinder.

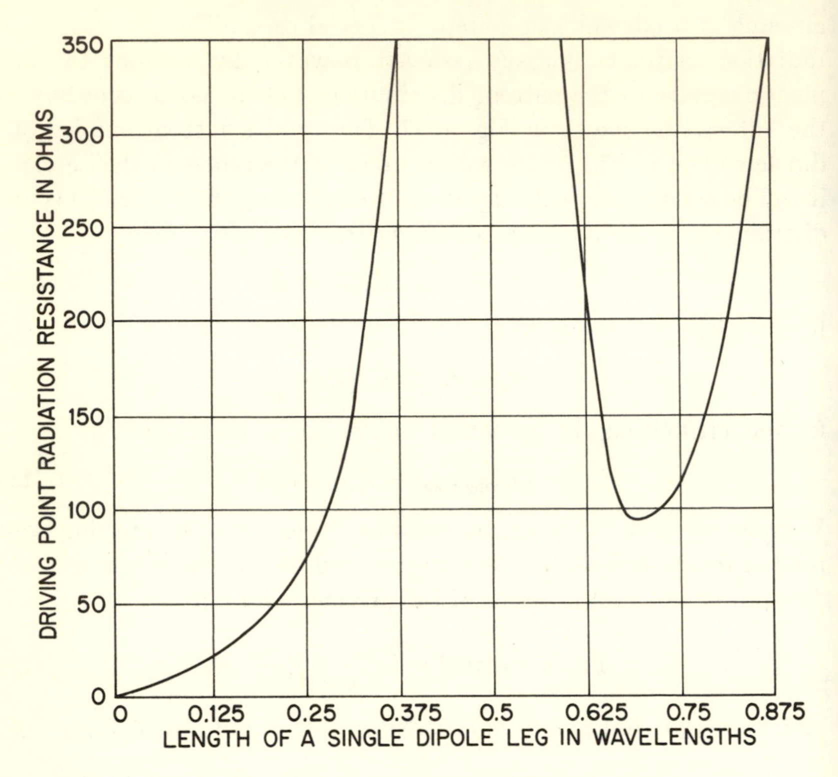

Fig. 13–9. Dipole radiation resistance referred to driving point.

13–6. PHYSICAL NATURE OF RADIATION FROM A DIPOLE

Let us briefly return to the physical meaning of the preceding analyses. In every case, we started with a sinusoidal current distribution assumption. This type of current distribution leads to a small but finite tangential electric field at the surface of the distribution. When we apply the results of the analyses to a wire antenna and compute the Poynting vector at the antenna surface for a sinusoidal current distribution, we would find a finite radial component. This would indicate that energy was emanating from the wire itself. Such a situation cannot exist, since the energy must always be in the field external to the wire. Actually the energy "flowers" outward from the driving point, being partially guided away from the driving point by the antenna wire itself. This is illustrated in Fig. 13–11. Note that no radiation occurs off the ends of the antenna.

Another way to view this phenomenon is to consider the outward radial component of the Poynting vector at the surface of the current

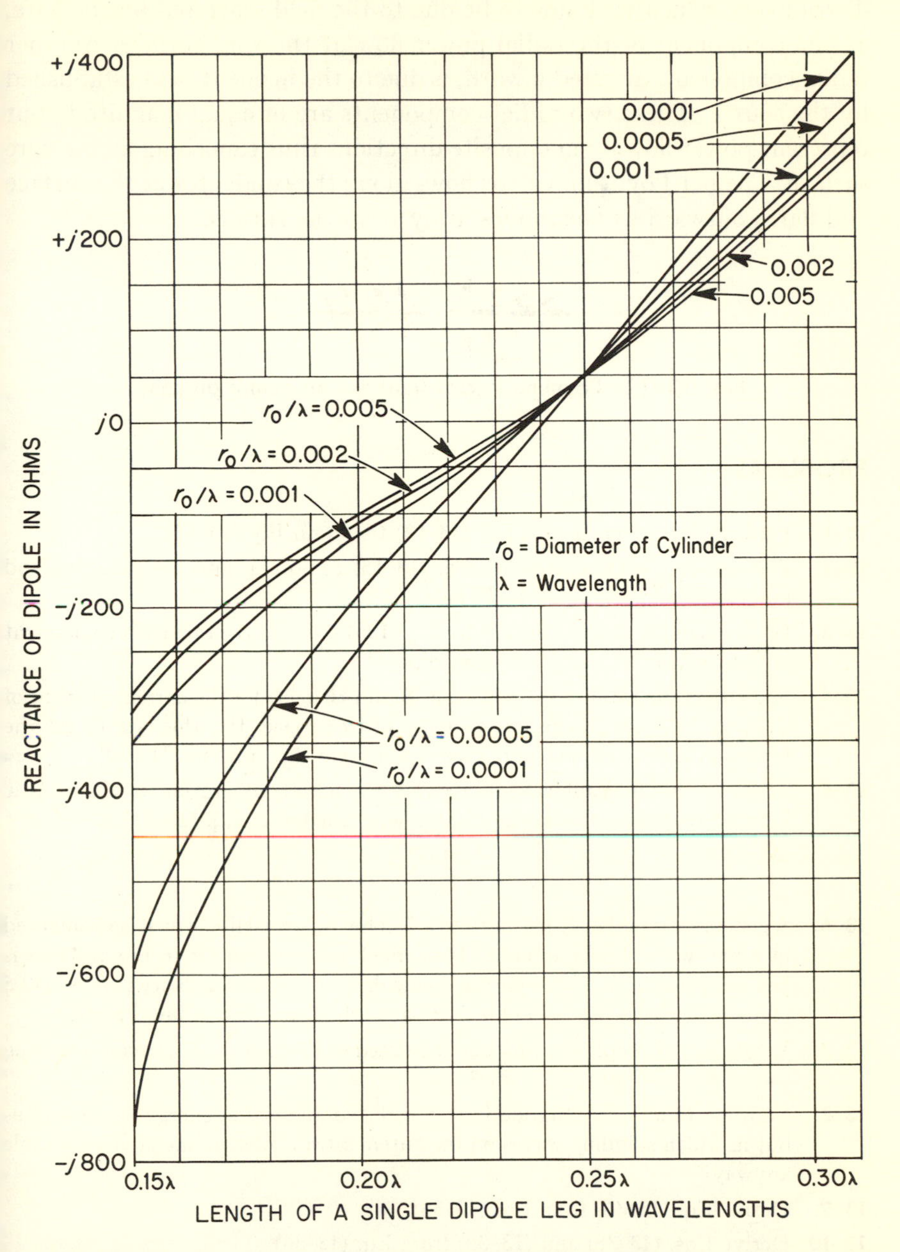

Fig. 13-10. Reactance of a cylindrical dipole referred to the driving point.

distribution, which we know to be due to the field scattered by the wire, as one component of the radial power flow at the wire surface. Another radial component, directed inward, is due to the incident field established by the source. These two radial components are of equal magnitude but represent power flowing in opposite directions thus cancelling at the wire surface. The net Poynting vector flows along the conductor at the surface and flares outward as one moves away from the surface.

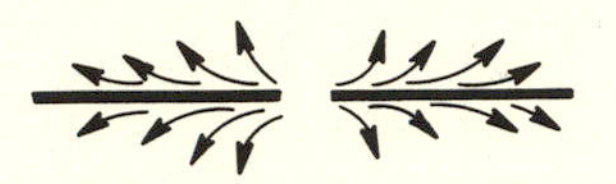

Fig. 13–11. Poynting vector field for an actual antenna.

PROBLEMS

13–1. Perform the integration of Eq. (13–12) to obtain Eq. (13–13).

13–2. Perform the curl operation on Eq. (13–15) to obtain both the radiation and induction components of the **H** field.

13–3. Go through the steps indicated by Eq. (13–17) to get the radiation component of the electric field of a half-wave dipole.

13–4. Show that the ratio of the radiation **E** and **H** fields for the current element of Chapter 12 is the intrinsic impedance of free space. Use this fact to get the radiation **E** field for the half-wave dipole from the radiation **H** field.

13–5. Show by either graphical methods or by actual integration of the elliptic integral that the value of the total power radiated in Eq. (13–21) is

$$\frac{1.218 I_{\rm rms}^2}{2\pi}$$

13–6. A generator is perfectly matched to a 73-ohm balanced line which is terminated in a half-wave dipole antenna. How much energy transfer to the antenna is lost by virtue of the fact that the dipole driving-point impedance is $73 + j42.5$ ohms rather than $73 + j0$ ohms? What is the transmission line VSWR?

13–7. What kind of simple network could be used in Problem 13–6 to avoid standing waves on the transmission line?

13–8. Can you think of a situation in which the simple loss of energy is not serious but in which standing waves on the antenna transmission line are undesirable anyway?

13–9. Derive Eq. (13–36) from Eq. (13–34).

13–10. Derive Eqs. (13–38) and (13–39) from Eq. (13–36).

13–11. Derive Eq. (13–44) from Eqs. (13–43) and (13–29).

13–12. A $3\lambda/2$ dipole ($3\lambda/4$ on each leg) is to be driven with a balanced transmission line. Ignoring reactive components, what is the radiation resistance if the antenna is to be center-fed?

14

General Consideration of Antenna Systems

14–1. INTRODUCTION

In Chapters 12 and 13, we were primarily concerned with the mathematical analysis of electromagnetic radiation and the concept of radiation resistance. The analysis was directed primarily at the dipole antenna and it may seem that this type of antenna was belabored. There are two reasons for giving such extensive consideration to the dipole. First, this is the antenna that has been studied the most completely, and therefore we know more about it than any other type of antenna. This, in turn, means that the correlation between mathematical analysis, physical concepts, and experimental results is more direct than it is in other types of antenna systems. A second and equally important reason is that once we have gone through the dipole in detail, we can extend the theory easily and rapidly to many other types of antenna systems, with a minimum of the complex mathematics that seems inevitably to accompany the study of radiating systems. The purpose of this chapter is to make such extensions so that the reader will have a reasonable idea of the nature of directional and other types of antennas, the relationship between receiving and transmitting antennas, the nature of the networks required to feed antenna systems, and standard methods and procedures used in measuring and describing such devices.

14–2. THE MONOPOLE ANTENNA AND THE METHOD OF IMAGES

In the discussion of the induced EMF method in Chapter 13, it was necessary to find the complete fields of the half-wave dipole antenna. They are given in Eqs. (13–36), (13–38), and (13–39) in cylindrical components and are repeated here for easy reference. Figure 13–6 shows the geometrical relationships among the variables.

$$H_\phi = -\frac{I}{4\pi j}\left[\frac{e^{-j\beta r_1}}{\rho} + \frac{e^{-j\beta r_2}}{\rho}\right] \tag{14–1}$$

$$E_\rho = \frac{I\beta}{4j\pi\rho\omega\epsilon_0}\left[\frac{L-z}{r_1}e^{-j\beta r_1} - \frac{L+z}{r_2}e^{-j\beta r_2}\right] \tag{14–2}$$

$$E_z = \frac{I\beta}{4j\pi\omega\epsilon_0}\left[\frac{e^{-j\beta r_1}}{r_1} + \frac{e^{-j\beta r_2}}{r_2}\right] \tag{14–3}$$

Now let us do a seemingly arbitrary thing to the dipole of Fig. 13–6. We are going to insert a perfectly conducting, thin infinite plane between the two dipole elements, perpendicular to their common axis. The antenna elements are insulated from the plane. The result is as shown in Fig. 14–1.

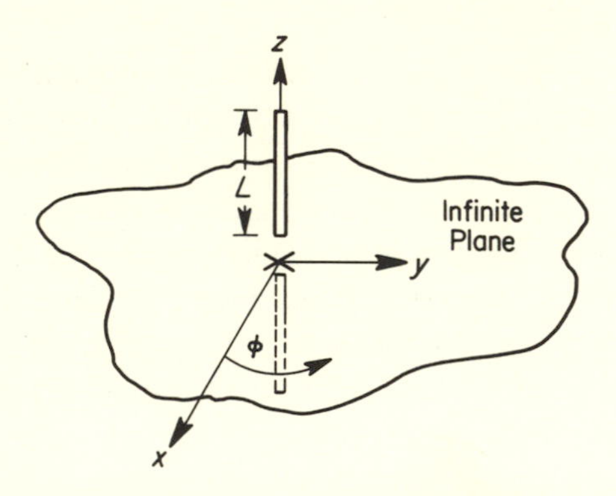

Fig. 14–1. Infinite conducting plane inserted between two dipole elements.

The first question that arises is, of course: In what way have we disturbed the electric and magnetic fields of the dipole? The magnetic intensity field of the dipole is given by Eq. (14–1). We note that there is a ϕ component of this field only. At the surface of the infinitely conducting plane, the magnetic intensity field is therefore entirely tangential. We would conclude that since there is no component of **H** perpendicular to the location of the conducting plane, the magnetic field would not be disturbed in the slightest manner unless the insertion of the plane in the electric field created a new and undesirable component of the **H** field. Now let us examine the electric field. As far as E_ρ is concerned [see Eq. (14-2)] everywhere on the conducting plane, z equals zero and r_1 equals r_2. The result is that everywhere on the conducting plane, E_ρ is zero whether the plane is there or not. The only possibility that this would not be true would be if some great disturbance was caused in E_z by the conducting plane. This could not be the case, however, because E_z is everywhere perpendicular to the conducting plane.

We thus come to a very interesting conclusion. The perpendicular insertion of a thin, infinitely conducting ground plane between two halves of a half-wave dipole antenna has no effect whatsoever on the fields created by that antenna for a given set of driving conditions. Had we solved for the general fields of a dipole of arbitrary length, we would have found the same condition to exist. Momentarily we may wonder just what became of the fields above the plane, formerly contributed by that portion of the antenna below the plane. We certainly would no longer expect such a contribution from the lower half of the antenna, since no electromagnetic wave can penetrate a perfectly conducting plane. The answer to this question is fairly obvious from examination of Fig. 14-2.

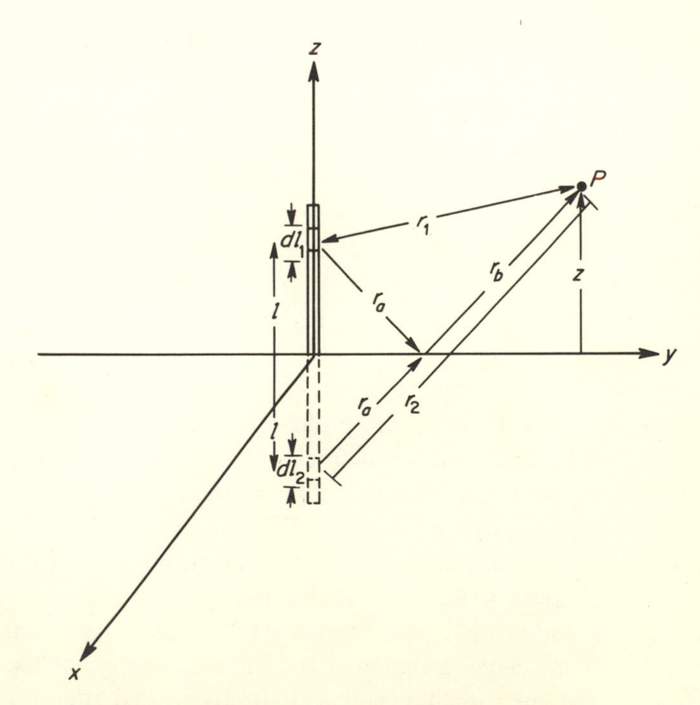

Fig. 14-2. Reflection effect of image plane.

At point P above the plane consider the fields due to two current elements, dl_1 and dl_2. These two current elements are "images" of each other. That is, they are equal in distance from the conducting plane and have currents equal in magnitude. The field components from dl_2 no longer reach the point P if the conducting plane is present, but they are exactly replaced by a component from dl_1 that cannot penetrate the plane and is therefore reflected to point P. The geometry is such that $r_a + r_b$ equals r_2, and therefore appropriate time phases are preserved.

Let us now make an obvious extension of the argument. We shall postulate a single straight antenna, one quarter-wavelength long and mounted perpendicularly to an infinitely conducting plane. We can immediately take stock of the geometric similarities and differences between this situation and a dipole twice as long with no conducting plane. When we do this, it is first apparent that the fields due to the monopole quarter-wave antenna are still given by Eqs. (14–1), (14–2), and (14–3). Further, we know from symmetry that the driving-point impedance of the quarter-wave monopole at the base is

$$Z_{11} = \frac{73 + j42.5}{2} = 36.5 + j21.25 \tag{14–4}$$

If our monopole antenna had been of some length other than quarter-wave, we could find the radiation resistance referred to a current loop simply by examination of Fig. 13–8 for the dipole, except that we would divide the resistance value obtained by two.

A little later on, we shall consider the mutual effects of two or more antennas on each other. If we should solve for the effects of two monopoles on each other, we would also have solutions for the effects of two dipoles on each other, provided the monopoles were working against an infinitely conducting ground plane and the dipoles were oriented parallel in free space in the proper manner.

Formally, the procedure we have just outlined is known as the *method of images* and is, of course, similar in philosophy to the method of images used in static fields. The method of images is a very useful tool in the design of multielement directional antenna arrays. Its mathematical justification rests on a theorem of differential equations which, in effect, says that if a solution is obtained that satisfies (1) the required differential equation and (2) a prescribed set of boundary conditions, then that solution is unique and is inherently correct. It does not matter that it was not obtained by some clever solution of the physical system at hand. If a solution can be obtained for one physical situation, and then boundary conditions can be fitted for that solution to a new physical situation, the solution is perfectly correct for the new situation also. This means that the method of images can be used to convert a boundary-value problem such as an antenna radiating in the presence of a perfectly conducting plane into a free-space problem involving the original radiator and its image. In free space the principle of linear superposition applies and the solutions for the original radiator and its image can be superimposed. The fields in the free-space portion of the original problem are the same as the fields in the corresponding portion of the free-space (source plus image) problem.

The image relationships for sources perpendicular to or parallel with a perfectly conducting plane are summarized in Fig. 14–3. In Fig. 14–3, the fields are identical in the region $z \geq 0$.

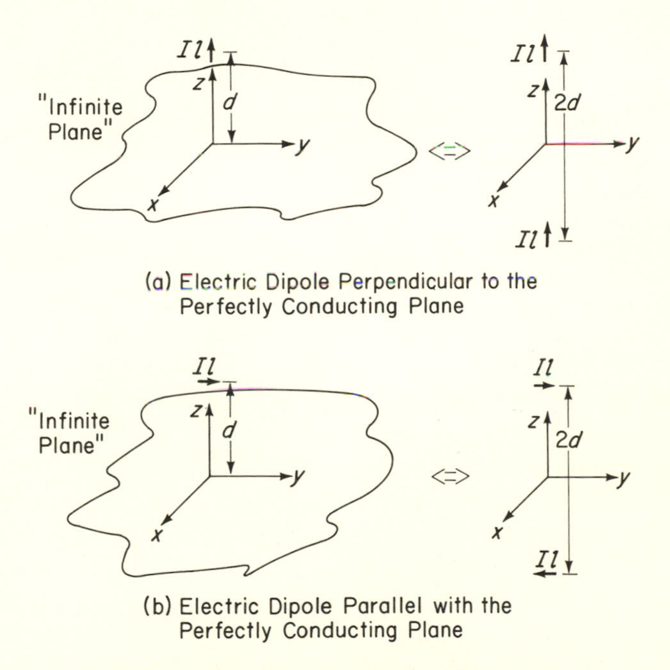

Fig. 14–3. Image relationships for electric dipoles.

14–3. DIRECTIONAL PATTERNS

In our discussion this far, we have not given consideration to the effectiveness of antennas in various directions. As a beginning, let us reexamine the nature of dipole radiation as a function of azimuth angle.

In particular, consider the half-wave dipole of Chapter 13. The Poynting vector for the far fields is written as

$$\begin{pmatrix}\text{Power radiated}\\ \text{in W/m}^2\end{pmatrix} = \frac{\eta|I|^2 \cos^2 [(\pi/2) \cos \theta]}{8\pi^2 r^2 \sin^2 \theta} \tag{14–5}$$

If one computes the Poynting vector as a function of θ, he finds that it is a maximum at 90 and 270 deg and zero at 0 and 180 deg. If one were to plot the directive characteristics as a function of θ, they would appear approximately as shown in Fig. 14–4. In this and subsequent antenna pattern

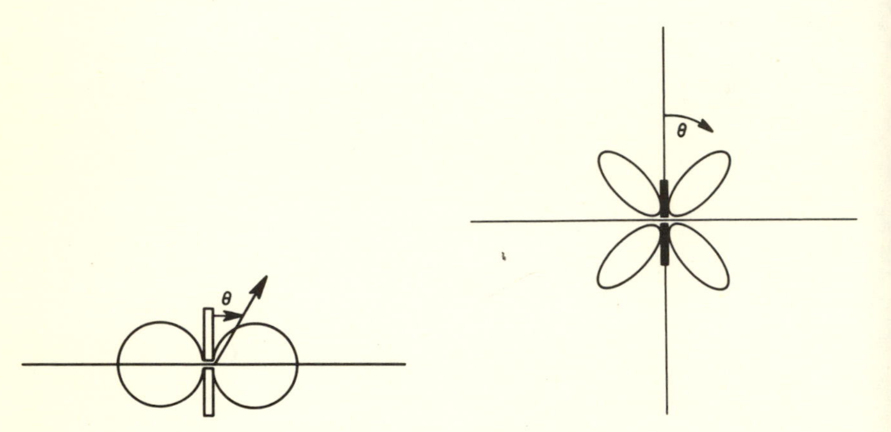

Fig. 14–4. Directional pattern of a half-wave dipole.

Fig. 14–5. Directional pattern of a dipole 2λ long.

figures, the distance from the origin to the pattern locus is proportional to the average Poynting vector in that direction. Thus it is seen that the dipole is most effective off the sides and least effective off the end. As a matter of interest, if the dipole had been extended so that the legs were each one wavelength long, the pattern would appear as shown in Fig. 14–5. In both cases the directions of major radiation are referred to as *lobes*. In some cases there are secondary radiation lobes, and these are often referred to as minor lobes.[1]

The vertical monopole obviously has an azimuth pattern that is circular regardless of its length. The angle θ in this case is measured from the vertical, and the pattern from this viewpoint is the same as for equivalent dipoles.

[1] A wide variety of antenna patterns can be found on antenna systems. For example, see R. E. Collin and F. J. Zucker, *Antenna Theory*, Parts I and II (New York, McGraw-Hill Book Co., Inc., 1969) or W. L. Weeks, *Antenna Engineering* (McGraw-Hill Book Co., Inc., 1968).

Let us suppose that we wish to build a directional antenna system with two driven elements. We would like to have maximum radiation in two directions, say, north and south, and minimum radiation in the other two directions, say, east and west. Two obvious possibilities immediately arise, as shown in Figs. 14–6(a) and 14–6(b). Figure 14–6(a) illustrates

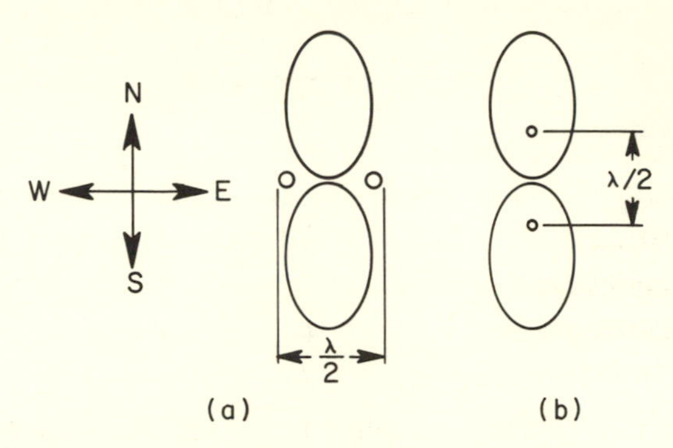

Fig. 14–6. (a) Two vertical elements placed one half-wavelength apart and oriented east and west. (b) Two vertical elements placed one half-wavelength apart and oriented north and south.

two vertical elements placed one half-wavelength apart and oriented east and west. These elements would have to be identical and have the same current distributions, which must be in phase with each other. Figure 14–6(b) illustrates two similar elements placed one half-wavelength apart and oriented north and south. These two elements would have to have similar current distributions but must be driven 180 degrees out of phase with each other. Many other possibilities exist, and little imagination is required to come up with them. We shall return to these two situations later.

14–4. THE RECIPROCITY LAW AND RECEIVING ANTENNAS

Let us imagine that we have a black box with two sets of terminals coming out of it. If we arbitrarily designate one set of terminals as port 1 and the other set as port 2, we can write the voltage and current equations:

$$V_1 = Z_{11}I_1 + Z_{12}I_2$$
$$V_2 = Z_{21}I_1 + Z_{22}I_2.$$

Let us apply a perfect voltage source V_1 at port 1 and a short circuit at port 2. The current flowing in port 2 is

$$I_2 = -\frac{Z_{21}}{Z_{22}Z_{11} - Z_{21}Z_{12}} V_1$$

If a perfect voltage source V_2 were applied to port 2 and a short circuit applied to port 1, the current flowing in port 1 would be

$$I_1 = -\frac{Z_{12}}{Z_{22}Z_{11} - Z_{21}Z_{12}} V_2$$

The usual statement of the law of reciprocity is that if $V_1 = V_2$ in the situations described above, then $I_2 = I_1$. This requires that $Z_{12} = Z_{21}$. Since this transfer impedance term is determined solely by what is in the box and is not dependent upon the excitation, we can write

$$\frac{V_2}{I_1} = \frac{V_1}{I_2} \tag{14–6}$$

where V_2 is the voltage source applied to port 2 which causes a current I_1 to flow in a load on port 1 and V_1 is the voltage source applied to port 1 which causes a current I_2 to flow in the load on port 2. The only restriction is that the generator impedance for excitation of port 1 be the same as the load impedance for excitation of port 2, and vice versa. Another possible relationship leading to Eq. (14–6) is that the generator impedances be the same in both cases of excitation as well as the load impedances being the same. A useful source-load combination which satisfies these conditions is current sources and open-circuited loads. We can then write Eq. (14–6) as

$$V_2\begin{pmatrix}\text{caused by}\\ \text{current at 1}\end{pmatrix} I_2\begin{pmatrix}\text{current}\\ \text{source}\\ \text{at 2}\end{pmatrix} = V_1\begin{pmatrix}\text{caused by}\\ \text{current at 2}\end{pmatrix} I_1\begin{pmatrix}\text{current}\\ \text{source}\\ \text{at 1}\end{pmatrix} \tag{14–7}$$

In particular, this is recognized as the situation which should exist when two current sources exist in free space. We can generalize Eq. (14–7) as volume integrals over the source distributions:

$$\iiint \mathbf{E}^{(1)} \cdot \mathbf{J}^{(2)}\, dv = \iiint \mathbf{E}^{(2)} \cdot \mathbf{J}^{(1)}\, dv \tag{14–8}$$

In Eq. (14–8) $\mathbf{J}^{(1)}$ and $\mathbf{J}^{(2)}$ are source current densities, $\mathbf{E}^{(1)}$ is the electric field intensity at current distribution (2) due to the current distribution (1), and $\mathbf{E}^{(2)}$ is the electric field intensity at current distribution (1) due to current distribution (2).

The law of reciprocity is useful in examining antenna characteristics. For example, we can use the law of reciprocity to prove that the directional pattern of an antenna is the same when the antenna is used to receive electromagnetic energy as it is when the antenna is used to transmit electromagnetic energy. To see that this is so, consider Fig. 14–7. In

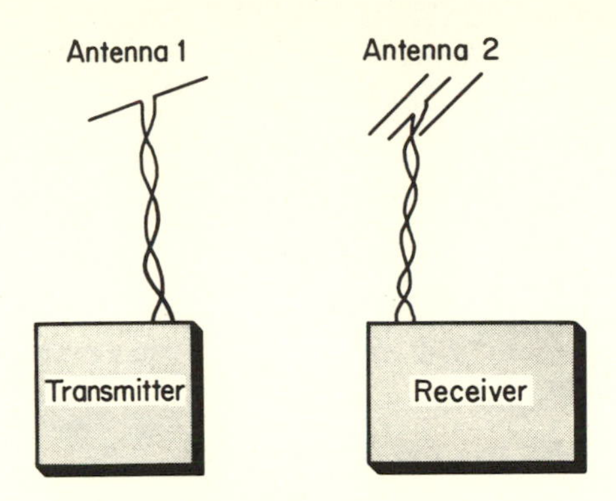

Fig. 14–7. Reciprocity example.

this figure, the transmitter supplies a current I to antenna 1 and the receiver (which is really a voltmeter connected to antenna 2) observes a signal V. The law of reciprocity states that if the transmitter is connected to antenna 2 and supplies a current I to it, and the receiver is connected to antenna 1, the voltmeter will again observe a signal V. Now return to the original situation with the transmitter connected to antenna 1 and the receiver connected to antenna 2. If we were to rotate antenna 1 about its axes (for example, change θ), the reading of the voltmeter on antenna 2 would be a measure of the transmitting pattern (the voltage pattern is proportional to the square root of the power pattern). Now, if the antennas were returned to their original orientations, the transmitter connected to antenna 2 and the receiver connected to antenna 1, reciprocity says that the voltmeter will read the same as it did in the original case (source connected to antenna 1 and receiver connected to antenna 2). In fact, as one rotates antenna 1 again (note that it is now the receiving antenna), for each orientation corresponding to the original situation, the voltmeter will read the same as it did in the original case. Again, the voltmeter reading is a measure of a directional pattern only this time it is the receiving pattern of antenna 1. Since the voltmeter readings for corresponding orientations are the same, one concludes that the directional pattern for a given antenna which is transmitting electromagnetic energy is the same as the directional pattern for that antenna when it is receiving electromagnetic energy.

The fact that the circuit models of a two-port such as the two-antenna situation we are considering are independent of the excitation leads us to conclude that the internal impedance of a given antenna receiving electromagnetic energy is equal to the driving point impedance of that antenna when it is radiating electromagnetic energy. To see that this is so, consider the two-port made up of two antennas such as those shown in Fig. 14–7. The voltage and current relationships are

$$V_1 = Z_{11}I_1 + Z_{12}I_2$$
$$V_2 = Z_{21}I_1 + Z_{22}I_2$$

If we consider antenna 1 as an example, we can start by asserting that its driving-point impedance, if it existed alone, would be Z_{11}. Now, suppose a current I_2 were made to flow in port 2. If an open circuit was the load at port 1, the circuit equations become

$$V_{1_{oc}} = Z_{12}I_2$$
$$V_2 = Z_{22}I_2$$

If, on the other hand, a short circuit were placed at port 1 we could write

$$0 = Z_{11}I_{1_{sc}} + Z_{12}I_2$$
$$V_2 = Z_{21}I_{1_{sc}} + Z_{22}I_2$$

Solving for the ratio of the open-circuit voltage to the short-circuit current at port 1 gives

$$\frac{V_{1_{oc}}}{I_{1_{sc}}} = -Z_{11}$$

The negative sign is a result of the current convention (positive current flows into the port). Thus the internal impedance of antenna 1 used as a receiving antenna is

$$Z_{\text{int}} = Z_{11}$$

This is equal to the driving-point impedance of the antenna radiating by itself.

14–5. COMPUTATION OF MUTUAL IMPEDANCE BETWEEN ANTENNA ELEMENTS

We know, of course, that two antennas will have an effect on each other. The question is: How shall we evaluate this effect? To begin with, let us attack the problem of the mutual impedance between two vertical monopoles working against a common ground plane. Then, through application of the method of images and circuit equivalents, we shall be able to extend the results directly to mutually driven dipoles and also to parasitic arrays.

Our main problem with the vertical monopoles is how to determine the driving-point impedances at the antenna bases when both elements are driven. We thus have the two-port network previously discussed. It can be assigned an equivalent tee configuration, as shown in Fig. 14-8.

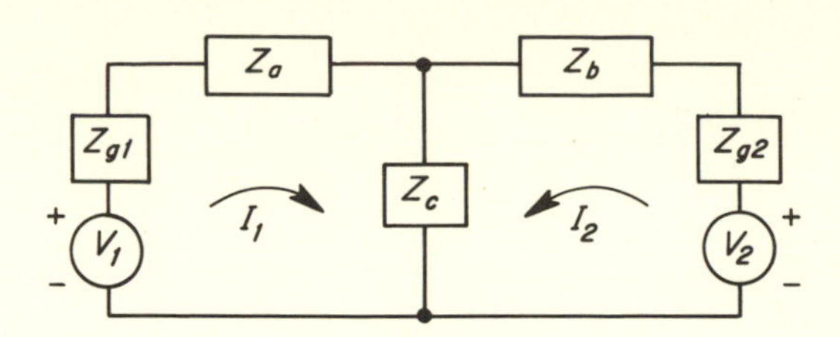

Fig. 14-8. Equivalent tee section of two antenna systems.

The next logical question is: What physical significance can we attach to Z_a, Z_b, and Z_c in Fig. 14-8? To begin to answer this question, let us write the equations of Fig. 14-8 in standard form:

$$\begin{aligned} V_1 &= (Z_{g_1} + Z_a + Z_c)\, I_1 + Z_c I_2 \\ V_2 &= Z_c I_1 + (Z_{g_2} + Z_b + Z_c)\, I_2 \end{aligned} \tag{14-9}$$

For simplicity, let $Z_{g_1} = Z_{g_2}$, and we shall hereafter write them both as Z_g. Now, if loop 2 could be opened in some way, the driving-point impedance seen by generator 1 (excluding Z_g) would be[2]

$$Z_{1(\mathrm{DP})}^{**} \text{ (loop 2 open)} = Z_a + Z_c \tag{14-10}$$

Similarly, if loop 1 were opened in some way, the driving-point impedance seen by generator 2 (excluding Z_g) would be

$$Z_{2(\mathrm{DP})}^{*} \text{ (loop 1 open)} = Z_b + Z_c \tag{14-11}$$

For all practical purposes, we may open loop 1 simply by disconnecting generator 1; or loop 2, by disconnecting generator 2. In other words, if we say we are opening loop 1, we mean antenna 1 is simply "floating"; similarly for loop 2. Under the condition that antenna 2 is floating, the driving-point impedance seen at antenna 1 is essentially the base impedance of this antenna, with nothing else in the vicinity. It is almost as

[2] For the purposes of this discussion only, one asterisk (*) in the superscript means loop 1 is open; two asterisks (**) mean loop 2 is open; one prime (′) in superscript means loop 1 is closed through Z_g but V_1 is shorted; and two primes (″) mean loop 2 is closed through Z_g but V_2 is shorted. The (*) and (**) impedances do not include Z_g, whereas the (′) and (″) impedances do include Z_g.

though antenna 2 were not there at all.[3] Presumably we know the self-impedance characteristics of the antennas, so we can write

$$Z^{**}_{1(\mathrm{DP})} = Z_1 \begin{pmatrix}\text{self-impedance characteristic}\\ \text{of antenna 1}\end{pmatrix} \approx Z_a + Z_c \qquad (14\text{–}12)$$

$$Z^{*}_{2(\mathrm{DP})} = Z_2 \begin{pmatrix}\text{self-impedance characteristic}\\ \text{of antenna 2}\end{pmatrix} \approx Z_b + Z_c \qquad (14\text{–}13)$$

It would seem that we now have a fair idea of the values of $Z_a + Z_c$ and $Z_b + Z_c$. Next we need to find either Z_a, Z_b, or Z_c alone so that all three may be evaluated. Suppose we open loop 2 at generator 2 and measure the voltage across the opened terminals. We shall call this voltage V_{21}. It is presumed, of course, that generator 2 is inoperative during this procedure. We shall then have

$$V_{21} = Z_c I^{**}_1(0) \qquad (14\text{–}14)$$

We would know $I^{**}_1(0)$, of course, because it is the base current of antenna 1 with loop 2 open. Therefore

$$Z_c = \frac{V_{21}}{I^{**}_1(0)} \qquad (14\text{–}15)$$

Similarly, if we had opened loop 1 and driven loop 2, we would have found

$$Z_c = \frac{V_{12}}{I^{*}_2(0)} \qquad (14\text{–}16)$$

where $I^{*}_2(0)$ is the base current of antenna 2 with loop 1 open.

We now wish to make use of the reciprocity rule, so let us postulate two situations.

Situation 1. Antenna 1 is connected to ground through Z_g. This is equivalent to shorting out V_1 in Fig. 14–8. The driving-point impedance seen at generator 2 is

$$Z'_{2(\mathrm{DP})} = Z_g + Z_b + \frac{Z_c(Z_a + Z_g)}{Z_a + Z_c + Z_g} \qquad (14\text{–}17)$$

Also under these conditions, if the voltage of generator 2 is V_2, the current at the base of antenna 2 is $I'_2(0)$ and

$$V_2 = I'_2(0) Z'_{2(\mathrm{DP})} \qquad (14\text{–}18)$$

[3] It will be noted that the statement is: "It is *almost* as though antenna 2 were not there at all." In other words, this is an approximation. If antenna 2 is very close to antenna 1, its proximity effect cannot be ignored. Also, if antenna 2 is a multiple of quarter-wavelengths (half-wavelengths for dipoles), the interaction is somewhat more pronounced, and the situation is a little more "almost" than otherwise. Nevertheless, these are working approximations and are used because they give reasonable results with reasonable effort. To attempt to work directional antenna problems from a boundary-value problem viewpoint leads to such complication that it is essentially useless at the current state of the art.

Further, this current in antenna 2 is a function of the position on the antenna. If it is assumed to vary sinusoidally, we can conveniently write the current in antenna 2 with loop 1 closed (and $V_1 = 0$) as $I_2'(l)$. Its value at the base is the $I_2'(0)$ of Eq. (14-18).

SITUATION 2. Generator V_1 is reconnected and generator V_2 is disconnected, but antenna 2 is grounded through Z_g. Under these conditions (assuming both antennas to be filamentary), a differential voltage will be induced in an element dl of antenna 2 by the electric field in the z direction of antenna 1. If this z component of electric field of antenna 1 adjacent to antenna 2 is symbolized as $E_{21}''(l)$,[4] then the differential voltage in question will be $E_{21}''(l)\, dl$. The current produced at the base of antenna 2 by this differential voltage is $dI_2''(0)$.

Now let us compare situations 1 and 2. The reciprocity rule allows us to write

$$\frac{V_2}{I_2'(l)} = \frac{E_{21}''(l)\, dl}{dI_2''(0)} \tag{14-19}$$

Substituting Eq. (14-18) into Eq. (14-19) gives

$$\frac{I_2'(0)Z_{2(\mathrm{DP})}'}{I_2'(l)} = \frac{E_{21}''(l)\, dl}{dI_2''(0)} \tag{14-20}$$

Rearranging gives

$$dI_2''(0) = \frac{1}{I_2'(0)Z_{2(\mathrm{DP})}'} E_{21}''(l)I_2'(l)\, dl \tag{14-21}$$

Let the length of antenna 2 be L_2. We shall integrate along the total length L_2 of antenna 2. The result is

$$I_2''(0) = \frac{1}{I_2'(0)Z_{2(\mathrm{DP})}'} \int_0^{L_2} E_{21}''(l)I_2'(l)\, dl \tag{14-22}$$

Now let us recall the situation under which Eqs. (14-14) and (14-15) were derived. In this case, loop 1 is driven normally but loop 2 is open-circuited. A Thévenin equivalent of this situation is illustrated in Fig. 14-9. Obviously, from Fig. 14-9 we can write

$$V_{21} = -I_2''(0)Z_{2(\mathrm{DP})}' \tag{14-23}$$

Substituting Eq. (14-22) into Eq. (14-23) gives

$$V_{21} = -\frac{1}{I_2'(0)} \int_0^{L_2} E_{21}''(l)I_2'(l)\, dl \tag{14-24}$$

[4] Note that at the surface of antenna 2, $z = l$.

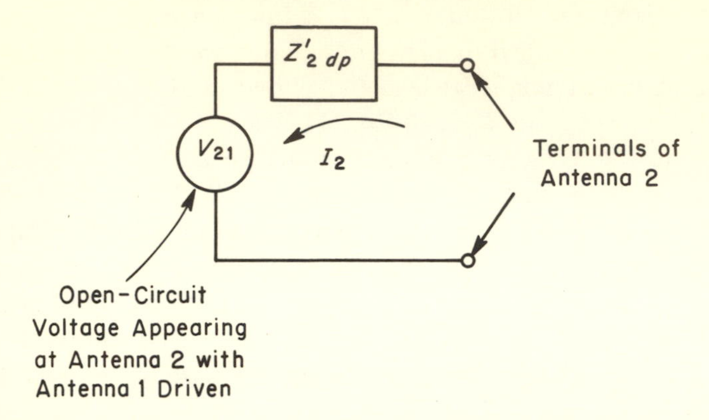

Fig. 14–9. Thévenin equivalent for loop 1 driven and loop 2 open.

Finally, let us recall Eq. (14–15). Substituting this into Eq. (14–24) allows us to write an expression directly for Z_c as follows:

$$Z_c = -\frac{1}{I_1^{**}(0)I_2'(0)} \int_0^{L_2} E_{21}''(l)I_2'(l)\,dl \tag{14–25}$$

To get Z_c, which has been our primary objective, simply requires that we evaluate the integral of Eq. (14–25) and multiply it by its coefficient. We are not quite out of the woods, however. Let us examine each mathematical function to see how we can obtain it.

1. $I_1^{**}(0)$ is the base current of antenna 1 when antenna 2 is floating. There is no problem here.
2. $I_2'(0)$ is the base current of antenna 2 when antenna 1 is connected to ground through Z_g. We do not really know what this function is.
3. $E_{21}''(l)$ is the z electric field at antenna 2 due to antenna 1 when antenna 2 is connected to ground through Z_g. Neither do we really know what this function is.
4. $I_2'(l)$ is the current in antenna 2 as a function of antenna 1 when it is connected to ground through Z_g. If we knew $I_2'(0)$ we would be fairly well off because we would make a sinusoidal current-distribution assumption in the standard manner. Unfortunately we do not know $I_2'(0)$.

All these dilemmas can be resolved by one rather simple assumption. Let us assume that the Z_g's are, for the moment, very high. Under these conditions, item (2) is the base current of antenna 2 when antenna 1 is floating, and therefore it is as easy to obtain as item (1). Once item (2) is obtained, item (4) follows immediately. Item (3) is simply the z electric field due to antenna 1 along antenna 2.[5] In other words, under the condition that Z_g is extremely large,

$$I_2'(0) \approx I_2^*(0) \tag{14-26}$$

and

$$E_{21}''(l) \approx E_{21}^{**}(l) \tag{14-27}$$

For this situation, we can then write

$$Z_c = -\frac{1}{I_1^{**}(0)I_2^*(0)} \int_0^{L_2} E_{21}^{**}(l)I_2^*(l)\,dl \tag{14-28}$$

Once we have found the mutual impedance for the two antennas in question, we can take advantage of the fact that this mutual impedance is really independent of the generator internal impedances, and therefore it can be used for any situation of generators and generator impedances.

It is very obvious that Z_c of Eqs. (14-25) and (14-28) is really independent of driving voltage or current amplitudes. $I_1^{**}(0)$ and $E_{21}''(l)$ would have the same constant multiplier representing the driving-current amplitude at antenna 1. Similarly, $I_2'(0)$ and $I_2'(l)$ would have the same multiplier [namely, $I_2'(0)$] representing the driving-current amplitude at antenna 2. Perhaps it is more meaningful simply to rewrite Eq. (14-25) [or Eq. (14-28)] as

$$Z_c = -\int_0^{L_2} f(l)g(l)\,dl \tag{14-29}$$

In Eq. (14-29), $f(l)$ represents the function inside the integral, multiplied by the driving-current amplitude of antenna 1; and $g(l)$ represents the function inside the integral, multiplied by the driving-current amplitude of antenna 2. The expression $f(l)$ is determined by the physical geometry of antenna 1 alone, and $g(l)$ is determined by the physical geometry of antenna 2 alone.

To see how this happens, let us compute the mutual impedance of two quarter-wave monopoles spaced one half-wavelength apart. The physical situation is illustrated in Fig. 14-10. Since the monopoles are a quarter-wavelength high, we can write

$$I_2^*(l) = I_2^*(0)\cos\beta l \tag{14-30}$$

From Eq. (14-3) we can write

$$E_{21}^{**}(l) = \frac{I_1^{**}(0)\beta}{4j\pi\omega\epsilon_0}\left(\frac{e^{-j\beta r_1}}{r_1} + \frac{e^{-j\beta r_2}}{r_2}\right) \tag{14-31}$$

[5] The measured z electric field at the surface of antenna 2 is, of course, zero. The field we are talking about, $E_{21}(l)$, is the component due to the current in antenna 1. The total z field, which must add up to zero at the antenna surface, is composed of $E_{21}(l)$ plus the self-field due to current in antenna 2.

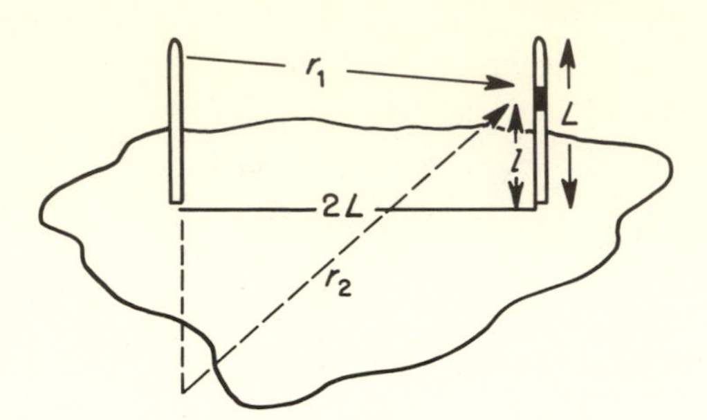

Fig. 14–10. Two quarter-wavelength monopoles spaced one half-wavelength apart.

Substituting Eqs. (14–30) and (14–31) into Eq. (14–28) and simplifying finally gives, for this particular situation,

$$Z_c = -\frac{1}{j4\pi v\epsilon_0}\int_0^L \left[\frac{e^{-j\beta r_1}}{r_1} + \frac{e^{-j\beta r_2}}{r_2}\right] \cos \beta l \, dl \qquad (14\text{–}32)$$

where

$$r_1 = \sqrt{4L^2 + (L - l)^2} \qquad (14\text{–}33)$$

and

$$r_2 = \sqrt{4L^2 + (L + l)^2} \qquad (14\text{–}34)$$

Equation (14–32) can be further simplified and broken into a real and imaginary integral as follows:

$$Z = +30 \int_0^L \left[\frac{\sin \beta r_1}{r_1} + \frac{\sin \beta r_2}{r_2}\right] \cos \beta l \, dl + j30 \int_0^L \left[\frac{\cos \beta r_1}{r_1} + \frac{\cos \beta r_2}{r_2}\right] \cos \beta l \, dl \qquad (14\text{–}35)$$

The first term of Eq. (14–35) represents the resistive component of Z_c, and the second term represents the reactive component. It is possible to find the value of Z_c by direct integration. The simplest way for this special case, however, is probably by graphical integration. The integrand functions of both integrals of Eq. (14–35) are plotted for illustrative purposes in Fig. 14–11. A planimeter evaluation of the curves of Fig. 14–11 leads to the conclusion that

$$Z_c = -6.2 - j14.9 \qquad (14\text{–}36)$$

It is worth while to give curves for some of the mutual impedances in other situations. Figure 14–12 is a family of curves representing the mutual resistances between a quarter-wave monopole (operating vertically above a ground plane) and a monopole parallel to it (also operating

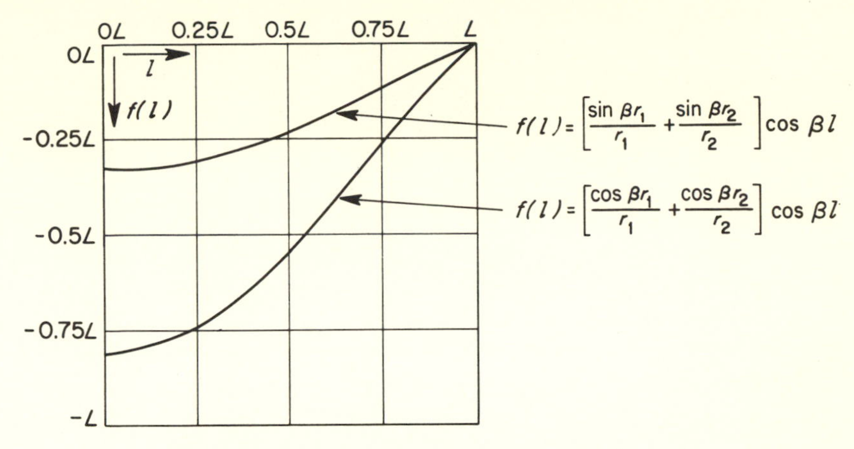

Fig. 14–11. Integrand functions for Eq. (14–35).

vertically above the ground plane) but of varying height. Since the mutual resistance is a function of the antenna spacing, a different curve is necessary for each spacing. Figure 14–13 is a family of curves representing the mutual reactances between a quarter-wave monopole and another monopole of varying height and spacing. It will be noted that the family of reactance curves exhibit a pseudoperiodicity as a function of spacing. This is because both the mutual resistance and reactance exhibit a periodic change of sign roughly every half-wavelength, although the points of zero mutual resistance or reactance do not necessarily come on exact half-wavelength values of spacing. Thus, had we given mutual resistance data for a wider variety of spacings, a similar periodic situation would have been evident. As an illustration of this, Fig. 14–14 is a plot of the mutual resistances and reactances between two thin quarter-wave monopoles as a function of spacing. The pseudoperiodicity characteristic is clearly evident.

The fact that mutual resistance can be a negative number, which is sometimes disturbing to a student, means that in our equivalent circuit for the antenna system, a negative resistance appears, and the solution of an antenna problem by simply building its electrical circuit analog may not be too practical. Those persons familiar with four-terminal network theory will recall that negative mutual impedances often occur in deriving an equivalent circuit to avoid a transformer. This situation is analogous. For any given antenna system, if the equivalent circuit is analyzed completely, one finds to his relief that the system really is conservative and totally passive even though a negative mutual resistance does

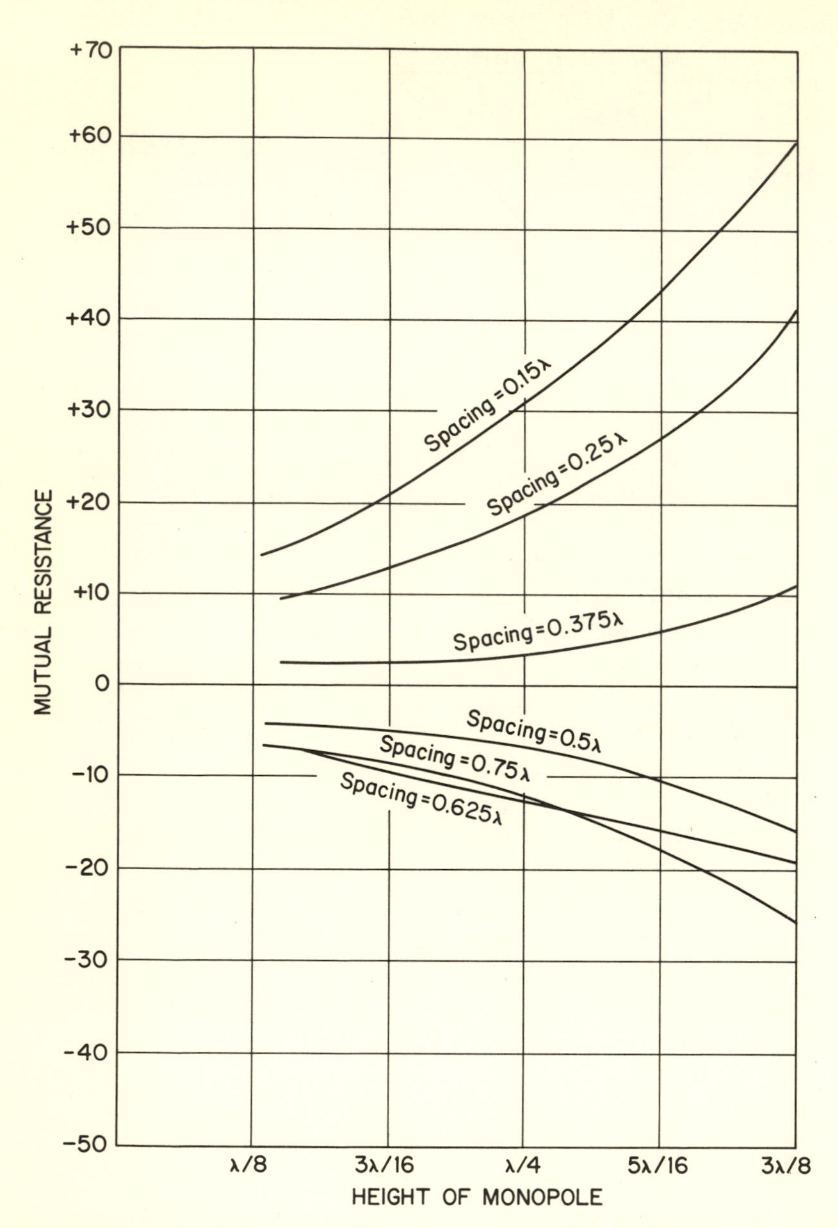

Fig. 14–12. Mutual resistances between a quarter-wavelength monopole and other-length monopoles for various spacings.

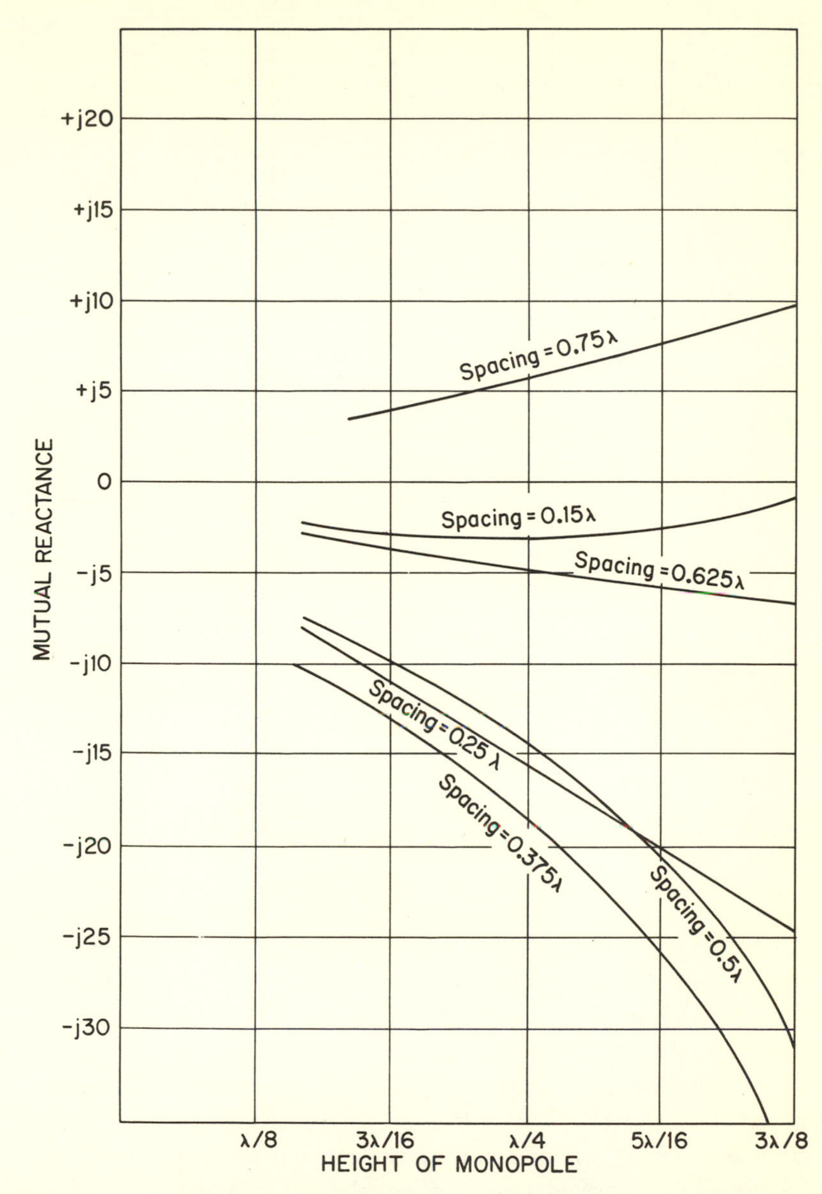

Fig. 14–13. Mutual reactances between a quarter-wavelength monopole and other-length monopoles for various spacings.

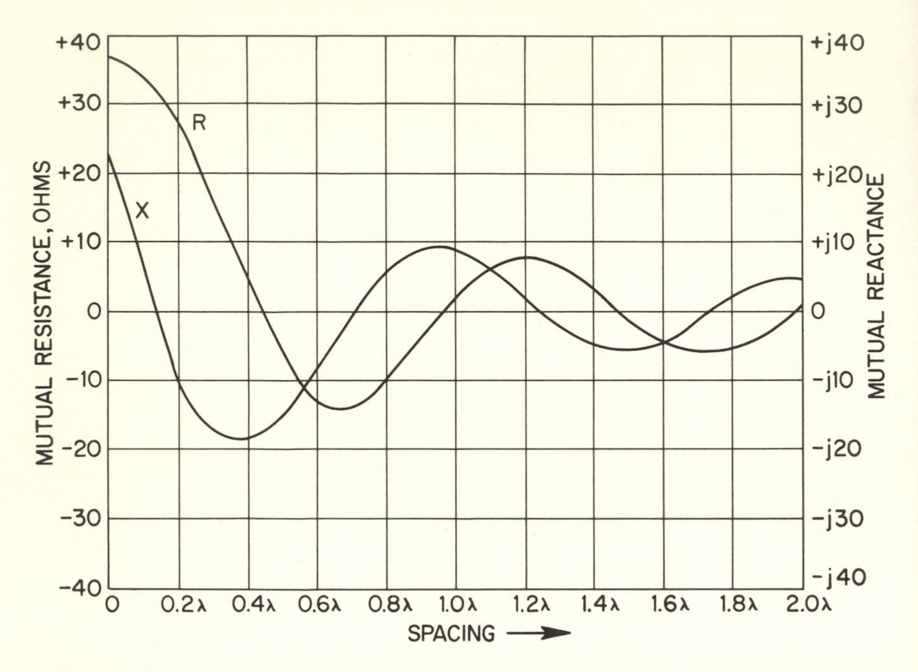

Fig. 14–14. Mutual resistance and reactance between two quarter-wave monopoles as a function of spacing.

occasionally occur. It should be remembered that in all work with analogies, the analogy is really a mathematical device used as a mental crutch and does not represent a real physical transformation.

It was found at the beginning of this chapter that the curves of radiation resistance (discussed in Chapter 13) for a dipole as a function of length hold equally well for monopoles through application of the method of images. In the same manner, Figs. 14–12, 14–13, and 14–14 give data that are equally valid for dipoles in free space. One merely has to double the resistance or reactance for any given spacing. For example, from Fig. 14–12 (or our own computation) we see that the mutual resistance between two quarter-wave monopoles spaced one half-wavelength is about -6.2 ohms. This means that the mutual resistance between two half-wave dipoles spaced one half-wavelength would be about -12.4 ohms.

The curves of Figs. 14–12, 14–13, and 14–14 were computed from Eq. (14–25), and we would say that as such they are impedances referred to the antenna bases or antenna driving point, depending on whether we are talking about monopoles or dipoles. In the literature, one often sees mutual impedances referred to current loops. This simply means that the expression for mutual impedance (Eq. 14–25) is multiplied by

$$\frac{I_1(0)I_2(0)}{I_{1m}I_{2m}} \tag{14–37}$$

Expression (14–37) is simply the ratio of the base currents (or driving-point currents in the case of the center-fed dipole) to the loop or maximum currents for both antennas. Since the $I(0)$ is usually expressed as a sinusoidal distribution with the corresponding I_m for a coefficient, the conversion is really a function of geometry and doesn't involve currents at all. Application of this function will be unnecessary for computations made from curves in this book but may be necessary if mutual impedance data are taken from other sources.

The curves in this chapter and in Chapter 13 can obviously be used to compute the characteristics of antennas with parasitic or undriven elements. The only difference between this situation and antenna systems in which all the elements are driven is that for the parasitic element, the driving generator is nonexistent and is considered to be shorted out in the equivalent circuit.

As an example of the manner in which simple directional antenna systems are designed, let us return to Fig. 14–6(a). It will be assumed that the monopoles are each one quarter-wavelength long. Our problem is really to find the driving generator characteristics necessary at each antenna base. In any directional antenna problem of this kind, the desired radiation characteristics are specified. Therefore the currents are usually known in both relative magnitude and phase. To get the radiation pattern of Fig. 14–6(a), the currents in both antennas must be of equal amplitude and in time phase. If currents are as assumed in Fig. 14–8, then we can write

$$\begin{pmatrix}\text{Voltage at terminals}\\ \text{of antenna 1}\end{pmatrix} = (V_1 - Z_{g_1}I_1) = I_1(Z_a + Z_c) + I_2Z_c \tag{14–38}$$

$$\begin{pmatrix}\text{Voltage at terminals}\\ \text{of antenna 2}\end{pmatrix} = (V_2 - Z_{g_2}I_2) = I_1Z_c + I_2(Z_b + Z_c) \tag{14–39}$$

Now, we know $(Z_a + Z_c)$ to be the self-impedance of antenna 1; $(Z_b + Z_c)$ is the self-impedance of antenna 2; and Z_c is the mutual impedance between the two antennas. In general the quantities $(Z_a + Z_c)$ and $(Z_b + Z_c)$ can be read directly from figures such as 13–10 and 13–11 (remembering that we are dealing with monopoles rather than dipoles, of course). In this case both quantities are equal to $36.5 + j21.25$. The mutual impedance is as given in Eq. (14–36). Putting these values into

Eqs. (14–38) and (14–39) yields Eq. (14–40). In this case, $I_1 = I_2$, and for simplicity we have let $Z_{g_1} = Z_{g_2}$.

$$V_1 = V_2 = (Z_{g_1} + 30.3 + j6.3)I_1 \tag{14–40}$$

Now, depending on the magnitude of I_1, V_1 and Z_{g_1} can be appropriately chosen. As a practical matter, the Z_g's will usually have an inserted reactance so that the generators are looking into a resistive load. Suppose that the desired radiated power is 1000 W. The current in each antenna is readily calculated to be 4.06 A.

Now let us slightly alter the problem. Suppose we wish to use the same antenna system to obtain the pattern of Fig. 14–6(b). In this case the currents in each antenna are of equal magnitude but of opposite phase. Substituting this information into Eqs. (14–38) and (14–39) (and letting Z_{g_1} again equal Z_{g_2}) gives

$$V_1 = -V_2 = (Z_{g_1} + 42.7 + j36.1)I_1 \tag{14–41}$$

The desired driving current in each antenna for a radiated power of 1000 W is now calculated to be 3.43 A.

Now let us postulate a somewhat different antenna. Suppose we have two half-wave dipoles in otherwise free space. The two dipoles are parallel and are spaced one half-wavelength, with one above the other. They are to be driven with identical currents because it is desired to radiate energy broadside but not up or down. It can be readily seen that this problem is essentially identical to the first example, using the two quarter-wave monopoles just discussed. All antenna impedances are twice their former values, and the solution method is exactly as before. If it had been our desire to place the dipoles parallel to each other and at the same height rather than one above the other, the analysis for broadside radiation would be identical to the second example, using the two quarter-wave monopoles except that all antenna impedances would be again doubled.

Finally, let us suppose that one of the dipoles is parasitic. The ends are simply connected together. In the monopole case this is equivalent to grounding the antenna base. Let us find the driving-point impedance of the driven element. The equivalent circuit is given in Fig. 14–15. The following values are apparent for the impedance elements:

$$Z_a + Z_c = Z_b + Z_c = 73 + j42.5 \tag{14–42}$$

$$Z_c = -12.4 - j29.8 \tag{14–43}$$

Thus

$$Z_a = Z_b = 85.4 + j72.3 \tag{14–44}$$

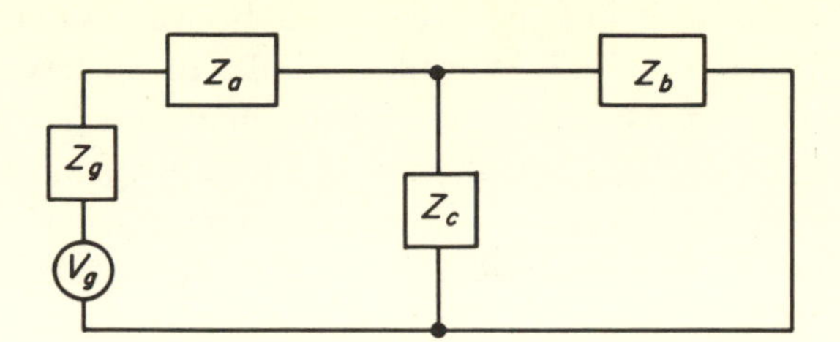

Fig. 14–15. Equivalent circuit of two parallel dipoles with one parasitic element.

The driving-point impedance of the active element is then given by

$$Z_{\rm DP} = Z_a + \frac{Z_b Z_c}{Z_b + Z_c} = 76 + j30 \tag{14–45}$$

The parasitic element made some adjustment in the driving-point impedance of the driven element. Since the mutual impedance is of fairly low magnitude, we would not have expected major changes in this driving-point impedance. It must be remembered, however, that the directional characteristics of this system are not the same as when both elements were driven because we no longer can independently specify the current in the second antenna element.

The question naturally arises as to the accuracy of the methods just discussed. It is perhaps sufficient to say that if one designed a large multi-tower array on this basis, he would not be wise to simply walk away from it after construction without checking all impedances with an r-f bridge. In general, such directional antenna systems require considerable adjusting after construction. Nevertheless the methods do give a good starting point, and one can be reasonably certain before construction that the system will be fairly close to the desired final result. In the case of monopole broadcast arrays, a major problem is that of obtaining a good conducting ground plane. Broadcast station sites are generally chosen for high ground conductivity, and the ground below the towers usually contains a wire network to approximate the desired ground plane.

Now suppose we have an antenna system of three or more elements. The mathematical method of handling such a situation follows directly. If we have three elements, we can write a symmetrical system of three equations. In these equations, Z_{11} will be the self-impedance of element 1, Z_{22} will be the self-impedance of element 2, and Z_{33} will be the self-impedance of element 3. Z_{12} will be the mutual impedance between elements 1 and 2 and can be read directly from Figs. 14–12, 14–13, and

14–14 since element 3 is not involved. That is, we assume that when element 3 is open, the system is essentially similar to one having elements 1 and 2 only. Analogous statements can be made for Z_{13} and Z_{23}. Of course, $Z_{12} = Z_{21}$, $Z_{13} = Z_{31}$, and $Z_{23} = Z_{32}$. The equivalent network can be drawn from these equations, although the computation of the actual elements of the equivalent network is involved and is not really necessary anyway.

14–6. DEFINITIONS OF DIRECTIVITY, GAIN, POLARIZATION, AND FIELD STRENGTH

In our discussion of directional antennas in Art. 14–3, it was implied that such antennas radiate more energy in one direction than another. If an antenna were used for receiving, the reciprocity rule says that it would be more sensitive to a wavefront from the first direction than from the second. Such antennas are often described in terms of *antenna directivity* and *antenna gain*. We shall define both terms.

Antenna Directivity. Let us postulate a perfectly spherical radiator, radiating W_r watts, such that the radial Poynting vector at a given distance from its center is equal to the radial Poynting vector at that same distance in any other direction. This is known as an isotropic radiator. The power density in watts per square meter is given by

$$P_{\text{ref}} = \frac{W_r}{4\pi r^2} \tag{14–46}$$

where r is the distance from the center of the radiator at which the reference point, P_{ref}, is measured. The distance r is usually considered large enough so that the antenna appears to be essentially a point source at the point of measurement whether the radiator is spherical or not. The power per unit of solid angle for the spherical radiator is

$$\text{Power per unit of solid angle} = U_{\text{ref}} = \frac{W_r}{4\pi} = r^2 P_{\text{ref}} \tag{14–47}$$

Now let us suppose that we have an antenna with some directional pattern. In particular, let us suppose that we have determined the direction of maximum radiation intensity of this antenna and that the power per unit solid angle in that direction is U_m. The directivity D of this antenna is then defined as

$$D = \frac{U_m}{U_{\text{ref}}} \tag{14–48}$$

Directivity is sometimes called *maximum directive gain.*

Antenna Gain. We shall define antenna gain, g, as

$$g = \frac{U_m}{U_a} \tag{14-49}$$

where U_a is the maximum radiation intensity of some accepted reference antenna when both antennas have the same power *input*. It is thus seen that directivity and gain differ in two important respects. First, the radiation efficiency of both antennas is considered in the definition of antenna gain where directivity is independent of power losses in the antenna itself. Secondly, the reference antenna is not necessarily a spherical radiator. If the reference antenna were a 100 per cent efficient spherical radiator and the unknown antenna were also 100 per cent efficient, then gain and directivity would be equivalent. More often, and particularly when discussing receiving antennas, the reference antenna is taken as a half-wave dipole oriented in the direction of maximum sensitivity to an incoming electromagnetic wave. Therefore, when one reads the words "antenna gain" in the literature, he should be careful to determine whether or not the writer really means "gain over a half-wave dipole." Furthermore, when one sees antenna gain expressed in decibels, this may mean decibel gain over a half-wave dipole or over a spherical radiator. Still another use of the word sometimes appears in the literature. One occasionally sees the word "gain" used with reference to some particular direction. This usually means in effect

$$\text{Gain in a given direction} = \frac{U_d}{U_a} \tag{14-50}$$

where U_d is the radiation intensity in the particular direction of interest.

Polarization. In the preceding discussion, we mentioned that the reference antenna must be oriented in a direction of maximum sensitivity to the incoming electromagnetic wave. Another way to say this would be that the antenna must be polarized with the incoming wave. In general, polarization of an electromagnetic wave is taken to mean the direction of the electric field. Thus a broadcast station antenna, which is usually one or more vertical monopoles, is vertically polarized, whereas a horizontal dipole is horizontally polarized. Generally speaking, once an electromagnetic wave is reflected from some object (or perhaps the ionosphere), its polarization may or may not be similar to that of the antenna from which it radiated.

Field Strength. The field strength of an electromagnetic wave is generally defined in terms of rms volts per meter in the direction of the electric field. This description is most appropriate for a plane wave front, which is generally encountered at reasonable distances from the radiating antenna. At higher frequencies, localized obstructions become important in the measurement of field strength because of the reflections and standing-wave patterns that are established. Just how one should go about measuring and describing such fields is not at all cut and dried and has occasionally been the source of much controversy.

14-7. ANTENNA APERTURE OR CAPTURE AREA, EFFECTIVE LENGTH, AND FREQUENCY FACTOR

Let us now concern ourselves with some of the more important aspects of the theory of receiving antennas. It is true that, owing to the reciprocity rule, we have specified such things as receiving directional patterns and internal impedances. However, this leaves a great deal unsaid. As an example, let us ask ourselves a rather obvious question. If we have an ordinary receiving antenna, say, a half-wave dipole, is the energy extracted from an oncoming electromagnetic wave simply equal to that intercepted by the physical structure of the dipole itself? If so, it would seem that a receiving dipole made of extremely small wire would be far less effective than one made of larger wire. Since this conclusion fits neither our intuition nor our experience, let us examine the question in greater detail.

Basically the problem is to examine the effects of inserting a half-wave dipole in an otherwise undisturbed plane wave field. This field will be oriented such that the direction of the **E** field is parallel to the axis of the receiving antenna, and the direction of the Poynting vector is, of course, perpendicular to the receiving antenna. Ordinarily we would describe this situation as adjusting the antenna to fit the polarization of the oncoming wave. If the **E** field of a plane wave is horizontal, we say that the wave is polarized horizontally; if the **E** field is vertical, the wave is polarized vertically, and so on. We shall describe the strength of the oncoming wave in a manner that has become customary among people working with such waves. The field will be expressed in *volts per meter*. This term describes the **E** field only, of course, but since the **E** field is related to the **H** field through the intrinsic impedance of the media through which the wave is traveling, we really know all about it when **E** is completely specified.

We can use the principle of reciprocity to determine how a typical receiving antenna extracts energy from the incident plane wave. Consider a situation where we have two elemental current sources of magnitude Il_1 and Il_2 respectively. The sources are located a distance d apart in space and oriented parallel to each other as represented in Fig. 14–16.

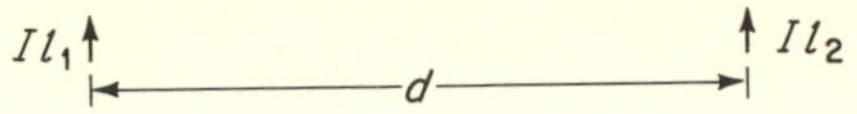

Il_1 Il_2 d

Fig. 14–16. Example of reciprocity using elemental antennas.

Application of the principle of reciprocity to this situation by means of Eq. (14–8) results in

$$V_1 I = \iiint_{\substack{\text{current}\\ \text{element 1}}} \mathbf{E}^{(2)} \cdot \mathbf{J}^{(1)}\, dv = V_2 I = \iiint_{\substack{\text{current}\\ \text{element 2}}} \mathbf{E}^{(1)} \cdot \mathbf{J}^{(2)}\, dv \tag{14–51}$$

V_1 is the voltage drop across current element 1 due to the electric field established by current element 2 and V_2 is the voltage drop across current element 2 due to the electric field established by current element 1.

For the case considered in Fig. 14–16, an expression for the electric field established by a current element is given by the far-field component of Eq. (12–39) (evaluated at $\theta = \pi/2$ and $r = d$). Equation (14–51) can now be written as

$$V_1 = E^{(2)} l_1 = \frac{-j\beta^2 I l_2 e^{-j\beta d}}{4\pi\omega\epsilon_0 d}\, l_1 = V_2 = \frac{-j\beta^2 I l_1 e^{-j\beta d}}{4\pi\omega\epsilon_0 d}\, l_2 \tag{14–52}$$

Now let us replace current element 2 with a half-wave dipole oriented parallel to current element 1. We can determine the voltage drop at current element 1 by using the electric field expression for the half-wave dipole given by Eq. (13–18) (evaluated at $\theta = \pi/2$ and $r = d$). Equation (14–51) can now be written as (assuming a current I at the input terminals of the half-wave dipole):

$$V_1 = \frac{-j\beta I e^{-j\beta d}}{2\pi\omega\epsilon_0 d}\, l_1 = V_2 = \iiint_{\substack{\text{half-wave}\\ \text{dipole}}} \mathbf{E}^{(1)} \cdot \mathbf{J}^{(2)}\, dv \tag{14–53}$$

The principle of reciprocity tells us that the voltage at the terminals of the half-wave dipole (V_2) will be equal to the voltage across the current element (V_1) provided a current source of I amperes is connected to the terminals of the dipole. Let us now define an effective length for the half-wave dipole in such a way that it could be considered to be a current element of magnitude Il_{eff} as far as its effect on the current element Il_1 is concerned. Comparison of Eqs. (14–52) and (14–53) indicates that the effective length for the half-wave dipole is

$$l_{\text{eff}} = \frac{2}{\beta} = \frac{\lambda}{\pi} \tag{14–54}$$

When the half-wave dipole is used as the receiving antenna the resultant open-circuit voltage appearing at the antenna terminals is

$$V_{\text{oc}} = +E_{\text{inc}}\left(\frac{\lambda}{\pi}\right) \tag{14–55}$$

Now let us terminate the half-wave dipole in conjugate-matching impedance 73 − j42.5 ohms as shown in Fig. 14–17. Half of the open-

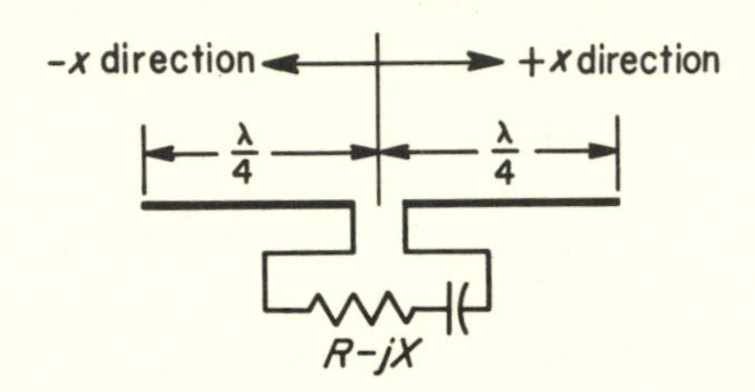

Fig. 14–17. Receiving half-wave dipole with terminating impedance of 73 − j42.5 ohms.

circuit voltage is dropped in the internal 73 ohms of the antenna itself. The power delivered to the 73-ohm load resistor is therefore equal to

$$W_{\text{rec}} = \frac{\left|\frac{E}{2}\right|^2 \lambda^2}{2 \times 73\pi^2} = \frac{|E|^2\lambda^2}{2 \times 292\pi^2} \tag{14–56}$$

Now, the power per unit area in the oncoming wave is equal to the electric field squared, divided by two times the intrinsic impedance of the propagation medium. This is the available power source for the power of Eq. (14–56).

$$|\mathbf{P}| = \begin{pmatrix}\text{Power per unit}\\ \text{of area available}\end{pmatrix} = \frac{|E|^2}{2\eta} = \frac{|E|^2}{2 \times 377} \tag{14–57}$$

The total area from which our antenna must take power, then, is Eq. (14–56) divided by Eq. (14–57). This is known as the capture area, or the antenna aperture.

$$\begin{pmatrix}\text{Capture area}\\ \text{of a half-wave dipole}\end{pmatrix} = \frac{277\lambda^2}{292\pi^2} = 0.131\lambda^2 \tag{14–58}$$

It is observed experimentally that an approximately sinusoidal current is induced in a half-wave dipole receiving antenna and that the power delivered is approximately that given by Eq. (14–56) when the antenna is properly matched. We conclude that a half-wave dipole actually captures energy from a rather large area surrounding it. This is illustrated in Fig. 14–18, in which the shaded area represents the equivalent aper-

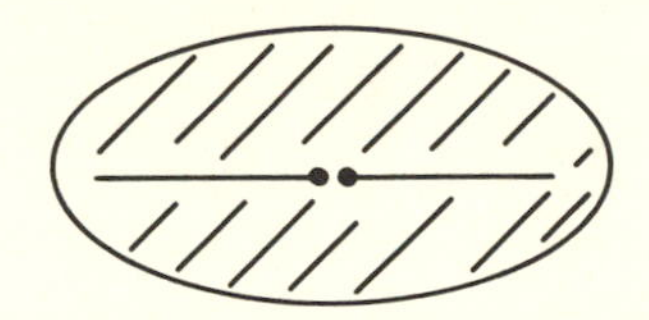

Fig. 14–18. Half-wave dipole capture area.

ture. For a dipole, this equivalent aperture is generally represented as an ellipse. Another way to state the results is as follows: The half-wave dipole is equally as effective as a perfectly matched, large elliptical horn which simply takes all the energy of the undisturbed plane wave striking within its boundary.

By using the effective length for a half-wave dipole, it is now fairly easy to compute the gain of a transmitting half-wave dipole over an isotropic, or spherical, radiator. From the work in previous chapters, we can show that the rms value of an electric far field for a current element, Il, is

$$E_{\text{far}} = 188.5 \frac{Il}{r\lambda} \tag{14–59}$$

where r is the distance from the antenna in meters. This E field is polarized in the same direction as the transmitting antenna. Now, if we consider our half-wave dipole in terms of its effective length, we see that it really is just a large current element. The saving grace of the situation is that the current all along the dipole length is essentially in phase, and therefore it may be considered to be in phase all along its effective length. This condition is inherent in the definition of a current element. If we get far enough away from the dipole, then, its far E field can be written as

$$E_{\text{far}} = 188.5 \times \frac{0.319\lambda I_0}{r\lambda} = 60 \frac{I_0}{r} \tag{14–60}$$

The power density in this far field is simply the square of the electric field divided by two times the intrinsic impedance of free space.

$$\begin{pmatrix}\text{Power density in far field of} \\ \text{half-wave transmitting dipole}\end{pmatrix} = P_m = \frac{3600 I_0^2}{2(377r^2)} \tag{14–61}$$

The total power radiated from the dipole W_r equals $73I_0^2/2$, so P_m can be written as

$$P_m = \frac{3600W_r}{377(73r^2)} \tag{14–62}$$

The power density radiated by a reference isotropic radiator is obviously

$$P_a = \frac{W_r}{4\pi r^2} \tag{14–63}$$

The half-wave dipole gain is

$$g_{\text{half-wave dipole}} = \frac{P_m}{P_a} = \frac{(4\pi)3600}{377(73)} = 1.64 \tag{14–64}$$

Thus the transmitting gain of a half-wave dipole (in its maximum direction) is 1.64 or 2.15 dB over a spherical, isotropic reference radiator.

We will find it useful in Chapter 15 to consider an isotropic receiving antenna as a reference or most basic receiving antenna. Such an antenna receives energy equally well from all directions. The capture area for such an antenna is defined as

$$(\text{C.A.})_{\text{reference antenna}} = \frac{(\text{C.A.})_{\text{half-wave dipole}}}{g_{\text{half-wave dipole}}} = \frac{\lambda^2}{4\pi} \tag{14–65}$$

This assumes that the isotropic radiator is properly matched. We will not be too much concerned about this since we will never use an isotropic radiator or receiver in a practical situation. The isotropic radiator or receiver is a mathematical abstraction which makes it easier to analyze antenna systems. We get back to the real world by multiplying received or transmitted signals calculated on the basis of an isotropic antenna by the gain (with respect to an isotropic antenna) of the actual antenna. One must bear in mind, however, that antenna gains are usually determined experimentally using a half-wave dipole as the reference antenna. These figures must be multiplied by 1.64 (or add 2.15 dB) to refer the gain to an isotropic radiator. There is one psychological trap in this procedure. The voltage produced at a 73-ohm terminating resistor by a half-wave dipole for any given oncoming field strength is a function of frequency. This is apparent from the capture area equation, Eq. (14–58). This frequency factor is too often forgotten in discussions on receiving-antenna gain.

14–8. ANTENNA ARRAYS

In order to achieve certain antenna pattern characteristics, it is necessary to use a number of antennas situated in some sort of an array configuration. Since the principle of linear superposition applies to the media in which these antennas will be located, the composite antenna pattern is determined by superimposing the fields due to the individual antennas. The simplest antenna configuration is that of two parallel identical half-wave dipoles located a distance d apart, as shown in Fig. 14-19.

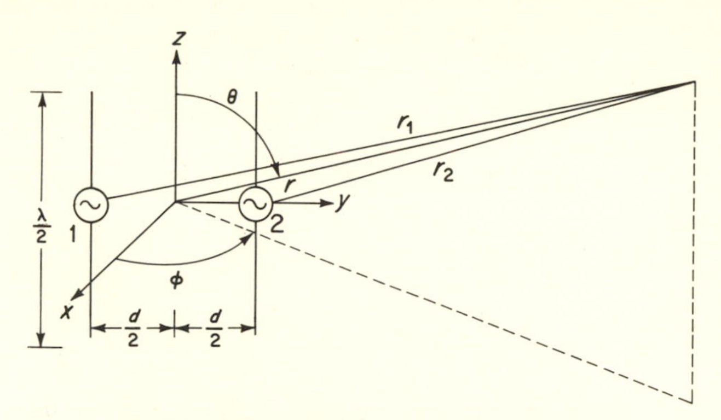

Fig. 14-19. A two-element antenna array.

The electric field at some point (r, θ, ϕ) is the sum of the electric field due to dipole 1 and the electric field due to dipole 2:

$$\mathbf{E} = \mathbf{E}^{(1)}(r_1, \theta, \phi) + \mathbf{E}^{(2)}(r_2, \theta, \phi) \tag{14-66}$$

The radiation component of the electric field expression for one of the dipoles is given by Eq. (13-18):

$$E_\theta = \frac{j\beta I e^{-j\beta r} \cos\left[(\pi/2)\cos\theta\right]}{2\pi\omega\epsilon_0 r \sin\theta}$$

For dipole 1, the distance from the center of the dipole to the field point is

$$r_1 = \sqrt{x^2 + (y + d/2)^2 + z^2}$$

and for dipole 2, the distance from the center of the dipole to the field point is

$$r_2 = \sqrt{x^2 + (y - d/2)^2 + z^2}$$

If $r = \sqrt{x^2 + y^2 + z^2}$ is very large, the following approximations can be made:

$$r_1 \approx r + \frac{y}{r}\frac{d}{2} = r + \frac{d}{2}\sin\theta\sin\phi \qquad \frac{1}{r_1} \approx \frac{1}{r}$$

$$r_2 \approx r - \frac{y}{r}\frac{d}{2} = r - \frac{d}{2}\sin\theta\sin\phi \qquad \frac{1}{r_2} \approx \frac{1}{r}$$

We can now approximate the composite electric field by

$$\mathbf{E} = \mathbf{E}^{(1)}(r, \theta, \phi)e^{-j\frac{\beta d}{2}\sin\theta\sin\phi} + \mathbf{E}^{(2)}(r, \theta, \phi)e^{+j\frac{\beta d}{2}\sin\theta\sin\phi} \tag{14-67}$$

The dipoles are identical, so we can write

$$\mathbf{E}^{(1)}(r, \theta, \phi) = \mathbf{a}_\theta I_1 b(r, \theta, \phi)$$

and

$$\mathbf{E}^{(2)}(r, \theta, \phi) = \mathbf{a}_\theta I_2 b(r, \theta, \phi)$$

where I_1 and I_2 are complex quantities (magnitude and phase angle), and

$$b(r, \theta, \phi) = \frac{j\beta e^{-j\beta r}}{2\pi\omega\epsilon_0 r}\frac{\cos[(\pi/2)\cos\theta]}{\sin\theta} \tag{14-68}$$

For the two dipoles of Fig. 14–19, the net electric field expression can now be written as

$$\mathbf{E}_\theta = b(r, \theta, \phi)[I_1 e^{-j\frac{\beta d}{2}\sin\theta\sin\phi} + I_2 e^{+j\frac{\beta d}{2}\sin\theta\sin\phi}] \tag{14-69}$$

Usually, the θ-dependence of the pattern of the array is determined by the θ-dependence of the individual dipoles, and the array is designed to produce a certain ϕ-dependence. Thus, we look at the array end on, as shown in Fig. 14–20. It is also convenient to specify an angle ψ such

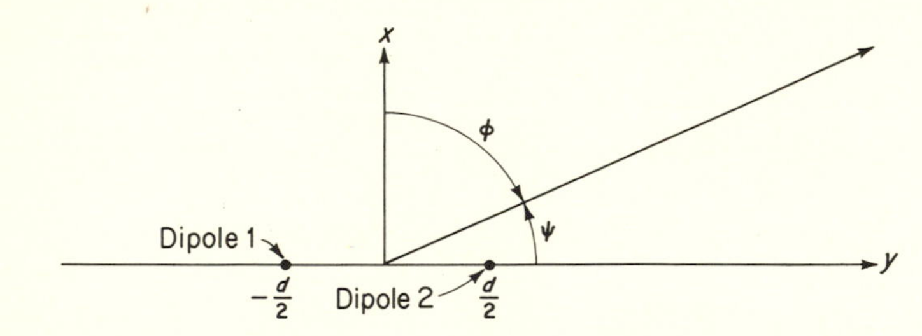

Fig. 14–20. End-on view of simple array.

that $\psi = \pi/2 - \phi$, as shown in Fig. 14–20. This means we are evaluating the electric field in the x-y-plane where the field of the individual dipoles is maximum. Now the electric field expression can be modified by setting $\theta = \pi/2$ and $\phi = \pi/2 - \psi$. The resultant expression is

$$\mathbf{E}_\theta = b\left(r, \frac{\pi}{2}, \frac{\pi}{2} - \psi\right)\left[I_1 e^{-j\frac{\beta d}{2}\cos\psi} + I_2 e^{+j\frac{\beta d}{2}\cos\psi}\right]$$

At this point let us develop far-field expressions for the specific cases of equal currents and opposite currents in the individual dipoles. If $I_1 = I_2 = I$

$$\mathbf{E}_\theta = b\left(r, \frac{\pi}{2}, \frac{\pi}{2} - \psi\right) I \left[e^{-j\frac{\beta d}{2}\cos\psi} + e^{+j\frac{\beta d}{2}\cos\psi}\right]$$

$$\mathbf{E}_\theta = 2Ib\left(r, \frac{\pi}{2}, \frac{\pi}{2} - \psi\right) \cos\left(\frac{\beta d}{2}\cos\psi\right)$$

If $I_1 = -I_2 = I$

$$\mathbf{E}_\theta = b\left(r, \frac{\pi}{2}, \frac{\pi}{2} - \psi\right) I \left[e^{-j\frac{\beta d}{2}\cos\psi} - e^{+j\frac{\beta d}{2}\cos\psi}\right]$$

$$\mathbf{E}_\theta = -2jIb\left(r, \frac{\pi}{2}, \frac{\pi}{2} - \psi\right) \sin\left(\frac{\beta d}{2}\cos\psi\right)$$

Remember, in the preceding examples I is, in general, complex. These equations could be used to predict the antenna patterns for the examples considered in Art. 14–3 (Figs. 14–6(a) and 14–6(b)),where $\beta d/2 = \pi/2$.

In general, if we express the currents in terms of arbitrary coefficients defined as $I_1 = a_1|I|$, $I_2 = a_2|I|$, we can then define an array factor

$$F(\psi) = a_1 e^{-j\frac{\beta d}{2}\cos\psi} + a_2 e^{+j\frac{\beta d}{2}\cos\psi} \tag{14–70}$$

The coefficients a_1 and a_2 are complex quantities which denote the relative amplitude and phase of the currents in the array.

Now let us consider some more complicated arrays. Suppose that there are N arbitrarily spaced antennas situated as shown in Fig. 14–21. The

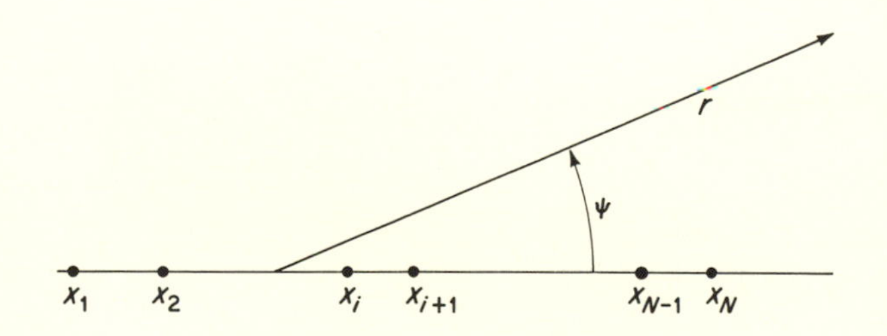

Fig. 14–21. The N-element array.

straightforward application of the foregoing theory results in the array factor expression

$$F(\psi) = \sum_{i=1}^{N} a_i e^{+j\beta x_i \cos\psi} \tag{14–71}$$

For example, if $N = 2$, $x_1 = -d/2$ and $x_2 = +d/2$, the resultant array factor is

$$F(\psi) = a_1 e^{-\frac{j\beta d}{2}\cos\psi} + a_2 e^{+j\frac{\beta d}{2}\cos\psi} \tag{14–72}$$

This is the same expression which was developed earlier. As a second example, consider the situation where there are five equally spaced elements driven by equal-magnitude currents phased in such a way that there is a progressive phase shift from element to element. That is,

$$a_i = e^{+j\alpha_i} = e^{ji\alpha}$$

where α is the phase shift between antennas. Say the antenna elements are d meters apart. The array factor for this situation is

$$F(\psi) = \sum_{i=1}^{5} e^{+j\beta id \cos\psi} e^{+ji\alpha} \tag{14-73}$$

$$F(\psi) = \sum_{i=1}^{5} e^{+j(\beta d \cos\psi + \alpha)i} \tag{14-74}$$

Since

$$\sum_{i=0}^{\infty} e^{-ui} = \frac{1}{1 - e^{-u}}$$

we can write

$$F(\psi) = \frac{e^{+j(\beta d \cos\psi + \alpha)}}{1 - e^{+j(\beta d \cos\psi + \alpha)}} - \frac{e^{+j6(\beta d \cos\psi + \alpha)}}{1 - e^{+j(\beta d \cos\psi + \alpha)}} \tag{14-75}$$

or

$$F(\psi) = \frac{e^{+j(\beta d \cos\psi + \alpha)}}{1 - e^{+j(\beta d \cos\psi + \alpha)}} \{1 - e^{+j5(\beta d \cos\psi + \alpha)}\} \tag{14-76}$$

This can be expressed as:

$$F(\psi) = {}^{+j3(\beta d \cos\psi + \alpha)} \left[\frac{\sin 5\left[\left(\frac{\pi d}{\lambda}\right)\cos\psi + \frac{\alpha}{2}\right]}{\sin\left[\left(\frac{\pi d}{\lambda}\right)\cos\psi + \frac{\alpha}{2}\right]} \right]$$

A plot of the magnitude of $F(\psi)$ versus ψ for this array with $\alpha = 0$ and $d = \lambda/2$ is shown in Fig. 14–22.

14–9. OTHER VARIETIES OF ANTENNA SYSTEMS

It has been one purpose of this and the previous chapters on antennas and antenna systems to develop some of the physical concepts necessary for rudimentary understanding of the subject. There are hundreds of different types of antenna systems in use today, and it would be a monumental task even to compile a complete list. However, we shall illustrate and briefly discuss a few of the more common systems, simply to give an idea of the variations that such systems can have.

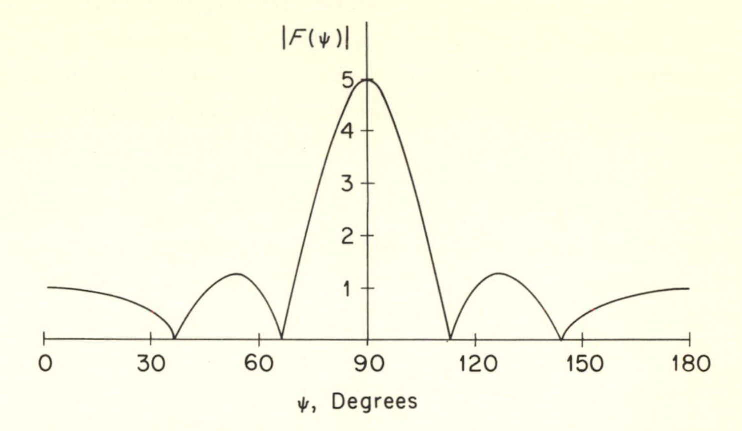

Fig. 14–22. The pattern for five identical elements spaced $\lambda/2$ apart and driven by equal in-phase currents.

Shunt-Fed Monopoles and Dipoles. The shunt-fed vertical monopole is similar in construction to the series-fed monopole we have been discussing except that the base is firmly bonded to the ground plane. The antenna is fed with a slant wire, as shown in Fig. 14–23.

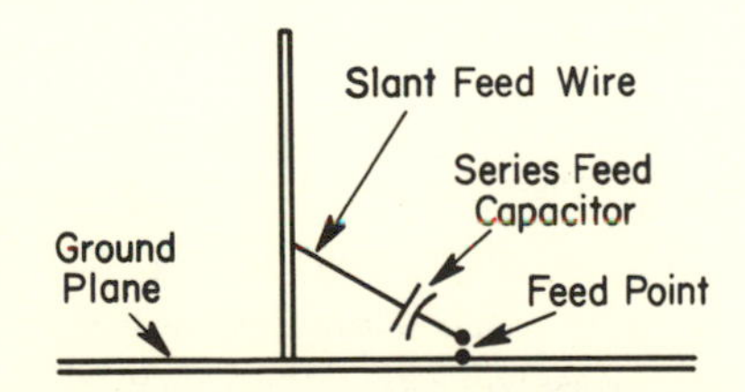

Fig. 14–23. Shunt-fed vertical monopole.

The current distribution in a shunt-fed monopole is not sinusoidal over its entire length, although reasonable analytical results are obtainable by assuming one sinusoidal distribution above the tap point and another sinusoidal distribution of different amplitude and space period below the tap point. The advantages of the shunt-fed monopole are that no base insulator is needed and that they are fairly easy to tune. The impedance observed at the bottom of the slant feed wire is resistive and inductively reactive. Therefore the slant wire is simply moved up and down until the resistive component equals the transmission-line characteristic impedance. The inductive reactance is then easily removed by putting a series capacitor of appropriate size in series with the slant wire at the base. Unfortunately it is not always practical to shunt-feed a vertical monopole, since for heights considerably in excess of a quarter-wave the impedances available for tap points of practical height are extremely unwieldy. Also, the feed wire has considerable, and often undesirable, effect on the radiation pattern, particularly when an attempt is made to use this type of antenna in a multiarray directional system.

A dipole can also be shunt-fed. The center points are tied together, and the two transmission-line wires are moved out and connected to each element at points symmetrically spaced and some distance from the center. Actually, such a dipole is the double image of a shunt-fed monopole working against an infinite ground plane. Therefore the analytical treatments of both are identical. The shunt-fed dipole is useful when it is desired to use a transmission line of much higher characteristic impedance than 73 ohms. Neither the shunt-fed monopole nor dipole is as commonly used as its series-fed counterpart.

Collinear Dipoles. One occasionally sees several dipoles spaced end on end in a collinear fashion, as illustrated in Fig. 14–24. This type of arrangement is generally situated such that the oncoming wave strikes each antenna perpendicularly. The signals from each antenna are then added in some sort of combining network. Generally speaking, the mutual effects between antennas are of less significance for a given spacing than would be the case if the antennas were parallel, as in a Yagi antenna. These mutual effects are readily calculable by means of the induced EMF method.

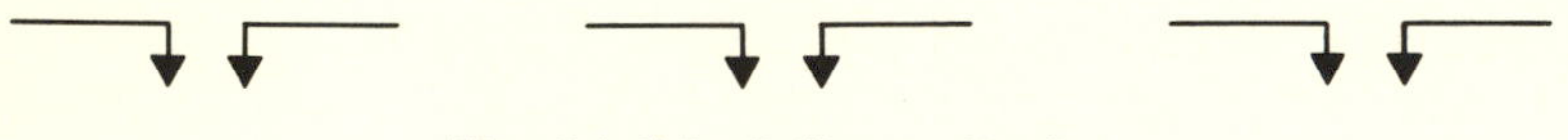

Fig. 14–24. Collinear dipoles.

Dipole with Reflector. Consider the possibility of putting a flat, metallic reflector behind a half-wave dipole, as shown in Fig. 14–25(a). If, for

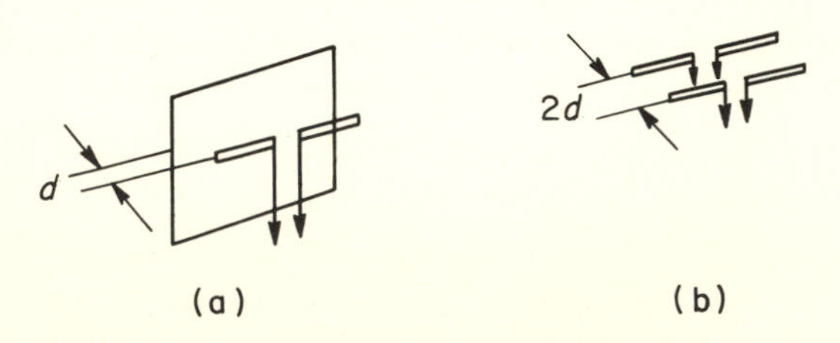

Fig. 14–25. Dipole with reflector and image equivalent.

example, the reflector is spaced one quarter-wavelength from the dipole, it is fairly easy to envision an approximately 6-dB gain in the direction away from the reflector toward the dipole. This is because the reflected wave that strikes the dipole is in phase with the initial wave which it intercepted, provided the system is oriented perpendicularly to the oncoming wave. The analysis of this antenna system is fairly straightforward through the use of the method of images, since it is equivalent to the two-active-element antenna illustrated in Fig. 14–25(b). In practice a variety of spacings are used, depending on the gain and directivity desired.

Yagi Antenna. The Yagi[6] antenna is generally considered to encompass a broad class of antennas which generally have one active dipole element, one reflector dipole element, and one or more director dipole elements. A typical arrangement is shown in Fig. 14–26. Most Yagi antennas

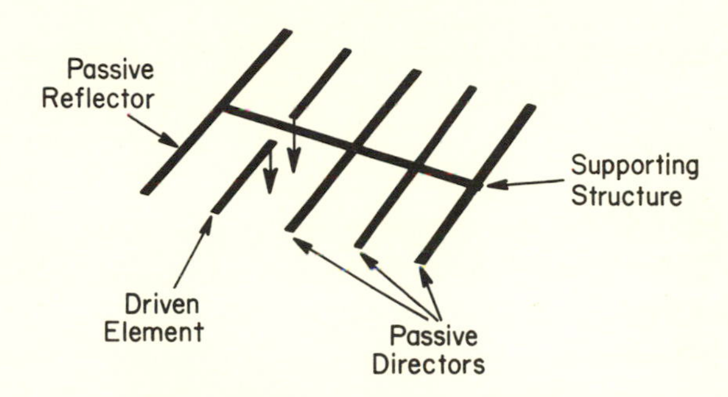

Fig. 14–26. A five-element Yagi.

have from three to five elements. However, antennas with 20 or more elements have been built. This design is an excellent antenna for narrow-band, high-gain work. The directors and reflector may be bolted directly to the metallic supporting structure, since the centers of these elements are at negligible potential. The active element must, of course, be isolated. The directional and band-width characteristics are determined by the element spacings and lengths. A common arrangement is for the reflector to be slightly longer than one half-wavelength, the active element about one half-wavelength, and the directors less than one half-wavelength. The element spacings are usually on the order of a quarter-

[6] Hidetsugu Yagi, "Beam Transmission of Ultra Short Waves," *Proc. I.R.E.*, vol. 16 (June 1928), pp. 715–40.

wavelength or less. One of the advantages of this type of antenna is that its characteristics are readily calculable with the induced EMF method.

Corner Reflector. There are other common ways to bias the directivity of a dipole. An obvious extension of the reflector sheet idea is the corner reflector illustrated in Figs. 14–27(a) and 14–27(b). This antenna has

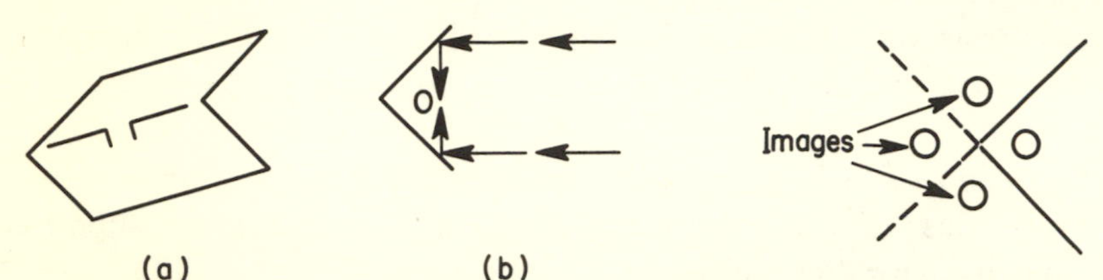

Fig. 14–27. Corner reflector.

Fig. 14–28. Images for the corner reflector.

considerably higher directivity than the dipole with a flat-sheet reflector. The corner is usually designed to be 90 degrees. The analysis is involved but nevertheless straightforward. The antenna may be attacked through the method of images. The image structure is shown in Fig. 14–28. There are three images for the two conducting planes. The directive characteristics are then computed through the induced EMF method.

Parabolic Reflector Systems. As the frequency of the oncoming wave increases, it begins to be possible to treat the wave much as a light wave is treated. In UHF systems, parabolic reflectors are often used as illustrated in Fig. 14–29. A receiving dipole or some other appropriate antenna

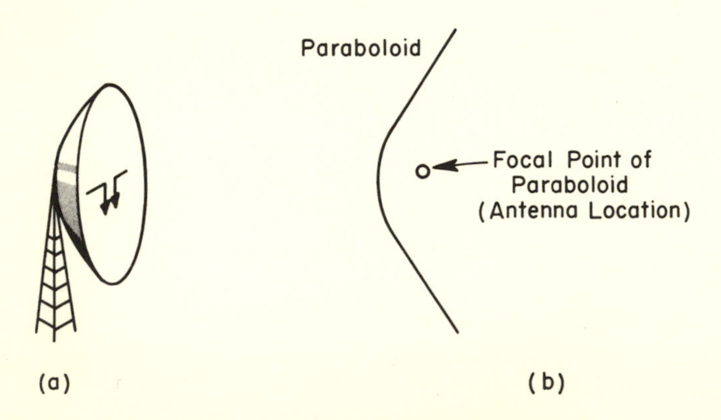

Fig. 14–29. Parabolic reflector systems.

is generally placed at the focal point of the parabola. Antenna gains of 10,000 or more and beam widths of a degree or less are obtainable with this type of system.

The operating characteristics of large-aperture antennas can be deduced by considering their behavior as receiving antennas. When receiving energy, the capture area of a large (with respect to wavelength) antenna is, for all practical purposes, equal to the area of the aperture, A_e. The gain of the antenna with respect to an isotropic radiator is simply the ratio of the capture areas:

$$g = \text{Power gain} = \frac{A_e}{\lambda^2/4\pi} = \frac{4\pi A_e}{\lambda^2}$$

Actually, this is an upper bound on the gain of aperture antennas since the capture area is usually somewhat less than the area of the aperture.

The parabolic reflector system has wide application in such things as radio astronomy, radar, and various varieties of tracking systems.

Dielectric Antennas. An antenna need not be an electrical conductor. Consider a plastic rod inserted in a waveguide. This rod can be reasonably matched to the guide such that energy flowing down the guide continues to flow down the plastic section. Once the wave has passed the end of the guide, however, it may strike a boundary between plastic and air. Part of the wave will be transmitted to the air, or radiated, and part of it will be reflected. The reflected part will ultimately strike another plastic-air boundary, whereupon part of it will be radiated, and so on. This phenomenon is illustrated in Fig. 14–30.

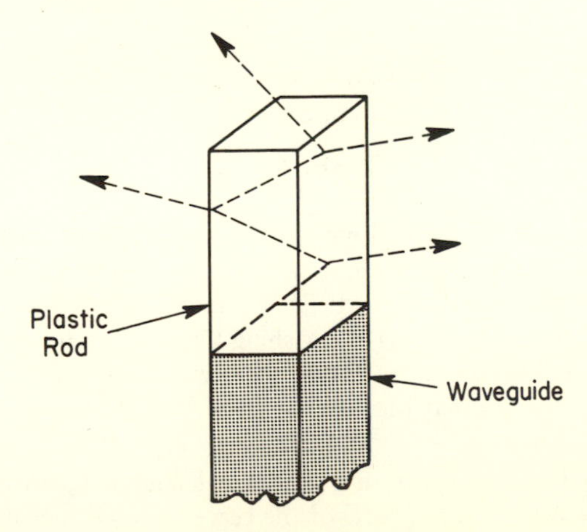

Fig. 14–30. Plastic antenna.

PROBLEMS

14–1. Prove the reciprocity theorem for a tee-section network as shown in the accompanying illustration.

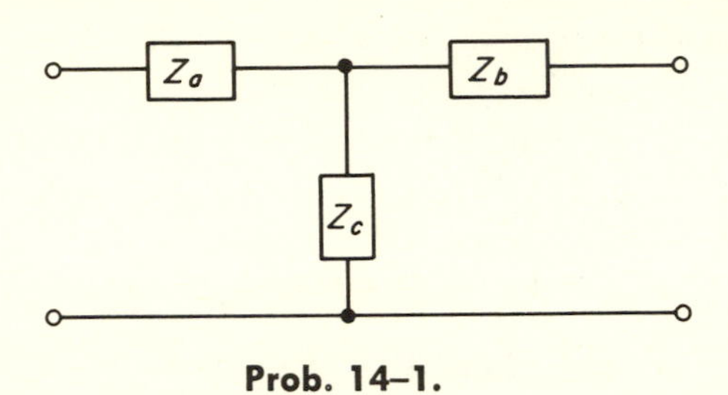

Prob. 14–1.

14–2. Design a two-element driven array to give an energy distribution similar to that of Figs. 14–6(a) and 14–6(b) except that the array is to be horizontally polarized.

14–3. Show that the driving-point impedances of two driven parallel antennas are functions of the ratio of the desired current in one antenna to that in the other.

14–4. A popular antenna for television reception of one channel is a three-element Yagi with dimensions as shown. A director of length 0.48λ is spaced 0.15λ in

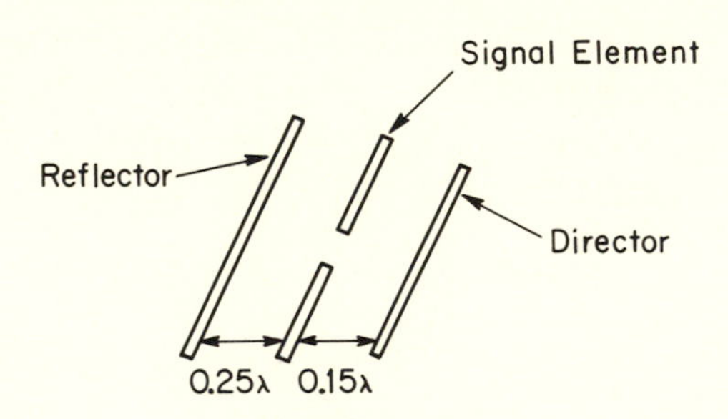

Prob. 14–4.

front of the half-wave signal dipole, and a reflector of length 0.51λ is spaced 0.25λ in back of it. This antenna usually has a 300-ohm twin lead connected directly to the signal element. How many decibels of gain improvement would be achieved by properly matching the antenna to the transmission line? The outside diameter of the tubing from which the antenna is made is $\frac{1}{2}$ in., and the frequencies that the antenna is designed to receive are TV Channel 8 (180 to 186 MHz).

14–5. Sketch the directional pattern of the two-element Yagi antenna of Fig. 14–15. Make your sketch on the basis of the currents in each element.

14–6. Plot a curve of half-wave dipole gain over an isotropic radiator as a function of frequency, when the dipole is used as a transmitting antenna.

14–7. Assume that a 100-MHz half-wave dipole is subjected to a plane wave of proper polarization and of 100-MHz frequency. Assume the dipole is terminated in $73 - j42.5$ ohms and that the power density of the oncoming wave is 1 mW/m^2. Compute the rms voltage across the 73-ohm resistor. Do the same for a 200-MHz dipole and a 200-MHz oncoming wave; for a 300-MHz antenna and wave; and so on every 100 MHz to 1000 MHz. Plot resistor voltage versus frequency.

14–8. Do the results of Problems 14–6 and 14–7 seem consistent with what you have been told about reciprocity? Explain.

14–9. Find the frequency at which the voltage produced across a 73-ohm terminating resistor on a half-wave dipole antenna exactly equals the field strength in volts per meter. Assume a conjugate-match condition.

14–10. Suppose our receiver has a 300-ohm input impedance instead of 73 ohms. We should, of course, go through an appropriate match transformer. If this transformer is lossless, at what frequency would the voltage at the receiver antenna terminals equal the field strength in volts per meter?

14–11. Repeat Problem 14–10 for a 50-ohm receiver input impedance.

14–12. Suppose a half-wave dipole is used to measure field strength. When it is necessary to change frequency, the dipole lengths are changed. The dipole is properly terminated, and the field strength is determined by reading the voltage across a 73-ohm resistor. Plot a curve of field strength over observed voltage as a function of frequency from 50 to 1000 MHz.

15

Elements of Radio-Wave Propagation

15–1. INTRODUCTION

The mechanisms responsible for the behavior of radio waves propagating in the earth's atmosphere have been a fascinating subject for study by radio scientists since the early 1900's. Experimental and theoretical work since that time has led the way to virtually complete understanding of at least the major features of the various propagation mechanisms. It is the purpose of this chapter to present the features of the propagation mechanisms in a form that emphasizes their role in determining the behavior of a communication path in the earth's atmosphere. Where practical, the theory will be developed and discussed; at times it will be necessary to simply state the results of an analysis and attempt to establish some feel for the phenomenon being described by means of an example. Since there exists a great body of literature (including books and journal articles) on the subject, references will be frequently cited so that the interested reader can go to the literature for additional information.

The chapter begins with a discussion of propagation in free space and the factors involved in defining propagation-path loss. Next, the basic mechanisms of reflection, refraction, scattering, and diffraction are discussed. Finally, applications of these basic mechanisms to practical propagation paths involving the earth and its atmosphere are presented.

15–2. PROPAGATION IN FREE SPACE

The simplest case of the transmission of electrical energy via the mechanism of radiation is propagation between two antennas situated in free space. Consider two isotropic antennas, one acting as the source and the other as the receiver, situated d meters apart in free space. If the transmitting antenna is radiating W_r watts, the power density ($\mathbf{P}_0$) in the field a distance d from the antenna is

$$|\mathbf{P}_0| = \begin{bmatrix} \text{Power per} \\ \text{unit area} \\ \text{(radially directed)} \end{bmatrix} = \frac{W_r}{4\pi d^2} \text{ watts/meter}^2$$

The rms magnitude of the electric field intensity at a point d meters from the isotropic source is readily determined to be

$$|\mathbf{E}_0| = \frac{\sqrt{30W_r}}{d}$$

The total power delivered to the load attached to the receiving antenna (W_{rec}) is simply the product of the power per unit area in the transmitting beam and the capture area of the isotropic receiving antenna. Since the isotropic receiving antenna has a capture area of

$$\text{Capture area}_{\text{iso}} = \frac{\lambda^2}{4\pi}$$

we can write

$$W_{\text{rec}} = W_r \left(\frac{\lambda}{4\pi d}\right)^2$$

This is not a very practical situation since the isotropic antenna is basically a mathematical concept which is useful in analytical work. The ideas presented are useful for more practical antennas, however, since we can multiply the transmitted energy density due to the isotropic radiator by the gain of the transmitting antenna, g_T; and the capture area of the istropic receiving antenna can be multiplied by the gain of the receiving antenna, g_R. The resultant expression is

$$W_{\text{rec}} = W_r \left(\frac{\lambda}{4\pi d}\right)^2 g_T g_R \qquad (15\text{–}1)$$

Example 15–1

Determine the power delivered to a matched load connected to a half-wave dipole which is receiving energy radiated by a second half-wave dipole located 10 km away in free space. The transmitting antenna is radiating 1 kW at a frequency of 3 MHz. For half-wave dipoles, $g_T = g_R = 1.64$. We are also given $W_r = 1$ kW, and $d = 10^4$ meters. In free space,

$$\lambda = \frac{3 \times 10^8}{3 \times 10^6} = 100 \text{ meters}$$

Substituting into Eq. (15–1),

$$W_{\text{rec}} = 10^3 \left(\frac{10^2}{4\pi \times 10^4}\right)^2 (1.64)^2 \text{ watts}$$
$$= 1.71 \times 10^{-3} \text{ watts}$$

15–3. TRANSMISSION LOSS

The concept of transmission loss is of great utility in designing radio communication systems. Transmission loss is defined in terms of the power radiated from the transmitting antenna (W_r) and the resulting signal power available from the loss-free receiving antenna (W_a):

$$\text{Transmission loss} = \frac{W_r}{W_a} \tag{15–2}$$

This definition includes the effects of the transmitting antenna, the receiving antenna, and the propagation path. The transmission loss, expressed in decibels, becomes

$$L_T = 10 \log_{10} W_r - 10 \log_{10} W_a \tag{15–3}$$

A very useful notion for analytical purposes is the transmission loss for the basic case of perfectly conducting isotropic antennas separated by a distance d in free space. Under such conditions,

$$W_a = \frac{W_r}{4\pi d^2}\left(\frac{\lambda^2}{4\pi}\right) = \left(\frac{\lambda}{4\pi d}\right)^2 W_r$$

The transmission loss for this case is denoted by L_b and can be written as

$$L_b = 10 \log_{10} \left(\frac{4\pi d}{\lambda}\right)^2 \tag{15–4}$$

In terms of frequency and distance, Eq. (15–4) becomes

$$L_b = [20 \log_{10} d + 20 \log_{10} f_{\text{MHz}} - 27.55] \text{ dB} \tag{15–5}$$

or, expressing d in miles,

$$L_b = [20 \log_{10} d_{\text{mi}} + 20 \log_{10} f_{\text{MHz}} + 36.58] \text{ dB} \tag{15–6}$$

This very simple case can be extended to more practical situations wherein the transmitting and receiving antennas have effective gains relative to isotropic antennas of g_T and g_R, respectively, by adding the effect of these gains to the transmission-loss expression:

$$L_T = L_b - G_T - G_R \tag{15–7}$$

Note that these "gains" act to reduce the transmission loss. If, in addition to antenna gains, we remove the constraint of free space between the transmitting and receiving antennas, we can take this into account by adding another term to the transmission-loss expression. This term is the propagation-path loss relative to free space, denoted by L_p. The final expression for the transmission loss becomes

$$L_T = L_b - G_T - G_R + L_p \tag{15–8}$$

The remainder of this chapter is devoted to the determination of L_p for a wide variety of situations which are frequently encountered in attempting to "engineer" a communication path. The propagation path loss is defined as

$$L_p = 10 \log_{10} \left[\frac{\text{Power at the receiving antenna due to an isotropic source transmitting in free space}}{\text{Power at the receiving antenna due to an isotropic source transmitting over the propagation path}} \right] \tag{15–9}$$

This loss can be formulated by calculating the power ratio or by calculating the magnitude of the electric field ratio and squaring it. In the latter case the electric field ratio is written as $|\mathbf{E}_0|/|\mathbf{E}|$, where $\mathbf{E}_0$ is the electric field strength due to the isotropic source in free space and $\mathbf{E}$ is the electric field strength of the signal due to the isotropic source transmitting over the propagation path. Later in the chapter we will formulate the ratio $|\mathbf{E}|/|\mathbf{E}_0|$ for use in determining the propagation path loss for specific propagation paths.

15–4. REFLECTION, REFRACTION, SCATTERING, AND DIFFRACTION

In most situations we are concerned with the transmission of energy through the atmosphere of the earth. We shall see that the earth and its atmosphere affect the propagation of radio waves by virtue of the dielectric constant of the atmosphere and the electrical nature of the earth. These effects result in several mechanisms which can be factors in the propagation of radio waves in or through the earth's atmosphere. These mechanisms are: (1) reflection, (2) refraction, (3) scattering, and (4) diffraction. We will consider these processes in some detail in this article and then relate them to the actual conditions in the earth's environment in the next article.

Reflection at a Material Interface. The reflection of a uniform plane wave obliquely incident on a material interface was considered in Chapter 8. Even though we are not, strictly speaking, dealing with uniform plane waves, the spherical waves which are radiated by a transmitting antenna can be approximated with little error by a uniform plane wave in the region where the reflection takes place. The terminology which is most useful in radio-wave propagation study is a little different from that used in Chapter 8, however, and should be discussed at this point. The angles of incidence and reflection are taken between the wave normal and the surface, as shown in Fig. 15–1.

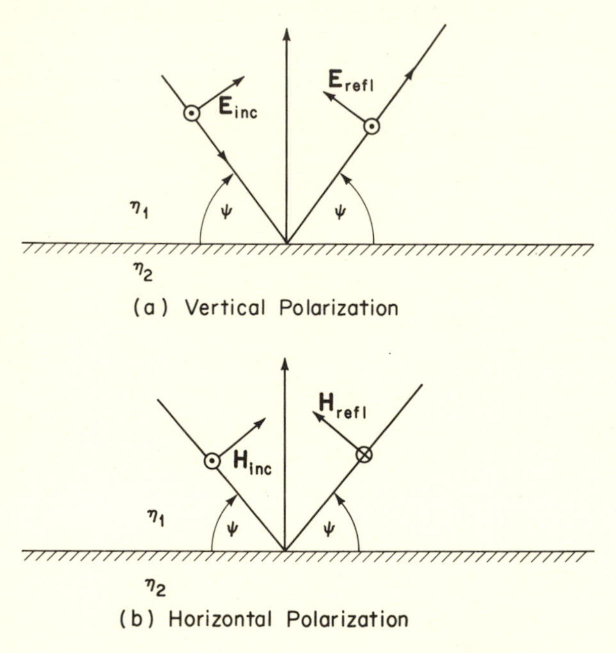

Fig. 15–1. Reflection at an interface.

Snell's law assures us that the angle of incidence is equal to the angle of reflection. In comparing the conditions in Figs. 15–1(a) and (b) with the results of Chapter 8 we see that we must substitute $\theta = (\pi/2) - \psi$. Referring to Fig. 15–1(a), a vertical polarization reflection coefficient R_v

will be so defined that the normal component of the reflected electric field is equal to the product of the vertical polarization reflection coefficient and the normal component of the incident electric field. That is,

$$E_{\text{normal (reflected)}} = R_v E_{\text{normal (incident)}}$$

This definition results in the positive tangential components being oppositely directed, but related in magnitude by the reflection coefficient. Since the positive-sign convention of Chapter 8 related the tangential components of the electric field at the interface, we see that the reflection coefficient for vertical polarization is the negative of the reflection coefficient introduced in Chapter 8 to facilitate the analysis of a uniform plane wave obliquely incident on a dielectric surface with its electric field in the plane of incidence. Thus we write

$$R_v = \frac{Z_{0z_1} - Z_{0z_2}}{Z_{0z_1} + Z_{0z_2}}$$

where

$$Z_{0z_1} = \eta_1 \cos\left(\frac{\pi}{2} - \psi\right) = \eta_1 \sin\psi$$

and

$$Z_{0z_2} = \eta_2 \sqrt{1 - \frac{\epsilon_1}{\epsilon_2}\cos^2\psi}$$

Substituting results in

$$R_v = \frac{\eta_1 \sin\psi - \eta_2 \sqrt{1 - \frac{\epsilon_1}{\epsilon_2}\cos^2\psi}}{\eta_1 \sin\psi + \eta_2 \sqrt{1 - \frac{\epsilon_1}{\epsilon_2}\cos^2\psi}} \tag{15-10}$$

For the case of horizontal polarization we can simply make the substitution for θ in the results of Chapter 8 to account for the changed reference for defining angles:

$$R_h = \frac{Z_{0z_2} - Z_{0z_1}}{Z_{0z_2} + Z_{0z_1}}$$

where

$$Z_{0z_1} = \eta_1 \sec\theta_1 = \frac{\eta_1}{\sin\psi}$$

and

$$Z_{0z_2} = \frac{\eta_2}{\sqrt{\left(1 - \frac{\epsilon_1}{\epsilon_2}\cos^2\psi\right)}}$$

Substituting results in

$$R_h = \frac{\eta_2 \sin\psi - \eta_1 \sqrt{1 - \frac{\epsilon_1}{\epsilon_2}\cos^2\psi}}{\eta_2 \sin\psi + \eta_1 \sqrt{1 - \frac{\epsilon_1}{\epsilon_2}\cos^2\psi}} \tag{15-11}$$

In the application of the reflection coefficient expressions to radio-wave propagation studies it will often be necessary to treat the case where one of the regions (usually region 2) is lossy. In such cases the loss must be incorporated in the dielectric constant. That is,

$$\epsilon_2 \longrightarrow \epsilon_2 - j\frac{\sigma}{\omega\epsilon_0}$$

where ϵ_2 is the relative permittivity and σ is the conductivity of region 2.

Refraction. Refraction is the bending of the path of a radio wave as it propagates through a medium characterized by a gradient of permittivity. In the regions where the permittivity is high, the phase velocity is reduced and the wave front moves more slowly than in the regions where the permittivity is lower. This causes a bending of the ray path as the wave normal changes direction.

The variation in permittivity in the medium is normally described as a variation in the index of refraction n. The index of refraction is defined as the ratio of the velocity of the wave in vacuum to the velocity of the wave in the medium:

$$n = \frac{c}{v} \tag{15-12}$$

where

$$c = \frac{1}{\sqrt{\mu_0\epsilon_0}} \qquad \text{and} \qquad v = \frac{1}{\sqrt{\mu\epsilon}}$$

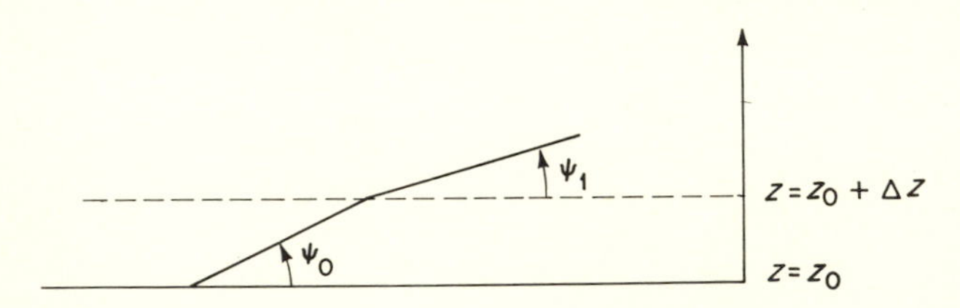

Fig. 15–2. Refraction geometry.

To see how the wave normal bends, consider Fig. 15–2. Assume that the refractive index is a function of z. Snell's law (Eq. (8–27)) can be written as

$$n_0 \cos\psi_0 = n_1 \cos\psi_1$$

This can be approximated by letting $\psi_1 = \psi_0 - \Delta\tau$, and $n_1 \simeq n_0 + \frac{dn}{dz}\Delta z$. We can then write

$$n_0 \cos \psi_0 = \left(n_0 + \frac{dn}{dz}\Delta z\right) \cos (\psi_0 - \Delta\tau)$$

This can be simplified by means of small argument approximations. That is, let $\cos \Delta\tau \approx 1$, and $\sin \Delta\tau \approx \Delta\tau$. This results in

$$\frac{\Delta\tau}{\Delta z} = -\frac{1}{n_0}\frac{dn}{dz} \cot \psi_0$$

This is usually written in terms of differentials as

$$d\tau = -\frac{dn}{n} \cot \psi \tag{15–13}$$

The total change in the direction of the wave normal as it passes from point 1 to point 2 is the integral of Eq. (15–13):

$$\tau_{12} = -\int_{n_1,\psi_1}^{n_2,\psi_2} \cot \psi \frac{dn}{n} \tag{15–14}$$

This integral is quite difficult to evaluate in most cases, but a number of techniques have been developed to evaluate it since it is fundamental to the pointing accuracy of a radar system operating in a medium characterized by a variation in refractive index. This aspect of refraction is beyond the scope of this text.[1]

The radius of curvature of a wave front propagating in a spherically stratified medium, characterized by $n(r)$, is of considerable utility in the analysis of radio-wave propagation around the earth. The geometry of such a situation is shown in Fig. 15–3. A wave front moves from AB to $A'B'$. If the phase velocity along BB' is $v + dv$, we can write

$$\frac{v}{R} = \frac{v + dv}{R + dR}$$

or

$$\frac{dv}{v} = \frac{dR}{R}$$

Since, by definition, $v = c/n$, we can differentiate to get $dv/v = -dn/n$. Combining expressions for dv/v results in

$$\frac{1}{R} = -\frac{1}{n}\frac{dn}{dR}$$

[1] B. R. Bean and E. J. Dutton, *Radio Meteorology*, National Bureau of Standards Monograph 92, Chapter 3, 1966.

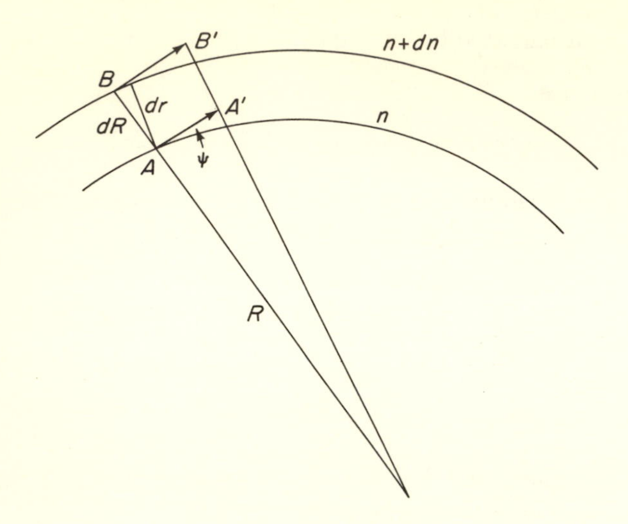

Fig. 15–3. Geometry of spherically stratified medium.

Consideration of Fig. 15–3 shows that $dr = dR \cos \psi$, and we can now write

$$\frac{1}{R} = -\frac{1}{n}\frac{dn}{dr}\cos \psi \tag{15–15}$$

R is the radius of curvature of the ray path traced by the wave normal as it propagates through the stratified atmosphere.

Scattering. Scattering differs from the other electromagnetic phenomena by virtue of the fact that a scattered wave is generally characterized by a non-uniform phase front which is the result of currents induced in the scatterer that act as secondary sources.

For wave-propagation purposes, the scatterer is characterized by a parameter called the *scattering cross-section* σ. The utility of the scattering cross-section concept can be seen by considering Fig. 15–4. θ is the angle between the incident wave normal and the reflected wave normal and X

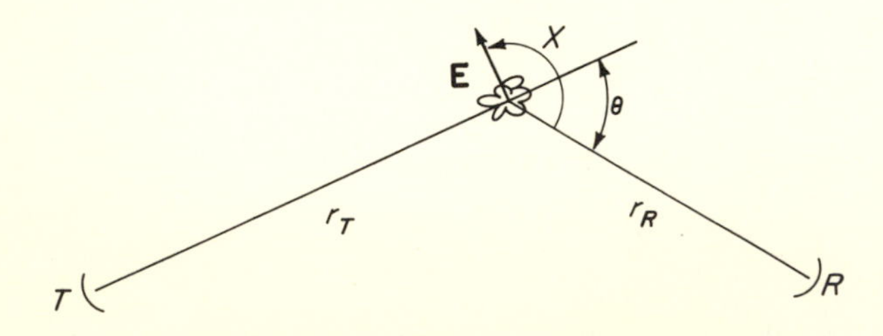

Fig. 15–4. The scattering geometry.

is the angle between the electric field vector and the scattered wave normal. The scattering cross-section is a function of these two variables. The scattering cross-section is defined as the scattered power per unit solid angle per unit incident power density. If the transmitter is radiating W_r watts by means of an antenna having a gain of g_T with respect to an isotropic radiator, the incident power density at the scatterer is

$$\begin{bmatrix}\text{Power per}\\ \text{unit area}\end{bmatrix}_{\text{inc}} = \frac{W_r g_T}{4\pi r_T^2} \text{ watts/meter}^2$$

From the definition of the scattering cross-section, the scattered power per unit solid angle is

$$\begin{bmatrix}\text{Power scattered per}\\ \text{unit solid angle}\end{bmatrix} = \frac{W_r g_T \sigma(\theta, X)}{4\pi r_T^2} \text{ watts/steradian}$$

The power density at the receiving antenna is

$$\begin{bmatrix}\text{Power per}\\ \text{unit area}\end{bmatrix}_{\text{rec}} = \frac{W_r g_T \sigma(\theta, X)}{4\pi r_T^2 r_R^2} \text{ watts/meter}^2$$

If the receiving antenna has a gain of g_R with respect to an isotropic radiation, its capture area is

$$\text{Capture area} = \frac{\lambda^2}{4\pi} g_R$$

The total power delivered to the receiver load is

$$W_{\text{rec}} = \frac{\lambda^2 P_r g_T g_R \sigma(\theta, X)}{(4\pi r_T r_R)^2} \tag{15-16}$$

In radio-wave-propagation work it is useful to define a differential scattering cross-section $\sigma_v(\theta, X)$. The differential scattering cross-section is defined as the scattered power per unit solid angle, per unit incident power density, and per unit element of volume. Using this definition, the total power delivered to the receiver load is

$$W_{\text{rec}} = \left(\frac{\lambda}{4\pi}\right)^2 \iiint \frac{\sigma_v(\theta, X) g_T g_R}{r_T^2 r_R^2} \, dv \tag{15-17}$$

Diffraction. When an electromagnetic wave is incident upon a material surface it induces currents in the material. These currents tend to flow along the surface of the object and guide electromagnetic energy over the surface. This process, which causes a radio wave to bend around curved surfaces and corners, is called diffraction.

There are two important cases of diffraction which are useful in the study of radio-wave propagation phenomena. These are: (1) diffraction caused by the curvature of the earth's surface, which bends the radio wave into the shadow region beyond the radio horizon; and (2) knife-edge diffraction, which directs the radio wave into the shadow region behind a sharp ridge.

The bending of radio waves by the curvature of the earth's surface, called "smooth-earth diffraction," is analyzed by considering a point source above the surface of a dielectric sphere. The solution of this mathematical model is accomplished in a straightforward manner with resultant fields being expressed as an infinite series of spherical Bessel functions and Legendre polynomials. For cases where the wavelength of the energy emitted by the source is much smaller than the radius of the sphere, the series converges very slowly and the solution is virtually useless. This situation is alleviated somewhat by means of a transformation that reduces the original series to another series representation which converges much more rapidly. In fact, a single term of this series is suitable for representing the fields in the diffraction region beyond the radio horizon. This series representation is written as[2]

$$\frac{|\mathbf{E}|}{|\mathbf{E}_0|} = \frac{\sqrt{d\lambda}}{h_0}\left|\sum_{s=1}^{\infty} G_S(h_1/h_0)G_S(h_2/h_0)F_S(d/d_0)\right|$$

where d = distance between antennas

h_1 = height of the transmitting antenna

h_2 = height of the receiving antenna

G_S = height gain function

F_S = range function

$$d_0 = \left(\frac{a_e^2\lambda}{\pi}\right)^{1/3}$$

$$h_0 = \frac{1}{2}\left(\frac{a_e\lambda^2}{\pi^2}\right)^{1/3}$$

a_e = effective earth radius

For the case where the first term of the series is valid, the field strength ratio can be written as

$$\frac{|\mathbf{E}|}{|\mathbf{E}_0|} = 2\sqrt{\pi X}|G_1(Z_1)G_1(Z_2)|e^{-2.02X} \tag{15-18}$$

where $X = d/d_0$ and $Z = h/h_0$. The main problem now is to determine $G_1(Z_1)$ and $G_1(Z_2)$. In general, the height-gain function is difficult to determine since it depends upon the polarization of the electrical signal and

[2] The format for this treatment of diffraction by a smooth dielectric sphere is that of J. E. Freehafer, *Propagation of Short Radio Waves*, edited by D. E. Kerr (McGraw-Hill Book Co., Inc., 1951, republished by Dover Publications, Inc., New York, 1965), Chapter 2.

the electrical properties of the earth. Graphs representing the height-gain function have been prepared and for the particular case of perfectly conducting earth and horizontal polarization are shown in Figs. 15-5(a) and 15-5(b). Equation (15-18) is valid as long as the receiver and the

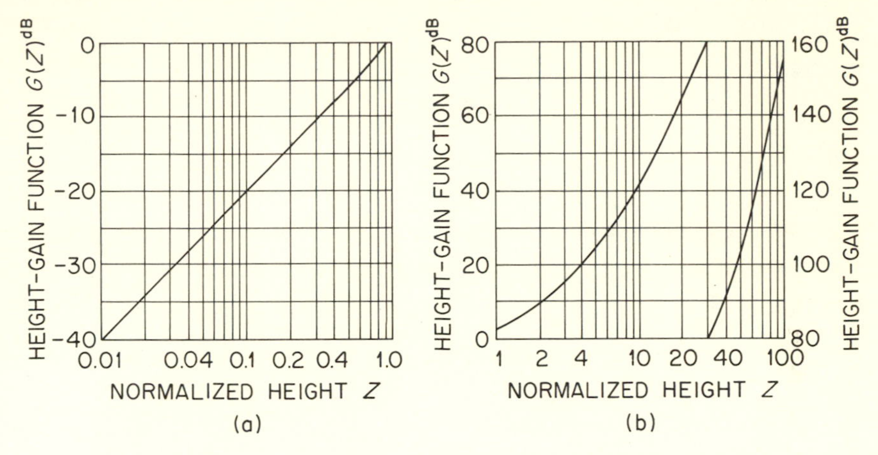

Fig. 15-5. Height-gain curves for smooth-earth diffraction.

transmitter are out of radio sight of each other. We will now show that the minimum distance between the transmitter and the receiver for which Eq. (15-18) is valid is formed by solving the equation

$$X_{\min} = \sqrt{Z_1} + \sqrt{Z_2} \tag{15-19}$$

This equation is derived by considering a transmitting antenna situated a height h above a spherical earth of radius a_e, as shown in Fig. 15-6.

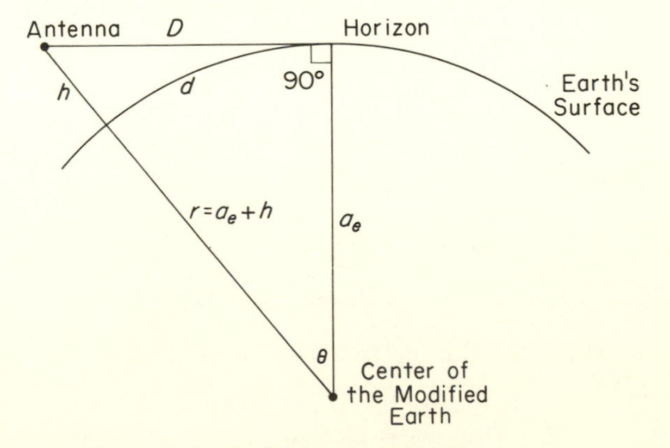

Fig. 15-6. Radio horizon geometry.

The earth's radius a_e includes any modification that may be caused by including the effect of a gradient of refractive index in the earth's atmosphere. Referring to Fig. 15–6, we can write

$$\begin{aligned}&\text{(a)} \quad (a_e + h)^2 = D^2 + a_e^2\\ &\text{(b)} \quad d = a_e\theta \qquad\qquad (15\text{–}20)\\ &\text{(c)} \quad D = a_e \tan\theta\end{aligned}$$

In most situations, $\theta \ll 1$, and

$$D \cong a_e\theta = d$$

Expanding Eq. (15–20a) results in

$$2a_eh + h^2 = D^2$$

For the antenna heights of concern here, $h \ll a_e$. All of the approximations and assumptions can be utilized to write

$$d = \sqrt{2a_eh} \qquad (15\text{–}21)$$

Adding d_1 (the distance to the horizon from the transmitting antenna) and d_2 (the distance to the horizon from the receiving antenna) yields the minimum distance

$$d_{\min} = \sqrt{2a_eh_1} + \sqrt{2a_eh_2}$$

This expression can be divided by d_0 to get Eq. (15–19).

As we consider typical propagation paths on the earth, we recognize that the earth's surface is anything but smooth. The terrain is often dominated by hills and ridges. We can usually get a reasonable approximation to the effect of a dominating ridge by calculating the diffraction over a "knife-edge" obstacle of the same height. The geometry of the "knife-edge" situation is shown in Fig. 15–7.

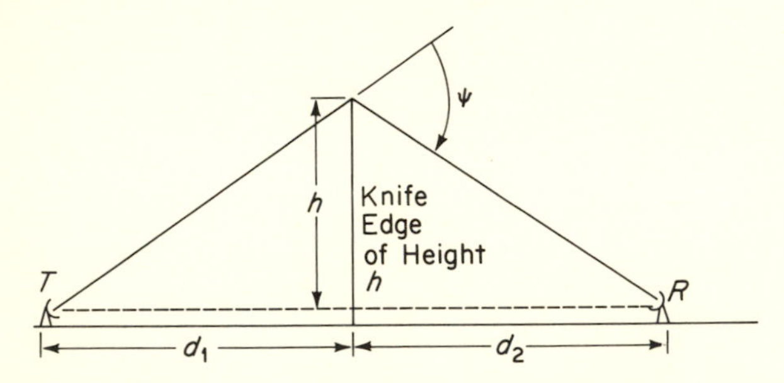

Fig. 15–7. Knife-edge diffraction path.

The magnitude of the electric field is expressed in terms of Fresnel integrals as[3]

[3] E. C. Jordan and K. G. Balmain, *Electromagnetic Waves and Radiating Systems*, 2d ed. (Englewood Cliffs, N.J.: Prentice-Hall, Inc., 1968), pp. 499–503.

$$\frac{|\mathbf{E}|}{|\mathbf{E}_0|} = \frac{1}{\sqrt{2}}\sqrt{\left(\frac{1}{2} - C(v)\right)^2 + \left(\frac{1}{2} - S(v)\right)^2} \tag{15-22}$$

where

$$C(v) = \int_0^v \cos\frac{\pi t^2}{2}\,dt,$$

$$S(v) = \int_0^v \sin\frac{\pi t^2}{2}\,dt,$$

and

$$v = h\sqrt{\frac{2}{\lambda}\left(\frac{1}{d_1} + \frac{1}{d_2}\right)} \tag{15-23}$$

The reference field strength $|\mathbf{E}_0|$ is the field which would exist in the absence of the obstacle. Thus, if one encountered a situation involving propagation over a curved surface dominated by a high ridge, he would determine the path loss by considering the path loss, in dB, due to knife-edge diffraction. A plot of the path loss as a function of the parameter v is shown in Fig. 15-8.

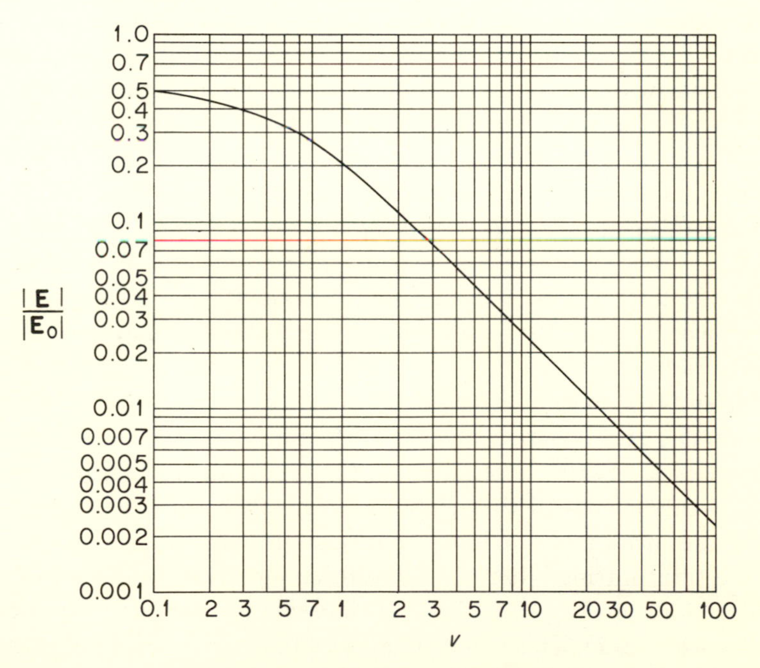

Fig. 15-8. Knife-edge shadow loss.

It is clear that one must plot a path profile in order to determine what value of effective earth radius to use in applying the smooth-earth diffraction theory and the height of a knife edge which is to be used to approximate any ridges present along the propagation path.

An example of knife-edge diffraction theory applied to a propagation path will be given in the next article.

15–5. RADIO-WAVE PROPAGATION AROUND THE EARTH

The propagation of radio waves around the earth is profoundly affected by the composition of the earth and its surrounding atmosphere. The earth is characterized by a conductivity σ, a relative permittivity ϵ_r, and a relative permeability μ_r. The earth's atmosphere is separated into two regions as far as its effect on radio-wave propagation is concerned. The lower layer of the atmosphere, which is called the *troposphere*, extends from the earth's surface to a height of about 11 kilometers at midlatitudes, ranging from about 18 kilometers at the equator to about 8 kilometers at the poles. The troposphere contains all of the earth's weather, nearly all of the atmospheric moisture, and most of the atmospheric weight; and is characterized by a gradual decrease in temperature with height. As we shall soon see, the dominant propagation mechanism in the troposphere is refraction, with scattering playing a lesser but nonetheless very important role. The second region of the earth's atmosphere which is of vital interest to radio scientists is a region of ionized gases above about 80 kilometers (with respect to sea level), called the *ionosphere.* The ionization mechanism is ultraviolet radiation from the sun. The dominant propagation mechanisms in the ionosphere are reflection and refraction. We will see, later in this article, that there exists a critical frequency f_c below which the ionosphere becomes completely reflective to electromagnetic radiation.

In attempting to establish a communication path between two points in the earth's atmosphere a number of factors must be considered. The simplest situations are those in which the transmitting antenna and the receiving antenna are within sight of each other and their directional patterns are such that there is only one possible path for a signal traveling from the transmitting antenna to the receiving antenna. This would be the direct wave of Fig. 15–9. The transmission loss for such situations is given by Eq. (15–7). This equation applies to a wide variety of practical communication paths, including well-designed microwave relay links, surface-to-aircraft links, air-to-air links, space communication links, etc.

The Ground-Wave Set. We will next consider the role of the earth's surface. When the receiving antenna is within sight of the transmitting

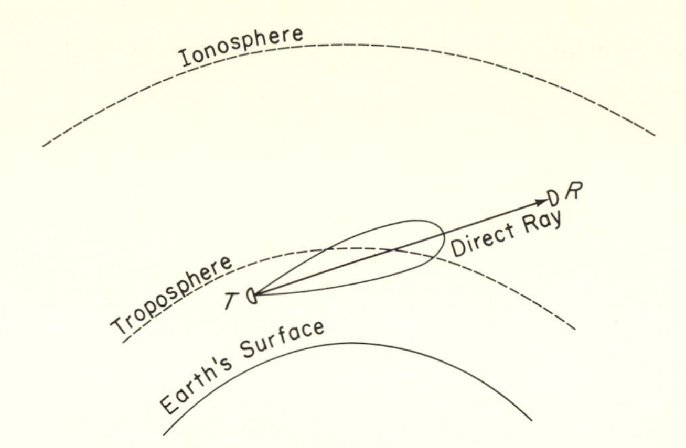

Fig. 15–9. Free-space propagation geometry.

antenna and the directional patterns of the antennas are such that energy can be reflected at the earth-atmosphere interface, a ground-reflected wave and a surface wave also become possible propagation mechanisms. The relative magnitudes of the three possible waves depend upon the electrical nature of the earth's surface and the height of the transmitting and receiving antennas. The signal at the receiver, called the *ground-wave set*, can be effectively treated by considering the earth to be plane. Bullington[4] has shown that the field components of the ground-wave set can be written as:

$$\frac{E}{E_0} = \underset{\text{direct wave}}{1} + \overset{\text{ground-reflected wave}}{Re^{j\Delta}} + \underset{\text{surface wave}}{(1 - R)Ae^{j\Delta}} + \dots \tag{15–24}$$

The relationship between the direct and the reflected wave is shown in

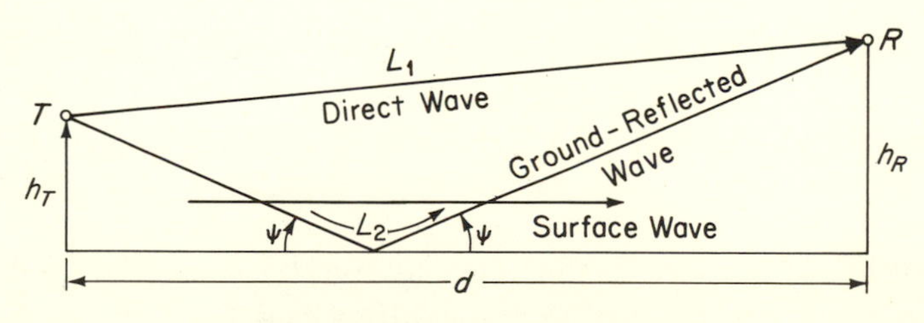

Fig. 15–10. Ground-wave set.

Fig. 15–10. The parameter R in Eq. (15–24) is the reflection coefficient of the earth's surface given by either Eq. (15–10) or Eq. (15–11), depend-

[4] K. Bullington, "Radio Propagation Fundamentals," *Bell System Technical Journal*, vol. 36, May 1957, pp. 593–626. Copyright © 1957, American Telephone and Telegraph Co., reprinted by permission.

ing upon the polarization of the source. The parameter Δ is the difference in phase between the direct wave and the reflected wave. That is,

$$\Delta = \beta(L_2 - L_1) = \frac{2\pi}{\lambda}(L_2 - L_1)$$

Consideration of Fig. 15–10 shows that

$$\Delta = \frac{2\pi}{\lambda}\{[h_T + h_R)^2 + d^2]^{1/2} - [(h_T - h_R)^2 + d^2]^{1/2}\} \qquad (15\text{–}25)$$

For d greater than about $5(h_R + h_T)$, Δ can be approximated by

$$\Delta \approx \frac{4\pi h_T h_R}{\lambda d} \qquad (15\text{–}26)$$

The surface-wave attenuation factor A in Eq. (15–24) is a function of both the polarization of the electrical signal and the ground constants. A plot of the attenuation factor A as a function of a numerical distance

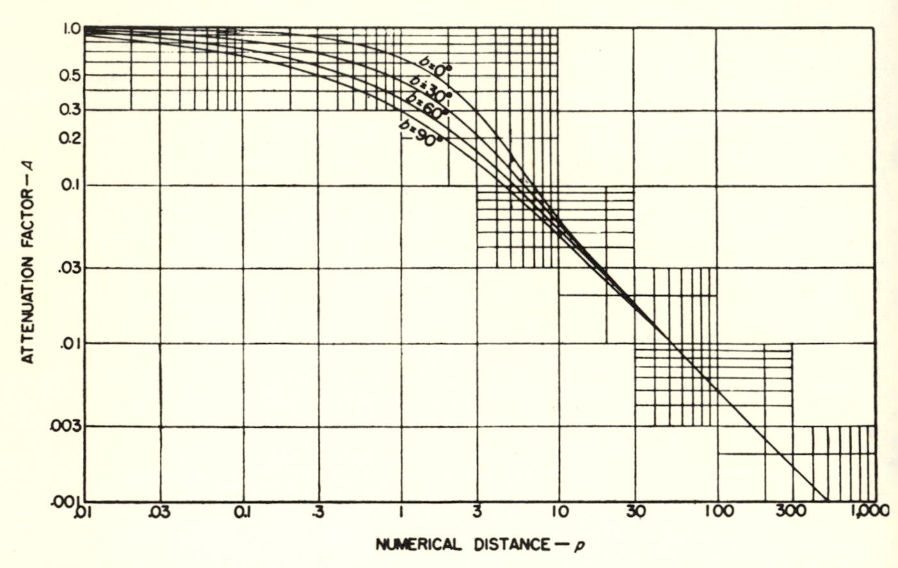

Fig. 15–11. Ground-wave attenuation factor as a function of the parameters p and b.

p has been presented by Norton[5] and is reproduced in Fig. 15–11. For vertical polarization, the numerical distance p is defined as

$$p = \frac{\pi d}{\lambda x}\frac{\cos^2 b''}{\cos b'} \approx \frac{\pi d}{\lambda x}\cos b \qquad (15\text{–}27)$$

[5] K. A. Norton, "The Calculation of Ground-Wave Field Intensity over a Finitely Conducting Spherical Earth," *Proceedings of the Institute of Radio Engineers*, vol. 29, pp. 623–639, 1941.

where

$$b = (2b'' - b') \approx \tan^{-1} \frac{\epsilon_r + 1}{x}$$

$$b'' = \tan^{-1} \frac{\epsilon_r}{x}, \qquad b' = \tan^{-1} \frac{\epsilon_r - \cos^2 \psi}{x} \approx \tan^{-1} \frac{\epsilon_r - 1}{x}$$

and

$$x = \frac{18 \times 10^3 \sigma}{f_{\mathrm{MHz}}} = 60\, \sigma \lambda_0$$

σ is the conductivity of the earth, ϵ_r is the relative permittivity of the earth, and ψ is the grazing angle (see Fig. 15-10). For horizontal polarization,

$$p \approx \frac{\pi\, xd}{\lambda} \frac{1}{\cos b'} \tag{15-28}$$

where $b = 180° - b'$.

For the situations where the transmitter and the receiver are at or very near the earth's surface, $\Delta \cong 0$, and $R \approx -1$. In this case

$$\frac{|\mathbf{E}|}{|\mathbf{E}_0|} \approx 2A \tag{15-29}$$

The only wave that is significant under these conditions is the surface wave. Equation (15-29) is applicable to standard broadcast transmitting antennas, which are normally sited on the surface of the earth. In order to apply this theory to specific situations involving propagation on the surface of the earth, we need typical values of the electrical constants. The electrical constants for several types of earth surface are tabulated in Table 15-1. All examples are nonmagnetic so $\mu_r = 1$.

TABLE 15-1

Electrical Constants for Typical Terrain Examples

Surface Type	ϵ_r	σ, mhos/m
Seawater	80	4
Fresh water	80	0.005
Dry sandy soil	10	0.002
Marshy soil	12	0.008
Fertile soil	15	0.01
Grazing land	13	0.005
Rocky terrain	10	0.002
Mountainous terrain	5	0.001

Example 15–2

To emphasize the role of the terrain in the excitation of the surface-wave component of the ground-wave set, we will compare the field strength ratio $|\mathbf{E}|/|\mathbf{E}_0|$ of a vertical monopole for two terrain conditions: case 1—the monopole is situated on a surface of dry sandy soil ($\sigma = 0.002$ mhos/m, $\epsilon_r = 10$); case 2—the monopole is situated on a surface of fertile soil ($\sigma = 0.01$ mhos/m, $\epsilon_r = 15$). The frequency of the radio wave is 1 MHz ($\lambda = 300$ m). The distance $d = 10$ km.

For case 1:

$$x = 60\,\sigma\lambda_0 = (60)(0.002)(300) = 36$$

$$b \cong \tan^{-1}\frac{11}{36} = \tan^{-1} 0.305 = 16.9°$$

$$p = \frac{\pi d}{\lambda x}\cos b = 2.86$$

From Fig. 15–11, $A \cong 0.27$.

For case 2:

$$x = (60)(0.01)(300) = 180$$

$$b = \tan^{-1}\frac{16}{180} = 5.1°$$

$$p = 0.58$$

From Fig. 15–11, $A \approx 0.75$.

For case 1, Eq. (15–29) shows that

$$\frac{|\mathbf{E}|}{|\mathbf{E}_0|} = 0.54$$

while for case 2, it is seen that

$$\frac{|\mathbf{E}|}{|\mathbf{E}_0|} = 1.5$$

The signal level over the fertile soil path is more than 9 dB greater than that over the dry sandy soil path. One of the reasons why standard broadcast band transmitters are often located in somewhat marshy areas with networks of good conducting cables located on or just below the surface is to achieve better excitation of the surface wave.

As the transmitting and receiving antennas are raised above the earth's surface, the phase factor Δ is no longer zero.

Generally speaking, the factor A can be neglected as long as both antennas are more than a wavelength above the earth or more than 5 to 10 wavelengths above seawater ($\epsilon_r = 80$, $\sigma = 4$ mhos/meter). The reflection coefficient is approximately -1 for either vertical or horizontal polarization if ψ is very small. Under these conditions we can write

$$\frac{E}{E_0} = 1 - e^{j\Delta} \tag{15–30}$$

Example 15–3

Determine the path loss for a horizontally polarized 100-MHz signal propagating over a 20-km path comprised of dry sandy soil ($\epsilon_r = 10$, $\sigma = 0.002$ mhos/m). The transmitter height is 100 meters. The receiver is also 100 meters above the surface.

Clearly, ψ is very small:

$$\psi \approx \frac{100 \text{ m}}{10 \text{ km}} = 10^{-2}$$

$$\cos\psi \approx 1$$
$$\sin\psi \approx \psi = 0.01$$

The complex permittivity of the earth, ϵ_2 of Eq. (15–11), is replaced by $\epsilon_2 - j\dfrac{\sigma}{\omega\epsilon_0}$:

$$\epsilon_2 - j\frac{\sigma}{\omega\epsilon_0} = 10 \frac{-j0.002}{(2\pi \times 10^8)\left(\dfrac{1}{36\pi} \times 10^{-9}\right)} = 10 - j0.36$$

From Eq. (15–11),

$$R_h = \frac{0.01 - \sqrt{9 - j0.36}}{0.01 + \sqrt{9 - j0.36}} = \frac{0.01 - 3\sqrt{1 - j0.04}}{0.01 + 3\sqrt{1 - j0.04}} \approx -1$$

Calculate Δ:

$$\Delta = \frac{4\pi}{3}\frac{(100)(100)}{(2 \times 10^4)} = \frac{2\pi}{3}$$

Now insert values into Eq. (15–30):

$$\frac{E}{E_0} = 1 - e^{j2\pi/3} = 1.5 - j\sqrt{3}/2$$

and

$$\frac{|\mathbf{E}|}{|\mathbf{E}_0|} = \sqrt{2.25 + 0.75} = 1.732$$

$$L_p = -20 \log_{10} 1.732$$
$$L_p = -2.4 \text{ dB}$$

We can manipulate Eq. (15–30) to obtain

$$\frac{E}{E_0} = -2je^{j\frac{\Delta}{2}} \sin\frac{\Delta}{2}$$

or, dealing in magnitudes,

$$\frac{|\mathbf{E}|}{|\mathbf{E}_0|} = 2 \sin\frac{\Delta}{2} \tag{15–31}$$

Equation (15–31) emphasizes the well-known lobe structure that is characteristic of radio-wave propagation over a plane (or approximately plane) surface.

A concept related to the lobe structure just discussed is the Fresnel zone. All points from which a wave can be reflected with a path difference ($L_2 - L_1$) of one half-wavelength form the boundary of the first Fresnel zone. Similarly, the locus of all points from which a wave could be reflected with a path difference of $n/2$ wavelengths forms the boundary of the nth Fresnel zone. This problem is usually concerned with establishing a parameter called the *Fresnel zone clearance.* This is the distance between the direct wave and the point of reflection H_n, as shown in Fig. 15–12. The boundary of the nth Fresnel zone is defined such that

$$\Delta = \frac{2\pi}{\lambda}(L_2 - L_1) = n\pi$$

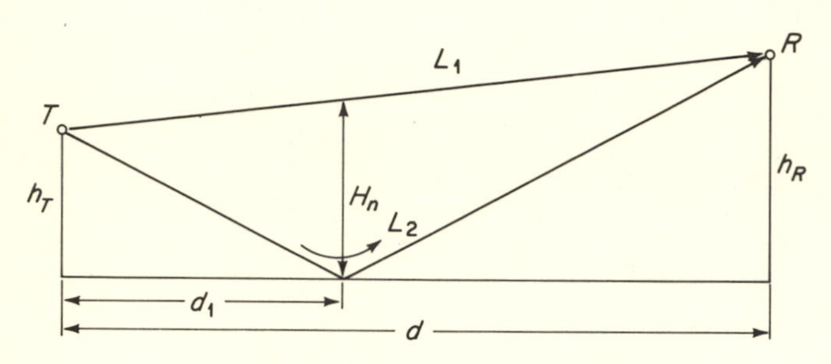

Fig. 15–12. Fresnel zone clearance.

The nth Fresnel zone clearance is then calculated to be

$$H_n = \sqrt{\frac{n\lambda\, d_1(d - d_1)}{d}} \tag{15–32}$$

Up to this point, the reflecting terrain has been considered to be perfectly smooth. If the terrain is rough (which it usually is), the reflection coefficient will be reduced. The reflection coefficient for rough ground is a difficult parameter to determine as it depends upon the electrical constants of the surface, the polarization of the incident energy, the angles of incidence and reflection (not necessarily equal), and the surface roughness statistics. A good treatment of this problem is given by Beckmann and Spizzichino.[6]

A simple way to treat this problem is to establish a criterion to determine whether a surface should be considered to be smooth or rough. The most frequently used criterion is the Rayleigh criterion. The essential features of the Rayleigh criterion are illustrated in Fig. 15–13. The surface is characterized by a distribution of irregularities of height h. A plane wave is incident on the surface at an angle ψ. The angle of

[6] P. Beckmann and A. Spizzichino, *The Scattering of Electromagnetic Waves* (New York: Pergamon Press, 1963).

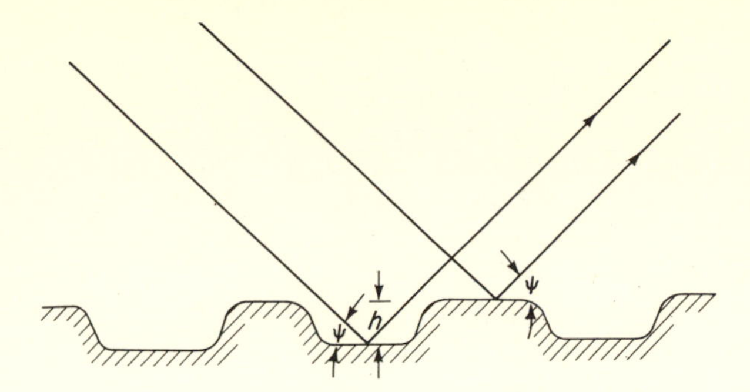

Fig. 15–13. Illustration of Rayleigh criterion.

reflection will also be assumed to be the angle ψ. The phase difference between the two rays shown in Fig. 15–13 is

$$\beta\Delta l = \frac{2\pi}{\lambda}(2h \sin \psi) = \frac{4\pi h}{\lambda} \sin \psi \tag{15–33}$$

If h/λ and ψ are such that $\beta\Delta l$ is very small, the surface looks smooth. However, if $\beta\Delta l = \pi$, these two rays cancel and no energy is reflected at the angle ψ. This means that the energy is redistributed in other directions and the classical reflection coefficient concept is not applicable. The Rayleigh criterion establishes an upper limit to Eq. (15–33) below which the surface is considered smooth. This is usually written as

$$\frac{4\pi h}{\lambda} \sin \psi \leq \frac{\pi}{4}$$

or

$$h \leq \frac{\lambda}{16 \sin \psi} \tag{15–34}$$

For values of h satisfying Eq. (15–34), the surface is considered smooth and all of the preceding theory applies to the ground-reflected wave. For h greater than the maximum value satisfying Eq. (13–34), the problem is more difficult analytically and is beyond the scope of this text.

A tabulation of the range of reflection coefficient magnitudes for several typical types of terrain is presented in Table 15–2 in the section on tropospheric wave propagation.

If the propagation path is such that the transmitter and receiver are not within sight of one another, the effects of earth curvature must be taken into account. The smooth earth diffraction theory of Article 15–4 can be used to predict the field strength and path loss for this case. We will now consider two examples of the application of diffraction theory.

Example 15–4

As an example of the application of the theory, let us determine the path loss for a basic communication path 80 statute miles long, utilizing transmitting and receiving antenna heights of 100 feet and a wavelength of 10 cm (3 GHz). We will also assume a 4/3 earth radius sphere to account for the gradient of the refractive index. This concept will be discussed in the next article. Thus, we take $a_e = 5280$ miles. Calculate d_0 and h_0, with the results

$$d_0 = \left[\frac{(5280\text{ mi})^2(10\text{ cm})(0.0328\text{ ft/cm})}{(3.14)(5280\text{ ft/mi})}\right]^{1/3} = 8.2\text{ mi}$$

$$h_0 = \frac{1}{2}\left[\frac{(5280\text{ mi})(5280\text{ ft/mi})(10\text{ cm})^2(0.0328\text{ ft/cm})^2}{(3.14)^2}\right]^{1/3}\text{ ft}$$

$$h_0 = 33.6\text{ ft}$$

Now check to see that the diffraction-zone assumption is valid:

$$X_{\min} = \sqrt{Z_1} + \sqrt{Z_2}$$

In this case,

$$Z_1 = Z_2 = \frac{100}{33.6} \approx 3 \quad \text{and} \quad X = \frac{80}{8.2} = 9.75$$

Substituting to determine $X_{\min}$:

$$X_{\min} = 2\sqrt{3} = 3.47$$

The diffraction-zone assumption is valid and we calculate

$$\begin{aligned} 2\sqrt{\pi X}e^{-2.02X} &= 2[30.6]^{1/2}e^{-21.4} = 11\,e^{-21.4} \\ &= 22 - 186\text{ dB} \\ &= -164\text{ dB} \end{aligned}$$

Using the value of Z determined above, and consulting Fig. 15–1(b), we see that the height-gain functions contribute +15 dB apiece; thus, the propagation path loss is

$$L_p = -20\log_{10}\frac{|\mathbf{E}|}{|\mathbf{E}_0|} = -[+15 + 15 - 164]\text{ dB} = 134\text{ dB}$$

Example 15–5

As an example of the use of the knife edge of diffraction theory, suppose that the propagation path of the smooth-earth diffraction example just considered was dominated by a relatively sharp mountain peak projecting 6000 ft above the terrain and located at midpath. Such a situation is shown in Fig. 15–14. We will model the mountain as a 6000-foot-tall knife edge.

Solve for v:

$$v = 6000\sqrt{\frac{2 \qquad 2}{(0.1)(3.25)(40)(5280)}} = \frac{6000}{1.31 \times 10^2} = 50.5$$

Using Fig. 15–8, $L = 0.0045$ or $|\mathbf{E}_0|/|\mathbf{E}| = 222$, and the loss in dB is given by

$$L_p = 20\log_{10}222 = 20(2.35) = 47\text{ dB}$$

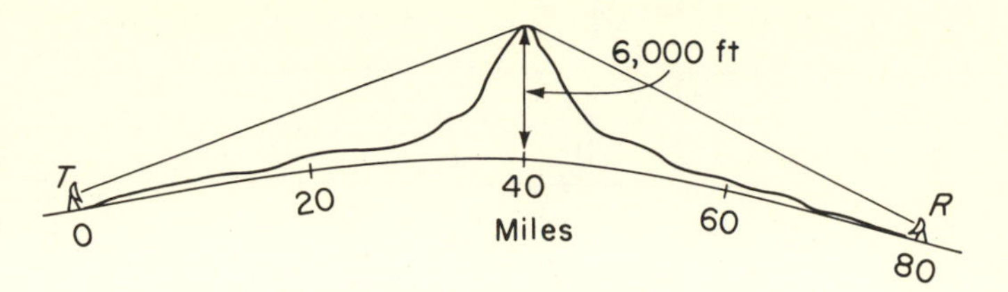

Fig. 15–14. Knife-edge diffraction example.

This is a substantial improvement over the situation for smooth-earth diffraction theory. Of course, this has been a highly idealized example; however, there are a number of successful propagation links in existence where knife-edge diffraction is the dominant propagation mechanism.

The most obvious idealization is that the dominating ridge is sharp enough to be considered a knife edge. There is a condition,[7] placed on an obstacle which has a radius of curvature R, which must be satisfied in order to use knife-edge diffraction theory to describe the fields in the shadow region with little error. This condition is

$$|\psi \text{ (radians)}| < \frac{1}{4}\left(\frac{\lambda}{R}\right)^{1/3}$$

where ψ is the angle between the edge of the shadow region and the ray from the obstacle to the receiver. In the example just considered,

$$\lambda = 0.325 \text{ ft}, \quad \psi \approx 2\left[\frac{6000}{5280(40)}\right] \approx 0.057 \text{ rad}$$

$$\frac{\lambda}{R_{\max}} = 0.012$$

$$R_{\max} = \frac{0.325}{0.012} = 27 \text{ ft}$$

This is a very sharp ridge, but the frequency is probably higher than one would expect to use in a situation like this. If the frequency were reduced to 300 MHz, for example, the maximum radius of curvature would be 270 ft, which is more realistic. The actual radius of curvature of an obstacle in a propagation path can only be determined by plotting a profile of the terrain over which the radio wave must propagate.

The Tropospheric Wave. Frequently, the antenna heights and directional patterns are such that the signal path is entirely within the troposphere and the effects of the earth are negligible. This wave is called the *tropospheric wave.*

The troposphere, which is that portion of the atmosphere next to the earth's surface extending 8 to 18 kilometers above the surface, is charac-

[7] S. O. Rice, "Diffraction of Plane Radio Waves by a Parabolic Cylinder," *Bell System Technical Journal,* vol. 33, no. 2, March 1954, pp. 417–504.

terized by a refractive index which is a function of temperature, pressure, and water vapor. This empirically derived expression is

$$(n - 1) \times 10^6 = \frac{77.6}{T}\left[p + \frac{4810e}{T}\right] \tag{15-35}$$

where p is the dry air pressure in millibars, T is temperature in degrees Kelvin, e is the partial pressure of water vapor in millibars, and n is the index of refraction. The water molecule is easily polarized and contributes substantially to the refractive index of the troposphere. The refractive index of the troposphere is considered to consist of a component which is fairly constant for a given locality but decreases with height, due to gravitational stratification of pressure. Superimposed on this is a fluctuating component caused by blobs or parcels of air of differing temperature and/or water vapor content. The fluctuating term is responsible for the scattering of UHF energy beyond the radio horizon via the propagation mechanism called tropospheric scatter.

The gravitationally stratified component of the refractive index is described by an exponential model:[8]

$$N = (n - 1) \times 10^6 = N_s \exp\{-C_e(h - h_s)\} \tag{15-36}$$

N is called the refractivity of the atmosphere, N_s is the surface refractivity, h is the height in kilometers, and h_s is the surface height in kilometers. C_e is defined as

$$C_e = \ln \frac{N_s}{N(1\ \text{km})}$$

Both heights are referred to sea level.

The average surface refractivity for the United States is $N_s = 301$ and $C_e = 0.139632$. The exponential atmosphere tends to bend radio waves downward, which complicates the analysis of radio waves propagating in such an atmosphere. This effect is usually taken into account by modifying the earth's radius so that the radio wave can propagate along straight rays over the modified radius earth. The relative curvature between the wave path and the earth's surface is maintained in the modification process. Define R_e' as the radius of curvature of the modified earth, R_{rw}' as the curvature of the ray over the modified earth, $R_e = -a$ as the radius of curvature of the actual earth, and R_{rw} as the radius of curvature of the radio wave in the actual earth's atmosphere. The relative

[8] B. R. Bean and E. J. Dutton, *Radio Meteorology*, National Bureau of Standards Monograph 92, 1966, page 65.

curvature between the ray and the earth will be preserved in both situations if

$$\frac{1}{R'_{rw}} - \frac{1}{R'_e} = \frac{1}{R_{rw}} - \frac{1}{R_e} \tag{15-37}$$

Notice that a positive radius of curvature is directed upward or outward radially. $1/R_{rw}$ is given by Eq. (15-15) with $\psi = 0$. If we want the ray path to be a straight line over the modified earth, $R'_{rw} = \infty$ and

$$-\frac{1}{R'_e} = \frac{1}{a} + \left.\frac{dN}{dh}\right|_{h=h_s} \times 10^{-6}$$

Using Eq. (15-36), we can write

$$\left.\frac{dN}{dh}\right|_{h=h_s} \times 10^{-6} = -301C_e \times 10^{-6} \approx -\frac{1}{4a}$$

since

$$a = 6.35 \times 10^3 \text{ km}$$

Thus,

$$-\frac{1}{R'_e} = \frac{1}{ka} = \frac{1}{a} - \frac{1}{4a}$$

where k is the factor by which the actual earth radius is modified. Clearly, $k = 4/3$.

This leads to the so-called 4/3 earth model. If one modifies the earth's radius by a factor of 4/3 and treats the resultant earth as airless, the effects of the exponential atmosphere have been automatically incorporated into the analysis.

At times the gradient of the refractive index can bend the ray path downward with a radius of curvature greater than the curvature of the earth. In such a case the radio wave is trapped near the surface of the earth, since it follows a path wherein it would reflect off the earth's surface and into the atmosphere only to be refracted back to the earth's surface. This is one form of "ducting," called a *surface duct*. One good way to examine this process is to consider the path of a ray over a plane earth under the conditions imposed by Eq. (15-37):

$$\frac{1}{R'_{rw}} = \frac{dN}{dh} \times 10^{-6} + \frac{1}{a}$$

If the first term, $(dN/dh) \times 10^{-6}$, is sufficiently negative, the radius of curvature of the ray over the plane surface will be negative and the ray path will bend downward and be reflected off the surface.

Another ducting possibility would occur when there is a large positive gradient of refractive index at one level with a large negative gradient at some height above it. Under such conditions energy may be trapped in the region between the gradient levels, since the ray path is forced to bend downward by the large negative gradient at the higher level, then forced to bend upward by the large positive gradient at the lower level. This situation is called an *elevated duct.* Figure 15–15 summarizes the refractive-

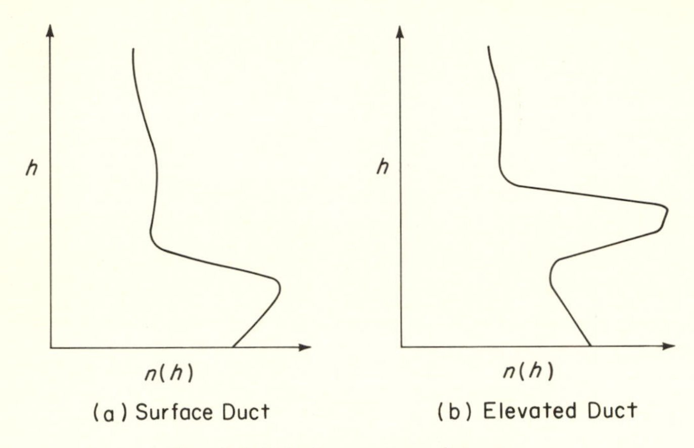

Fig. 15–15. Examples of ducts.

index conditions for surface and elevated ducts. Ducting usually results in occasional periods of abnormally high field strengths although it is possible for a duct to direct energy away from a desired receiving point.

The gradient of the refractive index is an important parameter to be considered in the design of radio relay links. These relay stations operate at microwave frequencies over propagation paths that are essentially line-of-sight. In such cases, the main consideration is to properly account for the refraction of the radio wave by the gradient of refractive index so as to avoid reflections from obstacles which will interfere with the main beam. To do this requires the preparation of a propagation-path profile and a knowledge of the gradient of the refractive index. An example of a 40-mile propagation-path profile is shown in Fig. 15–16. Plotted on the terrain profile are several different propagation paths, each characterized by a different radius of curvature of the main ray. The different radius-of-curvature paths correspond to different vertical gradients of refractive index over the propagation path. It is assumed that this gradient is the same over the entire path.

The problem in engineering radio relay links is to be sure that the main beam of the transmitting pattern is not bent too close to an obstacle such as a hill, a building, or some other source of reflection. The usual

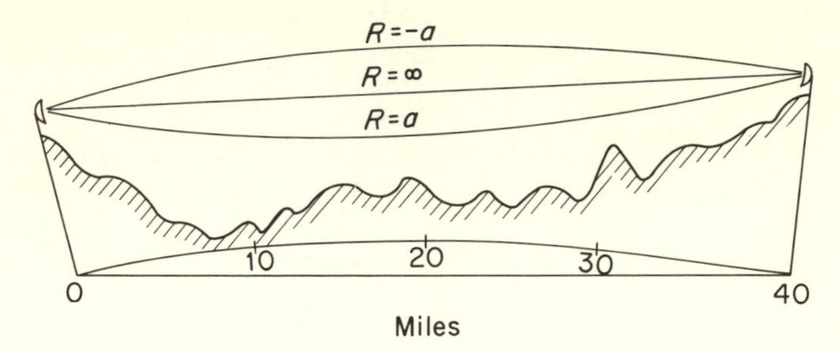

Fig. 15–16. Refraction in line-of-sight propagation paths.

rule is to ensure a clearance of six-tenths of the first Fresnel zone clearance distance as determined by Eq. (15–32). To calculate this distance one must know the worst gradient of refractive index condition for which reliable communication is expected. Data of this nature are difficult to get so the normal procedure is to provide ample design margin.

Even though there may be a reflected signal present in some cases, the reflected ray will not completely cancel the direct ray because the reflection coefficients are usually less than unity. Some typical reflection coefficient values and the depth of the null caused by destructive interference of the direct and reflected rays are presented in Table 15–2.[9]

TABLE 15–2

Typical Average Reflection Coefficient Magnitudes

Type of Terrain	*R*	*Depth of Fresnel-Zone Fade (dB)*
Heavily wooded forest land	0 to 0.1	0–2
Partially wooded forest land	0.1 to 0.4	2–5
Sagebrush and high grass	0.5 to 0.7	5–10
Rough seawater, low grassy areas	0.7 to 0.8	10–20
Smooth seawater, salt flats	0.9 +	20–40 +

The fluctuating component of the refractive index is taken into account by characterizing the variations in permittivity of the troposphere in terms of a scattering factor $\sigma_v(\theta, X)$. The scattering factor can then be used in Eq. 15–17 to determine the scattered power. Booker and Gordon

[9] R. U. Laine, "Microwave Propagation and Reliability," *Automatic Electric Technical Journal*, vol. 9, no. 4, October 1969, p. 149.

have developed an expression for the scattering factor in terms of parameters of the troposphere:[10]

$$\sigma_v(\theta, X) = \frac{\left(\frac{\overline{\Delta\epsilon}}{\epsilon}\right)^2 \left(\frac{2\pi l}{\lambda}\right) \sin^2 X}{\lambda \left[1 + \left\{\frac{4\pi l}{\lambda} \sin \frac{\theta}{2}\right\}^2\right]^2} \tag{15-38}$$

where $(\overline{\Delta\epsilon}/\epsilon)^2$ is the root-mean-square of the spatial fluctuations in the permittivity of the troposphere and l is the scale factor of the spatial variations.

The scattering from the variations in the refractive index of the troposphere is the mechanism for the propagation of UHF energy beyond the radio horizon. Through the use of powerful transmitters, sensitive receivers, and high-gain antennas one can reliably communicate at UHF frequencies over distances of several hundred miles. The propagation mechanism is such that it is difficult to develop analytical equations suitable for use in the design of tropospheric scatter links. A number of useful prediction formulas have been developed in an empirical manner, however, and one of these is given in Eq. (15–39):[11]

$$L_p = [30 \log f - 20 \log d + F(\theta d) + L_c - V(d_e)] \text{ dB} \tag{15-39}$$

where f = frequency in MHz
d = distance in km
θ = scattering angle in radians

$F(\theta d)$ is an attenuation function which is presented graphically and $V(d_e)$ is a climate variation factor which varies with geographical location and is also available in graphical form. L_c is the aperture-to-medium coupling loss, determined by

$$L_c = 0.07 \exp [0.055(g_T + g_R)]$$

The coupling-loss term is included as a result of the observation that in practice one does not realize the full plane-wave gain of a large antenna when it is utilized in a troposcatter link. This is felt to be due to the fact that the fields in the aperture of the receiving antenna are not plane waves but are more likely to be non-uniform and characterized by a correlation distance which is less than the aperture dimensions. The parameters $F(\theta d)$ and $V(d_e)$ are presented in a C.C.I.R. Report of 1966.

[10] H. G. Booker and W. E. Gordon, "A Theory of Radio Scattering in the Troposphree," *Proceedings of the Institute of Radio Engineers*, vol. 38, no. 4, April 1950, pp. 401–412.

[11] P. L. Rice, A. G. Longley, K. A. Norton, and A. P. Barsis, *Transmission Loss Prediction for Tropospheric Communication Circuits*, NBS Technical Note 101, May 1965.

An example of the sort of path loss expected for a troposcatter link is a 302-km propagation path from New Jersey to Massachusetts operating at 3670 MHz, which exhibited a measured path loss of 233.5 dB. Calculations using Eq. (15–39) and the appropriate graphs predicted a path loss of 233.7 dB.[12] This is remarkably close; such accuracy would not normally be expected. The example does serve to indicate the sort of path loss to be expected, however.

Due to the nature of the propagation mechanism, the received signal exhibits amplitude variations (fading) characterized by rapidly time-varying fluctuations superposed on a slowly varying component. Experimental observations of the received signal on tropospheric scatter links show that the statistical properties are approximated quite well by a Rayleigh distribution. Some of the long-term variability of the received signal can be overcome by employing a spatial array of receiving antennas. This is called *spatial diversity.*

Another characteristic of the troposphere which is becoming increasingly important in the design of communication circuits is the absorption of electromagnetic energy by constituent gases in the lower atmosphere. A plot of the absorption coefficient versus frequency is shown for the gaseous constituents H_2O (water vapor) and O_2 (oxygen) in Fig. 15–17.[13]

The lowest absorption frequency is seen to be a water vapor line at 22.5 GHz. There are also oxygen lines at 60 and 120 GHz plus another water vapor line at 188 GHz. These absorption bands impose practical limitations on the utilization of this portion of the electromagnetic spectrum in communication circuits. Advantages of operating in this portion of the spectrum, such as the ability to achieve very-high-gain antenna systems of moderate physical size, might well be overshadowed by the losses due to absorption.

Another factor of great concern, particularly to the designers of earth-space communication links, is the signal degradation due to rainfall. Plots of signal absorption versus frequency for various values of rainfall rate are also shown in Fig. 15–17. The only way to overcome the effects of rainfall on a communication path is to use spatial diversification so that at least one earth-based antenna is not receiving a signal which has been degraded by a local rainstorm.

[12] This example, plus a number of others, is presented in the article by R. Larsen, "A Comparison of Some Tropospheric Prediction Methods," *Tropospheric Wave Propagation, IEE Conference Publication, No. 48,* 1968.

[13] J. H. Van Vleck, E. M. Purcell, and H. Goldstein, *Propogation of Short Radio Waves,* edited by D. E. Kerr (New York: Dover Publications, Inc., 1965), Chapter 8.

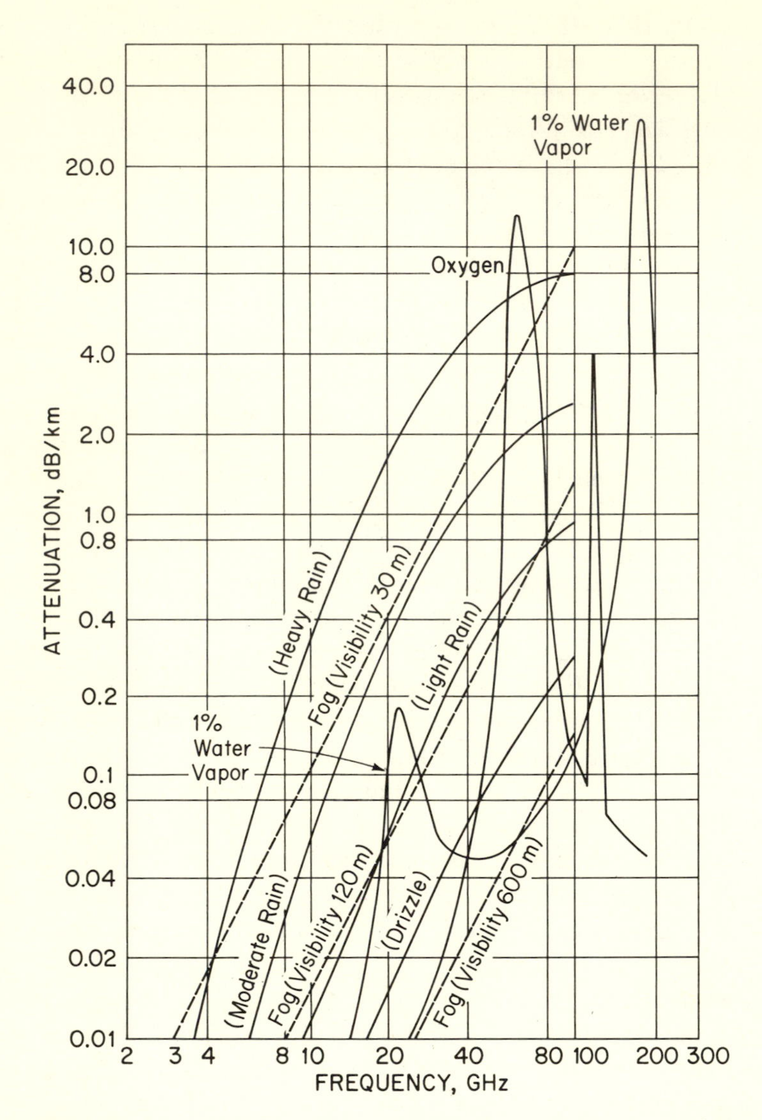

Fig. 15–17. Absorption due to atmospheric gases and rainfall.

A final comment on the troposphere concerns the propagation of optical frequencies. At these frequencies the water molecule acts as a scatterer of electromagnetic energy and does not contribute to the refractive index. The resultant expression for the refractivity is

$$N_v = (n - 1) \times 10^6 = \frac{77.6}{T} p \qquad (15\text{–}40)$$

$$N_v = 226\rho$$

where p = dry air pressure in millibars, T = temperature in degrees Kelvin, and ρ is the density of the atmosphere in kilograms per cubic meter.

Ionospheric Propagation. The ionosphere is the region of the earth's atmosphere which is characterized by the existence of an electrically neutral plasma formed by free electrons and protons. This region is approximately 80 km above the earth's surface and higher. The ionization mechanism is the radiation from the sun, and the concentration of electrons shows a pronounced variation with height as well as significant diurnal variation. The permittivity of the ionosphere is developed by assuming the ionosphere to be an electrically neutral plasma comprised of N electrons per unit volume and N protons per unit volume. The permittivity expression is developed in Appendix G. For purposes of studying the propagation of an electromagnetic wave in the ionosphere, the simplest expression for the permittivity of the ionosphere will be used. That is,

$$\epsilon = \epsilon_0 \epsilon_r = \epsilon_0 \left(1 - \frac{\omega_p^2}{\omega^2}\right) \tag{15–41}$$

The term $Ne^2/m\epsilon_0$ is designated as ω_p^2, where e is the charge of an electron, m is the mass of an electron, and ω_p is called the "plasma frequency." Equation (15–41) neglects collisions, which introduce a loss or dissipation term into the permittivity expression. The effects of the earth's magnetic field are also ignored. These effects make the ionosphere anisotropic, and the permittivity must be expressed as a tensor.

If one inserts known values for e, m, and ϵ_0 in Eq. (15–41), the relative permittivity expression for the ionosphere becomes

$$\epsilon_r = 1 - \frac{81N}{f_{\mathrm{kHz}}^2} \tag{15–42}$$

where N = number of electrons/cm^3. This expression for refractive index is interesting in that it is possible for ϵ_r to range from unity through zero to a very large negative number, depending upon the relationship between f and N.

A plot of electron density versus height is shown in Fig. 15–18.

A table of ionospheric regions and layers is presented by Davies[14] and is reproduced in Table 15–3. The layers are shown in Fig. 15–18.

When a radio wave is incident on the ionosphere, it will penetrate into the ionosphere and change direction in accordance with Snell's law:

$$n \sin \theta = n_0 \sin \theta_0$$

[14] K. Davies, *Ionospheric Radio Propagation*, National Bureau of Standards Monograph 80, 1965, Chapter 1.

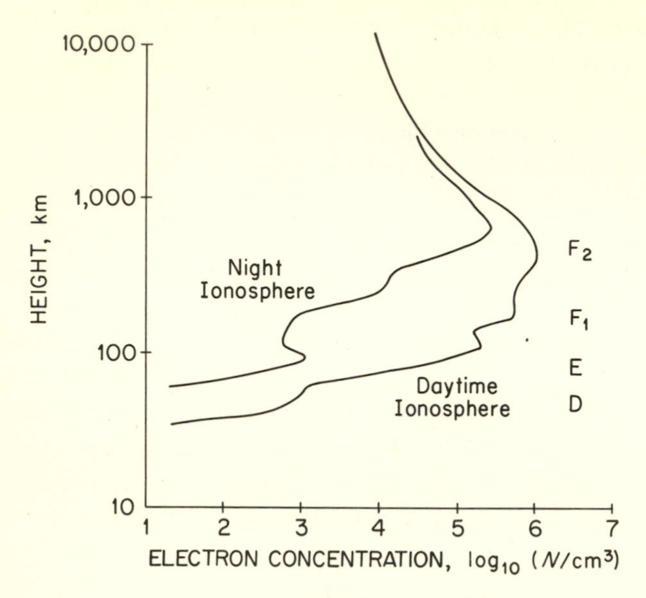

Fig. 15–18. Typical electron density profiles for the day and night ionosphere in the middle latitudes.

TABLE 15–3

Ionosphere Regions and Layers

Height (km)	*Region*	*Layers*
50–90	D	D
90 to 120–140	E	E_1, E_2, E_s
Above 120–140	F	F_1, F_2

where θ is the angle between the wave normal and the vertical. At some point the index of refraction may be such as to cause the wave to be traveling parallel to the earth's surface and then continue to turn back toward the earth. The turning point is where $\theta = \pi/2$ or $\sin\theta = 1$. This means that

$$n = n_0 \sin\theta_0$$

Using Eq. (15–42) with $n_0 = 1$ results in

$$\sqrt{1 - \frac{81N}{f_{\mathrm{kHz}}^2}} = \sin\theta_0 \tag{15–43}$$

If the incident wave is vertical, $\theta_0 = 0$ and

$$1 - \frac{81N}{f_{\mathrm{kHz}}^2} = 0$$

The frequency which satisfies this relationship is called the *critical frequency*, f_c. Clearly,

$$f_{c_{\mathrm{kHz}}} = 9N^{1/2} \tag{15–44}$$

If the wave is incident at some other angle, say θ_0, Eq. (15–43) can be solved for frequency, with the result

$$f_{\mathrm{kHz}} = 9N^{1/2} \sec \theta_0 = f_{c_{\mathrm{kHz}}} \sec \theta_0 \tag{15–45}$$

This is called the maximum usable frequency (MUF) since it is the highest frequency for which a wave can be reflected back for a given N and θ_0. At frequencies below the MUF reflections will take place from a given layer.

With this introduction, let us consider the characteristics of the ionosphere for a uniform plane wave. Again we begin with expressions for the constitutive parameters:

$$\epsilon = \epsilon_0 \epsilon_r = \epsilon_0 \left(1 - \frac{81N}{f_{\mathrm{kHz}}^2}\right)$$

$$\mu = \mu_0$$

$$\sigma = 0$$

The wave impedance and wave propagation factor become

$$\eta = \sqrt{\frac{\mu}{\epsilon}} = \frac{\eta_0}{\sqrt{1 - \dfrac{81N}{f_{\mathrm{kHz}}^2}}} \tag{15–46}$$

$$\gamma = \alpha + j\beta = j\omega\sqrt{\mu_0\epsilon_0}\sqrt{1 - \frac{81N}{f_{\mathrm{kHz}}^2}} = j\beta_0\sqrt{1 - \frac{81N}{f_{\mathrm{kHz}}^2}} \tag{15–47}$$

where the subscript 0 denotes free-space quantities. Consideration of Eq. (15–47) shows the velocity of propagation of the radio wave in the ionosphere to be

$$v_p = \frac{\omega}{\beta} = \frac{c}{\sqrt{1 - \dfrac{81N}{f_{\mathrm{kHz}}^2}}} \tag{15–48}$$

By reference to Appendix E, the group velocity v_g can be determined after some manipulation, to be

$$v_g = \frac{d\omega}{d\beta} = c\sqrt{1 - \frac{81N}{f_{\mathrm{kHz}}^2}} \tag{15–49}$$

This is the velocity of energy transmission in the ionosphere.

If an electromagnetic signal which is swept in frequency from a very low frequency to a frequency above the critical frequency for the highest level of electron concentration in the ionosphere is normally incident on the ionosphere, energy will be reflected back to the source at all frequencies for which the wave impedance is imaginary. A quantity called the *virtual height* is determined by measuring the time required for a signal to propagate to the reflection point and back, assuming that the wave had traveled at the speed of light. That is

$$h' = \text{virtual height} = \tfrac{1}{2}ct \tag{15-50}$$

where c is the velocity of light in vacuum and t is the time required for the signal to propagate to the reflection point in the ionosphere and back.

The wave velocity which determines the propagation time is the group velocity v_g since it is the time required for energy to travel to the point of reflection in the ionosphere and return. An instrument which measures the virtual height as a function of frequency is called an "ionosonde." This instrument utilizes a swept frequency source operating from 1 MHz to around 20 MHz. The output of the ionosonde is usually recorded on photographic film or facsimile. Such a record is called an "ionogram." The essential features of an ionogram are shown in Fig. 15–19.

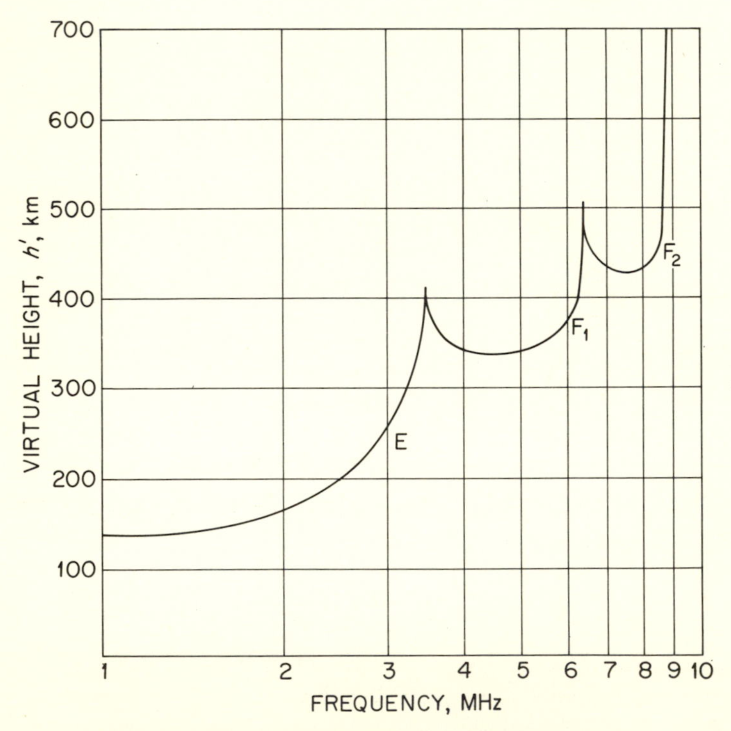

Fig. 15–19. Virtual height, h', versus frequency.

The reflection points for the E region, F_1 region, and F_2 region are indicated on the figure. The reason for the peaks between the regions is due to the fact that the wave group velocity decreases quite drastically with frequency in a given layer as the frequency approaches the critical frequency for the layer. After penetrating the layer the wave is reflected at a different layer having a higher critical frequency, so the group velocity quickly increases and the transit time drops. The process then repeats itself for the new layer. The concept of virtual height is quite important in the design of ionospheric communication paths. This is because of theorems due to Martyn and to Breit and Tuve.[15] The application of these theorems is illustrated in Fig. 15–20 for the case of a

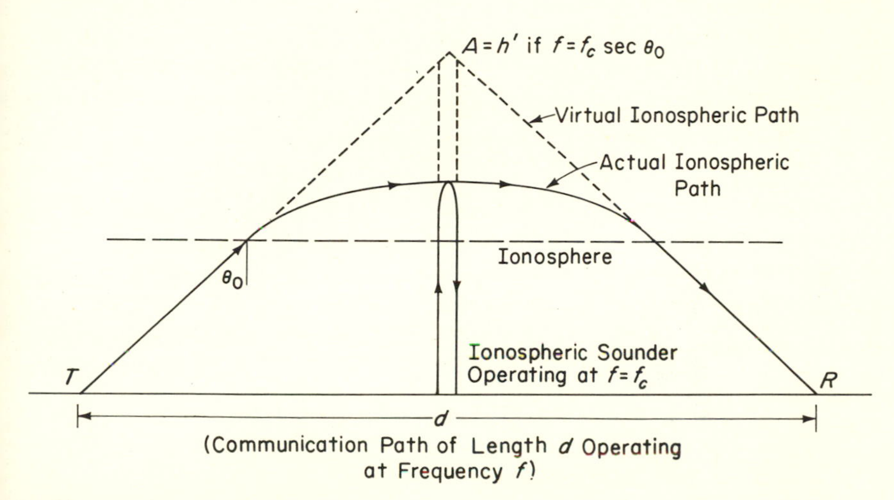

Fig. 15–20. Illustration of Martyn's theorem and Breit and Tuve's theorem as applied to an ionospheric propagation path.

model consisting of a plane earth and a plane ionosphere having no horizontal variations. Martyn's theorem states that if f and f_c are the frequencies of waves reflected obliquely and vertically, respectively, from the same real height, then the virtual height of reflection of the vertically incident signal (f_c) is equal to the height of the equivalent triangular path for the obliquely incident signal. That is, referring to Fig. 15–20, $A = h'$ if $f = f_c \sec \theta_0$. Breit and Tuve's theorem states that the effective or group length of actual ionospheric path is equal to the electrical length of the virtual path TAR in free space.

[15] K. Davies, *Ionospheric Radio Propagation*, National Bureau of Standards Monograph 80, 1965, Chapter 4.

We can use Martyn's theorem and an ionosonde record such as that shown in Fig. 15–19 to determine the operating frequency of a communication path such as is shown in Fig. 15–20. The frequency of operation is determined by solving the equation

$$f = f_c \sec \theta_0 = f_c \sqrt{1 + \left(\frac{d}{2h'}\right)^2} \tag{15–51}$$

A useful presentation can be made by solving Eq. (15–51) for h', given values of d (the distance between the transmitter and receiver), f (the operating frequency of the communication path), and f_c (the frequency of the ionospheric sounder). The result can be plotted on the ionosonde record for various values of f_c to generate a contour which intersects the critical frequency trace of the ionospheric sounder. The points of intersection represent critical frequencies which satisfy the condition $f = f_c \sec \theta_0$, and the values of h' which, for the given values of f and d, give the height of the equivalent triangular propagation path (*TAR* of Fig. 15–20). The solution of Eq. (15–51) for $f = 23$ MHz and $d = 3000$ km is plotted on the ionospheric record of Fig. 15–19 and presented in Fig. 15–21. Note that, at the frequency of 23 MHz, there are four intersections of the solution of Eq. (15–51) with the ionospheric sounder record. These occur at $h' = 340$ km, $h' = 435$ km, $h' = 450$ km, and $h' = 620$ km. This means that, at the frequency of 23 MHz, communication could be established in four different paths corresponding to four angles of incidence. Two paths involve reflection in the F_1 layer:

$$\left.\begin{aligned} \sec \theta_0 &= \frac{23}{5.2} \qquad \theta_0 \approx 77^\circ \\ \sec \theta_0 &= \frac{23}{6.35} \qquad \theta \approx 73.8^\circ \end{aligned}\right\} \quad F_1 \text{ layer}$$

The remaining two paths involve reflection in the F_2 layer:

$$\left.\begin{aligned} \sec \theta_0 &= \frac{23}{6.6} \qquad \theta_0 \approx 73.3^\circ \\ \sec \theta_0 &= \frac{23}{8.8} \qquad \theta_0 \approx 67.5^\circ \end{aligned}\right\} \quad F_2 \text{ layer}$$

The ray paths corresponding to these incident angles are shown in Fig. 15–22.

It should be noted that in the example just considered the frequency of the communication path was such that there was no reflection in the E layer. Had a lower frequency been used (in this example, lower than 20 MHz), reflections would have taken place in the E layer. If the fre-

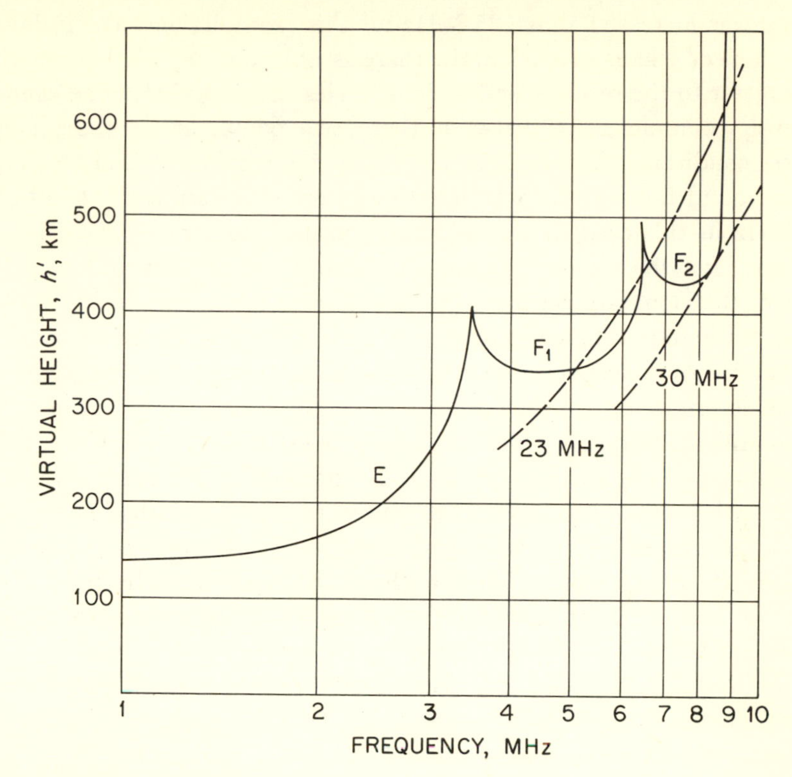

Fig. 15–21. Frequency versus h' plotted on an ionosonde record for a 3000-km propagation path.

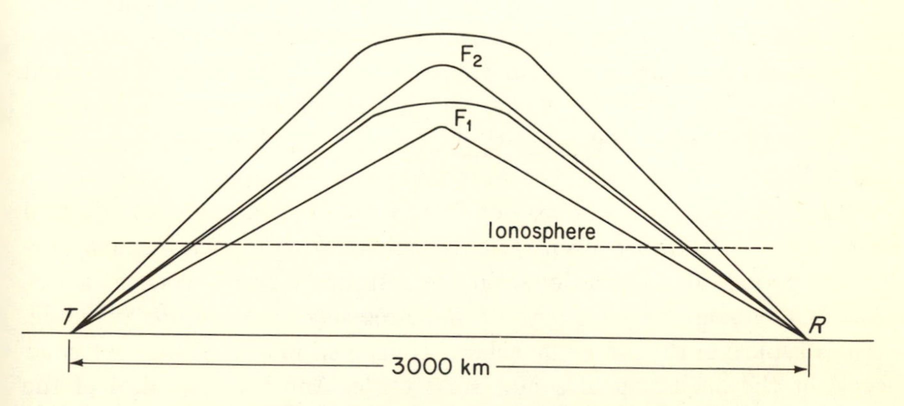

Fig. 15–22. Ray paths for ionospheric example.

quency was increased above 25 MHz for this example, only the F_2 layer would reflect energy. At 30 MHz there is only one ray path from the transmitter to the receiver and at frequencies above 30 MHz one cannot establish communication between these two points under these ionospheric conditions. The highest frequency for which a reflection point can be found for a given path length and set of ionospheric conditions (30 MHz in this case) is the maximum usable frequency (MUF). The distance d for which a given frequency curve is just tangent to the $h' - f_c$ curve of the ionogram is called the *skip distance* for that frequency. This is the minimum distance over which signals at this frequency can be transmitted by ionospheric reflection for these ionospheric conditions.

The determination of the path loss of an ionospheric propagation path is a complicated process which must include estimates of the absorption, defocusing, and depolarizing effects of the ionosphere. The initial estimate of the path loss is made as though one were working with a free-space path of length equal to the actual ionospheric propagation path (see Fig. 15–20, for example). The length of the equivalent triangle TAR could be used in the initial estimate without too much error, particularly if the path is long so that θ_0 is large.

The foregoing argument has been based on a plane-earth–plane-ionosphere model since this is a simple and easily understood physical configuration. One can do a great deal of useful engineering using this model, but for more accurate results the effects of the curvature of the earth and the ionosphere must be incorporated into the analysis.

The upper limit of the one-hop F_2-layer path length d is about 4000 km. To aid the designer of a long-path communication circuit via the sky wave, the United States Department of Commerce has published predictions of the monthly median values of critical frequency for normal incidence and the MUF for a 4000-km path for the F_2 layer as well as the E layer MUF for a 2000-km path.[16] These predicted median values are plotted as constant-value contours superimposed upon a map of the world. There are twelve maps per day to take into account the diurnal variation in electron concentration. Each map is applicable to a one-month period. The ionization, and the resulting electron concentration, is dependent on the solar activity so three sets of maps are available. One is typical of the minimum solar activity period at the close of a solar cycle or the beginning of a new solar cycle. Another is typical of the

[16] *Ionospheric Predictions*, U.S. Department of Commerce, Telecommunication Research and Engineering Report 13, Vols. 1 through 4, September 1971.

maximum solar activity period of an average solar cycle. The third set is typical of the maximum solar activity period of an above-average solar cycle. Tables of predicted values of indices of solar activity such as the Zurich smoothed mean sunspot number (R_{12}) used in the *Ionospheric Predictions* report are available from the same source. Examples of the maps are shown in Figs. 15–23(a) and (b). The critical frequency for normal incidence is labeled MUF (ZERO) F2 and the maximum usable frequency for a 4000-km path via reflection from the F_2 layer is labeled MUF (4000) F2.

In applying the maps to the design of a communication link, the designer plots a great-circle path from the transmitter site to the receiver site on the prediction maps appropriate to the month and time of day that the circuit will be established. The midpoint of the path is located and MUF (ZERO) and MUF (4000) values at that point determined. If the path is approximately 4000 km long, the MUF (4000) value would be used as the maximum operating frequency. If the path is less than 4000 km long, the MUF (ZERO) value is used in Eq. (15–51), where $h' = 320$ km, d is the length of the communication path, and $f_c =$ MUF (ZERO). At this point the design of a practical communication link becomes quite complicated and will not be further discussed here. The interested reader is referred to the excellent examples presented by Davies[17] or those contained in the *Ionospheric Predictions* report.

As a final example of the role of the ionosphere in radio-wave propagation phenomena, we will modify the permittivity of the ionosphere to include the effects of collisions of the electrons with molecules of constituent gases. As shown in Appendix G, the permittivity of the ionosphere, including the effects of collisions but ignoring the earth's magnetic field, is

$$\epsilon = \epsilon_0 \left[1 - \frac{\omega_p^2}{\omega^2 \left(1 - j\dfrac{\nu}{\omega}\right)} \right] \tag{15-52}$$

where ν is the collision frequency.

Equation (15–52) can be rewritten as

$$\epsilon = \epsilon_0 \left(1 - \frac{\omega_p^2}{\omega^2 + \nu^2} - j\,\frac{\omega_p^2 \nu}{\omega(\omega^2 + \nu^2)} \right) \tag{15-53}$$

[17] Kenneth Davies, *Ionospheric Radio Propagation,* National Bureau of Standards Monograph 80, 1965, Chapter 7.

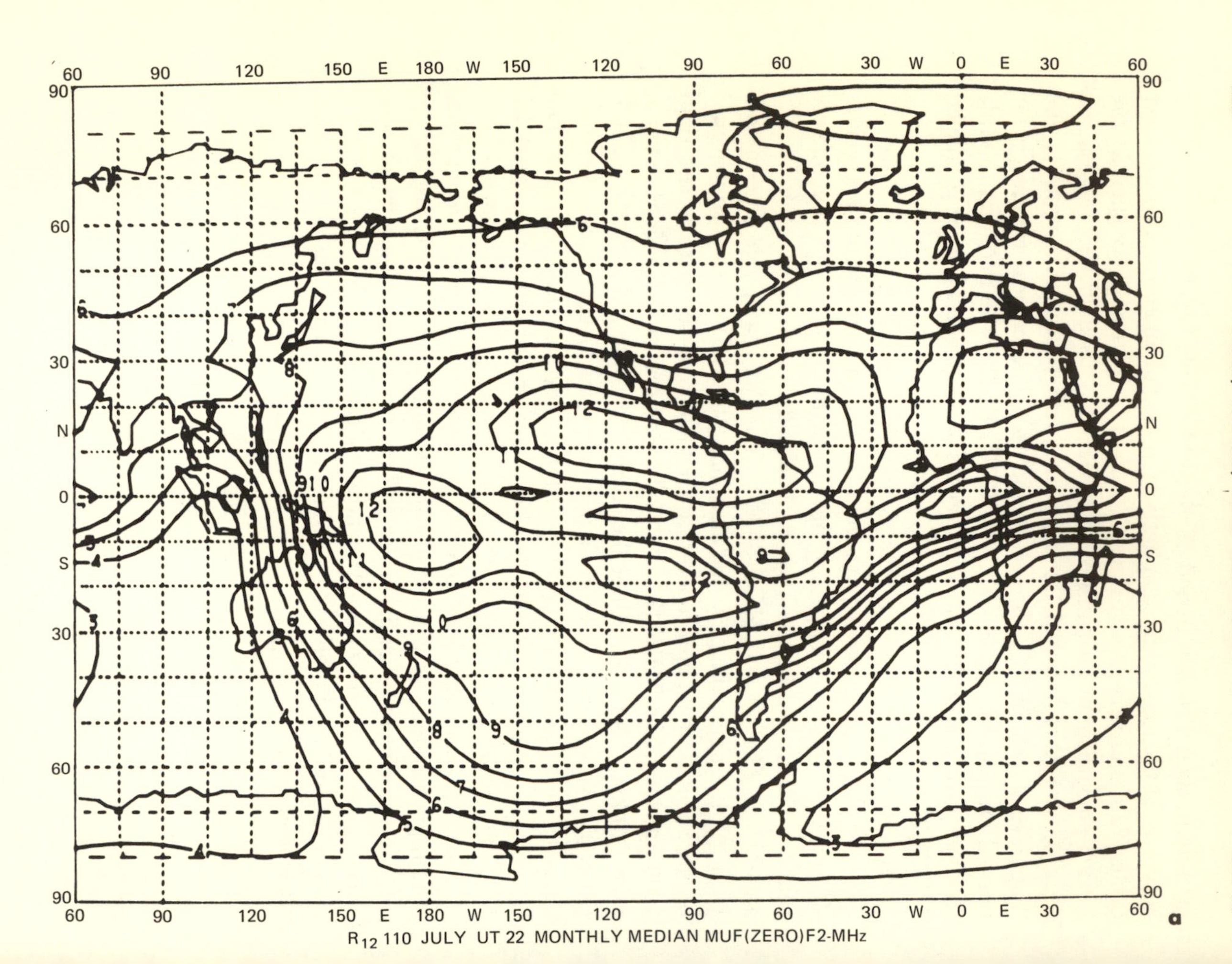
60 90 120 150 E 180 W 150 120 90 60 30 W 0 E 30 60
90 60 30 N 0 S 30 60 90
a
R_{12} 110 JULY UT 22 MONTHLY MEDIAN MUF(ZERO)F2-MHz

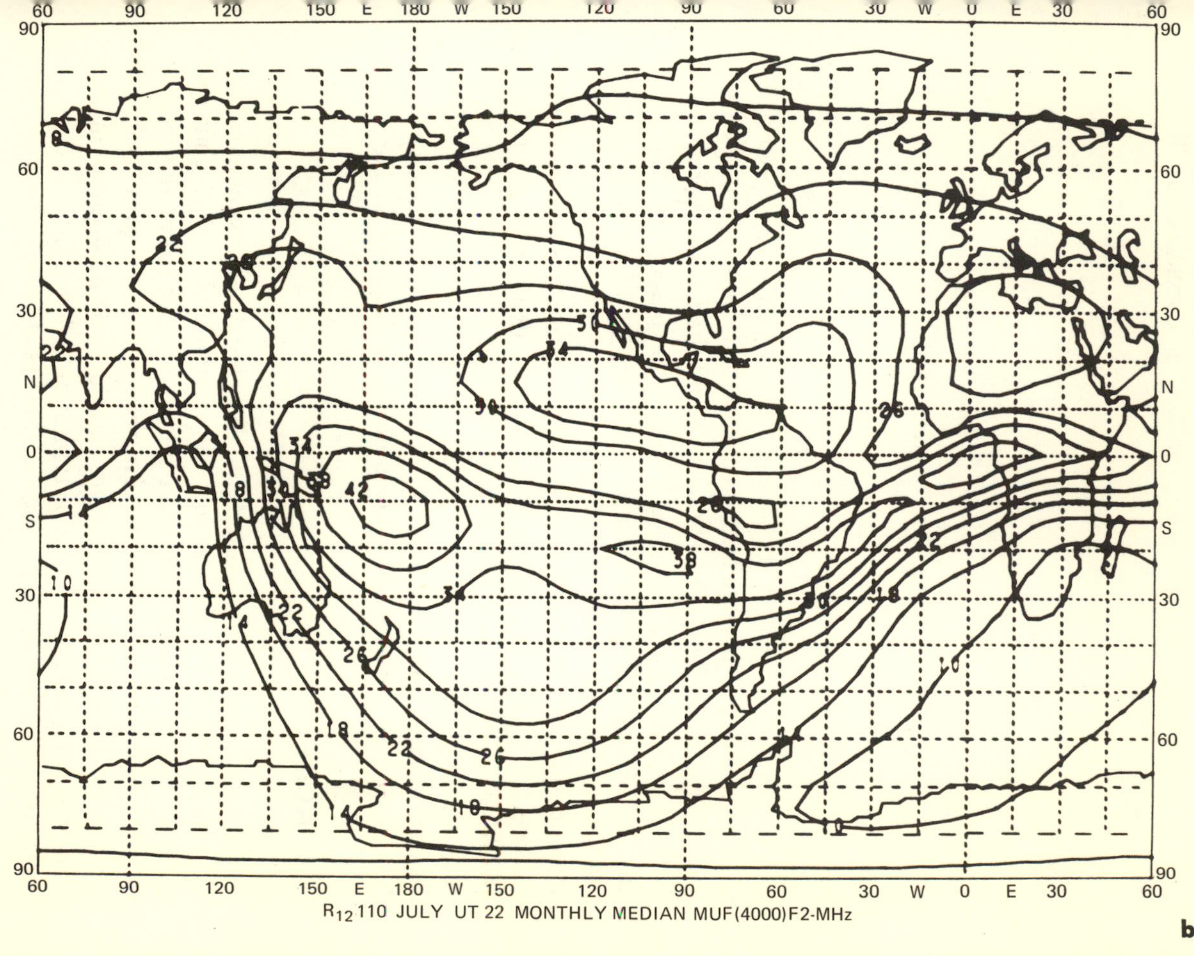

Fig. 15–23. Worldwide maps of predicted monthly median MUF values (d = 0 km and d = 4000 km) for the month of July (2200 UT), R_{12} = 110. (U.S. Department of Commerce, Office of Telecommunications.)

If we assume that $\omega_p^2/(\omega^2 + \nu^2) \ll 1$ and $\omega_p^2\nu/\omega(\omega^2 + \nu^2) < 1$, the small-loss formulas for propagation constant as developed in Chapter 8 can be applied:

$$\gamma = \alpha + j\beta = j\omega\sqrt{\mu_0\epsilon_0}\left(1 - j\,\frac{\omega_p^2\nu}{2\omega(\omega^2 + \nu^2)}\right)$$

Solving for α results in

$$\alpha = \frac{\omega_p^2\nu}{2c(\omega^2 + \nu^2)} = \frac{\eta_0 N e^2 \nu}{2m(\omega^2 + \nu^2)} \tag{15-54}$$

where η_0 is the intrinsic impedance of free space. For the situation where $\omega^2 \gg \nu^2$, Eq. (15-54) reduces to

$$\alpha = \frac{\eta_0 N e^2 \nu}{2m\omega^2} \tag{15-55}$$

The conditions imposed in deriving this equation are applicable to frequencies around the standard broadcast band propagating in the D region of the ionosphere. The D region, being the lowest region, has the highest concentration of gas molecules and, as a consequence, the highest collision frequency. It is the absorption of the sky wave by the D region which generally suppresses the sky wave and prevents it from being a factor in the propagation of broadcast-band transmissions. This limits the coverage of a broadcast station to that afforded by the ground-wave set. We are all aware that very often the sky wave is not severely attenuated, particularly at night, and broadcast stations are heard at large distances from the transmitting station. This is usually due to the fact that the D layer virtually disappears at night, as shown in Fig. 15–18.

15–6. NOISE

One of the main considerations in the design of a communication circuit is the power required of the transmitter. This is determined primarily by the transmission loss and signal level desired at the receiver. The minimum acceptable signal level at the receiver is in turn determined by the noise properties of the receiver and the radio noise power incident on the receiving antenna. A plot of typical noise levels in a 6-kHz bandwidth for several situations is shown in Fig. 15–24.

The atmospheric static is due to lightning and other natural electrical discharge phenomena. The values shown in Fig. 15–24 are middle-latitude values, which are lower than those encountered at the equator by 10 to 15 dB but higher than the normal values in arctic regions by 15 to 25 dB. The thermal noise curve is the noise power delivered by a resistor

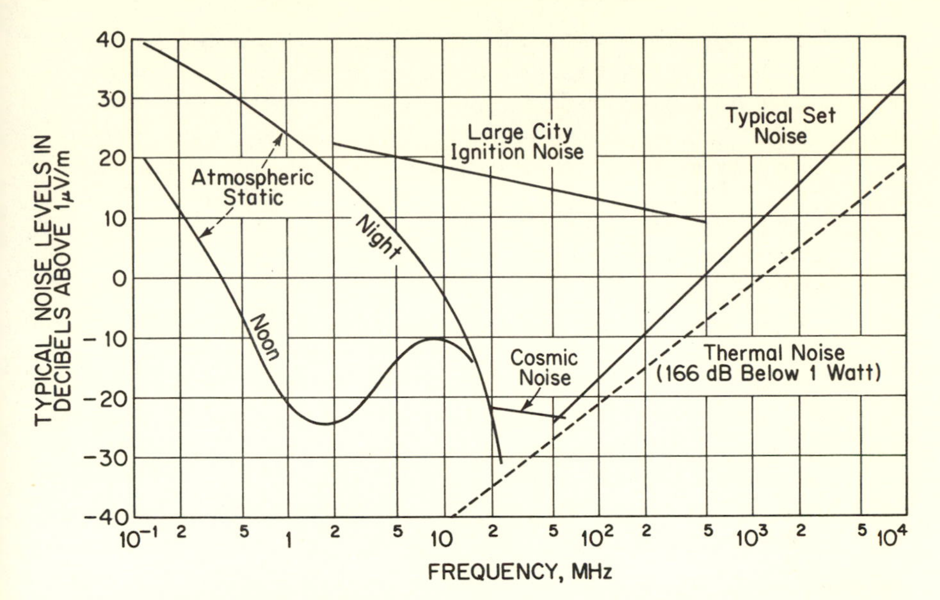

Fig. 15–24. Typical average noise levels in a 6-kHz band. (After K. Bullington, "Radio Propagation Fundamentals," *Bell System Technical Journal,* vol. 36, p. 624, May, 1957. Copyright © 1957, American Telephone and Telegraph Co.; reprinted by permission.)

at 295°K to a matched load over a 6-kHz band. The noise power delivered to the matched load is

$$P_n = kTB$$

where k is Boltzmann's constant (1.38×10^{-23} joule/°K), T is the temperature of the resistor in °K, and B is the bandwidth in Hz. Cosmic noise is due to noise sources (primarily thermal) outside the earth's atmosphere.

PROBLEMS

15–1. Determine the basic path loss L_b for a communication path from the moon to the earth operating at 3 GHz. The distance from the earth to the moon is 240,000 miles.

15–2. Determine the distance from the base of a 2000-foot tower to the radio horizon. Assume an effective earth radius of 5280 miles. What is the maximum distance between the 2000-foot tower and a 100-foot tower for which they will be within radio sight of each other?

15–3. Determine the ground-reflection coefficients for a 500-MHz horizontally polarized wave incident on a dry sandy soil surface. Assume an angle ψ of 30 degrees.

15–4. For vertical polarization one can find a Brewster angle where R_v goes to zero if both media are lossless. Find this angle for the case of $\epsilon_r = 10$. Use this angle to determine R_v for dry sandy soil ($\epsilon_r = 10$, $\sigma = 0.002$ ℧/m) at 500 MHz.

15–5. An AM broadcast station operating at 1400 kHz radiates 5 kW using a quarter-wavelength vertical monopole over fertile soil. At what distance from the transmitter is the rms signal strength equal to 500 μV/m? *Hint:* Use Eqs. (15–27) and (15–29) to develop expressions for A and p in terms of d. Plot these expressions on Fig. 15–11 to determine d.

15–6. Derive Eq. (15–32).

15–7. Determine the first Fresnel zone clearance for the propagation path shown in Fig. 15–16 for (a) an obstacle at midpath and (b) an obstacle located 15 miles from the transmitter. Assume a frequency of 7 GHz.

15–8. Repeat Example 15–4 for a spacing of 300 km between the transmitter and receiver and with the transmitter and receiver at the height $h=h_0$. Compare the result with the tropospheric scattering example.

15–9. Determine the critical frequencies for the F_2 layer of Fig. 15–18 for both nighttime and daytime cases.

15–10. Determine the MUF for a 3200-km path if f_c at the midpoint is 10 MHz. (Use the F_2 layer with $h' = 320$ km.)

15–11. It is desired to communicate with a mobile station in an urban environment at a frequency of 26 MHz using a bandwidth of 6 kHz. If the desired signal-to-noise ratio is 4:1, determine the signal strength required at the receiver.

APPENDIX

A

Transmission-Line Parameters

The inductance and capacitance of the coaxial line of Fig. A–1 may be found rather easily. For computation of the capacitance parameter C, suppose that the inner conductor has a linear charge density of $+\Lambda$ coulombs per meter,

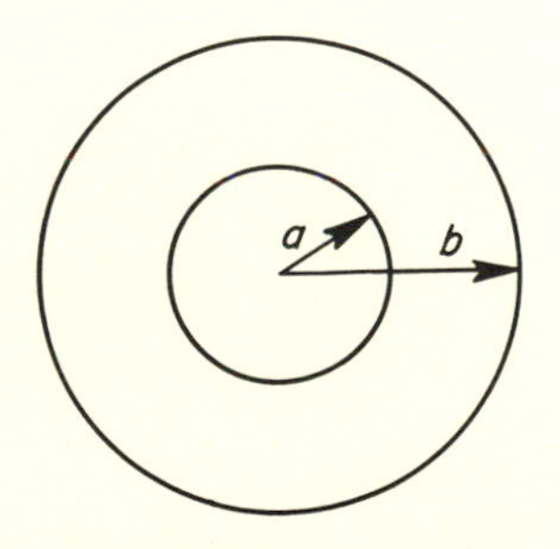

Fig. A–1. Coaxial line of thin shells.

while the outer conductor has a linear charge density of $-\Lambda$ coulombs per meter. Now apply Gauss's law to a 1-m length of the line. By symmetry the flux density is radial, and at a radius between a and b the flux density D is simply the total flux emanating, Λ, divided by the surface area of $2\pi\rho$ square meters.[1] Since the electric field intensity E is related to the flux density D by the expression $D = \epsilon E$, it follows that the electric field intensity E is radial

[1] In this Appendix, B, H, D, and E represent the magnitude of the corresponding vector quantities.

with the value $E = \Lambda/2\pi\rho\epsilon$. Now, by direct integration, the potential difference between the two conductors may be found as the negative of the line integral of the electric field intensity between the two conductors, with the specific value $\Phi_{ab} = [\Lambda \ln (b/a)]/2\pi\epsilon$. Using the conventional definition of capacitance, charge divided by voltage, we find that the capacitance per meter is

$$C = \frac{2\pi\epsilon}{\ln (b/a)} \quad \text{F/m}$$

In order to compute the inductance parameter for the coaxial line, we suppose that a current of I amperes flows down the center conductor and returns along the outer conductor. The magnetic field at some radius between the two conductors is circumferentially directed with a value of $I/2\pi\rho$. Since the relationship between the magnetic flux density B and the magnetic field intensity H is $B = \mu H$, we easily conclude that the magnetic flux density is $B = \mu I/2\pi\rho$. Now that the magnetic flux density B is known, the flux linkages can be computed by integration of the differential flux linkages over the range of radii from inner to outer conductor with a resultant value of $\phi = [\mu I \ln (b/a)]/2\pi$. Then, by using the conventional definition of inductance (flux linkages divided by current), we find the inductance per meter given by the expression

$$L = \frac{\mu}{2\pi} \ln (b/a) \quad \text{H/m}$$

In a similar fashion the inductance and capacitance parameters may be derived for other transmission-line configurations such as the parallel-wire line and the slab line. Idealized formulas for the inductance and capacitance of these configurations are given in Table A–1. In these formulas $\mu = \mu_0\mu_r$ and $\epsilon = \epsilon_0\epsilon_r$ and the values of the free-space constants μ_0 and ϵ_0 are conveniently given as

$$\mu_0 = 4\pi \times 10^{-7} \text{ H/m}$$

and

$$\epsilon_0 = \frac{10^{-9}}{36\pi} \quad \text{F/m}$$

The value of ϵ_0 is only approximate but is expressed in a convenient form for numerical computation and is satisfactorily accurate for many purposes.

The series resistance parameter may be calculated at low frequencies by the conventional dc formula for resistance. However, the resistance at high frequencies is considerably greater than at dc because the current tends to flow on the surface of the conductor and thus reduce the effective cross-sectional area. In fact, from field theory for a wave impinging on a flat conductor, it may be shown that the actual exponential current distribution in the conductor may be replaced, insofar as effect is concerned, by a uniform current distribution up to a thickness of

$$\delta = \frac{1}{\sqrt{\pi f \mu \sigma}} \quad \text{m}$$

TABLE A–1

Idealized Formulas for Inductance and Capacitance of Common Transmission-Line Configurations

	Coaxial Line *Inner Radius* = a *Outer Radius* = b	*Parallel-Wire Line* *Common Radius* = a *Spacing* = D	*Slab Line* *Width* = w *Space* = t
Value of C in F/m	$\dfrac{2\pi\epsilon}{\ln(b/a)}$	$\dfrac{\pi\epsilon}{\ln(D/a)}$	$\dfrac{w\epsilon}{t}$
Value of L in H/m	$\dfrac{\mu\ln(b/a)}{2\pi}$	$\dfrac{\mu\ln(D/a)}{\pi}$	$\dfrac{\mu t}{w}$

where the frequency is f and the conductivity of the conductor is σ. The conductivity of copper is 5.8×10^7 mhos/m.

This result may be applied with reasonable accuracy to curved conductors as long as the thickness so computed is much less than the radius of the conductor. Since resistance is inversely proportional to cross-sectional area, we have, at sufficiently high frequency for solid cylindrical conductors, the relationship

$$\frac{R_{\text{ac}}}{R_{\text{dc}}} = \frac{\text{area}_{\text{dc}}}{\text{area}_{\text{ac}}} = \frac{a}{2}\sqrt{\pi f \mu \sigma}$$

Probably the most important point to note from this formula is that the ac resistance increases *roughly* as the square root of the frequency.

The shunt conductance parameter may be calculated by conventional means, although it must be noted that integration is ordinarily required in the process. It is interesting and important to note that the conductance and capacitance for a transmission line are computed by using analogous laws. For example, the conductance parameter for the coaxial line of Table A–1 would be given by the expression

$$G = \frac{2\pi\sigma}{\ln(b/a)}$$

Or more generally, we may state that the conductance and capacitance are related for a given geometry by

$$\frac{G}{C} = \frac{\sigma}{\epsilon}$$

Now the power factor of a lossy dielectric is given by the expression

$$\text{Power factor} = \frac{\sigma}{\sqrt{\sigma^2 + (\omega\epsilon)^2}}$$

This expression for small power factors, a common occurrence, reduces to the very simple expression

$$\text{Power factor} \approx \theta = \frac{\sigma}{\omega\epsilon}$$

and consequently we can observe that the conductance parameter is related to the capacitance parameter and frequency by the relatively simple expression

$$G = 2\pi C\theta f$$

The power factor of polyethylene, a commonly used dielectric, is about 0.0004 and is essentially constant over an extremely broad frequency range. From the last formula for the conductance it may then be observed that the conductance varies *roughly* as frequency.

B

Line Characteristics

TABLE B–1

Radio Frequency Transmission Lines

Class of Cables	*Type*	*Inner Conductor*	*Nominal Overall Diam. (in.)*	*Nominal Capacitance (pF/ft)*	*Maximum Operating Voltage (rms)*
50 ohms	RG–8A/U	7/0.0296	0.405	29.5	4000
50 ohms	RG–58C/U	19/0.0071	0.195	28.5	1900
50 ohms	RG–122/U	27/0.005	0.160	29.3	1900
50 ohms	RG–55A/U	0.035	0.216	28.5	1900
50 ohms	RG–174/U	7/0.0063	0.100	30.0	1000
50 ohms	RG–196A/U	7/0.0040	0.080	29.0	1000
High attenuation	RG–126/U	7/0.0203	0.275	29.0	3000
High attenuation	RG–21A/U	0.053	0.332	29.0	2700
High delay	RG–65A/U	0.008*	0.405	44.0	1000
Twin conductor	RG–86/U	7/0.0285	0.30 × 0.65	7.8	—
Twin conductor	RG–130/U	7/0.0285	0.625	17.0	8000

* Formex F. Helix with diameter of 0.128 in.

TABLE B–2

Typical Power Line Characteristics at 60 Hz

Circuit Voltage (kV, L–L)	*Conductor Size (thousands of cir. mils, or AWG)*	*Equiv. Spacing (ft)*	*R (at 50°C—ohms per phase per mile)*	*X (ohms per phase per mile)*	*Shunt C (megohms per phase per mile)*	*Surge Imped. (ohms, L–N)*
69	2/0	8	0.481	0.784	0.182	378
69	336.4	19	0.306	0.808	0.191	393
115	336.4	13	0.306	0.762	0.180	370
115	336.4	22	0.306	0.826	0.196	402
138	397.5	15	0.259	0.764	0.181	371
138	397.5	24	0.259	0.821	0.195	399
161	397.5	17	0.259	0.779	0.185	379
161	397.5	25	0.259	0.826	0.196	402

C

Transmission-Line Charts

The equations needed for construction of the various charts will be derived in this Appendix. Only three cases need be considered: (1) the per-unit Smith chart labeled in rectangular values; (2) the per-unit Smith chart labeled in polar coordinates; and (3) the per-unit impedance chart.

The charts with bases other than 1 ohm are merely repetitions of these charts with all values scaled to the new base.

C–1. RECTANGULAR SMITH CHART (FIG. 3–1a)

The Smith chart is based upon a plot in the reflection coefficient plane of corresponding values of impedance. Let the complex value of K be expressed as $U + jV$. Then for $z = r + jx$ as a per-unit impedance, we have

$$z = \frac{1 + K}{1 - K} = r + jx = \frac{1 + U + jV}{1 - U - jV} = \frac{1 - U^2 - V^2 + j2V}{(1 - U)^2 + V^2}$$

By equating real and imaginary parts, we find that

$$r(1 - 2U + U^2 + V^2) = 1 - U^2 - V^2$$

and

$$x(1 - 2U + U^2 + V^2) = 2V$$

Now, by collecting common terms in U and V and completing the squares of the various terms, we have

$$\left(U - \frac{r}{r + 1}\right)^2 + (V - 0)^2 = \left(\frac{1}{r + 1}\right)^2$$

$$(U - 1)^2 + \left(V - \frac{1}{x}\right)^2 = \left(\frac{1}{x}\right)^2$$

Both equations are now in the canonical form of circles and may be readily plotted in terms of center and radius.

C-2. POLAR SMITH CHART (FIG. 3-1b)

The polar form of the Smith chart is found by expressing the per-unit impedance and reflection coefficients as $z = |z|e^{j\theta}$ and $K = U + jV$, with the results in the fashion shown before:

$$z = \frac{1+K}{1-K} = |z|e^{j\theta} = \sqrt{\frac{(1+U)^2+V^2}{(1-U)^2+V^2}}\, e^{j\left(\tan^{-1}\frac{2V}{1-U^2-V^2}\right)}$$

Now, by equating the magnitudes and angles, we find

$$|z|^2(1 - 2U + U^2 + V^2) = 1 + 2U + U^2 + V^2$$
$$(1 - U^2 - V^2)\tan\theta = 2V$$

By collecting common terms in U and V and completing the squares of the various terms, we find

$$U + \left(\frac{1+|z|^2}{1-|z|^2}\right)^2 + (V-0)^2 = \left(\frac{2|z|}{1-|z|^2}\right)^2$$

and

$$(U-0)^2 + \left(V + \frac{1}{\tan\theta}\right)^2 = \left(\frac{1}{\sin\theta}\right)^2$$

Both equations are now in the canonical form of circles and may be readily plotted in terms of center and radius.

C-3. RECTANGULAR CHART (FIG. 3-1c)

The impedance chart contours are found by essentially the reverse process. This time we let $z = r + jx$ and express K as $|K|e^{j\phi}$ to write

$$K = \frac{z-1}{z+1} = |K|e^{j\phi} = \frac{r+jx-1}{r+jx+1} = \sqrt{\frac{(r-1)^2+x^2}{(r+1)^2+x^2}}\, e^{j\left(\tan^{-1}\frac{2x}{r^2+x^2-1}\right)}$$

Then, by equating magnitudes and angles, we have

$$|K|^2(r^2 + 2r + 1 + x^2) = (r^2 - 2r + 1 + x^2)$$

and

$$(x^2 + r^2 - 1)\tan\phi = 2x$$

Now, by collecting terms in r and x, completing squares, and using the more appropriate expressions $|K| = (S-1)/(S+1)$ and $\phi = 2\beta d$, we find

$$\left(r - \frac{S^2+1}{2S}\right)^2 + (x-0)^2 = \left(\frac{S^2-1}{2S}\right)^2$$

and

$$(r - 0)^2 + \left(x + \frac{1}{\tan 2\beta d}\right)^2 = \left(\frac{1}{\sin 2\beta d}\right)^2$$

By using precisely the same techniques as illustrated for impedances, it may be shown that the same charts apply equally well for admittances.

D

Vector Analysis

Some of the more important relationships from vector analysis are presented here for reference purposes.

D–1. DOT PRODUCT

In rectangular coordinates the dot or scalar product of two vectors **A** and **B** is given by

$$\mathbf{A} \cdot \mathbf{B} = A_x B_x + A_y B_y + A_z B_z \tag{D–1}$$

where A_x, A_y, etc., refer to the rectangular components of the vectors. The dot product also has a geometric interpretation as the product of the two magnitudes times the cosine of the angle between the two vectors. Note that $\mathbf{A} \cdot \mathbf{B}$ is always a scalar quantity.

D–2. CROSS-PRODUCT

In rectangular coordinates the cross or vector product of two vectors **A** and **B** is given by

$$\mathbf{A} \times \mathbf{B} = \begin{vmatrix} \mathbf{a}_x & \mathbf{a}_y & \mathbf{a}_z \\ A_x & A_y & A_z \\ B_x & B_y & B_z \end{vmatrix} \tag{D–2}$$

where $\mathbf{a}_x$, $\mathbf{a}_y$, and $\mathbf{a}_z$ are unit vectors along the x, y, and z axes, respectively. The cross-product also has a geometric interpretation as a vector mutually perpendicular to **A** and **B**, with a magnitude equal to the product of the two times the sine of the angle between them. Note that $\mathbf{A} \times \mathbf{B}$ is always a vector quantity.

D–3. OPERATOR ∇

In rectangular coordinates the vector operator ∇ is given by

$$\nabla = \mathbf{a}_x \frac{\partial}{\partial x} + \mathbf{a}_y \frac{\partial}{\partial y} + \mathbf{a}_z \frac{\partial}{\partial z} \tag{D–3}$$

Note that ∇ is a vector operator.

D–4. OPERATOR ∇^2

The operator ∇^2 is defined as

$$\nabla^2 = \frac{\partial^2}{\partial x^2} + \frac{\partial^2}{\partial y^2} + \frac{\partial^2}{\partial z^2} \tag{D–4}$$

It may be thought of as arising from dotting ∇ into itself.

D–5. GRADIENT

The gradient of a scalar function Φ is given by

$$\text{grad}\,\Phi = \nabla\Phi = \mathbf{a}_x \frac{\partial\Phi}{\partial x} + \mathbf{a}_y \frac{\partial\Phi}{\partial y} + \mathbf{a}_z \frac{\partial\Phi}{\partial z} \tag{D–5}$$

This has a physical interpretation as a vector whose magnitude is equal to the maximum rate of change of Φ spacewise, and the vector direction is the direction in which this maximum rate of change takes place.

D–6. DIVERGENCE

The divergence of a vector $\mathbf{A}$ is the result of dotting ∇ into $\mathbf{A}$:

$$\text{div}\,\mathbf{A} = \nabla \cdot \mathbf{A} = \frac{\partial A_x}{\partial x} + \frac{\partial A_y}{\partial y} + \frac{\partial A_z}{\partial z} \tag{D–6}$$

This has a physical interpretation as the outward flow of the vector flux $\mathbf{A}$ per unit volume. That is, if one were to imagine an infinitesimal volume $\Delta\tau$, the divergence of $\mathbf{A}$ would be the integral of $\mathbf{A} \cdot \mathbf{da}$ over the surface of $\Delta\tau$ divided by $\Delta\tau$.

D–7. CURL

The curl of a vector $\mathbf{A}$ is the result of crossing ∇ into $\mathbf{A}$:

$$\text{curl}\,\mathbf{A} = \nabla \times \mathbf{A} = \begin{vmatrix} \mathbf{a}_x & \mathbf{a}_y & \mathbf{a}_z \\ \dfrac{\partial}{\partial x} & \dfrac{\partial}{\partial y} & \dfrac{\partial}{\partial z} \\ A_x & A_y & A_z \end{vmatrix} \tag{D–7}$$

This has a physical interpretation as the "circulation" of a vector field at a point. To be more specific, if one imagines an infinitesimal surface element $\Delta\sigma$, the component of curl **A** which is normal to the surface $\Delta\sigma$ would be the line integral of $\mathbf{A} \cdot \mathbf{dl}$ around the surface element, divided by $\Delta\sigma$. It can be seen that if this concept is applied to three mutually perpendicular surface elements at a point in space, the three rectangular components of curl **A** are obtained at the point.

D–8. VECTOR IDENTITIES

A number of useful vector identities are given below. Each may be verified by direct reduction of both sides of the equation. **A** and **B** denote vector quantities and Φ and ϕ denote scalars.

$$\nabla(\Phi + \phi) = \nabla\Phi + \nabla\phi \quad \text{(D–8)}$$

$$\nabla \cdot (\mathbf{A} + \mathbf{B}) = \nabla \cdot \mathbf{A} + \nabla \cdot \mathbf{B} \quad \text{(D–9)}$$

$$\nabla \times (\mathbf{A} + \mathbf{B}) = \nabla \times \mathbf{A} + \nabla \times \mathbf{B} \quad \text{(D–10)}$$

$$\nabla(\Phi\phi) = \Phi\nabla\phi + \phi\nabla\Phi \quad \text{(D–11)}$$

$$\nabla \cdot (\Phi\mathbf{A}) = \mathbf{A} \cdot \nabla\Phi + \Phi\nabla \cdot \mathbf{A} \quad \text{(D–12)}$$

$$\nabla \cdot (\mathbf{A} \times \mathbf{B}) = \mathbf{B} \cdot \nabla \times \mathbf{A} - \mathbf{A} \cdot \nabla \times \mathbf{B} \quad \text{(D–13)}$$

$$\nabla \times (\Phi\mathbf{A}) = \nabla\Phi \times \mathbf{A} + \Phi\nabla \times \mathbf{A} \quad \text{(D–14)}$$

$$\nabla \times (\mathbf{A} \times \mathbf{B}) = \mathbf{A}\nabla \cdot \mathbf{B} - \mathbf{B}\nabla \cdot \mathbf{A} + (\mathbf{B} \cdot \nabla)\mathbf{A} - (\mathbf{A} \cdot \nabla)\mathbf{B} \quad \text{(D–15)}$$

$$\nabla \cdot \nabla\Phi = \nabla^2\Phi \quad \text{(D–16)}$$

$$\nabla \cdot \nabla \times \mathbf{A} = 0 \quad \text{(D–17)}$$

$$\nabla \times \nabla\Phi = 0 \quad \text{(D–18)}$$

$$\nabla \times \nabla \times \mathbf{A} = \nabla(\nabla \cdot \mathbf{A}) - \nabla^2\mathbf{A} \quad \text{(D–19)}$$

D–9. STOKES' THEOREM

This integral theorem relates the surface integral of $\nabla \times \mathbf{A}$ to the line integral of **A** along a closed curve enclosing the surface. The theorem is

$$\oint_C \mathbf{A} \cdot \mathbf{dl} = \iint_S \nabla \times \mathbf{A} \cdot \mathbf{da} \quad \text{(D–20)}$$

D–10. DIVERGENCE THEOREM

This theorem relates the volume integral of $\nabla \cdot \mathbf{A}$ to the surface integral of **A** over the closed surface enclosing the volume. The theorem is

$$\iint_S \nabla \cdot \mathbf{da} = \iiint_V (\nabla \cdot \mathbf{A})\, dv \quad \text{(D–21)}$$

where S denotes the surface enclosing the volume V.

D–11. DEL OPERATIONS IN RECTANGULAR COORDINATES

The basic operations involving ∇ are summarized below in rectangular coordinates.

$$\nabla\Phi = \mathbf{a}_x \frac{\partial \Phi}{\partial x} + \mathbf{a}_y \frac{\partial \Phi}{\partial y} + \mathbf{a}_z \frac{\partial \Phi}{\partial z} \tag{D–22}$$

$$\nabla \cdot \mathbf{A} = \frac{\partial A_x}{\partial x} + \frac{\partial A_y}{\partial y} + \frac{\partial A_z}{\partial z} \tag{D–23}$$

$$\nabla \times \mathbf{A} = \begin{vmatrix} \mathbf{a}_x & \mathbf{a}_y & \mathbf{a}_z \\ \dfrac{\partial}{\partial x} & \dfrac{\partial}{\partial y} & \dfrac{\partial}{\partial z} \\ A_x & A_y & A_z \end{vmatrix} \tag{D–24}$$

$$\nabla^2\Phi = \frac{\partial^2 \Phi}{\partial x^2} + \frac{\partial^2 \Phi}{\partial y^2} + \frac{\partial^2 \Phi}{\partial z^2} \tag{D–25}$$

$$\nabla^2\mathbf{A} = \frac{\partial^2 \mathbf{A}}{\partial x^2} + \frac{\partial^2 \mathbf{A}}{\partial y^2} + \frac{\partial^2 \mathbf{A}}{\partial z^2} \tag{D–26}$$

D–12. DEL OPERATIONS IN CYLINDRICAL COORDINATES

The basic operations involving ∇ are given below in terms of cylindrical coordinates. The respective unit vectors are denoted as $\mathbf{a}_\rho$, $\mathbf{a}_\phi$, $\mathbf{a}_z$, and the coordinate system is shown in Fig. D–1.

$$\nabla\Phi = \mathbf{a}_\rho \frac{\partial \Phi}{\partial \rho} + \mathbf{a}_\phi \frac{1}{\rho}\frac{\partial \Phi}{\partial \phi} + \mathbf{a}_z \frac{\partial \Phi}{\partial z} \tag{D–27}$$

$$\nabla \cdot \mathbf{A} = \frac{1}{\rho}\frac{\partial}{\partial \rho}(\rho A_\rho) + \frac{1}{\rho}\frac{\partial A_\phi}{\partial \phi} + \frac{\partial A_z}{\partial z} \tag{D–28}$$

$$\nabla \times \mathbf{A} = \begin{vmatrix} \dfrac{1}{\rho}\mathbf{a}_\rho & \mathbf{a}_\phi & \dfrac{1}{\rho}\mathbf{a}_z \\ \dfrac{\partial}{\partial \rho} & \dfrac{\partial}{\partial \phi} & \dfrac{\partial}{\partial z} \\ A_\rho & \rho A_\phi & A_z \end{vmatrix} \tag{D–29}$$

$$\nabla^2\Phi = \frac{1}{\rho}\frac{\partial}{\partial \rho}\left(\rho \frac{\partial \Phi}{\partial \rho}\right) + \frac{1}{\rho^2}\frac{\partial^2 \Phi}{\partial \phi^2} + \frac{\partial^2 \Phi}{\partial z^2} \tag{D–30}$$

$$\nabla^2\mathbf{A} = \nabla(\nabla \cdot \mathbf{A}) - \nabla \times \nabla \times \mathbf{A} \tag{D–31}$$

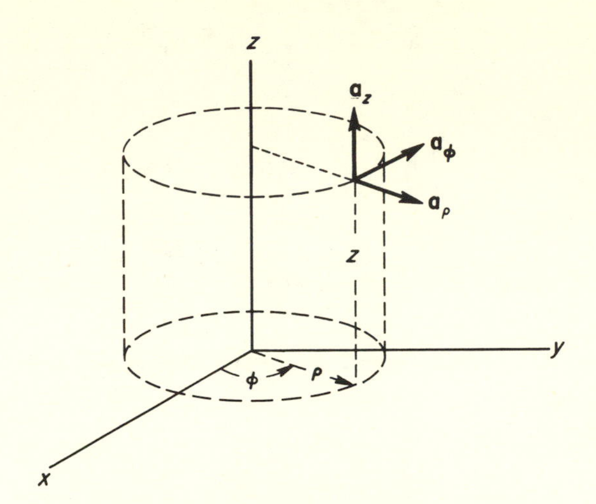

Fig. D–1. Cylindrical coordinate system.

D–13. DEL OPERATIONS IN SPHERICAL COORDINATES

The basic operations involving ∇ are given below in terms of spherical coordinates. The respective unit vectors are denoted as $\mathbf{a}_r$, $\mathbf{a}_\theta$, $\mathbf{a}_\phi$, as shown in Fig. D–2.

$$\nabla\Phi = \mathbf{a}_r \frac{\partial \Phi}{\partial r} + \mathbf{a}_\theta \frac{1}{r}\frac{\partial \Phi}{\partial \theta} + \mathbf{a}_\phi \frac{1}{r \sin\theta}\frac{\partial \Phi}{\partial \phi} \tag{D–32}$$

$$\nabla \cdot \mathbf{A} = \frac{1}{r^2}\frac{\partial}{\partial r}(r^2 A_r) + \frac{1}{r\sin\theta}\frac{\partial}{\partial\theta}(\sin\theta A_\theta) + \frac{1}{r\sin\theta}\frac{\partial A_\phi}{\partial \phi} \tag{D–33}$$

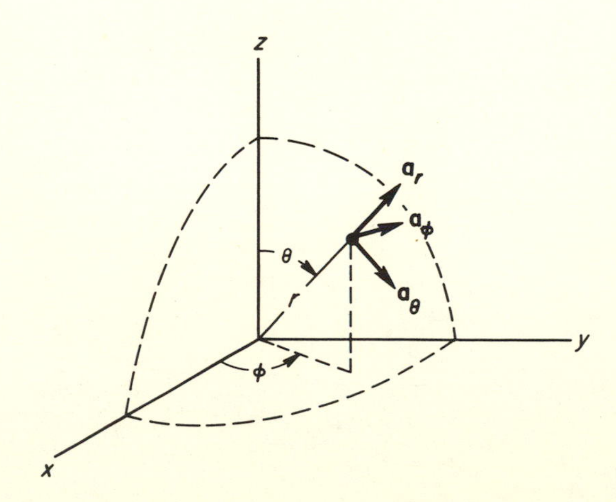

Fig. D–2. Spherical coordinate system.

$$\nabla \times \mathbf{A} = \begin{vmatrix} \dfrac{1}{r^2 \sin\theta}\mathbf{a}_r & \dfrac{1}{r\sin\theta}\mathbf{a}_\theta & \dfrac{1}{r}\mathbf{a}_\phi \\ \dfrac{\partial}{\partial r} & \dfrac{\partial}{\partial \theta} & \dfrac{\partial}{\partial \phi} \\ A_r & rA_\theta & r\sin\theta A_\phi \end{vmatrix} \tag{D-34}$$

$$\nabla^2\Phi = \frac{1}{r^2}\frac{\partial}{\partial r}\left(r^2\frac{\partial\Phi}{\partial r}\right) + \frac{1}{r^2\sin\theta}\frac{\partial}{\partial\theta}\left(\sin\theta\frac{\partial\Phi}{\partial\theta}\right) + \frac{1}{r^2\sin^2\theta}\frac{\partial^2\Phi}{\partial\phi^2} \tag{D-35}$$

$$\nabla^2\mathbf{A} = \nabla(\nabla\cdot\mathbf{A}) - \nabla\times\nabla\times\mathbf{A} \tag{D-36}$$

E

Phase and Group Velocities

Consider a simple, single-frequency wave which has the mathematical form

$$e = A \cos (\omega t - \beta z) \tag{E-1}$$

For purposes of illustration this may be thought of as a voltage wave along a transmission line, but the following remarks are perfectly general and apply to any wave exhibiting this mathematical form. The wave is shown in Fig. E-1

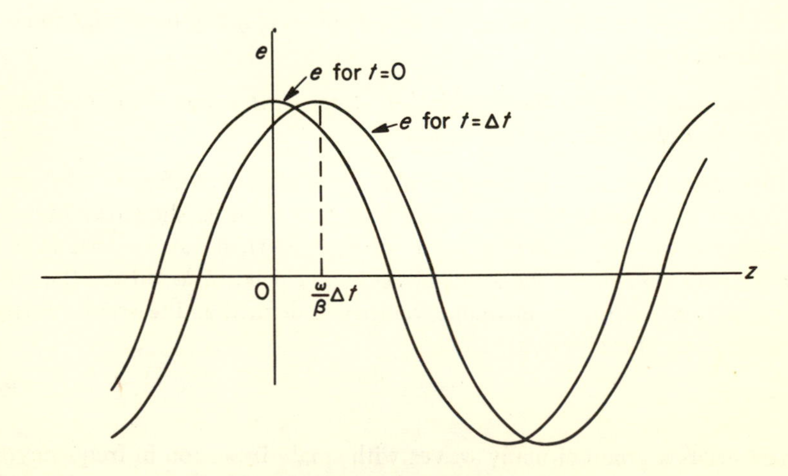

Fig. E-1. Single-frequency wave.

as a function of z for $t = 0$ and $t = \Delta t$. It can be seen that the wave moves through a distance $(\omega/\beta)\,\Delta t$ in a time Δt, and thus

$$\text{Velocity} = \frac{(\omega/\beta)\,\Delta t}{\Delta t} = \frac{\omega}{\beta} \tag{E-2}$$

This velocity, defined as the *phase velocity* v_p, is the velocity at which the whole wave picture moves in the z direction. Note that this is apparent motion and no motion of physical particles is implied.

Next consider a "wave" made up of two frequency components of equal amplitude, and let one of these be at a frequency slightly greater than ω_0 and one slightly less than ω_0. Also, let the phase-shift constant β be related to

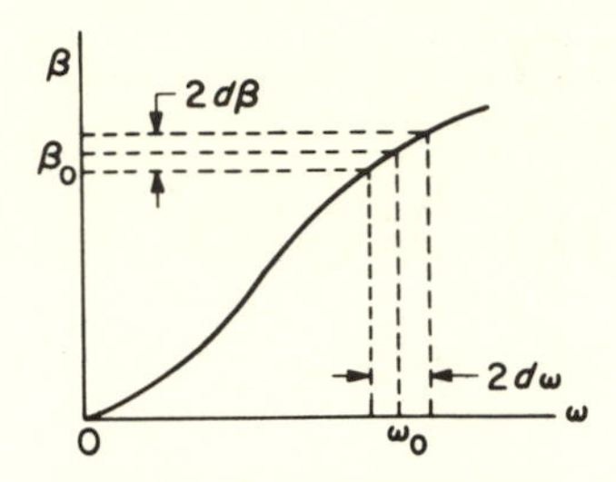

Fig. E–2. Plot of β versus ω.

frequency ω by some general curve such as that shown in Fig. E–2. Such a wave could be expressed mathematically as

$$e = A\cos\left[(\omega_0 - d\omega)t - (\beta_0 - d\beta)z\right] + A\cos\left[(\omega_0 + d\omega)t - (\beta_0 + d\beta)z\right] \tag{E-3}$$

By rearranging terms and taking advantage of an identity from trigonometry, the equation becomes

$$\begin{aligned} e &= A\cos\left[(\omega_0 t - \beta_0 z) - (d\omega t - d\beta z)\right] + A\cos\left[(\omega_0 t - \beta_0 z) + (d\omega t - d\beta z)\right] \\ &= 2A\cos(d\omega t - d\beta z)\cos(\omega_0 t - \beta_0 z) \end{aligned} \tag{E-4}$$

The first cosine term may be thought of as a modulation coefficient, modulating a "carrier" at a frequency ω_0. A sketch of this wave is shown in Fig. E–3. By inspection of the respective terms of Eq. (E–4), it can be seen that the modulation or envelope of the wave moves at a velocity of $d\omega/d\beta$ and the carrier at a velocity of ω_0/β_0. The envelope velocity is defined as the *group velocity* v_g and is given by the equation

$$v_g = \frac{d\omega}{d\beta} \tag{E-5}$$

If there exists a group of many waves with small dispersion in frequency, the resultant envelope will move at the group velocity. Thus, when one considers

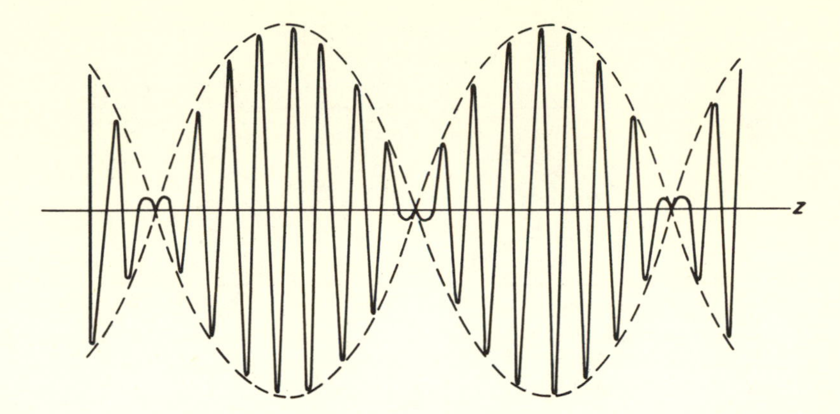

Fig. E–3. Sketch of two-frequency wave.

a modulated RF system, the "signal" travels at the group velocity rather than the phase velocity.

It can be seen from Eqs. (E–2) and (E–5) that the phase and group velocities are equal when β is a linear function of ω. This is the case in lossless transmission-line theory (or any distortionless system), and thus there is no distinction between phase and group velocities in this problem. This is not so in the waveguide problem, though, since the phase-shift function β is given by

$$\beta = \omega\sqrt{\mu\epsilon}\sqrt{1 - \left(\frac{\omega_c}{\omega}\right)^2} \tag{E–6}$$

Thus the phase and group velocities are

$$v_p = \frac{\omega}{\beta} = \frac{v}{\sqrt{1 - (\omega_c/\omega)^2}} \tag{E–7}$$

$$v_g = \frac{d\omega}{d\beta} = \frac{1}{(d\beta/d\omega)} = v\sqrt{1 - \left(\frac{\omega_c}{\omega}\right)^2} \tag{E–8}$$

Note that the geometric mean of v_g and v_p is always the free velocity v; i.e.,

$$v_g v_p = v \tag{E–9}$$

Therefore the group or signal velocity is always less than the free velocity in waveguide transmission.

The group velocity also has an interesting interpretation as the velocity of energy flow in a waveguide system. This concept can be readily verified for the TE_{10} rectangular mode, in which case the propagating wave can be resolved into two uniform plane waves bouncing off the side walls obliquely as shown in Fig. E–4. To show this mathematically, consider a superposition of two equal-amplitude plane waves which are polarized in the y-direction and which

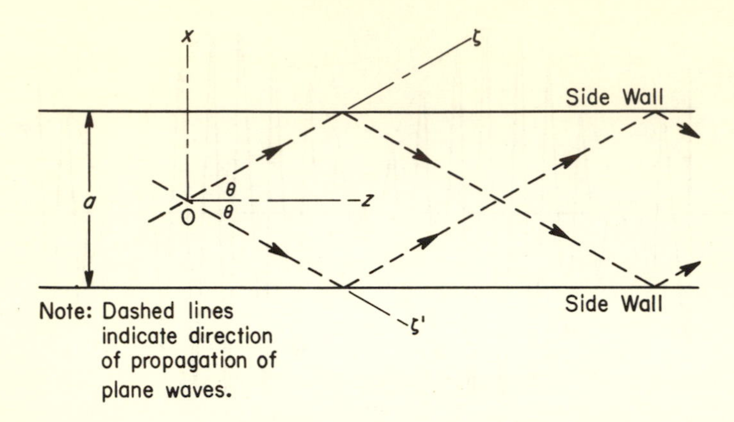

Fig. E–4. Top view showing plane waves bouncing off side walls for TE_{10} mode.

are traveling in the ζ and ζ' directions, respectively. The total electric field for such a combination may be written in phasor form as

$$E_y = E_0 e^{-jk\zeta} + E_0 e^{-jk\zeta'} \tag{E–10}$$

where

$$k = \omega\sqrt{\mu\epsilon} \tag{E–11}$$

and ζ and ζ' are coordinates, as shown in Fig. E–4. Now ζ and ζ' may be expressed in terms of x and z as

$$\begin{aligned} \zeta &= x \sin\theta + z\cos\theta \\ \zeta' &= -x\sin\theta + z\cos\theta \end{aligned} \tag{E–12}$$

and substitution of these in Eq. (E–10) leads to

$$\begin{aligned} E_y &= E_0 e^{-jkx \sin\theta} e^{-jkz\cos\theta} + E_0 e^{jkx\sin\theta} e^{-jkz\cos\theta} \\ &= 2E_0 \cos(kx\sin\theta) e^{-j(k\cos\theta)z} \end{aligned} \tag{E–13}$$

Next we let

$$\cos\theta = \sqrt{1 - \left(\frac{\omega_c}{\omega}\right)^2} \tag{E–14}$$

which is always possible if $\omega > \omega_c$. Then

$$k\cos\theta = \omega\sqrt{\mu\epsilon}\sqrt{1 - \left(\frac{\omega_c}{\omega}\right)^2}$$

$$kx\sin\theta = \omega\sqrt{\mu\epsilon}x\frac{\omega_c}{\omega} = \frac{\pi x}{a} \tag{E–15}$$

and the expression for E_y becomes

$$E_y = 2E_0 \cos\frac{\pi x}{a} e^{-j\beta z} \tag{E–16}$$

This distribution of E_y is precisely the same as that determined in Chapter 9, and thus resolving the total wave into two plane waves is valid.[1]

Now if we think of the energy as being associated with the plane waves which are bouncing off the side walls obliquely, the component of energy velocity in the z-direction is

$$\text{Energy velocity (in } z\text{-direction)} = v \cos\theta = v\sqrt{1 - \left(\frac{\omega_c}{\omega}\right)^2}$$

which is simply the group velocity previously discussed from a different viewpoint.

[1] Note that the origin of the coordinate system in this case was chosen at the center of the guide rather than at one edge, as in Chapter 9. This gives rise to cosinusoidal rather than sinusoidal variation in the x-direction, but the physical result is the same for both cases.

F

Constants for Common Materials

TABLE F–1

Good Conductors

Material	*Conductivity (mhos/m)*
Aluminum	3.72×10^7
Brass	1.57×10^7
Copper	5.8×10^7
Gold	4.1×10^7
Silver	6.17×10^7
Solder	0.71×10^7

TABLE F–2

Dielectrics*

Material	*Dielectric Constant*			*Loss Tangent*		
	$f = 10^6$ *Hz*	$f = 10^8$ *Hz*	$f = 10^{10}$ *Hz*	$f = 10^6$ *Hz*	$f = 10^8$ *Hz*	$f = 10^{10}$ *Hz*
Bakelite	4.4	4.0	3.7	0.03	0.04	0.04
Fused quartz	3.78	3.78	3.78	2×10^{-4}	1×10^{-4}	1×10^{-4}
Glass, Corning 7070	4.0	4.0	4.0	8×10^{-4}	12×10^{-4}	21×10^{-4}
Micarta	4.5	3.8	3.2	0.04	0.05	0.04
Nylon	3.4	3.2	3.0	0.02	0.02	0.01
Paper, Royalgrey	3.0	2.8	2.6	0.04	0.07	0.04
Polyethylene	2.25	2.25	2.25	4×10^{-4}	4×10^{-4}	4×10^{-4}
Polystyrene	2.56	2.55	2.54	0.7×10^{-4}	1×10^{-4}	4×10^{-4}
Ruby mica	5.4	5.4	—	3×10^{-4}	2×10^{-4}	—
Teflon	2.1	2.1	2.1	2×10^{-4}	2×10^{-4}	4×10^{-4}
Water	78	78	55	0.04	0.005	0.5

* The values listed are typical. Data for dielectric materials vary considerably from one published source to another. One of the reasons for apparent discrepancies between sources is that there are many varieties of most materials. For example, there are literally dozens of varieties of polystyrene, all of which have somewhat different characteristics. Also, both the dielectric constant and loss tangent are sensitive to temperature, moisture content, impurities, etc. For serious engineering work, one should consult manufacturers' data or an elaborate reference such as A. R. Von Hippel, *Dielectric Materials and Applications* (Cambridge, Mass.: The Technology Press of Massachusetts Institute of Technology, 1954).

G

Tensor Properties of Ferrites and Plasmas

A ferrite is a member of the class of materials called *ferrimagnetics*. The chemical composition of ferrites includes iron and oxygen plus small fractions of metals such as zinc, nickel, or manganese. Ferrites have very low conductivity and a relative dielectric constant in the range of 10 to 20. The dominant contribution to the magnetic moment of a ferrite molecule is that of bound electron spin.

A magnetic material is characterized by an array or lattice of elemental magnetic dipoles. These elemental dipoles are associated with molecules of the ferrite compound. The effect of the distribution of elemental dipoles is contained in the macroscopic field quantity $\mathfrak{M}$, called the magnetic dipole moment per unit volume. Mathematically,

$$\mathfrak{M} = \frac{1}{V} \sum_{\substack{\text{all dipoles} \\ \text{in } V}} \mathbf{m}_i \tag{G-1}$$

where $\mathbf{m}_i$ is the ith elemental dipole in the volume V.

When a magnetic field is applied to a ferrite, the field interacts with the elemental dipoles. The model used to describe the interaction of an electromagnetic field with a ferrite is based on that of Polder.[1] The model assumes that the spinning electrons possess magnetic dipole moment $\mathbf{m}$ and angular momentum $\mathfrak{J}_m$, which are antiparallel and related by

$$\mathbf{m} = \gamma \mathfrak{J}_m \tag{G-2}$$

[1] D. Polder, "On the Theory of Electromagnetic Resonance," *Philosophical Magazine* vol. 40, pp. 95–115, 1949.

The quantity γ is called the *gyromagnetic ratio* and is equal to -1.759×10^{11} m^2/weber-sec in the rationalized MKS system of units. Losses are neglected in this development.

The torque acting on an elemental magnetic dipole in a magnetic field is

$$\mathcal{T} = \mathbf{m} \times \mathcal{B} \tag{G-3}$$

By Newton's law, the torque is equal to the time rate of change of the angular momentum. Thus,

$$\frac{d\mathcal{J}_m}{dt} = \mathbf{m} \times \mathcal{B} \tag{G-4}$$

Using Eq. (G–2) and Eq. (G–4), we can write

$$\frac{d\mathbf{m}}{dt} = \gamma(\mathbf{m} \times \mathcal{B}) = \mu_0\gamma(\mathbf{m} \times \mathcal{H}) \tag{G-5}$$

If all of the magnetic dipoles in the material are identical and parallel, Eq. (G–1) can be rewritten as

$$\mathcal{M} = N_0\mathbf{m} \tag{G-6}$$

where N_0 is the number of dipoles per unit volume. Equation (G–6) can be solved for $\mathbf{m}$ and substituted into Eq. (G–5) with the result

$$\frac{d\mathcal{M}}{dt} = \gamma(\mathcal{M} \times \mathcal{B}) = \mu_0\gamma(\mathcal{M} \times \mathcal{H}) \tag{G-7}$$

Define $\mathcal{M}$ and $\mathcal{H}$ to consist of dc components oriented parallel to the z-axis plus ac components varying sinusoidally with frequency ω. That is:

$$\begin{aligned} \mathcal{H} &= \mathbf{a}_z H_0 + \text{Re}\,(\mathbf{H}e^{j\omega t}) \\ \mathcal{M} &= \mathbf{a}_z M_0 + \text{Re}\,(\mathbf{M}e^{j\omega t}) \end{aligned} \tag{G-8}$$

Substituting into Eq. (G–7) and retaining terms of frequency ω result in

$$\begin{aligned} j\omega M_x &= \mu_0\gamma[M_yH_0 - M_0H_y] = -\omega_0 M_y + \omega_M H_y \\ j\omega M_y &= \mu_0\gamma[M_0H_x - M_xH_0] = -\omega_M H_x + \omega_0 M_x \\ j\omega M_z &= 0 \end{aligned} \tag{G-9}$$

where $\omega_0 = -\gamma\mu_0 H_0$ is the angular gyromagnetic frequency, and $\omega_M = -\gamma\mu_0 M_0$. Solving for the components of M:

$$\begin{aligned} M_x &= \left[\frac{\omega_0\omega_M}{\omega_0^2 - \omega^2}\right] H_x + \left[\frac{j\omega\omega_M}{\omega_0^2 - \omega^2}\right] H_y \\ M_y &= \left[\frac{-j\omega\omega_M}{\omega_0^2 - \omega^2}\right] H_x + \left[\frac{\omega_0\omega_M}{\omega_0^2 - \omega^2}\right] H_y \\ M_z &= 0 \end{aligned} \tag{G-10}$$

Since $\mathbf{B} = \mu_0(\mathbf{H} + \mathbf{M})$, we can write

$$B_x = \mu_0\left\{\left[1 + \frac{\omega_0\omega_M}{\omega_0^2 - \omega^2}\right]H_x + \left[\frac{j\omega\omega_M}{\omega_0^2 - \omega^2}\right]H_y\right\}$$

$$B_y = \mu_0\left\{\left[\frac{-j\omega\omega_M}{\omega_0^2 - \omega^2}\right]H_x + \left[1 + \frac{\omega_0\omega_M}{\omega_0^2 - \omega^2}\right]H_y\right\} \tag{G-11}$$

$$B_z = \mu_0 H_z$$

This can be simplified by defining $\mu = \left(1 + \frac{\omega_0\omega_M}{\omega_0^2 - \omega^2}\right)$ and $\left(\frac{\omega\omega_M}{\omega_0^2 - \omega^2}\right) = K$. Substituting into Eq. (G–11):

$$\begin{aligned} B_x &= \mu_0\mu H_x + j\mu_0 K H_y \\ B_y &= -j\mu_0 K H_x + \mu_0\mu H_y \Leftrightarrow \\ B_z &= \mu_0 H_z \end{aligned} \quad \begin{bmatrix} B_x \\ B_y \\ B_z \end{bmatrix} = \mu_0 \begin{bmatrix} \mu & jK & 0 \\ -jK & \mu & 0 \\ 0 & 0 & 1 \end{bmatrix} \begin{bmatrix} H_x \\ H_y \\ H_z \end{bmatrix} \tag{G-12}$$

Clearly the tensor permeability of a ferrite biased with a z-directed dc magnetic field is

$$\bar{\bar{\mu}}_r = \begin{bmatrix} \mu & jK & 0 \\ -jK & \mu & 0 \\ 0 & 0 & 1 \end{bmatrix} \tag{G-13}$$

If the direction of the bias field is changed, one can deduce the permeability tensor by a cyclic interchange of subscripts on the elements of the permeability tensor. In Eq. (G–13), $\mu_{xx} = \mu_{yy} = \mu$, $\mu_{xy} = -\mu_{yx} = jK$, $\mu_{xz} = \mu_{zx} = \mu_{yz} = \mu_{zy} = 0$, and $\mu_{zz} = 1$. If we shift the dc bias field so that it is parallel to the y-axis, we simply change the subscripts so that $z \to y$, $y \to x$, $x \to z$. The resultant permeability tensor is

$$\bar{\bar{\mu}} = \begin{bmatrix} \mu & 0 & -jK \\ 0 & 1 & 0 \\ jK & 0 & \mu \end{bmatrix} \tag{G-14}$$

The permittivity of a plasma is developed by assuming the plasma to consist of an electrically neutral medium comprised of N electrons per unit volume and N protons per unit volume. The electrons have a charge $-e = -1.6 \times 10^{-19}$ coulomb and are free to move under the influence of an applied electric field. The protons have a charge of $+1.6 \times 10^{-19}$ coulomb and will be considered to remain stationary. If an electric field $\mathcal{E}$ is incident upon the distribution of electrons, a force

$$\mathfrak{F} = -e\mathcal{E}$$

is exerted on each electron. This force causes the electron to accelerate according to Newton's law:

$$\mathfrak{F} = m\mathfrak{a} = m\frac{d\mathfrak{v}}{dt}$$

Equating the force expressions and assuming a time dependence of $e^{j\omega t}$, we can write

$$-e\mathbf{E} = j\omega m\mathbf{v} \tag{G-15}$$

If we consider a surface, 1 meter square, through which the electrons are moving, there will be a convection current density of

$$\mathbf{J}^c = -Ne\mathbf{v} \tag{G-16}$$

Substituting results in

$$\mathbf{J}^c = \frac{Ne^2}{j\omega m}\mathbf{E} \tag{G-17}$$

In addition to the convection current density, there will be a displacement current density flowing through the region.

$$\mathbf{J}^d = j\omega\epsilon_0\mathbf{E}$$

The total current density is

$$\mathbf{J} = \mathbf{J}^c + \mathbf{J}^d = j\omega\epsilon_0\left(1 - \frac{Ne^2}{\omega^2 m\epsilon_0}\right)\mathbf{E} \tag{G-18}$$

The term $Ne^2/m\epsilon_0$ is designated as ω_p^2; ω_p is called the "plasma frequency." Thus, the plasma is accounted for by modifying the permittivity to

$$\epsilon = \epsilon_0\epsilon_r = \epsilon_0\left(1 - \frac{\omega_p^2}{\omega^2}\right) \tag{G-19}$$

This analysis neglects collisions which introduce a loss or dissipation term into the permittivity expression.

If losses are included, a term which represents the time rate of momentum loss of the electron is added to the force equation:

$$\mathbf{F} = m\mathbf{a} = -e\mathbf{E} - m\nu\mathbf{v} \tag{G-20}$$

where ν is the collision frequency. This results in an expression which corresponds to Eq. (G–15):

$$-e\mathbf{E} = j\omega m\left(1 - j\frac{\nu}{\omega}\right)\mathbf{v} \tag{G-21}$$

The convection current density is

$$\mathbf{J}^c = \frac{Ne^2}{j\omega m(1 - j\nu/\omega)}\mathbf{E} \tag{G-22}$$

The total current density is the sum of the convection current density, Eq. (G–22), and the displacement current density:

$$\mathbf{J} = j\omega\epsilon_0\left(1 - \frac{Ne^2}{\omega^2\epsilon_0 m(1 - j\nu/\omega)}\right)\mathbf{E} \tag{G-23}$$

The permittivity of a plasma, including collisions, becomes:

$$\epsilon = \epsilon_0\left(1 - \frac{\omega_p^2}{\omega^2(1 - j\nu/\omega)}\right) \tag{G-24}$$

To this point the situations considered resulted in scalar permittivity expressions. If a dc magnetic field is applied to the plasma, a tensor permittivity will result. The dc magnetic field adds an additional term to the force equation. The force equation becomes

$$\mathfrak{F} = m\mathbf{a} = -e[\mathcal{E} + \mathbf{v} \times \mathcal{B}] - m\nu\mathbf{v} \tag{G–25}$$

Introducing the $e^{j\omega t}$ time-dependence results in

$$j\omega m\mathbf{v} = -e[\mathbf{E} + \mathbf{v} \times \mathbf{B}] - m\nu\mathbf{v} \tag{G–26}$$

Equation (G–26) can be solved for the components of $\mathbf{v}$, which in turn can be substituted into Eq. (G–16) to determine the convection current density. The convection current is then added to the displacement current to get the total current density. The result is a set of three equations similar to Eq. (G–11). The permittivity tensor of the plasma can be deduced from this set of equations. For the case where there are no losses ($\nu = 0$) and the dc magnetic field is z-directed, the resultant permittivity tensor is

$$\bar{\bar{\epsilon}}_r = \begin{bmatrix} \left(1 + \dfrac{\omega_p^2}{\omega_0^2 - \omega^2}\right) & -j\dfrac{\omega_p^2\omega_0}{\omega(\omega_0^2 - \omega^2)} & 0 \\ +\dfrac{j\,\omega_p^2\omega_0}{\omega(\omega_0^2 - \omega^2)} & \left(1 + \dfrac{\omega_p^2}{\omega_0^2 - \omega^2}\right) & 0 \\ 0 & 0 & 1 - \dfrac{\omega_p^2}{\omega^2} \end{bmatrix} \tag{G–27}$$

where $\omega_0 = B_0e/m$ is the angular gyromagnetic frequency. B_0 is the dc bias field.

H

List of Symbols

Symbol	*Meaning*	*Unit*
$\mathcal{A}$, **A**	Vector magnetic potential (time and phasor forms, respectively)	webers/meter
A_e	Effective aperture area	weber^2
a	Earth's radius (6.36×10^6)	meter
a_e	Effective earth radius	meter
$\mathfrak{a}$, **a**	Acceleration vector (time and phasor forms, respectively)	meters/second^2
$\mathbf{a}_x$, $\mathbf{a}_y$, . . .	Unit vectors in x, y, . . . directions	
a_1, a_2, . . .	Scattering matrix representation variables (incident wave)	watts$^{1/2}$
$\mathcal{B}$, **B**	Magnetic flux density (time and phasor forms, respectively)	webers/meter^2
B_0	Dc magnetic flux density (bias field applied to ferrite materials)	webers/meter^2
b_1, b_2, . . .	Scattering matrix representation variables (reflected wave)	watt$^{1/2}$
C	Capacitance per unit length	farads/meter
c	Velocity of light (3×10^8)	meters/second
c_1, c_2, . . .	Transmission matrix representation variables	watt$^{1/2}$
$\mathcal{D}$, **D**	Electric flux density (time and phasor forms, respectively)	coulombs/meter^2
da	Differential surface vector	meter^2
dl	Differential length vector	meter
D	Antenna directivity	

Symbol	*Meaning*	*Unit*
D	Ionospheric layer	
d	Fixed distance (or length of line)	meter
$\mathcal{E}$, **E**	Electric field intensity (time and phasor forms, respectively)	volts/meter
$\mathcal{E}_x, \mathcal{E}_y, \ldots; E_x, E_y, \ldots$	Components of $\mathcal{E}$ and **E** in respective directions	volts/meter
E'_x, E'_y, E'_z	Components of **E** with z-dependence removed (i.e., these are functions of x and y only)	volts/meter
E	Ionospheric layer	
e	Partial pressure of water vapor	newtons/meter^2
e	Electronic charge (1.602×10^{-19}, negative)	coulomb
$\mathcal{F}$, **F**	Force (time and phasor forms, respectively)	newton
F	Antenna array factor	
F	Ionospheric layer	
F_s	Range-gain function	
f	Frequency	hertz
f_c	Critical frequency in a plasma	hertz
f_c	Cut-off frequency in waveguide	hertz
f_N	Plasma frequency ($\sqrt{Ne^2/\epsilon_0 m}/2\pi$)	second^{-1}
f_0	Center frequency	hertz
G	Antenna gain	decibel
G	Conductance per unit length	mhos/meter
G_s	Height-gain function	
g	Antenna gain	
$\mathcal{H}$, **H**	Magnetic field intensity (time and phasor forms, respectively)	amperes/meter
$\mathcal{H}_x, \mathcal{H}_y, \ldots; H_x, H_y, \ldots$	Components of $\mathcal{H}$ and **H** in respective directions	amperes/meter
H'_x, H'_y, H'_z	Components of **H** with z-dependence removed	amperes/meter
h	Antenna height above the earth	meter
h'	Virtual height ($\frac{1}{2}ct$)	meter
$\mathcal{I}$, I	Current (time and phasor forms, respectively)	
I^+, I^-	Incident and reflected currents, respectively	ampere
i^+, i^-	Normalized incident and reflected currents, respectively	ampere-ohm$^{1/2}$
$\mathcal{J}_m$, $\mathbf{J}_m$	Angular momentum (time and phasor forms, respectively)	kilogram-meters/second

Symbol	*Meaning*	*Unit*
$\mathcal{J}$, $\mathbf{J}$	Current density (time and phasor forms, respectively)	amperes/meter^2
$\mathcal{J}^c$, $\mathbf{J}^c$	Conduction current density (time and phasor forms, respectively)	amperes/meter^2
$\mathcal{J}^d$, $\mathbf{J}^d$	Displacement current density (time and phasor forms, respectively)	amperes/meter^2
$\mathcal{J}^i$, $\mathbf{J}^i$	Impressed current density (time and phasor forms, respectively)	amperes/meter^2
$\mathcal{J}_s$, $\mathbf{J}_s$	Surface current (time and phasor forms, respectively)	amperes/meter
J_n	Bessel function of first kind and nth order	
J'_n	Derivative of J_n	
j	Unit imaginary number ($\sqrt{-1}$)	
K	Reflection coefficient	
k	Boltzmann's constant (1.380×10^{-23})	joule/°K
k_c^2	$\gamma^2 + \omega^2\mu\epsilon$ in waveguide work	meter^{-2}
L	Inductance per unit length	henrys/meter
L_b	Transmission loss (isotropic antennas)	decibel
L_p	Propagation-path loss	decibel
L_t	Transmission loss	decibel
l	Length	meter
$\mathcal{M}$, $\mathbf{M}$	Magnetic dipole moment per unit volume (time and phasor forms, respectively)	amperes/meter
MUF	Maximum usable frequency	hertz
$\mathbf{m}$	Elemental magnetic dipole moment	ampere-meter^2
m	Electronic mass (9.107×10^{-31})	kilogram
N	Electron number density	electrons/meter^3
N	Refractivity $(n - 1) \times 10^6$	
N_n	Bessel function of second kind and nth order	
N'_n	Derivative of N_n	
n	Index of refraction	
$\mathcal{P}$, $\mathbf{P}$	Poynting vector (instantaneous function of time and average value forms, respectively)	watts/meter^2
P^+, P^-	Incident and reflected powers, respectively	watt
P_L	Power dissipated per unit length	watt
P_n	Total average noise power	watt
P_r	Total average power radiated	watt
P_T	Total average power transmitted	watt

Symbol	*Meaning*	*Unit*
p	Atmospheric dry air pressure	newtons/meter^2
Q	Quality factor	
Q_e	External quality factor	
Q_L	Loaded quality factor	
Q_0	Unloaded quality factor	
R	Resistance per unit length	ohms/meter
R	Real (resistive) part of impedance Z	ohm
R_h	Reflection coefficient (horizontal polarization)	
R_s	Surface resistance ($\sqrt{\omega\mu/2\sigma}$)	ohm
R_v	Reflection coefficient (vertical polarization)	
R_0	Characteristic resistance (real number)	ohm
R_{12}	Zurich smoothed mean sunspot number	
$\mathbf{r}$	Field point position vector	meter
$\mathbf{r}'$	Source point position vector	meter
r	Per-unit value of resistance	
r	Radial distance (spherical polar coordinate system)	meter
s	Distance between source point and field point	meter
S	Voltage standing wave ratio	
S	Richards' transformation ($j \tan \theta$)	
$S_{11}, S_{12}, \ldots$	Elements of the scattering matrix	
$\boldsymbol{\mathcal{T}}$, $\mathbf{T}$	Torque (time and phasor forms, respectively)	kilogram-meter
T	Temperature	degree Kelvin
$T_{11}, T_{12}, \ldots$	Elements of the transmission matrix	
t	Time	second
t	Tangent βl	
U	Power gain per unit solid angle	
$\mathcal{V}$, V	Voltage (time and phasor forms, respectively)	volt
V^+, V^-	Incident and reflected voltages, respectively	volt
$\boldsymbol{\upsilon}$, $\mathbf{v}$	Velocity vector (time and phasor forms, respectively)	meters/second
v	Free-wave velocity	meters/second
v_g	Group velocity	meters/second
v_p	Phase velocity	meters/second
v^+, v^-	Normalized incident and reflected voltages, respectively	volts/ohm$^{1/2}$

Symbol	*Meaning*	*Unit*
W	Energy	joule
W_a	Power available from a loss-free receiving antenna	watt
W_r	Radiated power	watt
W_{rec}	Received power	watt
w_e	Electric field energy density	joules/meter^3
w_m	Magnetic field energy density	joules/meter^3
X	Imaginary (reactive) part of impedance Z	ohm
x	Per-unit reactance	
x	Distance	meter
x	Loss factor ($\sigma/\omega\epsilon_0$)	
Y	Admittance	mho
$Y_{11}, Y_{12}, \ldots$	Elements of admittance matrix	mho
y	Per-unit admittance	
y	Distance	meter
$y_{11}, y_{12}, \ldots$	Normalized admittance matrix elements	
Z	Impedance	ohm
Z_0	Characteristic impedance	ohm
$Z_{0(TE)}, Z_{0(TM)}$	Characteristic impedance for TE and TM modes	ohm
$Z_{11}, Z_{12}, \ldots$	Elements of impedance matrix	ohm
z	Per-unit impedance	
z	Distance	meter
$z_{11}, z_{12}, \ldots$	Normalized impedance matrix elements	
α	Attenuation constant	nepers/meter
β	Phase shift constant	radians/meter
β	Resonator coupling factor	
γ	Propagation constant ($\alpha + j\beta$)	meter^{-1}
γ	Gyromagnetic ratio	meters^2/weber-second
Δ	Faraday rotation angle	radian
Δ	Phase factor (ground-wave set)	radian
δ	Depth of penetration ($\sqrt{2/\omega\mu\sigma}$)	meter
δ	Frequency difference ($\omega - \omega_0$)	radians/second
ϵ	Permittivity ($\epsilon = \epsilon_r\epsilon_0$)	farads/meter
ϵ_r	Relative permittivity (dielectric constant)	
ϵ_0	Permittivity of free space ($10^{-9}/36\pi$)	farads/meter
η	Intrinsic impedance ($\sqrt{\mu/\epsilon}$)	ohm
θ	Spherical polar coordinate	radian
θ	Loss tangent ($\sigma/\omega\epsilon$)	

Symbol	*Meaning*	*Unit*
θ	Angle between incident wave normal and scattered wave normal	radian
θ	Electrical angle (βl)	radian
κ	Off-diagonal element of permeability tensor	
λ	Wavelength (v/f)	meter
λ_g	Guide wavelength (v_p/f)	meter
μ	Permeability ($\mu = \mu_r\mu_0$)	henrys/meter
μ	On-diagonal element of permeability tensor	
μ_r	Relative permeability	
μ_0	Permeability of free space ($4\pi \times 10^{-7}$)	henrys/meter
$\mu_{xx}, \mu_{xy}, \ldots$	Elements of the permeability tensor	
ν	Collision frequency	second^{-1}
ρ	Charge density	coulomb
ρ	Radial distance (cylindrical coordinate system)	meter
ρ	Density of the atmosphere	kilograms/meter^3
σ	Conductivity	mhos/meter
σ	Damping constant (with reference to exponentially decaying phenomena)	
σ_v	Differential scattering cross-section	steradian^{-1} meter^{-1}
τ	Total angular refraction	radian
Φ	Scalar electric potential	volt
ϕ	Magnetic flux	weber
ϕ	Cylindrical coordinate	radian
χ	Angle between electric field vector and scattered-wave normal	radian
ψ	Grazing angle	radian
$\boldsymbol{\omega}_0$	Precessional frequency	second^{-1}
ω	Angular frequency	
ω_c	Cut-off angular frequency in waveguide	radians/second
ω_p	Plasma frequency	radians/second
ω_r	Resonant angular frequency of cavity	radians/second
ω_0	Center angular frequency	radians/second
ω_0	Angular gyromagnetic frequency	radians/second
$\boldsymbol{\nabla}$	Vector operator "del"	
∇^2	Laplacian operator	
∇_t^2	Transverse (two-dimensional) Laplacian operator	

Index